Omics Approaches to Immune-Mediated Inflammatory Diseases: Towards Novel Biomarkers and Potential Therapeutic Targets

Editor

Maria-Ioanna (Marianna) Christodoulou

MDPI

Basel • Beijing • Wuhan • Barcelona • Belgrade • Novi Sad • Cluj • Manchester

Editor
Maria-Ioanna (Marianna) Christodoulou
European University Cyprus
Nicosia, Cyprus

Editorial Office
MDPI
St. Alban-Anlage 66
4052 Basel, Switzerland

This is a reprint of articles from the Special Issue published online in the open access journal *Biomedicines* (ISSN 2227-9059) (available at: https://www.mdpi.com/journal/biomedicines/special_issues/Omics_Inflammatory_Dieases).

For citation purposes, cite each article independently as indicated on the article page online and as indicated below:

Lastname, A.A.; Lastname, B.B. Article Title. *Journal Name* **Year**, *Volume Number*, Page Range.

ISBN 978-3-0365-9276-3 (Hbk)
ISBN 978-3-0365-9277-0 (PDF)
doi.org/10.3390/books978-3-0365-9277-0

Contents

biomedicines

MDPI

Article

Comprehensive Profiling of Early Neoplastic Gastric Microenvironment Modifications and Biodynamics in Impaired BMP-Signaling FoxL1^{+}-Telocytes

Alain B. Alfonso, Véronique Pomerleau, Vilcy Reyes Nicolás, Jennifer Raisch, Carla-Marie Jurkovic, François-Michel Boisvert † and Nathalie Perreault *,†

Département d'Immunologie et Biologie Cellulaire, Faculté de Médecine et des Sciences de la Santé, Université de Sherbrooke, Sherbrooke, QC J1E 4K8, Canada

* Correspondence: nathalie.perreault@usherbrooke.ca

† Co-senior authors.

Abstract: FoxL1^{+}telocytes (TC^{FoxL1+}) are novel gastrointestinal subepithelial cells that form a communication axis between the mesenchyme and epithelium. TC^{FoxL1+} are strategically positioned to be key contributors to the microenvironment through production and secretion of growth factors and extracellular matrix (ECM) proteins. In recent years, the alteration of the bone morphogenetic protein (BMP) signaling in TC^{FoxL1+} was demonstrated to trigger a toxic microenvironment with ECM remodeling that leads to the development of pre-neoplastic gastric lesions. However, a comprehensive analysis of variations in the ECM composition and its associated proteins in gastric neoplasia linked to TC^{FoxL1+} dysregulation has never been performed. This study provides a better understanding of how TC^{FoxL1+} defective BMP signaling participates in the gastric pre-neoplastic microenvironment. Using a proteomic approach, we determined the changes in the complete matrisome of *BmpR1a*$^{\triangle FoxL1+}$ and control mice, both in total antrum as well as in isolated mesenchyme-enriched antrum fractions. Comparative proteomic analysis revealed that the deconstruction of the gastric antrum led to a more comprehensive analysis of the ECM fraction of gastric tissues microenvironment. These results show that TC^{FoxL1+} are key members of the mesenchymal cell population and actively participate in the establishment of the matrisomic fraction of the microenvironment, thus influencing epithelial cell behavior.

Keywords: FoxL1^{+}-telocytes; epithelial–mesenchymal interaction; BMP signaling; extracellular matrix; mechanical microenvironment; matrisome

Citation: Alfonso, A.B.; Pomerleau, V.; Nicolás, V.R.; Raisch, J.; Jurkovic, C.-M.; Boisvert, F.-M.; Perreault, N. Comprehensive Profiling of Early Neoplastic Gastric Microenvironment Modifications and Biodynamics in Impaired BMP-Signaling FoxL1^{+}-Telocytes. *Biomedicines* **2023**, *11*, 19. https://doi.org/10.3390/biomedicines11010019

Academic Editor: Marianna Christodoulou

Received: 18 November 2022
Revised: 13 December 2022
Accepted: 16 December 2022
Published: 22 December 2022

1. Introduction

The extracellular matrix (ECM) is a complex assembly of large fibrous proteins, glycoproteins, proteoglycans, and ECM-associated proteins, such as growth factors, whose composition varies from one tissue to another [1]. The ECM represents the insoluble fraction of the microenvironment, and although it was long believed to be a passive component, it is in fact highly dynamic and influences the behavior of neighboring cells through mechanosensing and signaling [2,3]. Thus, the architecture and homeostasis of a tissue, such as the stomach, are maintained in part by tight regulation of ECM dynamics. Dysregulation of the ECM composition in the microenvironment creates a disbalance in the physical (force, porosity, stiffness) and biochemical (growth factor density, cell adhesion, signaling) stimuli, providing an abnormal cell response to these biomechanical forces and leading to the development of diseases such as gastric neoplasia [4–8]. In gastric cancer, pre-malignant lesions already show dysregulation in ECM dynamics and will also influence the prognosis outcome and therapeutic strategies at later stages of the disease [2,5,9].

In mammals, the ECM is composed of approximately 300 proteins. This represents the core matrisome, which is mainly composed of proteins, such as collagens (CLs) and proteoglycans, with structural and fibrillar glycoproteins [10–13]. The biochemical properties of these proteins, such as their size, insolubility, and cross-linking, have made attempts to systematically characterize the entire tissue ECM composition challenging [14]. Recently, Naba et al. developed a proteomics-based approach to identify, quantify, and compare the matrisome of whole tissues, partially resolving the limitations of in vivo analysis of ECM dynamics [14]. This approach allows for comprehensive evaluation of the proteins from the core matrisome, as well as the components of matrisome-associated proteins such as ECM regulators (ECM-remodeling enzymes, cross-linkers, proteases) and secreted factors such as growth factors and cytokines binding the ECM [13,14].

As the microenvironment plays an essential role in tissue homeostasis and in the development of pathologies such as gastric cancer [4–8], mesenchymal cells have attracted considerable attention in recent years [15–17]. Mesenchymal cells, more precisely myofibroblasts as well as FoxL1^{+}telocytes (TC^{FoxL1+}), are better known for their contribution to the sub-epithelial microenvironment. Both myofibroblasts and TC^{FoxL1+} are capable secretors of cytokines, chemokines, growth factors, and ECM proteins [16–22]. In addition, TC^{FoxL1+} are advantageously positioned directly underlying the epithelium, forming a 3D nexus between the epithelium and the rest of the stroma [17,23]. TC^{FoxL1+} contribute to the stem cell niche microenvironment by secreting soluble factors such as WNT5a, R-spondin3, and gremlin, which has been documented in recent years [15,17,20,23,24]. However, the precise role of TC^{FoxL1+} in the insoluble fraction of the gastrointestinal (GI) microenvironment is poorly defined. Considering the effect of TC^{FoxL1+} on GI epithelial cells [17–19,22,23], there is a critical need to rigorously characterize the role of the ECM biodynamic microenvironment on GI epithelial cell behavior in vivo and determine the contribution of TC^{FoxL1+}.

To date, there have been limitations to the study of the various roles of TC^{FoxL1+} in the in vivo microenvironment because of the limited models available [17,20,23,25,26]. A previous study, using a murine model with TC^{FoxL1+} impaired BMP signaling pathway, demonstrated the importance of these cells and this pathway in inducing gastric neoplastic lesions and polyps in 90-day-old mice [22]. *BmpR1a*$^{\triangle FoxL1+}$ mice did not develop chronic inflammation or a malignant phenotype; however, disturbed TC^{FoxL1+} led to early precancerous events with important disorganized gastric glands architecture, intestinal metaplasia, and spasmolytic polypeptide-expressing metaplasia (SPEM), in addition to remodeling of the ECM into a reactive microenvironment [22]. Consequently, *BmpR1a*$^{\triangle FoxL1+}$ mice represent an excellent model to investigate the TC^{FoxL1+} contribution instructing the microenvironment ECM biodynamics, leading to gastric neoplasia. Using this model, we can perform a matrisomic investigative of the stomach of control and *BmpR1a*$^{\triangle FoxL1+}$ mice, and better understand the contribution of TC^{FoxL1+} to this aspect of the microenvironment [13,14].

In the present study, we evaluated the contribution of TC^{FoxL1+} to the matrisomic microenvironment in mice with early gastric neoplasia. This matrisomic investigative approach, used in concert with the TC^{FoxL1+} signaling impaired gastric pre-neoplastic mouse model, revealed a detailed inventory of dysregulated core-matrisome and matrisome-associated proteins in early events of gastric neoplasia. We identified important and subtle changes in the ECM biology that occur during the etiology of gastric neoplasia associated with Bmp-signaling impaired TC^{FoxL1+}.

2. Materials and Methods

2.1. Animals

The transgenic mouse line C57BL/6J *FoxL1*Cre was provided by Dr. Kaestner [27] and 129 SvEv-*BmpR1*afx/fx mice were supplied by Dr. Mishina [28]. *BmpR1a*$^{\Delta FoxL1+}$ conditional knockout mice were generated as previously described [18,21,22]. Male and female 90-day-old age-matched mice were used for the study. All experiments were performed in accordance with our animal welfare protocol (approval number: FMSS-2019-2370).

2.2. Deconstruction of Mouse Ex Vivo Stomach Tissues

Tissue deconstruction was performed stepwise to enrich each compartment (the epithelial, mesenchymal, and muscular layers). First, stomachs were opened along the greater curvature and rinsed with cold 1× PBS, and the antrums were isolated from the corpus and fundus sections of the total tissue. Mouse antrums were cut with a razor blade into 5 mm tissue sections and the muscle layer was mechanically dissociated using forceps under a stereomicroscope. Leftover tissues (mesenchyme and epithelium) were subsequently incubated in 4 mL sterile Corning™ Cell Recovery Solution without agitation (Corning Life Science, Corning, NY, USA) at 4 °C for 24 h. The following day, dissociation of the epithelial layer was performed with a 30 min incubation of the tissue on ice followed by vigorous manual shaking for 15 s. The mesenchymal tissue was incubated once again in 6 mL of sterile Corning™ Cell Recovery Solution (Corning Life Science, Corning, NY, USA) on ice with gentle shaking for 30 min followed by further dissociation by vigorous manual shaking for 15 s. Finally, mesenchymal tissues were washed four times with 1× PBS while all remaining epithelial cells were pooled and kept on ice. Deconstructed tissue sections were either snap-frozen for immunoblotting and proteomic analysis or fixed in 4% paraformaldehyde (PFA) (Thermo Fisher Scientific, Waltham, MA, USA) and paraffin-embedded for histological analysis. Total tissue samples were also collected to allow for a more comprehensive comparison of the matrisome content.

2.3. Histological Analysis

The total stomach antrum or deconstructed fractions were fixed overnight at 4 °C in 4% PFA (Thermo Fisher Scientific, Waltham, MA, USA) and subsequently processed for tissue embedding as previously described [18,21]. To avoid the diffusion of cells in paraffin, the epithelial layer from the deconstructed tissue was embedded in HistoGelTM (Thermo Fisher Scientific, Waltham, MA, USA) and wrapped in lens paper prior to embedding. Histological staining (H&E) on tissue sections was performed as previously described [18,21]. Virtual images were acquired with a slide scanner (Nanozoomer; Hamamatsu, Japan) and visualized using the NDP.view2 software (version 2.8.24).

2.4. In-Solution Digestion of Proteins to Peptides for Mass Spectrometry Analysis

Frozen samples of either the total stomach antrum or mesenchymal-enriched stomach antrum fractions were thawed on ice and homogenized directly in 8 M urea (Sigma Aldrich, St. Louis, MO, USA) dissolved in 10 mM HEPES pH 8.0 (Wisent, Saint-Jean-Baptiste, QC, Canada) (100 µL/10 mg wet tissue weight), using the QIAGEN TissueLyser LT (Hilden, Germany). Prior to protein quantification by BCA assay (Pierce Thermo Scientific, Waltham, MA, USA), samples were centrifuged following their homogenization to remove urea-insoluble materials. Following the protocol described by Naba et al., proteins were reduced, alkylated, deglycosylated, and digested, except for the Lys-C digestion, which was omitted [14,29]. Solutions were prepared using MS-grade water and low protein binding tubes were used for these experiments.

2.5. Purification and Desalting of the Peptides on C18 Columns

Trifluoroacetic acid (TFA) was added following incubation with the proteases to a final concentration of 0.2%, and the samples were desalted using C18 tips (Pierce Thermo Scientific, Waltham, MA, USA). Acetonitrile was first aspirated in the C18 tip initially and then equilibrated with 0.1% TFA. Each peptide sample was bound to the C18 tip by 10 successive up-and-down until the entire sample was loaded. The tip was then washed with a solution containing 0.1% TFA, and the peptides were eluted in a separate low-bind tube using a 50% acetonitrile/1% formic acid solution. The eluted peptides were lyophilized using a centrifugal evaporator at 60 °C and the dry peptides were resuspended in 1% formic acid. The peptide concentration was measured using a NanoDrop spectrophotometer (Thermo Fisher Scientific, Waltham, MA, USA) at 205 nm absorbance. The peptide samples were transferred to autosampler vials and stored at −20 °C until analyzed by mass spectrometry.

2.6. LC-MS/MS Analysis

Analysis of the purified peptides was carried out at the Université de Sherbrooke proteomics facility using the following parameters: Each sample (was injected into an HPLC system (NanoElute, Bruker Daltonics, Billerica, MA, USA) for LC-MS/MS. A total of 250 ng of peptides were loaded onto a trap column at a constant flow of 4 μL/min (Acclaim PepMap100 C18 column, 0.3 mm id × 5 mm, Dionex Corporation, Sunnyvale, CA, USA) and eluted onto the C18 analytical column (1.9 μm beads size, 75 μm × 25 cm, PepSep) over a 2 h gradient of acetonitrile (5–37%) in 0.1% FA at 500 nL/min into a TimsTOF Pro ion mobility mass spectrometer equipped with a Captive Spray nanoelectrospray source (NanoElute, Bruker Daltonics, Billerica, MA, USA). The data were acquired in data-dependent MS/MS mode with a 100–1700 m/z mass range, and the number of PASEF scans was set at 10 (1.27 s duty cycle) with a dynamic exclusion m/z isolation window of 0.4 min. The collision energy was set at 42.0 eV, and the target intensity was 20,000 with an intensity threshold of 2500.

2.7. Protein Identification Using MaxQuant Analysis

MaxQuant software version 1.6.17 (Munich, Bavaria, Germany), was used to analyze the raw files using the Uniprot mouse proteome database (25 March 2020, 55,366 entries). The analysis was performed under TIMS-DDA type in group-specific parameters, and included the following parameters: two miscleavages were allowed; fixed modification was carbamidomethylation of cysteine; the enzyme selected was trypsin (not before a proline). The following variable modifications were included in the analysis: methionine oxidation, N-terminal protein acetylation, and protein carbamylation (K, N-terminal). The limit for mass tolerance was set at 10 ppm for the precursor ions and at 20 ppm for the fragment ions. The identification values "PSM FDR", "Protein FDR", and "Site decoy fraction" were set to 0.05. The minimum peptide count was set to 1. Label-free quantification (LFQ) was performed using an LFQ minimal ratio count of 2. Both the "Second peptides" and "Match between runs" were allowed.

2.8. Differential and Statistical Analyses of Mass Spectrometry Data

Following the MaxQuant analysis, LFQ intensities were sorted according to several parameters using the Prostar software version 1.28.1 (Grenoble, France) [30]. Filtered proteins positive for the "Reverse", "Only.identified.by.site", or "Potential.contaminant" categories were eliminated, as were proteins identified from only one unique peptide. Data were normalized with quantile centering set to 0.5 for the intensity distribution. The non-detection of a protein was considered biologically relevant in the following cases: 75% (3 of 4) of the control or mutant mice group with respect to the other for total antrum (TA) and in 83% (5 of 6) of the control or mutant mice group with respect to the other for enriched mesenchyme (EM). Considering the aforementioned conditions, for all data corresponding to the matrisome, the partially observed value (POV) imputation was revised according to the following cases, followed by recalculation of Log2FC and *p*-value in ProStar. For TA data, the imputed POV was removed and replaced by the minimum POV when three out of four mice presented an LFQ intensity = 0 for a given protein. If two out of four mice presented LFQ intensity = 0, the Log2FC and the *p*-value recalculated in ProStar were considered non-conclusive (NC). For the EM data, the imputed POV was removed and replaced by the minimum POV when five out of six mice in one of the two groups presented an LFQ intensity = 0 for a given protein. If four out of six mice presented an LFQ intensity = 0 in one or both groups, the Log2FC and *p*-value recalculated by ProStar were considered NC. Structured least square adaptation (SLSA) and detQuantile imputation were performed for POV and missing values in the entire condition (MEC), respectively. The results were ranked to preserve the proteins present in at least three of the four (in TA) and three of the six (in MS EM) biological replicates for each condition. For hypotheses testing, a Limma statistical test was used, with a fold-change threshold of 1.5 and a *p*-value of 0.05, to determine the list of differentially abundant proteins. A "st.boot" calibration plot was chosen for *p*-value distribution.

2.9. Matrisome Identification

The Matrisome Annotator webtool (matrisomeproject.mit.edu) was used to annotate the list of differentially abundant proteins as previously described [13]. Matrisome divisions (core matrisome or matrisome-associated) and categories (ECM glycoproteins, collagens (CLs) and proteoglycans, ECM-affiliated, ECM regulators, and secreted factors) were used according to Naba et al. [13].

2.10. Indirect Immunofluorescence

Indirect immunofluorescence of stomach sections from 90-day old control and *BmpR1a*$^{\triangle FoxL1+}$ mice was performed as previously described [18,21,22,31–33]. Antigen blocking was performed with a solution of 2% bovine serum albumin (BSA), 0.1% fish gelatin, and 0.2% Triton X-100 in 1× PBS for 1 h at room temperature. The following primary antibodies were used in this study: S100A9 (Cell Signaling Technology, Danvers, MA, USA; Cat#73425; RRID:AB_2799839), fibronectin (Millipore; Burlington, MA, USA, Cat# AB2033, RRID:AB_2105702), and tenascin C (Millipore, Burlington, MA, USA; Cat# AB19013, RRID:AB_2256033). The following day, slides were incubated with anti-rabbit IgG Alexa-488 labeled secondary antibody (Cell Signaling Technology; Danvers, MA, USA; Cat# 4412; RRID:AB_1904025). Slides were examined under a Zeiss Axioscope 5 (Oberkochen, Germany) equipped with a Zeiss Axiocam 705 mono CMOS camera. Images were analyzed using ImageJ v.1.53j (RRID:SDR_003070).

2.11. Picro-Sirius Red Staining

Tissue sections of 90-day old mouse stomach were stained with picrosirius red following a previously published protocol [34] and CL content and fibers were analyzed under bright-field and polarized light. Images from four mice in each group were taken using a Zeiss Axioscope 5 equipped with a linear polarizer and analyzer. Multiple representative regions of interest (ROI) were assessed per image to characterize the alignment properties of CL fibers. ROI were selected in both the top and middle antrum glands of *BmpR1a*$^{\triangle FoxL1+}$ mice to better assess tissue complexity. Each ROI was the same dimension. The distribution of CL fiber angles and coherency was determined using ImageJ software (Madison, WI, USA) package Orientation J (version 2.0.5; RRID:SCR_014796). Statistical analysis was performed using Prism v9.4.1 (San Diego, CA, USA, RRID:SCR_002798). To test the normal distribution of the samples, we used D'Agostino-Pearson omnibus normality test and for group analyses we used nested ANOVA.

2.12. Immunoblot Analysis

The same 8 M urea proteins extracts from total antrum tissues used for proteomic analyses were also assessed to validate the potential proteins of interest (n = 4). Samples (10 µg each) were separated on NuPage 4–12% Bis-Tris gels (Thermo Fisher Scientific, Waltham, MA, USA) with MES buffer and transferred onto a PVDF membrane. Membranes were probed with the following antibodies: S100A8 (Proteintech, Rosemont, IL, USA; Cat# 15792-1-AP, RRID:AB_10666315), S100A9 (Cell Signaling Technology; Danvers, MA, USA; Cat#73425; RRID:AB_2799839), SPARCL1 (R&D Systems, Minneapolis, MN, USA; Cat# AF2836, RRID:AB_2195097), and ADAM9 (Cell Signaling Technology; Danvers, MA, USA; Cat# 4151, RRID:AB_1903892). GAPDH (Cell Signaling Technology; Danvers, MA, USA; Cat# 2118, RRID:AB_561053) was used as a loading control. Anti-rabbit (Cat#7074; RRID:AB_2099233) HRP-labeled secondary antibodies were purchased from Cell Signaling Technology; Danvers, MA, USA; and anti-goat HRP-labeled antibodies (Cat#705-035-003; RRID:AB_2340390) were from Jackson ImmunoResearch Laboratories (West Grove, PA, USA). Immunoreactive bands were detected using the Amersham ECL Western blotting Detection System (GE Healthcare Life Sciences/Cytiva, Chicago, IL, USA) with an Azure Biosystems c280 digital imager (Azure Biosystems, Dublin, CA, USA). Quantification was performed using ImageJ v1.53j ($n = 4$ mice/group). The Mann–Whitney U test was used to determine data significance.

3. Results

To study the contribution of TC^{FoxL1+} in instructing the microenvironment ECM biodynamic leading to gastric neoplasia through a matrisomic investigative approach, we compared and analyzed two methods for tissue preparation of the stomach antrum of 90-day-old control and $BmpR1a^{\triangle FoxL1+}$ mice (Figure 1A). In the first approach, an 8 M urea extraction of total proteins was performed on the stomach antrum of the control and $BmpR1a^{\triangle FoxL1+}$ mice. Proteins from the total antrum were identified using LC-MS/MS as previously described [14]. For the second method, we investigated whether other cell compartments in the tissue caused unwanted interference during the protein identification and quantification within the proteomic analysis. As the bulk of ECM/matrisome proteins is located in the mesenchymal compartment, we decided to deconstruct the stomach antrum to obtain an enriched mesenchymal compartment (Figure 1B–E). First, the stomach antrum was isolated from control and $BmpR1a^{\triangle FoxL1+}$ mice (Figure 1B), and the muscle layers (Figure 1C) were mechanically separated from the antrum using tweezers. Next, the remaining epithelium/mesenchymal fraction (Figure 1D) was incubated with a non-enzymatic cell recovery solution that dissociated the epithelial fraction (Figure 1E) from the underlying mesenchyme, as previously described [18,32,33,35]. The 8 M protein extraction was carried out for the isolated enriched mesenchymal fraction, and the analysis was performed as described above for the total tissue.

3.1. Analysis of the Matrisome from Total Antrum of $BmpR1a^{\triangle FoxL1+}$ Mouse

To evaluate the changes in ECM composition in our pre-neoplastic gastric $BmpR1a^{\triangle FoxL1+}$ mouse model, we calculated the fold change in matrisome proteins between the total antrum of mutant and control mice. The ratio ($BmpR1a^{\triangle FoxL1+}$/control) of relative expression of total proteins between both groups was compared. Among the 3803 proteins detected, 279 were shown to be upregulated, while 484 were downregulated (Figure 2A). The analysis identified, from the total antrum, the presence of 36 overexpressed matrisome proteins (dark red spots, FC > 1.5) and 37 downregulated proteins (dark blue spots, FC < −1.5) in $BmpR1a^{\triangle FoxL1+}$ mice compared to those observed in the control group (Figure 2A). Matrisome proteins were identified using the Matrisome Annotator analytical tool (http://matrisomeproject.mit.edu/; accessed on 29 September 2020) [13,14,36]. A total of 169 proteins were identified, 70 of them belonging to the core matrisome and 99 to matrisome-associated proteins. Of the proteins belonging to the core matrisome, we identified 11 proteoglycans, 10 CLs, and 49 glycoproteins, whereas we identified 28 ECM-affiliated proteins, 54 ECM regulators, and 17 secreted factors among the matrisome-associated proteins (Figure 2B). Surprisingly, except for two the CL chains (CL1α2, CL4α1, and α2; CL6α1, α2, and α5; CL12α1 and CL14α1) in $BmpR1a^{\triangle FoxL1+}$ mice, all were downregulated compared to those observed in controls (Table 1). Only CL15α1 and CL18α1 were upregulated in the mutant mice compared to those in the controls (Table 1). Similarly, most proteoglycans (HSPG2, perlecan; ASPN, asporin; DCN, decorin; LUM, lumican; and VCAN, versican) were observed to be negatively modulated in $BmpR1a^{\triangle FoxL1+}$ mice compared to those in the controls. Only biglycan (BGN) and bone marrow proteoglycan (PRG2) were upregulated in the mutant mice compared to those in the controls (Table 1) Glycoproteins such as Agrin (AGRN), fibronectin I (FNI), tenascin C (TNC), vitronectin (VTN), and periostin (POSTN) were upregulated in mutant mice compared to those in the controls, whereas others such as microfibrillar-associated proteins (MFAP2, 4, and 5), Nidogen1 and 2 (NID1 and NID2), as well as SPARC-like protein-1 (SPARCL-1) were downregulated (Table 1). Among the matrisome-associated proteins, the analysis revealed that ECM-affiliated proteins such as proteins of the annexin family including annexin 10 (ANXA10) and different galectins, such as galectin-4 (LGALS4) and mucin 4 (MUC4), were upregulated, whereas annexin 6 (ANXA6) and chondroitin sulfate proteoglycan 4 (CSPG4) were downregulated in *BmpR1a*$^{\triangle FoxL1+}$ mice compared to those in the controls (Table 1). Analysis of ECM regulators revealed that disintegrin, metalloproteinase family members (ADAM9 and 10), and various serpins (SERPINB1a, SERPINB5, and SERPINB12) were overexpressed, whereas α-1-microglobulin/bikunin (AMBP) and transglutaminase 2 (TGM2) were downregulated in mutant mice compared to those in the controls (Table 1). For the secreted factors, proteomic analyses showed that most members of the S100 protein group (S100A1, A2, A4, A6, A8, A9, A11, A13, A14, A16, and G) were overexpressed, except for S100B, which was downregulated in *BmpR1a*$^{\triangle FoxL1+}$ mice compared to that measured in controls (Table 1).

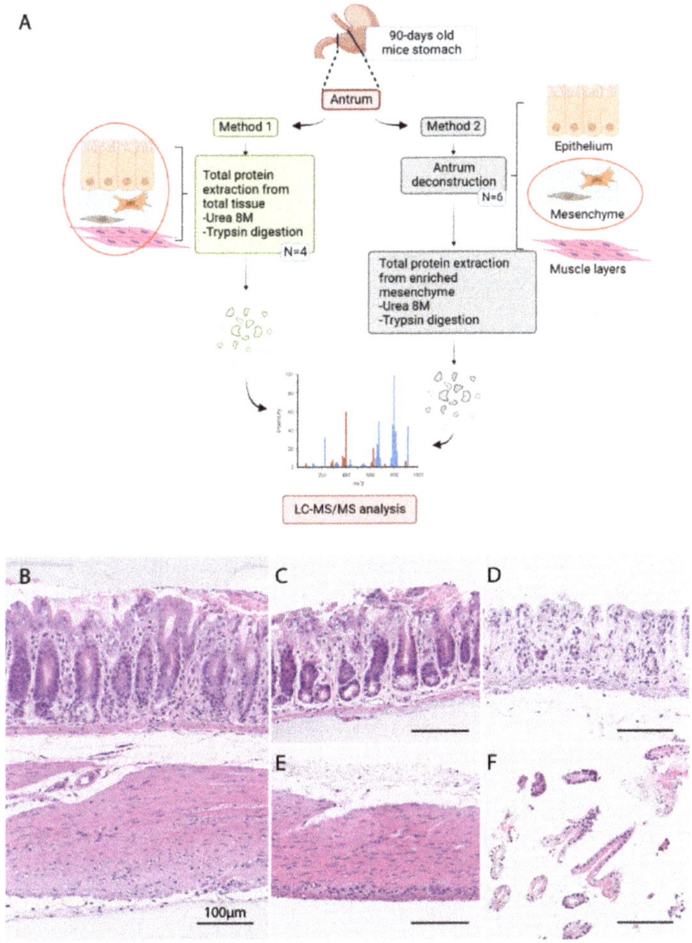

Figure 1. Methods for tissue preparation of stomach antrum for proteomic analysis. (**A**) Schematic representation of the experimental pipeline to assess the gastric matrisome profile in the *BmpR1a*$^{\triangle FoxL1+}$ mouse model. Created with BioRender.com. (**B–F**) Histological assessment of the deconstructed antrum tissue. Total antrum tissue (**B**) was deconstructed in a stepwise manner, where the muscle layers (**E**) were first dissociated from the other two compartments (**C**). Epithelial/mesenchymal tissue (**C**) was further dissociated, yielding the mesenchyme compartment (**D**) and the epithelium (**F**). Scale bar = 100 µm.

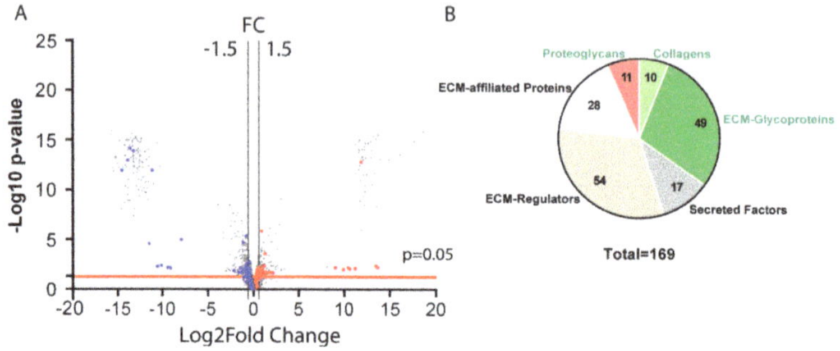

Figure 2. Total antrum matrisome in mice upon deletion of telocyte BMP-associated signaling. (**A**) Proteomic data from total antrum tissue isolated from control and $BmpR1a^{\triangle FoxL1+}$ mice (n = 4) were analyzed using ProStar to determine which proteins were significantly modulated. The volcano plot shows all differentially regulated proteins identified following mass spectrometry, highlighting significant matrisome proteins with at least a 1.5-fold change (plotted as log2FC) and a p-value lower than 0.05. Blue dots represent downregulated matrisome proteins; red dots represent upregulated matrisome proteins. The horizontal line represents the threshold p-value of 0.05. Vertical lines represent the 1.5-fold change threshold (in log2). Volcano plot was generated using GraphPad Prism version 9.4.1. (**B**). Pie chart indicates the number of matrisome proteins identified in total antrum tissue according to categories (core matrisome proteins in green and matrisome-associated proteins in black).

3.2. Analysis of the Matrisome from Enriched Mesenchymal Antrum of $BmpR1a^{\triangle FoxL1+}$ Mouse

Next, we evaluated changes in the ECM composition of antrum-enriched mesenchyme extracts from both mutant and control mice. We detected 37.5% fewer proteins in the enriched mesenchyme (2377) compared to those in the total antrum (3803); however, we discovered that a greater number of proteins were modulated, with 827 being upregulated and 492 being downregulated (Figure 3A). The analysis of the enriched mesenchymal antrum revealed the presence of 34 overexpressed matrisome proteins (dark red spots, FC > 1.5) and 59 downregulated proteins (dark blue spots, FC < −1.5) in $BmpR1a^{\triangle FoxL1+}$ mice compared to those in the control group (Figure 3A). As described above, matrisome proteins were identified using the Matrisome Annotator analytical tool (access date: 15 December 2020). A total of 135 proteins were identified, of which 68 belonged to the core matrisome and 67 to the matrisome-associated proteins. Of the proteins belonging to the core matrisome, we identified 10 proteoglycans, 12 CLs, and 46 glycoproteins, whereas among the matrisome-associated proteins, 21 ECM-affiliated proteins, 34 ECM regulators, and 12 secreted factors were identified (Table 2). As observed for the total tissue extract, most CL chains (CL1α1, CL4α1, CL6α1, α2, α3 and α5, and CL15α1) and most proteoglycans (perlecan, asporin, decorin, lumican, and versican) in the antrum enriched mesenchyme were downregulated in $BmpR1a^{\triangle FoxL1+}$ mice compared to those in the controls (Table 2). We observed that, unlike the total antrum extract, biglycan was downregulated in the enriched mesenchymal antrum extract from mutant mice compared to that from controls (Table 2). Similar results were obtained with the enriched mesenchymal antrum extract for glycoproteins. FN1, TNC, and VTN were upregulated, whereas MFAP2, 4, and 5, NID1 and NID2, and SPARCL-1 were downregulated in mutant mice compared to those measured in controls (Table 2). However, in the enriched mesenchymal antrum extract, Agrin was downregulated, in contrast to our observations for the total antrum extract. Finally, our analysis of the matrisome-associated proteins, ECM-affiliated proteins, ECM regulators, and secreted factors revealed variations in mostly similar proteins identified in the total tissue extract (Table 2). When we compared both analyses, we discovered that the matrisomic variations obtained from the enriched mesenchymal antrum extracts were more robust than those obtained from the total antrum extract.

Table 1. Total antrum tissue.

Core Matrisome

ECM-Glycoproteins

Name	FC	p-Value
Agrn	2372	0.0076
Dmbt1	3.60	0.0188
Fgb	2.96	0.0200
Fgg	2.67	0.0206
Fga	2.54	0.0175
Vtn	2.49	0.0191
Thbs1	2.10	0.0049
Mfge8	1.51	0.071
Tnc	1.40	0.0253
Igfbp7	1.39	0.1144
Creld2	1.35	0.1078
Fn1	1.33	0.0094
Vwa1	1.28	NC/0.0437
Fbln2	1.24	0.174
Sparc	1.23	0.1481
Postn	1.22	0.0344
Lrg1	1.14	0.2343
Vwa5a	1.13	0.1869
Tgfbi	1.12	0.1154
Aebp1	1.03	0.8022
Efemp1	1.02	0.7931
Ltbp4	1.01	0.906
Fbln1	−1.01	0.8928
Thbs4	−1.12	NC/0.5325
Nid2	−1.20	0.1529
Fbln5	−1.23	0.1405
Lamb1	−1.27	0.0300
Pcolce	−1.31	NC/0.0097
Lamb3	−1.32	NC/0.0020
Sbspon	−1.34	NC/0.0423
Tsku	−1.44	NC/0.0188
Dpt	−1.58	0.0069
Mfap4	−1.60	0.0210
Adipoq	−1.61	0.0623
Lama4	−1.67	0.0145
Lamc1	−1.69	0.0269
Mfap5	−1.69	0.0429
Nid1	−1.70	0.0116
Lama2	−1.70	0.0377
Tinagl1	−1.74	0.0283
Lama5	−1.77	0.0121
Emilin1	−1.79	NC/4 × 10^{-6}
Lamb2	−2.33	0.0140
Tnxb	−2.38	0.0193
Fbn1	−243	1×10^{-5}
Abi3bp	−1117	0.0038
Mfap2	−2262	NC/1 × 10^{-12}
Sparcl1	−15,377	1 × 10^{-13}
Spp1	−24,277	NC/1 × 10^{-12}

Collagen Chains

Name	FC	p-Value
Col18a1	1.94	0.0099
Col15a1	1.19	NC/0.1689
Col6a5	−1.15	0.4277
Col4a1	−1.36	0.1037
Col12a1	−1.73	0.0030
Col14a1	−2.08	0.0558
Col6a1	−3.10	0.0178
Col6a2	−4.52	0.0124
Col4a2	−553	0.0071
Col1a2	−1496	0.0049

Proteoglycans

Name	FC	p-Value
Bgn	1.31	0.0160
Prg2	1.09	0.7556
Vcan	−1.27	0.0841
Hspg2	−1.39	0.0302
Prelp	−1.41	0.0513
Lum	−1.47	0.0188
Aspn	−1.56	0.0104
Ogn	−1.85	0.0065
Dcn	−2.05	0.0094
Podn	−2835	3×10^{-5}
Fmod	−12,933	7×10^{-15}

Matrisome-Associated

ECM-Regulators

Name	FC	p-Value
Adam9	510	0.0063
Prss3	2.51	NC/0.0528
Leprel1	2.38	NC/0.0002
Serpinb1a	2.29	0.0044
Serpinb5	2.07	0.0183
Ctss	2.05	0.0043
Ctsc	1.85	0.0134
Serpinb12	1.84	NC/0.0554
Ctsh	1.73	0.0113
F2	1.56	0.0280
Try10	1.53	NC/0.0394
Hrg	1.51	0.0357
Plg	1.39	0.0466
Ctse	1.38	0.0548
Serpinf2	1.37	0.0572
Itih3	1.36	0.0472
Ctsa	1.35	0.0365
Serpini2	1.34	NC/0.3921
Serping1	1.33	0.0997
Serpinc1	1.30	0.1275
Itih2	1.26	0.1419
Kng1	1.24	0.0814
F13a1	1.23	0.0278
Cst3	1.23	0.0257
Adam10	1.21	0.1819
Ctsb	1.17	0.1297
Ctsl	1.15	0.1851
Ctsz	1.14	0.1260
A2m	1.13	0.3687
Itih1	1.13	0.2482
Serpinb9	1.10	0.4599
Serpina1e	1.08	0.8936
Serpinh1	1.06	0.5231
Ctsd	1.05	0.4297
Cstb	1.04	0.5912
Ngly1	1.03	0.8976
Serpinf1	−1.00	0.9864
Serpind1	−1.02	0.8891
Cela2a	−1.05	0.9419
Fam20b	−1.06	0.5655
St14	−1.09	0.4949
Cela3b	−1.11	0.8716
Serpina3k	−1.18	0.5193
Prss2	−1.20	0.7794
Serpina1d	−1.25	0.1443
Tgm2	−1.29	0.0344
Serpina1c	−1.30	0.3113
Cela1	−1.30	0.6149
F12	−1.38	0.0016
P4ha1	−1.41	NC/0.0071
Serpina1b	−1.47	0.1053
P4ha2	−1.63	NC/0.0159
Serpina6	−1.76	0.1680
Ambp	−671	0.0064

ECM-Affiliated Proteins

Name	FC	p-Value
Muc4	1594	0.0084
Mbl2	1382	0.0066
Lgals7	4.40	NC/0.0210
Lgals4	2.09	0.0276
Lgals9	1.63	0.0293
Anxa3	1.54	0.0224
Anxa10	1.51	0.0454
Anxa1	1.48	0.0121
Lgalsl	1.43	0.0206
Reg1	1.35	0.6373
Hpx	1.31	0.0656
Reg2	1.29	NC/0.1715
Anxa2	1.26	0.0574
Anxa7	1.25	0.0696
Lman1	1.12	0.4453
Anxa5	1.10	0.1127
Anxa4	1.08	0.2235
Muc6	1.05	0.7641
Sema4b	1.01	0.9016
Lgals3	−1.03	0.7888
Anxa11	−1.05	0.3729
Plxnb2	−1.06	0.3421
Lgals1	−1.10	0.2588
Anxa6	−1.47	0.0357
Muc5ac	−1.50	0.0477
Sdc1	−1.55	NC/0.0175
Lgals2	−2.76	0.0067
Cspg4	−9977	1 × 10^{-14}

Secreted factors

Name	FC	p-Value
S100a9	13058	0.0059
S100a8	11412	0.0043
Sfrp1	3720	NC/1 × 10^{-13}
Il1rn	948	0.0100
S100a6	1.99	0.0389
Rptn	1.86	NC/1 × 10^{-13}
S100g	1.58	0.0410
S100a4	1.52	0.0059
S100a1	1.46	0.0119
S100a13	1.38	0.0825
S100a11	1.36	0.0774
S100a14	1.32	0.0266
Il18	1.20	0.2779
S100a16	1.18	0.1155
Hcfc1	−1.12	0.0848
S100a10	−1.23	0.0344
S100b	−2.20	1 × 10^{-5}

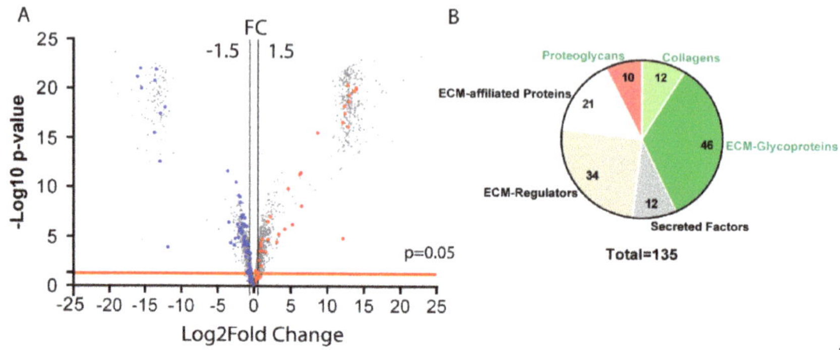

Figure 3. Enriched mesenchyme antrum matrisome in mice upon deletion of telocyte BMP-associated signaling. (**A**) Proteomic data from mesenchyme-enriched antrum tissue isolated from control and *BmpR1a*$^{\triangle FoxL1+}$ mice ($n = 6$) were analyzed using ProStar to determine which proteins were significantly modulated. The volcano plot shows all differentially regulated proteins identified following mass spectrometry, highlighting significant matrisome proteins with at least a 1.5-fold change (plotted as log2FC) and a p-value lower than 0.05. Blue dots represent downregulated matrisome proteins; Red dots represent upregulated matrisome proteins. The horizontal line represents the threshold p-value of 0.05. Vertical lines represent the 1.5-fold change threshold (in log2FC). Volcano plot was generated using GraphPad Prism version 9.4.1. (**B**) Pie chart indicates the number of matrisome proteins identified in mesenchyme-enriched antrum tissue according to categories (core matrisome proteins in green and matrisome-associated proteins in black).

Table 2. Total Enriched mesenchyme from antrum tissue.

Core Matrisome — ECM-Glyucoproteins

Name	FC	p-Value
Fbln1	5535	3×10^{-18}
Dmbt1	74,2	5×10^{-12}
Fgb	18,4	2×10^{-6}
Fgg	10,4	6×10^{-6}
Fga	8.91	3×10^{-5}
Tnc	1.96	9×10^{-5}
Vtn	1.93	NC/4×10^{-5}
Mfge8	1.78	0.0003
Fn1	1.46	0.0760
Fbln5	1.41	NC/0.0454
Igfbp7	1.30	0.1291
Ecm1	1.10	NC/0.3423
Creld2	1.09	0.6556
Ltbp4	−1.04	0.8548
Tgfbi	−1.06	0.5927
Agrn	−1.08	0.4679
Vwf	−1.32	0.0853
Vwa1	−1.32	NC/0.0617
Postn	−1.58	0.0054
Vwa5a	−1.61	0.0076
Mfap4	−1.61	0.0311
Lamb1	−1.73	0.0002
Emilin1	−1.79	0.0432
Adipoq	−1.95	NC/0.0357
Papln	−1.96	NC/9×10^{-7}
Aebp1	−2.00	0.0004

Core Matrisome — Collagen Chains

Name	FC	p-Value
Col18a1	2.28	2×10^{-5}
Col4a1	−1.17	0.4016
Col6a4	−1.49	0.0575
Col4a2	−1.72	0.0288
Col15a1	−2.72	2×10^{-5}
Col6a2	−2.84	6×10^{-7}
Col6a1	−2.96	3×10^{-7}
Col6a3	−3.08	1×10^{-7}
Col1a2	−6.44	7×10^{-5}
Col1a1	−9.15	4×10^{-5}
Col6a5	−10.97	3×10^{-7}
Col4a6	−12,771	NC/2×10^{-21}

Core Matrisome — Proteoglycans

Name	FC	p-Value
Prg2	2.27	0.0003
Hspg2	−1.4	0.0134
Bgn	−1.50	0.1212
Podn	−2.23	0.0010
Prelp	−3.01	5×10^{-7}
Aspn	−3.42	1×10^{-7}
Dcn	−3.62	3×10^{-9}
Lum	−3.78	9×10^{-10}
Ogn	−4.29	7×10^{-10}
Vcan	−12.30	2×10^{-12}

Matrisome-Associated — ECM-Regulators

Name	FC	p-Value
Serpinb5	15,966	7×10^{-21}
Plg	8304	4×10^{-20}
Mmp9	8201	2×10^{-19}
Loxl2	7570	NC/6×10^{-17}
Fam20b	5265	5×10^{-19}
Adam10	4586	2×10^{-5}
P4ha2	4433	NC/2×10^{-17}
Ctse	3.81	3×10^{-7}
Cst3	2.08	NC/0.0088
Serpinc1	1.88	0.1149
F13a1	1.66	0.1621
Serpinb9	1.64	NC/0.0025
Serpinb1a	1.27	0.0296
A2m	1.20	0.1905
Ctsc	−1.05	0.6744
Itih1	−1.08	0.7179
Ctsh	−1.12	0.2113
Ctsb	−1.13	0.2275
Ctsa	−1.20	0.1233
Serpina1c	−1.22	0.1968
Ctsz	−1.26	0.0404
Itih3	−1.30	0.2241
Ctsd	−1.40	0.0005
Itih2	−1.49	0.2789
Cstb	−1.52	0.0248
Serpinh1	−1.66	0.0003

Matrisome-Associated — ECM-Affiliated Proteins

Name	FC	p-Value
Muc4	7357	4×10^{-21}
Muc5ac	409	NC/3×10^{-16}
Anxa10	90.47	8×10^{-9}
Lgals4	25.55	1×10^{-10}
Lgals9	4.89	9×10^{-8}
Lgals2	3.87	2×10^{-5}
Anxa3	1.74	0.0003
Anxa11	1.35	0.0461
Anxa7	1.16	0.1634
Lman1	1.15	0.2975
Plxnb2	1.12	0.4665
Anxa4	−1.07	0.4975
Anxa1	−1.17	0.2537
Muc6	−1.22	0.2380
Cspg4	−1.25	NC/0.0437
Lgals3	−1.28	0.0832
Anxa2	−1.31	0.0187
Sema3d	−1.44	NC/0.0015
Anxa5	−2.05	5×10^{-5}
Lgals1	−2.78	8×10^{-5}
Anxa6	−3.53	2×10^{-6}

Matrisome-Associated — Secreted factors

Name	FC	p-Value
S100a16	16,177	1×10^{-20}
S100a14	12,545	2×10^{-20}

Table 2. *Cont.*

Core Matrisome						Matrisome-Associated					
ECM-Glyucoproteins			Collagen Chains			ECM-Regulators			ECM-Affiliated Proteins		
Name	FC	*p*-Value	Name	FC	*p*-Value	Name	FC	*p*-Value	Name	FC	*p*-Value
Nid2	−2.09	9×10^{-7}				Serpina3k	−1.68	NC/7×10^{-5}	S100a9	85.2	NC/3×10^{-12}
Lamc1	−2.33	9×10^{-6}				Serping1	−1.93	NC/3×10^{-5}	S100a8	37.9	6×10^{-7}
Lama4	−2.34	1×10^{-7}				P4ha1	−2.01	8×10^{-5}	S100a1	3.1	NC/0.0003
Nid1	−2.50	5×10^{-7}				Ctss	−2.13	5×10^{-5}	S100a4	1.3	0.2283
Sbspon	−2.54	NC/3×10^{-5}				Tgm2	−2.88	6×10^{-8}	Angptl2	1.22	NC/0.0289
Lama5	−2.57	2×10^{-5}				Cela1	−3.38	4×10^{-6}	Hcfc1	1.20	0.2782
Tinagl1	−3.45	1×10^{-6}				Ambp	−13,221	3×10^{-16}	S100a6	1.20	0.1979
Lamb2	−3.98	1×10^{-6}				Adamts20	−68,487	NC/6×10^{-22}	S100a11	1.12	0.3482
Mfap5	−4.09	3×10^{-6}							S100a13	−1.12	0.4822
Tnxb	−4.37	1×10^{-5}							S100a10	−2.16	0.0004
Lama2	−4.66	6×10^{-7}									
Dpt	−5.49	3×10^{-11}									
Fbn1	−5.99	2×10^{-5}									
Sparc	−3620	0.0001									
Spp1	−4710	NC/8×10^{-19}									
Mmrn2	−7450	3×10^{-18}									
Mfap2	−7588	3×10^{-13}									
Abi3bp	−11,200	1×10^{-22}									
Fbn2	−46,287	1×10^{-20}									
Spon1	−51,711	NC/9×10^{-23}									

Data from both types of tissue extracts analyzed were further processed to remove irrelevant data, which led to the identification of 184 matrisome proteins between both experiments (Figure 4). Venn diagrams of the different protein categories, core matrisome (in green), and matrisome-associated proteins (in black), revealed that mesenchymal enrichment did not lead to heavy loss of matrisomal proteins in relation to the total tissue extract, except for the ECM regulators, which were more affected by the tissue treatment. Next, we performed a functional association network using the STRING database and the 116 matrisome proteins that were identified to be significantly modulated in both experiments to obtain a signature profile of proteins indicative of biological processes occurring in the microenvironment of our mouse model. The STRING analysis revealed changes in proteins involved in immune regulation, fibrosis, and tumor microenvironment in *BmpR1a*$^{\triangle FoxL1+}$ mice compared to those in controls (data not shown).

3.3. Loss of BMP Signaling in Gastric TC^{FoxL1+} Induces Dysregulations in ECM Biodynamics Associated with Inflammation

The tissue microenvironment can play an important role in cellular behavior, and ECM proteins influence the biodynamics as well as cell biology of tissues [37–39]. The core matrisome proteins' influence on the microenvironment through biomechanical and biochemical sensing is evident. However, it is important to take into consideration that the ECM can act as a reservoir for secreted growth factors, chemokines, and cytokines also affecting the microenvironment and impacting cell behavior [37,39]. Histopathologically, *BmpR1a*$^{\triangle FoxL1+}$ mice have been shown to be more prone to gastric neoplasia with mild inflammation [22]. Here, a part of the functional network analysis suggested a protein signature profile linked to immune regulation. S100A8 and S100A9, both secreted factors associated with the ECM, have been associated with acute and chronic inflammatory conditions and autoimmune diseases [40–42]. Matrisomic profiling revealed a significant increase in S100A8 and S100A9 between *BmpR1a*$^{\triangle FoxL1+}$ mice and controls in total antrum (FC = 11,412 and 13058, respectively; Table 1) as well as in the enriched mesenchymal antrum (FC = 37.9 and 85.2, respectively; Table 2). S100A9 expression in mutant mice was confirmed through immunofluorescence, with strong expression in the *BmpR1a*$^{\triangle FoxL1+}$ mouse mesenchyme, whereas controls showed no expression of the protein (Figure 5A). In addition, immunoblot analysis against secreted factors S100A8 and A9 revealed de novo expression of both proteins in the mutant mice but not controls, where these proteins were not detected (fold change = 20.34 and 20.48, respectively; Figure 5B,C).

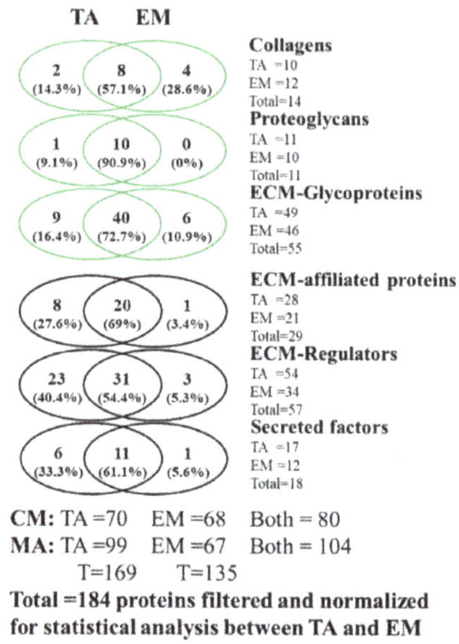

Figure 4. Mesenchymal enrichment of the antrum does not lead to notable ECM protein loss. Venn diagrams illustrating the overall *BmpR1a*$^{\triangle FoxL1+}$ mouse gastric matrisome proteins identified using the two methods combined, indicating a wide overlap between the two approaches. Core matrisome proteins are presented in green and matrisome-associated proteins are shown in black. TA, total antrum; EM, enriched mesenchyme; CM, core matrisome; MA, matrisome-associated.

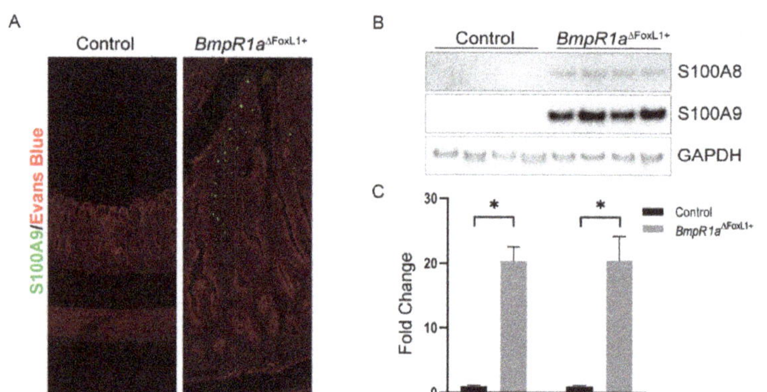

Figure 5. S100A8 and A9 proteins are upregulated secreted factors in *BmpR1a*$^{\triangle FoxL1+}$ mice, indicating an inflammatory response. (**A**) Immunostaining against S100A9 (shown in green) revealed an increase in its expression in the mesenchyme-enriched area of the antrum tissue of BmpR1a$^{\triangle FoxL1+}$ mice compared to that in controls. (**B**) Immunoblot analysis of the total antrum tissue indicates strong expression of both S100A8 and S100A9 proteins in BmpR1a$^{\triangle FoxL1+}$ mice compared to that in controls. (**C**) Quantification of immunoblots confirmed a significant increase in both S100A8 and S100A9 in the mutant animals (FC = 20.34 and 20.48, respectively) compared to that in controls. Statistical analysis was assessed using the Mann–Whitney test with * $p < 0.05$. Evans blue was used as a counterstain (red signal in (**A**)). Scale bar = 100 µm.

3.4. Disruption of the CL Network in Mice with Impaired Gastric BMP Signaling in TC^{FoxL1+}

CL is a dominant and important element in the pathological microenvironment and has a significant influence on the initiation and development of pathologies such as cancer [10]. Furthermore, its expression is generally increased in gastric cancers [43]. However, as shown in Tables 1 and 2, the expression of almost all CL chains was negatively modulated in $BmpR1a^{\triangle FoxL1+}$ mice compared to that in controls (CL1α2, CL4α1, and α2; CL6α1, α2 and α5; CL12α1 and CL14α1). Only a few examples were observed to be positively modulated in mutant mice using both tissue preparation methods (Tables 1 and 2). These results differ from previously published work with this mouse model [22], in which marked expression and accumulation of CLI and IV in the gastric glands of $BmpR1a^{\triangle FoxL1+}$ mice were observed. Therefore, we decided to perform further analyses of the CL network in both mouse groups. Collagen deposition, fiber orientation, and spatial distribution were analyzed using picrosirius red staining under bright and polarized light microscopy in both control and mutant mice (Figure 6). The loss of BMP signaling in TC^{FoxL1+} mice affected the sub-epithelial CL fiber network in mutant mice, mainly towards the upper part of the gland, compared to controls, as shown following picrosirius red staining under bright field (Figure 6A, left panels). Visualization of CL fibers orientation and alignment was performed with polarized light, where fibrillar CL appeared in a range of colors from red, yellow, orange, and green (Figure 6A middle panels). Heterogeneous organization of CL fibers was observed in $BmpR1a^{\triangle FoxL1+}$ mice, with areas of increased alignment of fibrillar collagen towards the top of the gland compared to that in controls (Figure 6A middle and right panels). Analysis using the OrientationJ plugin in ImageJ indicates a similar distribution of fiber angles between the control and $BmpR1a^{\triangle FoxL1+}$ mice in the middle part of the glands (Figure 6B). However, the upper gland of the mutant mice revealed a divergent spatial organization of CL fibers with respect to the organization observed in the controls (Figure 6C). The coherency factor was significantly higher in the top of the gland in $BmpR1a^{\triangle FoxL1+}$ mice (CF = 0.338), indicating that the CL fibers tended to be in a predominant direction and had an increased alignment compared to that observed in control mice (CF = 0.245; Figure 6D).

3.5. Loss of BMP Signaling in Gastric TC^{FoxL1+} Causes Remodeling of ECM Glycoproteins Associated with Early Gastric Neoplasia

ECM glycoproteins and ECM regulators are other matrix components essential for proper tissue function, including the stomach [10,44]. In addition, part of the functional annotation analysis also suggested a protein signature profile linked to the tumor microenvironment. Over the years, several ECM glycoproteins and ECM regulators have been associated with every stage of gastric cancer [45–47]. Matrisomic profiling revealed a significant increase in ECM glycoproteins such as FN1 between $BmpR1a^{\triangle FoxL1+}$ and control enriched mesenchymal antrum (FC = 1.46; Table 2) and TNC in total antrum (FC = 1.4; Table 1) as well as in enriched mesenchymal antrum (FC = 1.95; Table 2). A significant decrease in SPARCL-1 in total antrum (FC = −15377; Table 1) was also observed. Finally, we identified a significant increase in the ECM regulator, ADAM9, only in the in total antrum (FC = 510; Table 1). FN1 (Figure 7A) and TNC (Figure 7B) exhibited increased expressions in $BmpR1a^{\triangle FoxL1+}$ mice compared to those in controls, as confirmed by immunofluorescence of stomach sections (Figure 7A,B). The immunoblot analysis against SPARCL-1 confirmed a significant decrease in this ECM glycoprotein in mutant mice compared to that measured in controls (fold change = 0.48; Figure 7C,D). Immunoblot analysis against ADAM9 confirmed a significant increase in this ECM regulator in $BmpR1a^{\triangle FoxL1+}$ mice compared to that in controls (fold change = 1.976; Figure 7C,D).

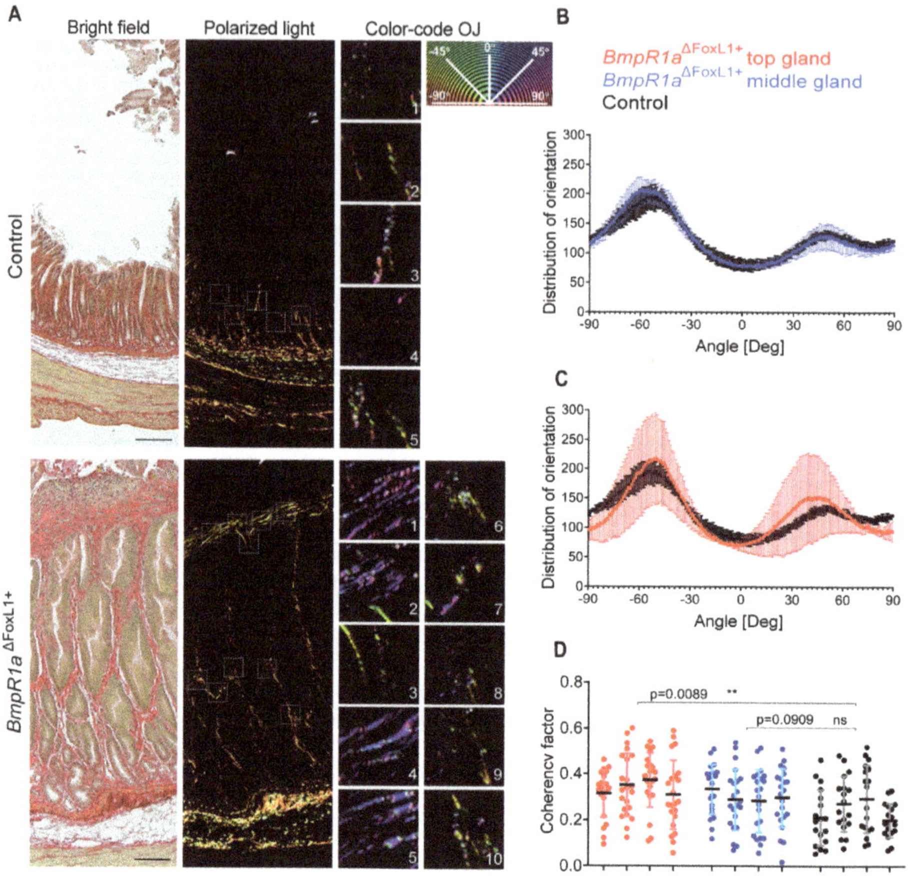

Figure 6. Loss of BMP signaling in gastric TC^{FoxL1+} disrupts the collagen network. (**A**) Picrosirius red staining was performed on stomach sections from both control and *BmpR1a*$^{\triangle FoxL1+}$ mice. Collagen fiber organization and alignment was evaluated under bright field (left panels) and polarizing light (middle panels). Imaging was performed using a Zeiss Axioscope 5 equipped with an analyzer and a linear polarizer. ROI (dotted squares) were converted to grayscale 16-bit images and color-coded where pixel hue corresponds to the angle of local fiber orientation, which ranges from −90° to +90°. Representative ROI are shown with their color-coded fiber orientation (right panels) and color-coded orientation legend is shown. (**B**) Distribution of fiber orientations was compiled for each ROI in all analyzed images, to compare control tissue with middle of the gland in *BmpR1a*$^{\triangle FoxL1+}$ mice antrum. Data are shown as means of distribution ± SD, for four individual mice in each group. (**C**) Distribution of fiber orientations was compiled for each ROI in all analyzed images, to compare control tissue with top of the gland in *BmpR1a*$^{\triangle FoxL1+}$ mice antrum. Data are shown as means of distribution ± SD, for four individual mice in each group. (**D**) Coherency factor was computed for all ROI and data were plotted showing a significant increase of fiber alignment in the top part of antrum gland in *BmpR1a*$^{\triangle FoxL1+}$ mice, with a mean coherency factor of 0.338 compared to 0.245 observed in control mice. No significant difference was observed between the middle part of antrum gland in control and that in *BmpR1a*$^{\triangle FoxL1+}$ mice. Statistical analyses were performed using Prism, with a table and group nested ANOVA. Scale bar = 50 μm. ** $p < 0.01$. ROI: representative region of interests.

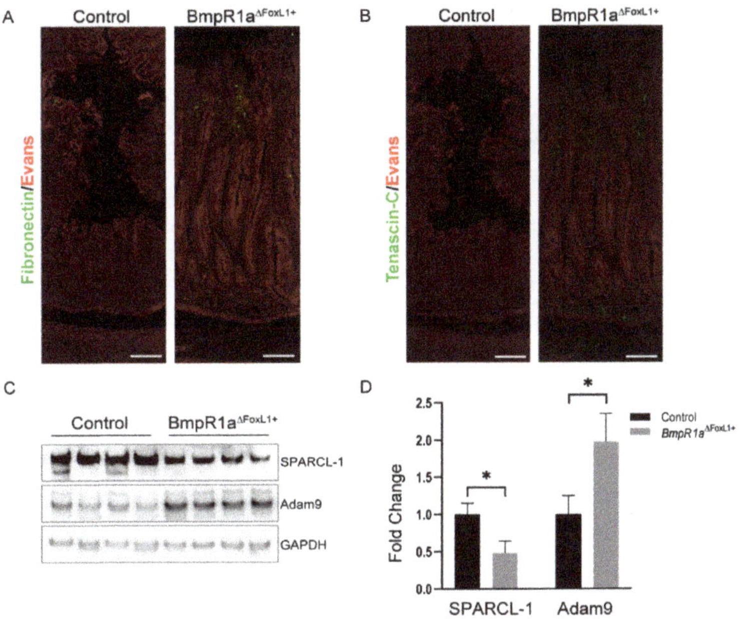

Figure 7. Modulations in ECM glycoproteins and ECM regulator correlate with a neoplasia phenotype in stomach of *BmpR1a*$^{\triangle FoxL1+}$ mice. (**A**). Immunostaining against ECM glycoprotein fibronectin (shown in green) revealed an increased expression in the enlarged mesenchymal area of the antrum tissue in *BmpR1a*$^{\triangle FoxL1+}$ mice compared to that in controls. (**B**) Immunostaining against ECM glycoprotein Tenascin C (shown in green) revealed an increased expression in the antrum mesenchyme of *BmpR1a*$^{\triangle FoxL1+}$ mice compared to that in controls. (**C**). Immunoblot analysis showed a decrease of the ECM glycoprotein SPARCL-1 expression and an increase of the ECM regulator ADAM9 in total antrum samples of *BmpR1a*$^{\triangle FoxL1+}$ mice compared to that in controls. GAPDH was used as a loading control. (**D**) Quantification of immunoblots revealed a significant modulation of SPARCL-1 and ADAM9 between both group (FC = 0.48 and 1.98, respectively). All quantifications were performed using ImageJ and statistical analyses were performed using Prism. All immunoblot quantification data are presented as the mean ± SD (n = 4). Statistical analysis was assessed using the Mann–Whitney test with * $p < 0.05$. Evans blue was used as a counterstain (red signal in A and B). Scale bar = 100 µm.

4. Discussion

Due to the complexity and extremely low solubility of the ECM, exhaustive biochemical characterization of tissues has long been a challenge. In recent years, mass spectrometry has been used to characterize ECM proteins in various tissues [14,48–50]. In addition, the developments brought forward by Naba et al. of an in silico definition of the matrisome provide a possibility for a detailed characterization of the biochemistry and composition of the ECM in normal and diseased tissues [13,14,48,51]. Similar to other diseases, ECM deregulation has been shown to play a role in gastric neoplasia by creating a favorable microenvironment for the transformed cells to thrive from pre-neoplastic lesions to metastatic stages [5,52]. Recent studies have demonstrated that TC^{FoxL1+} are strong contributors to the GI microenvironment [15,17,20,23,24]; however, their precise contribution to the ECM fractions of the microenvironment is less clear. Qualitative analysis of ECM proteins in the *BmpR1a*$^{\triangle FoxL1+}$ mouse, where TC^{FoxL1+} are impaired in BMP signaling, suggests a potential role for this mesenchymal cell population in contributing to the ECM fraction of the microenvironment [18,21,22]. In addition, the pathophysiological phenotype of the *BmpR1a*$^{\triangle FoxL1+}$ mouse model is characterized by the development of gastric pre-neoplastic lesions [22]. Together, we discovered that *BmpR1a*$^{\triangle FoxL1+}$ mice represent an adequate model for understanding how TC^{FoxL1+} participates in an aberrant gastric pre-neoplastic ECM microenvironment.

As part of our study was to characterize the ECM contribution of BMP-signaling impaired TC^{FoxL1+} to the pre-neoplastic gastric microenvironment, we explored the validity of using enriched mesenchyme over total tissue extract for targeting matrisomic proteins. Tissue deconstruction into minimal mesenchymal compartment, where TC^{FoxL1+} and the microenvironment are observable, allows for the possibility of circumventing the complexity of the total tissue protein content. As expected, we observed an important decrease in the presence of ECM regulator proteins when we used enriched mesenchymal extract in comparison to the total tissue extract because these proteins are not bound to the ECM. Thus, they are easily lost during purification processes [48]. Deconstruction of the gastric antrum provides a more comprehensive analysis of the matrisome in *BmpR1a*$^{\triangle FoxL1+}$ mice compared to controls, with the removal of background noise from non-matrisomic proteins. In addition, the mesenchymal-enriched extract allows for improved identification of proteins with low expression levels that could be easily lost in a larger pool of proteins.

In a previous study, the gastric pathophysiological aspects of the *BmpR1a*$^{\triangle FoxL1+}$ mouse model showed that disruption of BMP signaling in TC^{FoxL1+} led to the creation of a toxic microenvironment with an increase in CLI, fibronectin, HGF, and FSP1/S100A4, pressuring the epithelium to initiate pre-malignant lesions [22]. Correa's cascade of gastric carcinogenesis shows that a normal gastric epithelium gradually transitions from initial gastritis to chronic gastritis, mucosal atrophy, metaplasia, dysplasia, and carcinoma [53,54]. Early steps of this cascade prior to carcinoma involve the presence of inflammatory processes [54,55] and a reorganization of the nurturing microenvironment into a tumor microenvironment [5]. Interestingly, some protein profiles, such as immune regulation, fibrosis, and tumor microenvironment, were noticeably modulated in the *BmpR1a*$^{\triangle FoxL1+}$ matrisome analysis. Thus, the present protein profile, in combination with our previous phenotypic analysis of *BmpR1a*$^{\triangle FoxL1+}$ mice, allows for a better understanding of the sequence of events occurring in the ECM microenvironment of these mice with BMP-impaired TC^{FoxL1+} with regard to early events in gastric neoplasia.

Consequently, the overexpression of S100A8 and A9 in the matrisomic analysis, as secreted factors associated with the ECM, supports these profiles. Both proteins have been associated with numerous human disorders, including acute and chronic inflammatory conditions, autoimmune diseases, and cancer [40,56,57]. They are also reported to represent highly potent biomarkers of a wide range of inflammatory processes, including rheumatoid arthritis and inflammatory bowel disease [41,58]. In tumor biology, both proteins play a fundamental role, and their levels are elevated in numerous tumors, including gastric cancer, which is in line with our model [57,59–63]. Although there are signs of inflammation in mice with infiltration of lymphocytes (CD3) and macrophages (F4/80), no chronic inflammation was observed [22]. This could partially explain the overexpression of S100A8/A9 in the gastric microenvironment of the *BmpR1a*$^{\triangle FoxL1+}$ mice.

As for the tumor microenvironment profile identified in this study, ECM glycoproteins and ECM regulators are known to play key roles in the microenvironment for proper tissue function including the stomach [2,5,10,45,64–66]. For example, matricellular proteins such as FN1, TNC, and ADAM9 were upregulated, while SPARCL-1/Hevin was downregulated. In addition, these ECM glycoproteins and ECM regulators have been linked to the tumor microenvironment in various stages of gastric cancer [67–70]. Deregulation of protein expression, such as FN1 and ADAM9 (upregulated) or SPARCL-1 (downregulated), has been shown to affect cell growth and tissue proliferation in gastric cancer [70–73]. The hyperplasia seen in the gastric glands of *BmpR1a*$^{\triangle FoxL1+}$ mice [22] could be, in part, explained by the modification of these proteins in the microenvironment. TNC is generally absent or suppressed in most normal adult tissues, while it is markedly overexpressed in some pathological conditions, such as wound healing, inflammation, and in a variety of neoplasms [74]. This expression pattern was observed in the stomachs of *BmpR1a*$^{\triangle FoxL1+}$ mice when compared to that of controls. Thus, similar to gastrointestinal stromal tumors [67], whereas TNC is used as a potential marker, it can also be used as an indicator of gastric premalignancies, according to the results shown in this study.

CL is a polymeric protein present in greater quantities in the ECM under physiological conditions [75,76], as well as in the tumor microenvironment, where its extensive deposition is one of the pathological characteristics of cancers, such as gastric neoplasia [43,77]. As collagens play an important structural role in the ECM and contribute to its mechanical properties by influencing cellular behavior [78], any changes in CL organization, expression, and/or crosslinking will directly affect optimal tissue function [79]. Unexpectedly, in this study, we discovered that almost all CL chains analyzed using MS were downregulated in the *BmpR1a*$^{\triangle FoxL1+}$ pre-neoplastic model. This is in contrast to previous findings, especially regarding what is known from descriptive studies on ECM in gastric cancer, as well as previous studies with *BmpR1a*$^{\triangle FoxL1+}$ [5,22,43]. Other proteomic analyses have shown the difficulties of optimal CL protein extraction from tissues, especially when fibrotic [36,80,81]. We hypothesize that the extraction method used in this study was not optimal for CL protein analysis [81]. However, the choice of another method favoring CL protein extraction could be detrimental to the analysis of other matrisomic proteins [81]. Considering that CL chain expression, as well as its mechanical and biochemical organization, could be validated through other techniques, proteomic analyses would not be the preferred technique for studying fibrotic tissues. In this study, Sirius red staining under bright field was used for the visualization of total CL deposition in tissue, while under polarized light microscopy it provided more relevant information regarding the CL network, such as its organization, stiffness, and fiber alignment.

Altogether, the present study provides a more comprehensive representation of the evolving ECM fraction from the microenvironment in pre-neoplastic gastric lesions associated with BMP signaling-impaired TC^{FoxL1+}. These findings support the importance of TC^{FoxL1+} and BMP signaling in the maintenance of a healthy microenvironment to maintain gastric homeostasis and prevent the development of pathologies such as neoplasia.

Author Contributions: Conceptualization, N.P. and F.-M.B.; methodology, A.B.A.; software, J.R., C.-M.J.; validation, A.B.A., V.P. and V.R.N.; formal analysis, A.B.A., V.P., F.-M.B. and NP.; investigation, A.B.A.; resources, N.P.; data curation, A.B.A.; writing—original draft preparation, N.P.; writing—review and editing, N.P. and F.-M.B.; visualization, A.B.A.; supervision, N.P. and F.-M.B.; project administration, N.P.; funding acquisition, N.P. and F.-M.B. All authors have read and agreed to the published version of the manuscript.

Funding: This research was funded by the Natural Sciences and Engineering Research Council of Canada (NSERC) RGPIN-2018-05414 to F.M.B. and RGPIN 2018-06115 to N.P.

Institutional Review Board Statement: The study protocol was approved by the Animal Welfare Research Committee of the Faculty of Medicine and Health Sciences of the Université de Sherbrooke (FMSS-2019-2370), and all experiments were conducted in strict adherence to the standards and policies of the Canadian Council on Animal Care in Sciences.

Data Availability Statement: Raw files, databases, and MaxQuant results have been deposited in ProteomeXchange with the accession number PXD038603.

Acknowledgments: NP and FMB are members of the Fonds de Recherche du Québec-Santé-funded "Centre de Recherche CHUS". FMB is an FRQS senior scholar (award number 281824). The authors thank KHK for providing the *FoxL1*Cre transgenic line, Ariane De Castro for her technical assistance with the mouse colony and genotyping, and Electron Microscopy and Histology Research Core of the Faculté de Médecine et des Sciences de la Santé at the Université de Sherbrooke for their histology, electron microscopy, and phenotyping services.

Conflicts of Interest: The authors declare no conflict of interest.

References

1. Mouw, J.K.; Ou, G.; Weaver, V.M. Extracellular matrix assembly: A multiscale deconstruction. *Nat. Rev. Mol. Cell Biol.* **2014**, *15*, 771–785. [CrossRef] [PubMed]
2. Walker, C.; Mojares, E.; Del Rio Hernandez, A. Role of Extracellular Matrix in Development and Cancer Progression. *Int. J. Mol. Sci.* **2018**, *19*, 3028. [CrossRef]
3. Filipe, E.C.; Chitty, J.L.; Cox, T.R. Charting the unexplored extracellular matrix in cancer. *Int. J. Exp. Pathol.* **2018**, *99*, 58–76. [CrossRef] [PubMed]
4. Pickup, M.W.; Mouw, J.K.; Weaver, V.M. The extracellular matrix modulates the hallmarks of cancer. *EMBO Rep.* **2014**, *15*, 1243–1253. [CrossRef] [PubMed]
5. Moreira, A.M.; Pereira, J.; Melo, S.; Fernandes, M.S.; Carneiro, P.; Seruca, R.; Figueiredo, J. The Extracellular Matrix: An Accomplice in Gastric Cancer Development and Progression. *Cells* **2020**, *9*, 394. [CrossRef]
6. Gerarduzzi, C.; Hartmann, U.; Leask, A.; Drobetsky, E. The Matrix Revolution: Matricellular Proteins and Restructuring of the Cancer Microenvironment. *Cancer Res.* **2020**, *80*, 2705–2717. [CrossRef]
7. Brauchle, E.; Kasper, J.; Daum, R.; Schierbaum, N.; Falch, C.; Kirschniak, A.; Schaffer, T.E.; Schenke-Layland, K. Biomechanical and biomolecular characterization of extracellular matrix structures in human colon carcinomas. *Matrix Biol.* **2018**, *68–69*, 180–193. [CrossRef]
8. Xu, S.; Xu, H.; Wang, W.; Li, S.; Li, H.; Li, T.; Zhang, W.; Yu, X.; Liu, L. The role of collagen in cancer: From bench to bedside. *J. Transl. Med.* **2019**, *17*, 309. [CrossRef]
9. Yang, Z.; Xue, F.; Li, M.; Zhu, X.; Lu, X.; Wang, C.; Xu, E.; Wang, X.; Zhang, L.; Yu, H.; et al. Extracellular Matrix Characterization in Gastric Cancer Helps to Predict Prognosis and Chemotherapy Response. *Front. Oncol.* **2021**, *11*, 753330. [CrossRef]
10. Bonnans, C.; Chou, J.; Werb, Z. Remodelling the extracellular matrix in development and disease. *Nat. Rev. Mol. Cell Biol.* **2014**, *15*, 786–801. [CrossRef]
11. Hynes, R.O.; Naba, A. Overview of the matrisome—An inventory of extracellular matrix constituents and functions. *Cold Spring Harb. Perspect. Biol.* **2012**, *4*, a004903. [CrossRef] [PubMed]
12. Karamanos, N.K.; Theocharis, A.D.; Piperigkou, Z.; Manou, D.; Passi, A.; Skandalis, S.S.; Vynios, D.H.; Orian-Rousseau, V.; Ricard-Blum, S.; Schmelzer, C.E.H.; et al. A guide to the composition and functions of the extracellular matrix. *FEBS J.* **2021**, *288*, 6850–6912. [CrossRef] [PubMed]
13. Naba, A.; Clauser, K.R.; Ding, H.; Whittaker, C.A.; Carr, S.A.; Hynes, R.O. The extracellular matrix: Tools and insights for the "omics" era. *Matrix Biol.* **2016**, *49*, 10–24. [CrossRef] [PubMed]
14. Naba, A.; Clauser, K.R.; Hoersch, S.; Liu, H.; Carr, S.A.; Hynes, R.O. The matrisome: In silico definition and in vivo characterization by proteomics of normal and tumor extracellular matrices. *Mol. Cell. Proteom.* **2012**, *11*, M111.014647. [CrossRef]
15. McCarthy, N.; Manieri, E.; Storm, E.E.; Saadatpour, A.; Luoma, A.M.; Kapoor, V.N.; Madha, S.; Gaynor, L.T.; Cox, C.; Keerthivasan, S.; et al. Distinct Mesenchymal Cell Populations Generate the Essential Intestinal BMP Signaling Gradient. *Cell Stem Cell* **2020**, *26*, 391–402.e5. [CrossRef]
16. Powell, D.W.; Pinchuk, I.V.; Saada, J.I.; Chen, X.; Mifflin, R.C. Mesenchymal cells of the intestinal lamina propria. *Annu. Rev. Physiol.* **2011**, *73*, 213–237. [CrossRef]
17. Shoshkes-Carmel, M.; Wang, Y.J.; Wangensteen, K.J.; Toth, B.; Kondo, A.; Massasa, E.E.; Itzkovitz, S.; Kaestner, K.H. Subepithelial telocytes are an important source of Wnts that supports intestinal crypts. *Nature* **2018**, *557*, 242–246. [CrossRef]
18. Allaire, J.M.; Roy, S.A.; Ouellet, C.; Lemieux, E.; Jones, C.; Paquet, M.; Boudreau, F.; Perreault, N. Bmp signaling in colonic mesenchyme regulates stromal microenvironment and protects from polyposis initiation. *Int. J. Cancer* **2016**, *138*, 2700–2712. [CrossRef]
19. Langlois, M.J.; Servant, R.; Reyes Nicolas, V.; Jones, C.; Roy, S.A.B.; Paquet, M.; Carrier, J.C.; Rivard, N.; Boudreau, F.; Perreault, N. Loss of PTEN Signaling in Foxl1(+) Mesenchymal Telocytes Initiates Spontaneous Colonic Neoplasia in Mice. *Cell. Mol. Gastroenterol. Hepatol.* **2019**, *8*, 530–533.e5. [CrossRef]
20. McCarthy, N.; Kraiczy, J.; Shivdasani, R.A. Cellular and molecular architecture of the intestinal stem cell niche. *Nature* **2020**, *22*, 1033–1041. [CrossRef]
21. Reyes Nicolas, V.; Allaire, J.M.; Alfonso, A.B.; Pupo Gomez, D.; Pomerleau, V.; Giroux, V.; Boudreau, F.; Perreault, N. Altered Mucus Barrier Integrity and Increased Susceptibility to Colitis in Mice upon Loss of Telocyte Bone Morphogenetic Protein Signalling. *Cells* **2021**, *10*, 2954. [CrossRef] [PubMed]
22. Roy, S.A.; Allaire, J.M.; Ouellet, C.; Maloum-Rami, F.; Pomerleau, V.; Lemieux, E.; Babeu, J.P.; Rousseau, J.; Paquet, M.; Garde-Granger, P.; et al. Loss of mesenchymal bone morphogenetic protein signaling leads to development of reactive stroma and initiation of the gastric neoplastic cascade. *Sci. Rep.* **2016**, *6*, 32759. [CrossRef] [PubMed]
23. Aoki, R.; Shoshkes-Carmel, M.; Gao, N.; Shin, S.; May, C.L.; Golson, M.L.; Zahm, A.M.; Ray, M.; Wiser, C.L.; Wright, C.V.; et al. Foxl1-expressing mesenchymal cells constitute the intestinal stem cell niche. *Cell. Mol. Gastroenterol. Hepatol.* **2016**, *2*, 175–188. [CrossRef] [PubMed]
24. Greicius, G.; Kabiri, Z.; Sigmundsson, K.; Liang, C.; Bunte, R.; Singh, M.K.; Virshup, D.M. PDGFRalpha(+) pericryptal stromal cells are the critical source of Wnts and RSPO3 for murine intestinal stem cells in vivo. *Proc. Natl. Acad. Sci. USA* **2018**, *115*, E3173–E3181. [CrossRef]

25. Kaestner, K.H. The Intestinal Stem Cell Niche: A Central Role for Foxl1-Expressing Subepithelial Telocytes. *Cell. Mol. Gastroenterol. Hepatol.* **2019**, *8*, 111–117. [CrossRef]
26. Kondo, A.; Kaestner, K.H. Emerging diverse roles of telocytes. *Development* **2019**, *146*, dev175018. [CrossRef]
27. Sackett, S.D.; Fulmer, J.T.; Friedman, J.R.; Kaestner, K.H. Foxl1-Cre BAC transgenic mice: A new tool for gene ablation in the gastrointestinal mesenchyme. *Genesis* **2007**, *45*, 518–522. [CrossRef]
28. Mishina, Y.; Hanks, M.C.; Miura, S.; Tallquist, M.D.; Behringer, R.R. Generation of Bmpr/Alk3 conditional knockout mice. *Genesis* **2002**, *32*, 69–72. [CrossRef]
29. Pomerleau, V.; Nicolas, V.R.; Jurkovic, C.M.; Faucheux, N.; Lauzon, M.A.; Boisvert, F.M.; Perreault, N. FOXL1+ Telocytes in mouse colon orchestrate extracellular matrix biodynamics and wound repair resolution. *J. Proteom.* **2023**, *21*, 104755. [CrossRef]
30. Wieczorek, S.; Combes, F.; Lazar, C.; Giai Gianetto, Q.; Gatto, L.; Dorffer, A.; Hesse, A.M.; Coute, Y.; Ferro, M.; Bruley, C.; et al. DAPAR & ProStaR: Software to perform statistical analyses in quantitative discovery proteomics. *Bioinformatics* **2017**, *33*, 135–136. [CrossRef]
31. Allaire, J.M.; Darsigny, M.; Marcoux, S.S.; Roy, S.A.; Schmouth, J.F.; Umans, L.; Zwijsen, A.; Boudreau, F.; Perreault, N. Loss of Smad5 leads to the disassembly of the apical junctional complex and increased susceptibility to experimental colitis. *Am. J. Physiol. Gastrointest. Liver Physiol.* **2011**, *300*, G586–G597. [CrossRef] [PubMed]
32. Gagne-Sansfacon, J.; Allaire, J.M.; Jones, C.; Boudreau, F.; Perreault, N. Loss of Sonic hedgehog leads to alterations in intestinal secretory cell maturation and autophagy. *PLoS ONE* **2014**, *9*, e98751. [CrossRef] [PubMed]
33. Maloum, F.; Allaire, J.M.; Gagne-Sansfacon, J.; Roy, E.; Belleville, K.; Sarret, P.; Morisset, J.; Carrier, J.C.; Mishina, Y.; Kaestner, K.H.; et al. Epithelial BMP signaling is required for proper specification of epithelial cell lineages and gastric endocrine cells. *Am. J. Physiol. Gastrointest. Liver Physiol.* **2011**, *300*, G1065–G1079. [CrossRef] [PubMed]
34. Lattouf, R.; Younes, R.; Lutomski, D.; Naaman, N.; Godeau, G.; Senni, K.; Changotade, S. Picrosirius red staining: A useful tool to appraise collagen networks in normal and pathological tissues. *J. Histochem. Cytochem.* **2014**, *62*, 751–758. [CrossRef] [PubMed]
35. Perreault, N.; Beaulieu, J.F. Primary cultures of fully differentiated and pure human intestinal epithelial cells. *Exp. Cell Res.* **1998**, *245*, 34–42. [CrossRef]
36. Naba, A.; Pearce, O.M.T.; Del Rosario, A.; Ma, D.; Ding, H.; Rajeeve, V.; Cutillas, P.R.; Balkwill, F.R.; Hynes, R.O. Characterization of the Extracellular Matrix of Normal and Diseased Tissues Using Proteomics. *J. Proteome Res.* **2017**, *16*, 3083–3091. [CrossRef] [PubMed]
37. Hynes, R.O. Stretching the boundaries of extracellular matrix research. *Nat. Rev. Mol. Cell Biol.* **2014**, *15*, 761–763. [CrossRef]
38. Gjorevski, N.; Sachs, N.; Manfrin, A.; Giger, S.; Bragina, M.E.; Ordonez-Moran, P.; Clevers, H.; Lutolf, M.P. Designer matrices for intestinal stem cell and organoid culture. *Nature* **2016**, *539*, 560–564. [CrossRef]
39. Lu, P.; Weaver, V.M.; Werb, Z. The extracellular matrix: A dynamic niche in cancer progression. *J. Cell Biol.* **2012**, *196*, 395–406. [CrossRef]
40. Grevers, L.C.; de Vries, T.J.; Vogl, T.; Abdollahi-Roodsaz, S.; Sloetjes, A.W.; Leenen, P.J.; Roth, J.; Everts, V.; van den Berg, W.B.; van Lent, P.L. S100A8 enhances osteoclastic bone resorption in vitro through activation of Toll-like receptor 4: Implications for bone destruction in murine antigen-induced arthritis. *Arthritis Rheum.* **2011**, *63*, 1365–1375. [CrossRef]
41. Ehrchen, J.M.; Sunderkotter, C.; Foell, D.; Vogl, T.; Roth, J. The endogenous Toll-like receptor 4 agonist S100A8/S100A9 (calprotectin) as innate amplifier of infection, autoimmunity, and cancer. *J. Leukoc. Biol.* **2009**, *86*, 557–566. [CrossRef] [PubMed]
42. Foell, D.; Wittkowski, H.; Roth, J. Mechanisms of disease: A 'DAMP' view of inflammatory arthritis. *Nat. Clin. Pr. Rheumatol.* **2007**, *3*, 382–390. [CrossRef] [PubMed]
43. Zhou, Z.H.; Ji, C.D.; Xiao, H.L.; Zhao, H.B.; Cui, Y.H.; Bian, X.W. Reorganized Collagen in the Tumor Microenvironment of Gastric Cancer and Its Association with Prognosis. *J. Cancer* **2017**, *8*, 1466–1476. [CrossRef] [PubMed]
44. Kim, S.; Min, S.; Choi, Y.S.; Jo, S.H.; Jung, J.H.; Han, K.; Kim, J.; An, S.; Ji, Y.W.; Kim, Y.G.; et al. Tissue extracellular matrix hydrogels as alternatives to Matrigel for culturing gastrointestinal organoids. *Nat. Commun.* **2022**, *13*, 1692. [CrossRef] [PubMed]
45. Abbaszadegan, M.R.; Mojarrad, M.; Moghbeli, M. Role of extra cellular proteins in gastric cancer progression and metastasis: An update. *Genes Environ.* **2020**, *42*, 18. [CrossRef] [PubMed]
46. Li, L.; Zhu, Z.; Zhao, Y.; Zhang, Q.; Wu, X.; Miao, B.; Cao, J.; Fei, S. FN1, SPARC, and SERPINE1 are highly expressed and significantly related to a poor prognosis of gastric adenocarcinoma revealed by microarray and bioinformatics. *Sci. Rep.* **2019**, *9*, 7827. [CrossRef] [PubMed]
47. Sung, J.Y.; Cheong, J.H. The Matrisome Is Associated with Metabolic Reprograming in Stem-like Phenotypes of Gastric Cancer. *Cancers* **2022**, *14*, 1438. [CrossRef]
48. Naba, A.; Clauser, K.R.; Hynes, R.O. Enrichment of Extracellular Matrix Proteins from Tissues and Digestion into Peptides for Mass Spectrometry Analysis. *J. Vis. Exp.* **2015**, e53057. [CrossRef]
49. Socovich, A.M.; Naba, A. The cancer matrisome: From comprehensive characterization to biomarker discovery. *Semin. Cell Dev. Biol.* **2019**, *89*, 157–166. [CrossRef]
50. Taha, I.N.; Naba, A. Exploring the extracellular matrix in health and disease using proteomics. *Essays Biochem.* **2019**, *63*, 417–432. [CrossRef]
51. Shao, X.; Taha, I.N.; Clauser, K.R.; Gao, Y.T.; Naba, A. MatrisomeDB: The ECM-protein knowledge database. *Nucleic Acids Res.* **2020**, *48*, D1136–D1144. [CrossRef] [PubMed]
52. Hanahan, D.; Weinberg, R.A. Hallmarks of cancer: The next generation. *Cell* **2011**, *144*, 646–674. [CrossRef] [PubMed]

53. Correa, P. A human model of gastric carcinogenesis. *Cancer Res.* **1988**, *48*, 3554–3560. [PubMed]
54. Piazuelo, M.B.; Correa, P. Gastric cancer: Overview. *Colomb. Med.* **2013**, *44*, 192–201.
55. De Re, V. Molecular Features Distinguish Gastric Cancer Subtypes. *Int. J. Mol. Sci.* **2018**, *19*, 3121. [CrossRef] [PubMed]
56. Hiratsuka, S.; Watanabe, A.; Aburatani, H.; Maru, Y. Tumour-mediated upregulation of chemoattractants and recruitment of myeloid cells predetermines lung metastasis. *Nat. Cell Biol.* **2006**, *8*, 1369–1375. [CrossRef]
57. Salama, I.; Malone, P.S.; Mihaimeed, F.; Jones, J.L. A review of the S100 proteins in cancer. *Eur. J. Surg. Oncol.* **2008**, *34*, 357–364. [CrossRef]
58. Foell, D.; Wittkowski, H.; Vogl, T.; Roth, J. S100 proteins expressed in phagocytes: A novel group of damage-associated molecular pattern molecules. *J. Leukoc. Biol.* **2007**, *81*, 28–37. [CrossRef]
59. Emberley, E.D.; Murphy, L.C.; Watson, P.H. S100 proteins and their influence on pro-survival pathways in cancer. *Biochem. Cell Biol.* **2004**, *82*, 508–515. [CrossRef]
60. Gebhardt, C.; Nemeth, J.; Angel, P.; Hess, J. S100A8 and S100A9 in inflammation and cancer. *Biochem. Pharmacol.* **2006**, *72*, 1622–1631. [CrossRef]
61. Hermani, A.; De Servi, B.; Medunjanin, S.; Tessier, P.A.; Mayer, D. S100A8 and S100A9 activate MAP kinase and NF-kappaB signaling pathways and trigger translocation of RAGE in human prostate cancer cells. *Exp. Cell Res.* **2006**, *312*, 184–197. [CrossRef] [PubMed]
62. Ott, H.W.; Lindner, H.; Sarg, B.; Mueller-Holzner, E.; Abendstein, B.; Bergant, A.; Fessler, S.; Schwaerzler, P.; Zeimet, A.; Marth, C.; et al. Calgranulins in cystic fluid and serum from patients with ovarian carcinomas. *Cancer Res.* **2003**, *63*, 7507–7514. [PubMed]
63. Bresnick, A.R.; Weber, D.J.; Zimmer, D.B. S100 proteins in cancer. *Nat. Rev. Cancer* **2015**, *15*, 96–109. [CrossRef] [PubMed]
64. Efthymiou, G.; Saint, A.; Ruff, M.; Rekad, Z.; Ciais, D.; Van Obberghen-Schilling, E. Shaping Up the Tumor Microenvironment With Cellular Fibronectin. *Front. Oncol.* **2020**, *10*, 641. [CrossRef]
65. Liu, A.Y.; Zheng, H.; Ouyang, G. Periostin, a multifunctional matricellular protein in inflammatory and tumor microenvironments. *Matrix Biol.* **2014**, *37*, 150–156. [CrossRef]
66. Sirica, A.E. Matricellular proteins in intrahepatic cholangiocarcinoma. *Adv. Cancer Res.* **2022**, *156*, 249–281. [CrossRef]
67. Shen, C.; Wang, C.; Yin, Y.; Chen, H.; Yin, X.; Cai, Z.; Chen, Z.; Zhang, B.; Zhou, Z. Tenascin-C expression is significantly associated with the progression and prognosis in gastric GISTs. *Medicine* **2019**, *98*, e14045. [CrossRef]
68. Sun, Y.; Zhao, C.; Ye, Y.; Wang, Z.; He, Y.; Li, Y.; Mao, H. High expression of fibronectin 1 indicates poor prognosis in gastric cancer. *Oncol. Lett.* **2020**, *19*, 93–102. [CrossRef]
69. Wang, J.; Deng, L.; Huang, J.; Cai, R.; Zhu, X.; Liu, F.; Wang, Q.; Zhang, J.; Zheng, Y. High expression of Fibronectin 1 suppresses apoptosis through the NF-kappaB pathway and is associated with migration in nasopharyngeal carcinoma. *Am. J. Transl. Res.* **2017**, *9*, 4502–4511.
70. Jakharia, A.; Borkakoty, B.; Singh, S. Expression of SPARC like protein 1 (SPARCL1), extracellular matrix-associated protein is down regulated in gastric adenocarcinoma. *J. Gastrointest. Oncol.* **2016**, *7*, 278–283. [CrossRef]
71. Pankov, R.; Yamada, K.M. Fibronectin at a glance. *J. Cell Sci.* **2002**, *115*, 3861–3863. [CrossRef] [PubMed]
72. Wang, J.; Zhou, Y.; Fei, X.; Chen, X.; Yan, J.; Liu, B.; Zhu, Z. ADAM9 functions as a promoter of gastric cancer growth which is negatively and post-transcriptionally regulated by miR-126. *Oncol. Rep.* **2017**, *37*, 2033–2040. [CrossRef] [PubMed]
73. Wang, H.; Zhang, J.; Li, H.; Yu, H.; Chen, S.; Liu, S.; Zhang, C.; He, Y. FN1 is a prognostic biomarker and correlated with immune infiltrates in gastric cancers. *Front. Oncol.* **2022**, *12*, 918719. [CrossRef] [PubMed]
74. Midwood, K.S.; Orend, G. The role of tenascin-C in tissue injury and tumorigenesis. *J. Cell Commun. Signal.* **2009**, *3*, 287–310. [CrossRef] [PubMed]
75. Bella, J.; Hulmes, D.J. Fibrillar Collagens. *Subcell. Biochem.* **2017**, *82*, 457–490. [CrossRef]
76. Ricard-Blum, S. The collagen family. *Cold Spring Harb. Perspect. Biol.* **2011**, *3*, a004978. [CrossRef]
77. Fang, M.; Yuan, J.; Peng, C.; Li, Y. Collagen as a double-edged sword in tumor progression. *Tumor Biol.* **2014**, *35*, 2871–2882. [CrossRef]
78. Taufalele, P.V.; VanderBurgh, J.A.; Munoz, A.; Zanotelli, M.R.; Reinhart-King, C.A. Fiber alignment drives changes in architectural and mechanical features in collagen matrices. *PLoS ONE* **2019**, *14*, e0216537. [CrossRef]
79. Kirkness, M.W.; Lehmann, K.; Forde, N.R. Mechanics and structural stability of the collagen triple helix. *Curr. Opin. Chem. Biol.* **2019**, *53*, 98–105. [CrossRef]
80. Krasny, L.; Paul, A.; Wai, P.; Howard, B.A.; Natrajan, R.C.; Huang, P.H. Comparative proteomic assessment of matrisome enrichment methodologies. *Biochem. J.* **2016**, *473*, 3979–3995. [CrossRef]
81. McCabe, M.C.; Schmitt, L.R.; Hill, R.C.; Dzieciatkowska, M.; Maslanka, M.; Daamen, W.F.; van Kuppevelt, T.H.; Hof, D.J.; Hansen, K.C. Evaluation and Refinement of Sample Preparation Methods for Extracellular Matrix Proteome Coverage. *Mol. Cell. Proteom.* **2021**, *20*, 100079. [CrossRef] [PubMed]

biomedicines

MDPI

Article

Aberrant Expression and Prognostic Potential of IL-37 in Human Lung Adenocarcinoma

Panayiota Christodoulou [1,2,†], Theodora-Christina Kyriakou [3,†], Panagiotis Boutsikos [2], Maria Andreou [1], Yuan Ji [4], Damo Xu [5], Panagiotis Papageorgis [3] and Maria-Ioanna Christodoulou [1,4,*]

1 Tumor Immunology and Biomarkers Laboratory, Basic and Translational Cancer Research Center, Department of Life Sciences, School of Sciences, European University Cyprus, Nicosia 2404, Cyprus
2 School of Medicine, European University Cyprus, Nicosia 2404, Cyprus
3 Tumor Microenvironment, Metastasis and Experimental Therapeutics Laboratory, Basic and Translational Cancer Research Center, Department of Life Sciences, School of Sciences, European University Cyprus, Nicosia 2404, Cyprus
4 School of Infection and Immunity, University of Glasgow, Glasgow G12 8TA, UK
5 State Key Laboratory of Respiratory Disease for Allergy at Shenzhen University, Shenzhen Key Laboratory of Allergy and Immunology, Shenzhen University School of Medicine, Shenzhen 518055, China
* Correspondence: mar.christodoulou@euc.ac.cy; Tel.: +357-22-713044
† These authors contributed equally to this work.

Abstract: Interleukin-37 (IL-37) is a relatively new IL-1 family cytokine that, due to its immunoregulatory properties, has lately gained increasing attention in basic and translational biomedical research. Emerging evidence supports the implication of this protein in any human disorder in which immune homeostasis is compromised, including cancer. The aim of this study was to explore the prognostic and/or diagnostic potential of IL-37 and its receptor SIGIRR (single immunoglobulin IL-1-related receptor) in human tumors. We utilized a series of bioinformatics tools and -omics datasets to unravel possible associations of IL-37 and SIGIRR expression levels and genetic aberrations with tumor development, histopathological parameters, distribution of tumor-infiltrating immune cells, and survival rates of patients. Our data revealed that amongst the 17 human malignancies investigated, IL-37 exhibits higher expression levels in tumors of lung adenocarcinoma (LUAD). Moreover, the expression profiles of IL-37 and SIGIRR are associated with LUAD development and tumor stage, whereas their high mRNA levels are favorable prognostic factors for the overall survival of patients. What is more, *IL-37* correlates positively with a LUAD-associated transcriptomic signature, and its nucleotide changes and expression levels are linked with distinct infiltration patterns of certain cell subsets known to control LUAD anti-tumor immune responses. Our data indicate the potential value of IL-37 and its receptor SIGIRR to serve as biomarkers and/or immune-checkpoint therapeutic targets for LUAD patients. Further, the data highlight the urgent need for further exploration of this cytokine and the underlying pathogenetic mechanisms to fully elucidate its implication in LUAD development and progression.

Keywords: interleukin (IL-)37; lung adenocarcinoma; biomarker; survival; infiltration rates

Citation: Christodoulou, P.; Kyriakou, T.-C.; Boutsikos, P.; Andreou, M.; Ji, Y.; Xu, D.; Papageorgis, P.; Christodoulou, M.-I. Aberrant Expression and Prognostic Potential of IL-37 in Human Lung Adenocarcinoma. *Biomedicines* **2022**, *10*, 3037. https://doi.org/10.3390/biomedicines10123037

Academic Editor: Manoj K. Mishra

Received: 20 October 2022
Accepted: 21 November 2022
Published: 24 November 2022

1. Introduction

Interleukin-37 (IL-37) is one of the latest members included in the IL-1 family of cytokines, known to suppress innate immune responses and modulate acquired ones. Thus, this cytokine possesses a pivotal role in inflammation related to the pathophysiology of various human disorders, including autoimmune diseases, inflammatory systemic conditions, infections, and cancer [1]. It is produced by immune and non-immune cells and acts via inhibition of the production of pro-inflammatory cytokines and activation of anti-inflammatory signals [2]. Similar to other immune-regulatory cytokines (e.g., TGF-β and IL-10), IL-37 has attracted notable interest both from a basic biological but also from a translational research perspective [2].

The human *IL-37* gene is located on chromosome 2q12-13, very close to the regulatory regions of the genes encoding the IL-1a and IL-1b cytokines [3]. The gene encodes for five protein isoforms, a, b, c, d, and e; however, their specific functions as well as their relative abundance are not yet fully elucidated [2]. Among all isoforms, IL-37b is the longest one (consisting of five of the six exons; all except exon 3) and the most well-studied [4,5]. In humans, IL-37 is reported to be constitutively expressed by circulating monocytes, tissue macrophages, dendritic cells (DCs), tonsil B cells, and plasma cells [2,6]. Upon pro-inflammatory stimuli, its expression is significantly augmented by tissue cells and peripheral blood mononuclear cells (PBMCs), predominantly monocytes, thus possessing a prevailing immunoregulatory role [1,6,7].

IL-37 exerts its effects on target cells via two distinct mechanisms: (i) Extracellularly, it binds to interleukin 18 receptor 1 (IL-18R1), which recruits interleukin 1 receptor 8 (IL-1R8, also known as single Ig IL-1-related receptor—SIGIRR), essential for the anti-inflammatory actions of the cytokine and forms a complex that transduces the signal intracellularly [8–11]; the activity of the cytokine lies mainly in the suppression of pro-inflammatory signaling factors, including mTOR (mammalian target of rapamycin), STAT1 (signal transducer and activator of transcription 1), AKT (Ak strain transforming), p53, p38, SHP-2 (SH2 domain-containing protein tyrosine phosphatase-2), Syk (Spleen Associated Tyrosine Kinase) [2], and also on the enhancement of anti-inflammatory signaling factors, such as PTEN (phosphatase and tensin homolog) phosphatase, to further inhibit inflammation through PI_3K (Phosphoinositide 3-kinases) kinase, mTOR, MAPK (mitogen-activated protein kinase), and FADK (focal adhesion kinase) pathways [12]. (ii) Intracellularly, upon its cleavage by caspase-1 at aspartic acid (D20) residue and binding to Smad3, IL-37 can be translocated into the nucleus, where it dampens the expression of inflammatory genes [13–16].

Evidence of the role of IL-37 in cancer, emerging during the last years, support its tumor-protective properties exerted through the enhancement of anti-tumor immunity, specifically within the tumor microenvironment (TME). At this point, it is worthwhile to highlight the importance of the local microenvironment, consisting of distinct immune and non-immune, cellular and non-cellular components (growth factors, chemokines, cytokines), in the development and progression of human tumors, as well as response-to-treatment [17]. TME interplays are orchestrated by tumor-infiltrating lymphocytes (TILs), natural killer (NK) cells, tumor-infiltrating dendritic cells (TIDCs), tumor-associated macrophages (TAMs), tumor-associated neutrophils (TANs), cancer-associated fibroblasts (CAFs), and myeloid-derived suppressor cells (MDSCs). Within these cell types, T helper (Th) 1, cytotoxic T, NK, B cells, M1 macrophages (MΦ), and mature DCs represent partners of immune control against malignant cells, and Th2, regulatory T cells (Tregs), M2 MΦ, neutrophils, CAFs, immature DCs, and MDSCs promote immune escape. Essentially, certain elements of TME have been targeted by therapeutic drugs (antibodies against immune checkpoints, such as PD-1/PD-L1, as well as anti-angiogenic factors, such as anti-VEGF-A) that are associated with good clinical outcomes [18].

IL-37 overexpression by TAMs derived from patients with human hepatocellular carcinoma (HCC) inhibits M2 polarization via regulation of the IL-6/STAT3 pathway, to suppress tumor growth in vivo [19]. High IL-37 expression by HCC tumor cells is associated with upregulated levels of CCL3 and CCL20 and increased recruitment of $CD1a^{+}$ dendritic cells (DCs) in tumor infiltrations [20]. What is more, IL-37 secreted from HCC cells enhances the expression of MHC-II, CD86, and CD40 surface molecules and the secretion of IL-2, IL-12, IL-12p70, interferon-a (IFN-α), and IFN-γ cytokines by DCs, which is, in turn, associated with an increased proportion of $IFN\gamma^{+}CD8^{+}$ T cells [20]. Additional in vivo experiments showed that overexpression of IL-37 in HCC cells resulted in increased recruitment of $CD11c^{+}$ DCs in the tumor microenvironment and tumor growth delay [20]. On the other hand, a very recent study using an experimental colorectal cancer (CRC) model reported that IL-37 transgenic mice are highly prone to developing colitis-associated CRC, which is characterized by severely increased tumor burdens and dysfunction of infiltrating $CD8^{+}$ T cells, dependent on SIGIRR [21].

Apart from immune-related effects, IL-37 also exerts its anti-cancer activity on other aspects of tumor development. First, it acts as an anti-angiogenic factor; its expression by cancer cells suppresses the tubule formation of human umbilical vein endothelial cells (HUVEC) cells in vitro, decreases the expression of matrix metallopeptidase 2 (MMP2) and vascular endothelial growth factor (VEGF) in SK-Hep-1 cells, and inhibits tumor angiogenesis in a murine model of HCC [22]. Second, it suppresses migration through the inhibition of Rac1 activation in various tumor cell types; indeed, intracellular IL-37 binds to the C-terminal region of the protein, preventing its membrane translocation and downstream signaling [22,23]. It has been observed that decreased expression of IL-37 in human lung adenocarcinoma (LUAD) biopsies is associated with tumor metastasis [23]. Lastly, the cytokine can act against tumor progression through the modulation of N6-methyladenosine (m6A) activity and inhibition of the Wnt5a/5b pathway in lung cancer cells [24].

Clinical observations over the last years have shed light on the potential of this cytokine to serve as a possible biomarker in various human malignancies. In CRC patients, serum IL-37 levels were found to be significantly elevated and positively correlated with the levels of CEA (carcinoembryonic antigen), a commonly used diagnostic biomarker for the disease [21]. In these patients, a negative correlation between IL-37 levels in the serum and $CD8^+$ T cell infiltration in the tumor was also observed [21]. Importantly, IL-37 expression in CRC tumors was found (a) to be linearly correlated with their stage, with the highest expression detected in stage I and the lowest in stage IV tumors; and (b) to be associated with survival rates, with higher levels predicting longer disease-free (DFS) and overall (OS) survival [25]. It is of note that intratumoral IL-37 levels, together with the incidence of $CD66b^+$ neutrophils, as well as mismatch repair (MMR) status, have been proven to be independent prognostic factors and are included in nomograms predicting DFS and OS in CRC, which could facilitate individualized patients' management [25].

Elevated serum IL-37 levels were also detected in patients with transitional cell carcinoma of the bladder (TCC) [26]. In melanoma, high levels of IL-37 expressed by peripheral Tregs were found to mirror the secretion of IL-1β mediators, especially TGFβ, by the tumor, suggesting it could be used as a possible biomarker for tumor-induced immunosuppression [27]. Furthermore, in HCC tumor infiltration high prevalence of $IL\text{-}37^+CD1a^+$ DCs biopsies was linked to higher survival rates of patients [20]. On the other hand, the ratio of IL-18-to-IL-37 levels was higher in the serum and PBMCs of patients with oral squamous cell carcinoma (OSCC) compared to non-cancer individuals and associated with shorter OS and DFS [28]. Low levels of IL-37 in the sera of patients with acute myeloid leukemia (AML) were shown to be associated with poor prognosis of the disease, but they were restored to normal in complete remission [29]. Finally, in breast cancer, peripheral blood *IL-37* mRNA levels and $CD8^+$ T cell numbers were decreased in patients compared to healthy individuals, and they were correlated with $ER^+/PR^+/HER2^+$ status [30].

In this study, we aimed at the investigation of the possible prognostic potential of IL-37 in patients with cancer utilizing bioinformatics tools and publicly available databases. Since our initial results indicated that among various human malignancies, *IL-37* exerts its highest expression in lung adenocarcinoma (LUAD), the study was subsequently focused on this cancer type. IL-37 levels were found to be correlated with tumor development, stage, grade, and the improved overall survival of patients, and mutations and gene expression levels were associated with the differential distribution of immune cells infiltrating the tumor.

2. Materials and Methods

2.1. Study Design

We first investigated the possible differential distribution of *IL-37* and *SIGIRR* expression levels in various human cancers using the Tissue Atlas tool of the Human Protein Atlas website [31]. Lung cancer, and more specifically lung adenocarcinoma (LUAD), was selected for further analysis. The TNMplot web tool [32] was used to compare the mRNA expression levels in LUAD versus non-LUAD specimens, the UALCAN portal [33] to explore the differential distribution among tumors of various histology, stage, nodal metastasis, or TP53 status, and the Kaplan–Meier plotter tool [34] to assess the effect of mRNA levels on survival rates of LUAD patients. Protein expression profiles as well as associations with various parameters of the pathology of the tumor were investigated through the UALCAN portal [33] and the Pathology tool of the Human Protein Atlas website [31]. To explore the expression distribution of *IL-37* and *SIGIRR* genes in different cell types of the human lung, the Single-Cell Type Atlas part of the Human Protein Atlas was used [35]. Finally, the effect of *IL-37* nucleotide changes or aberrations in expression levels on the differential distribution of various immune cell subsets infiltrating the LUAD tumor was analyzed using the TIMER2.0 webserver [36].

2.2. Study of the Expression Levels of IL-37 and SIGIRR in Various Human Cancers

To explore the expression patterns of *IL-37* and *SIGIRR* in various human cancers, the Tissue Atlas tool [31] of the Human Protein Atlas website was used (www.proteinatlas.org, accessed on 24 August 2022). The program processes data from RNA-sequencing experiments on tumor samples of various origin (17 different types of cancer, n = 7932 total samples). The mean of the FPKM levels of the genes in each cancer type and in the total cohort of cancer patients as well as its standard deviation (SD) was estimated. Non-parametric Mann–Whitney U test were applied for the evaluation of differences in gene expression levels.

2.3. Investigation of IL-37 and SIGIRR Expression Levels in LUAD versus Non-LUAD Lung Tissue

The TNMplot tool (www.tnmplot.com, accessed on 3 September 2022) [32] was used to explore impaired expression levels of the *IL-37* and *SIGIRR* genes in LUAD tumors. Comparative analysis processed RNA-sequencing data deposited in The Cancer Genome Atlas (TCGA) database from (a) 524 LUAD versus 486 non-LUAD individuals and (b) 57 pairs of LUAD versus adjacent normal tissue biopsies. Fold-changes of the median expression levels between the groups and the *p*-values assessed utilizing the non-parametric Mann–Whitney U test are reported; significant changes were considered those with a p-value < 0.05 and a fold-change > 2 or <0.5.

2.4. Exploration of Associations between IL-37 or SIGIRR Expression Levels with Certain Pathological Characteristics of the LUAD Tumor

To investigate possible associations between the gene-expression levels of *IL-37* and *SIGIRR* with certain parameters of the pathology of the tumor, data from the UALCAN portal were processed http://ualcan.path.uab.edu/ [33] (accessed on 3 September 2022). Gene expression levels were analyzed in correlation with histological type, cancer stage, nodal metastasis, and TP53 mutation status. Fold-changes compared to control groups >2 or <0.5 and *p*-values < 0.05 were considered significant.

2.5. Assessment of the Effect of IL-37 and SIGIRR Expression Levels on Survival Rates of Patients with LUAD

The Kaplan–Meier (KM) Plotter tool (www.kmplot.com, accessed on 24 August 2022) [34] was used to evaluate the ability of *IL37* expression to serve as prognostic factor for overall or relapse-free survival (OS or RFS, respectively) in LUAD patients. The tool analyzed RNA-sequencing data from 504 LUAD participants (deposited in Gene Expression Omnibus (GEO), EGA (European Genome-Phenome Archive), and TCGA databases) categorized based on *IL-37* or *SIGIRR* levels of expression ranging from high to low and calculated the hazard ratio (HR) and logrank *p*-values for the probability of survival at 250 months.

2.6. Development of a List of Positively/Negatively Correlated Genes of IL-37 in LUAD Biopsies

To obtain a list with genes whose expression levels are positively or negatively correlated with those of *IL-37*, the UALCAN tool (http://ualcan.path.uab.edu/ (accessed on 20 September 2022)) was utilized [33]. The tool processed RNAseq data from LUAD tumor biopsies of TCGA. A heatmap was generated, and Pearson's correlation test was applied to assess the significance of the data ($p < 0.05$ were considered significant). Correlations of *IL-37* with genes encoding IC proteins were specifically explored via the Correlation analysis option of the TNM plotter tool [32]. Data from RNA seq experiments on LUAD tumors were processed; Spearman's rho and *p*-values were obtained, and a correlation coefficient cut-off = 0 was set.

2.7. Investigation of the IL-37 and SIGIRR Protein Expression Levels

Protein expression in LUAD tumors was first explored using the Pathology tool of the Human Protein Atlas website, www.proteinatlas.org [35] (accessed on 20 September 2022). Deposited pictures of immunohistochemically stained sections of paraffin-embedded LUAD tissues were observed. Staining had been performed using polyclonal antibodies against human IL-37 (HPA054371) and SIGIRR (HPA023188). Levels of protein expression (z-values) in paired LUAD primary vs. normal tissues, as well as in tumors of different histological subtype, stage, grade, status of the HIPPO, WNT, mTOR, NRF2, RTK, or p53/Rb-related pathways, SWI-SNF complex, MYC/MYCN or chromatin modifier, were analyzed through the UALCAN web portal http://ualcan.path.uab.edu/ (accessed on 1 October 2021) [33]. *p*-Values calculated utilizing the non-parametric Mann–Whitney U test for differences in between-two group analyses are reported; *p*-values < 0.05 were considered significant.

2.8. Blood and Immune Single-Cell Analysis in Lung Tissue

To visualize single-cell RNA-seq (scRNAseq) data from human lung tissue, the Single-Cell Type Atlas was used. *IL-37* and *SIGIRR* expression profiles of blood and immune cells including macrophages, alveolar cells type 1 and 2, T cells, granulocytes, fibroblasts, club cells, ciliated cells, and endothelial cells are depicted in colored clusters at UMAP plots and in bar charts. Elevated expression levels (read counts normalized to transcripts per million protein coding genes, pTPM) of *IL-37* and *SIGIRR* in different blood and immune cell groups can categorize genes as cell type-enriched (at least four-fold higher mRNA level in a certain cell type compared to any other cell type), group-enriched (at least four-fold higher average mRNA level in a group of 2–10 cell types compared to any other cell type), and cell type-enhanced (at least four-fold higher mRNA level in a cell certain cell type compared to the average level in all other cancer types).

2.9. Analysis of Possible Correlations between IL-37 Gene Alterations or Expression Levels with Immune Cell Infiltration Patterns

To analyze the effect of *IL-37* nucleotide changes on or the association of *IL-37* expression levels with immune cell infiltration of lung tumors, the TIMER2.0 webserver was used [36]. The "mutation" module was utilized for the investigation of possible differential distribution of macrophages, $CD4^+$ or $CD8^+$ T cells, Tregs, dendritic cells (DCs), neutrophils, B cells, monocytes, NK cells, MDSCs, and endothelial cells in LUAD tumors of patients with *IL-37* somatic mutations vs. those without. The Wilcoxon *p*-value and $\log_2$fold-change of infiltration levels between the groups were estimated. The "gene" module was utilized for the exploration of possible correlations between the expression levels of *IL-37* and the levels of infiltration of the tumors by $CD4^+$ or $CD8^+$ T cells, Tregs, $\gamma\delta$ T cells B cells, neutrophils, monocytes, macrophages, DCs, NK, mast cells, cancer-associated fibroblasts, lymphoid, myeloid or granulocyte-lymphocyte progenitor cells, endothelial cells, eosinophils, hematopoietic stem cells, and MDSCs. All data were filtered for tumor purity. Spearman's rho and *p*-values were calculated for the evaluation of the linear positive or negative correlation.

2.10. Analysis of Possible Correlations between IL-37 Gene Alterations or Expression Levels with Immune Cell Infiltration Patterns

The STRING (Search Tool for the Retrieval of Interacting Genes/Proteins) database (https://string-db.org) [37] (accessed on 1 October 2022) was used to explore the cancer/LUAD molecular/cellular networks in which IL-37 signaling is implicated. The Biological Process (Gene Ontology), Molecular Function (Gene Ontology), Cellular Component (Gene Ontology), Reference publications (PubMed), Local network cluster (STRING), KEGG Pathways, Reactome Pathways, WikiPathways, Tissue expression (TISSUES), Subcellular localization (COMPARTMENTS), Annotated Keywords (UniProt), Protein Domains (Pfam), Protein Domains and Features (InterPro), and Protein Domains (SMART) depositories were searched. The level of confidence for the minimum interaction score was 0.4. Pathways with a false-discovery rate (FDR) < 0.05 were considered to be significantly affected.

2.11. Statistical Analysis

Part of the statistical analysis of the data had been performed by the aforementioned web portals. Additional statistical tests included (as required): (a) for between two-groups analysis, non-parametric Mann–Whitney *U* or unpaired t-test with Welch's correction; (b) for differences among means, an ordinary one-way ANOVA test; (c) for linear correlations, Pearson's *r* test. The analysis was performed using the GraphPad Prism 8.4.2 software (GraphPad Software, San Diego, CA, USA). *p*-Values < 0.05 were considered significant.

3. Results

3.1. Among Human Cancers, IL-37 Exhibits the Highest Expression Levels in Lung Adenocarcinoma (LUAD)

The first aim of this study was to investigate the expression patterns of *IL-37* and its receptor, *SIGIRR*, in tumor biopsies of various human cancer types, utilizing the Tissue Atlas tool of the Human Protein Atlas database [31]. Analysis revealed that the *IL-37* expression pattern significantly differs in various types of human cancers (one-way ANOVA $p < 0.0001$), whereas among the 17 types examined, lung cancer exhibited the highest *IL-37* mRNA levels (mean ± SD: 6.5 ± 31.88, n = 994) (Table 1, Figure 1A). Lung cancer specimens were further analyzed based on their specific type: either lung adenocarcinoma (LUAD) or lung squamous carcinoma (LUSC); the former exhibited increased *IL-37* expression levels compared to the latter (12.85 ± 44.05, n = 500 vs. 0.25 ± 1.79, n = 494, respectively; Mann–Whitney $p = 0.0001$). In contrast, *SIGIRR* was expressed at similar levels in almost all cancer types, besides glioma and head-and-neck tumors, in which its expression levels were decreased compared to the rest (Table 1, Figure 1B). However, within lung cancer samples, LUAD specimens exhibited higher *SIGIRR* levels compared to LUSC specimens (41.6 ± 4.46 vs. 13.64 ± 7.03, respectively; Mann-Whitney $p < 0.0001$). It is also noteworthy that, overall, all human tumor biopsies express lower *IL-37* compared to *SIGIRR* mRNA levels (1.029 ± 11.68 vs. 13.04 ± 8.911, respectively, n = 7932, $p < 0.0001$).

Table 1. *IL-37* and *SIGIRR* mRNA expression levels (FPMKs) in various human cancer types. Data analyzed were obtained from www.proteinatlas.org [31] (accessed on 24 August 2022). *p*-Values compared to total lung or LUAD specimens using non-parametric Mann-Whitney U test are reported.

Human Cancer Type	No of Patients	Gene Expression Levels			
		IL-37		*SIGIRR*	
		Mean FPKM ± SD	*p*-Value Compared to Lung/LUAD	Mean FPKM ± SD	*p*-Value Compared to Lung/LUAD
Breast	1075	0.07 ± 0.58	<0.0001/<0.0001	14.04 ± 10.16	<0.0001/NS
Cervical	291	0.17 ± 0.19	<0.0001/<0.0001	12.77 ± 8.29	<0.0001/0.0006
Colorectal	597	0.68 ± 2.38	0.09/<0.0001	17.64 ± 7.15	<0.0001/<0.0001

Table 1. *Cont.*

Human Cancer Type	No of Patients	Gene Expression Levels			
		IL-37		*SIGIRR*	
		Mean FPKM ± SD	*p*-Value Compared to Lung/LUAD	Mean FPKM ± SD	*p*-Value Compared to Lung/LUAD
Endometrial	541	0.19 ± 1.45	<0.0001/<0.0001	22.52 ± 10.64	<0.0001/<0.0001
Glioma	153	0.05 ± 0.07	<0.0001/<0.0001	2.56 ± 1.68	<0.0001/<0.0001
Head-and-neck	499	0.18 ± 0.49	<0.0001/<0.0001	5.04 ± 4.21	<0.0001/<0.0001
Liver	365	0.03 ± 0.07	<0.0001/<0.0001	14.02 ± 12.76	<0.0001/0.0008
Lung	994	6.59 ± 31.88	NA/NA	10.30 ± 6.78	NA/NA
LUAD	500	12.85 ± 44.05	NA/NA	13.64 ± 7.03	NA/NA
LUSC	494	2.56 ± 17.89	NA/0.0001	6.92 ± 4.46	NA/<0.0001
Melanoma	102	0.83 ± 2.11	0.0002/<0.0001	10.96 ± 6.44	NS/<0.0001
Ovarian	373	0.10 ± 0.15	<0.0001/<0.0001	16.70 ± 8.41	<0.0001/<0.0001
Pancreatic	176	0.34 ± 0.54	NS/<0.0001	15.54 ± 7.74	<0.0001/0.005
Prostate	494	0.01 ± 0.05	<0.0001/<0.0001	15.57 ± 6.26	<0.0001/<0.0001
Renal	877	0.02 ± 0.06	<0.0001/<0.0001	9.94 ± 6.93	0.05/<0.0001
Stomach	354	0.35 ± 1.21	0.0001/<0.0001	15.17 ± 5.94	<0.0001/<0.0001
Testis	134	1.08 ± 2.27	NS/NS	12.25 ± 5.31	<0.0001/0.08
Thyroid	501	0.25 ± 1.24	<0.0001/<0.0001	11.02 ± 4.66	<0.0001/<0.0001
Urothelial	406	0.67 ± 8.58	<0.0001/<0.0001	11.40 ± 7.3	0.006/<0.0001

NA: not applicable, NS: non-significant.

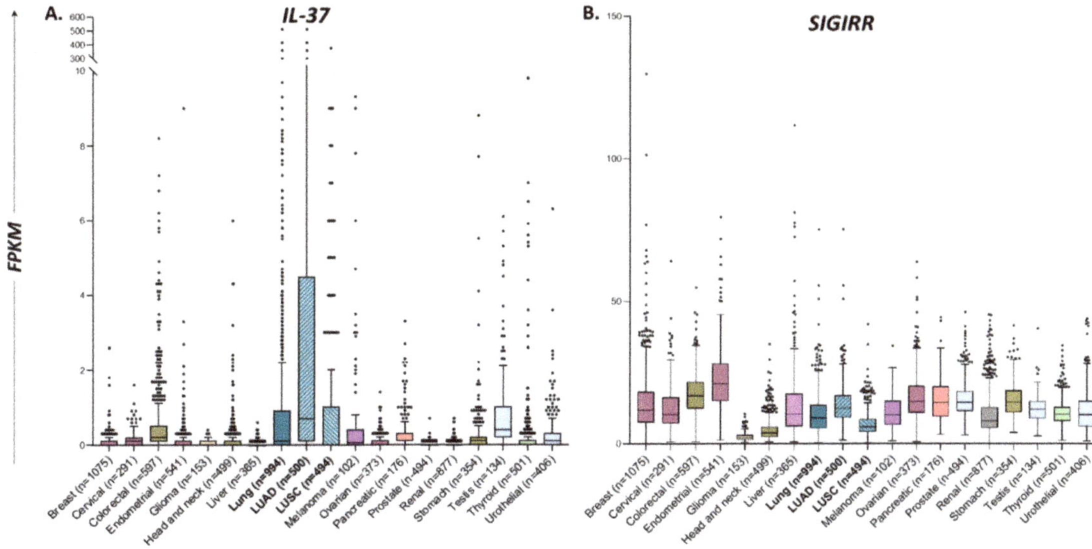

Figure 1. Box and Tukey whiskers diagrams showing the differential distribution of *IL-37* (**A**) and *SIGIRR* (**B**) expression levels (FPKMs), as analyzed by RNA sequencing in human tumor samples of

various cancer types. Especially for lung cancer, samples were separately analyzed as lung adenocarcinoma (LUAD) or squamous carcinoma (LUSC). Data were obtained from www.proteinatlas.org [31] (accessed on 24 August 2022) and further processed for statistical analysis. Left and right sides of the boxes correspond to the lower and upper quartiles; the box covers the interquartile interval including 50% of the data. The vertical line in the box represents the median. Whiskers outside the box expand from the minimum to the lower quartile and from the upper quartile to the maximum of data range. Each dot represents the value of a single sample.

3.2. IL-37 Levels Are Increased in LUAD versus Non-LUAD Lung Tissues

Following the observation that LUAD exhibits the highest levels of *IL-37* expression among solid tumors, we investigated its differential expression pattern in cancerous (LUAD) vs. non-cancerous lung biopsies (non-LUAD) using the TNMplot web tool [32]. *IL-37* mRNA levels were found to be increased by 12-fold in samples from LUAD (n = 524) compared to those from non-LUAD individuals (n = 486, Mann–Whitney $p = 3.83 \times 10^{-59}$), and by 11-fold in tumor compared to paired adjacent normal tissues from LUAD individuals (n = 57; $p = 3.81 \times 10^{-8}$) (Figure 2A). *SIGIRR* mRNA expression was slightly decreased (fold-change = 0.85; $p = 5.63 \times 10^{-5}$) between LUAD and non-LUAD individuals; however, no change was observed between tumor and adjacent normal tissues in LUAD patients (fold-change = 1.05; $p = 6.28 \times 10^{-1}$) (Figure 2A).

3.3. IL-37 and SIGIRR Levels Are Associated with Histological Type and Tumor Grade of LUAD Tumors

Based on the UALCAN webtool analysis [33], *IL-37* and *SIGIRR* expression levels derived from RNA-sequencing experiments are associated with the histological type of LUAD tumors. The one-way ANOVA test revealed significant differential expression among groups ($p = 0.0001$). More specifically, LBC (lung bronchioloalveolar carcinoma)-mucinous and mucinous-coloid tumors express significantly higher *IL-37* levels (TPM median (range), fold-change, *p*-value, respectively; for LBC-mucinous (n = 5): 1.07 (0.20–1.96), 13.38, 0.0072 and for mucinous-coloid (n = 10): 0.26 (0–0.76), 3.25, 0.0153) compared to normal lung tissue (n = 59, TPM median (range) 0.08 (0–0.38)) (Figure 2B, Table 2). Regarding *SIGIRR* levels, the observed changes were modest (fold-changes: between 0.5–2; one-way ANOVA $p > 0.05$). *IL-37* levels also correlated with LUAD tumor stage. Compared to controls, stage 1 biopsies (n = 277) expressed 7.63 times higher mRNA levels (TPM median (range) = 28 (0–23.62)) (Figure 2B, Table 2). Compared to stage 1 biopsies, expression levels were significantly decreased in stage 2 (n = 125; 0.37 (0–13.55)) and stage 4 (n = 28; 0.35 (0–6.17)) specimens, being 4.63- and 4.38-fold, respectively, higher than controls. In stage 3, *IL-37* levels were 2.13-fold increased compared to non-LUAD biopsies (n = 85; 0.17 (0–19.55)). A notable trend of association was observed between *SIGIRR* levels and the stage of the LUAD tumor; none of the stage subgroups exhibited significant change compared to controls (n = 59; 35.06 (17.01–55.55)) (fold-changes ranged between 1.21 and 0.73). However, *SIGIRR* levels displayed a statistically significant linear decrease among subgroups, following the stage 1-to-4 order (one-way ANOVA test: $r^2 = 0.6909$, $p < 0.0001$). Between-group analyses revealed that *SIGIRR* levels were significantly lower in biopsies of stage 2 compared to those of stage 1 (36.16 (12.59–75.62) vs. 42.53 (3.97–89.73), unpaired t-test $p < 0.0001$) in biopsies of stage 3 (33.03 (3.84–74.01)) compared to those of stage 1 ($p < 0.0001$), in biopsies of stage 3 compared to those of stage 2 ($p = 0.038$), and in biopsies of stage 4 (25.65 (7.76–69.02)) compared to those of stage 2 ($p = 0.022$). When compared to the control group, only stage 1 biopsies were found to express significantly different (higher) *SIGIRR* levels ($p < 0.0001$).

Table 2. Differential distribution of *IL-37* and *SIGIRR* mRNA levels in LUAD vs. non-LUAD tissues or among LUAD tumors of different histological type or stage. Number of patients in each group, the median levels of expression and their range (where available), fold-change over non-LUAD/normal samples, *p*-values, and the webtool used for the analysis (TNMplot [32], UACLAN [33]) are reported.

	IL-37				*SIGIRR*				
				Non-LUAD vs. LUAD					
	n	Levels (Median; Range)	Fold-Change	*p*-Value	n	Levels (Median; Range)	Fold-Change	*p*-Value	Platform
Non-LUAD	486	1			486	959.5			TNMplot
LUAD	524	11	12	3.83×10^{-59}	524	811	0.85	5.63×10^{-5}	TNMplot
Normal	59	0.08 (0–0.38)			59	35.06 (17.01–55.55)			UACLAN
Primary tumor	515	0.43 (0–19.6)	5.38	5.97×10^{-11}	515	38.71 (3.84–84.81)	1.11	9.20×10^{-7}	UACLAN
				Tumors vs. adjacent Normal					
Adjacent normal	57	1			57	713			TNMplot
Tumor	57	11	11	3.81×10^{-8}	57	749	1.05	6.28×10^{-1}	TNMplot
				LUAD histological subtype					
Normal	59	0.08 (0–0.38)			59	35.06 (17.01–55.55)			UACLAN
NOS	320	0.42 (0–16.99)	5.25	NS	320	37.08 (3.84–84.81)	1.06	NS	UACLAN
Mixed	107	0.7 (0–42.55)	5.6	NS	107	42.19 (9.77–71.23)	1.20	0.0003	UACLAN
Clear cell	2	22.40 (0–44.79)	280	NS	2	30.54 (29.11–31.98)	0.87	0.0022	UACLAN
LBC-Nonmucinous	19	1.08 (0–14.61)	13.5	NS	19	31.81 (24.68–59.76)	0.90	NS	UACLAN
Solid pattern predominant	5	0.10 (0–3.13)	1.25	NS	5	35.81 (27.55–62.29)	1.02	NS	UACLAN
Acinar	18	0.34 (0–9.66)	4.25	NS	18	48.39 (20.35–122.59)	1.38	NS	UACLAN
LBC-Mucinous	5	1.07 (0.20–1.96)	13.38	0.0072	5	52.32 (37.87–60.69)	1.49	0.0008	UACLAN
Mucinous colloid	10	0.26 (0–0.76)	3.25	0.0153	10	47.72 (35.68–70.80)	1.36	0.0011	UACLAN
Papillary	23	0.30 (0–19.55)	3.75	NS	23	43.01 (23.36–70.09)	1.23	NS	UACLAN
Mucinous	2	0.09 (0–0.17)	1.13	NS	2	56.86 (52.91–60.81)	1.62	NS	UACLAN
Micropapillary	3	11.24 (3.76–31.37)	140.05	NS	3	23.67 (21.42–73.27)	0.67	NS	UACLAN
Signet ring	1	0.82 (0.82–0.82)	10.25	NA	1	45.86 (45.86–45.86)	1.30	NS	UACLAN
				Cancer stage					
Normal	59	0.08 (0–0.38)			59	35.06 (17.01–55.55)			UACLAN
Stage 1	277	28 (0–23.62)	7.63	NS	277	42.53 (3.97–89.73)	1.21	<0.0001	UACLAN
Stage 2	125	0.37 (0–13.55)	4.63	NS	125	36.16 (12.59–75.62)	1.03	NS	UACLAN
Stage 3	85	0.17 (0–19.55)	2.13	NS	85	33.03 (3.84–74.01)	0.94	NS	UACLAN
Stage 4	28	0.35 (0–6.17)	4.38	NS	28	25.65 (7.76–69.02)	0.73	NS	UACLAN
Linear trend Stage 1→4				NS				<0.0001	

NOS: not otherwise specified, LBC: lung bronchioloalveolar carcinoma, NS: non-significant, NA: not applicable. Other between-two-groups comparisons; for *IL-37*; NOS vs. mixed: NS, NOS vs. clear cell: NA, NOS vs. LBC-nonmucinous: NS, NOS vs. solid pattern predominant: NS, NOS vs. acinar: $p = 0.0192$, NOS vs. mucinous: $p = 0.000003$, NOS vs. mucinous colloid: NA, NOS vs. papillary: NS, NOS vs mucinous: $p = 0.000003$, NOS vs. micropapillary: NS, NOS vs. signet ring: NA, mixed vs. clear cell: NA, mixed vs. LBC-nonmucinous: NS, mixed vs. solid pattern predominant: NS, mixed vs. acinar: $p = 0.011$, mixed vs. LBC-mucinous: $p = 0.0006$, mixed vs. mucinous colloid: NA, mixed vs. papillary: NS, mixed vs. mucinous: NA, mixed vs. micropapillary: NS, mixed vs. signet ring: NA, clear cell vs. LBC-nonmucinous: NA, clear cell vs. solid pattern predominant: NA, clear cell vs. acinar: NA, clear cell vs. LBC-mucinous: NA, clear cell vs. mucinous colloid: NA, clear cell vs. papillary: NA, clear cell vs. mucinous: NA, clear cell vs. micropapillary: NA, clear cell vs. signet ring: NA, LBC-nonmucinous vs. solid pattern predominant: NS, LBC-nonmucinous vs. acinar: NS, LBC-nonmucinous vs. LBC-mucinous: NS, LBC-nonmucinous vs. mucinous colloid: NA, LBC-nonmucinous vs. papillary: NS, LBC-nonmucinous vs. papillary: NS, LBC-nonmucinous vs. nucinous:, LBC-nonmucinous vs. micropapillary: NS, LBC-nonmucinous vs. signet ring: NA, solid pattern predominant vs. acinar: NS, solid pattern predominant vs. LBC-mucinous: NS, solid pattern predominant vs. mucinous colloid: NS, solid pattern predominant vs. papillary: NS, solid pattern predominant vs. mucinous: NS, solid pattern predominant vs. micropapillary: NS, solid pattern predominant vs. signet ring: NA, acinar vs. LBC-mucinous: NS, acinar vs. mucinous colloid: NA, acinar vs. papillary: NS, acinar vs. mucinous: NA, acinar vs. micropapillary: NS, acinar vs. signet ring: NA, LBC-mucinous vs. mucinous colloid: NA, LBC-mucinous vs. papillary: NS, LBC-mucinous vs. mucinous: NS, LBC-mucinous vs. micropapillary: NS, LBC-mucinous vs. signet ring: NA, for *SIGIRR*; NOS vs. mixed: NS, NOS vs. clear cell: NS, NOS vs. LBC-nonmucinous: NS, NOS vs. solid pattern predominant: NS, NOS vs. acinar: $p = 0.0066$, NOS vs. mucinous: NS,

NOS vs. mucinous colloid: NS, NOS vs. papillary: NS, NOS vs mucinous: NS, NOS vs. micropapillary: NS, NOS vs. signet ring: NA, mixed vs. clear cell: NS, mixed vs. LBC-nonmucinous: $p = 0.0422$, mixed vs. solid pattern predominant: NS, mixed vs. acinar: $p = 0.0307$, mixed vs. LBC-mucinous: NS, mixed vs. mucinous colloid: NS, mixed vs. papillary: NS, mixed vs. mucinous: NS, mixed vs. micropapillary: NS, mixed vs. signet ring: NA, clear cell vs. LBC-nonmucinous: NS, clear cell vs. solid pattern predominant: NS, clear cell vs. acinar: NS, clear cell vs. LBC-mucinous: NS, clear cell vs. mucinous colloid: NS, clear cell vs. papillary: NS, clear cell vs. mucinous: $p = 0.0245$, clear cell vs. micropapillary: NS, clear cell vs. signet ring: NA, LBC-nonmucinous vs. solid pattern predominant: NS, LBC-nonmucinous vs. acinar: $p = 0.0140$, LBC-nonmucinous vs. LBC-mucinous: NS, LBC-nonmucinous vs. mucinous colloid: $p = 0.0414$, LBC-nonmucinous vs. papillary: NS, LBC-nonmucinous vs. papillary: NS, LBC-nonmucinous vs. mucinous, LBC-nonmucinous vs. micropapillary: NS, LBC-nonmucinous vs. signet ring: NA, solid pattern predominant vs. acinar: NS, solid pattern predominant vs. LBC-mucinous: NS, solid pattern predominant vs. mucinous colloid: NS, solid pattern predominant vs. papillary: NS, solid pattern predominant vs. mucinous: NS, solid pattern predominant vs. micropapillary: NS, solid pattern predominant vs. signet ring: NA, acinar vs. LBC-mucinous: NS, acinar vs. mucinous colloid: NA, acinar vs. papillary: NS, acinar vs. mucinous: NA, acinar vs. micropapillary: NS, acinar vs. signet ring: NA, LBC-mucinous vs. mucinous colloid: NS, LBC-mucinous vs. papillary: NS, LBC-mucinous vs. mucinous: NS, LBC-mucinous vs. micropapillary: NS, LBC-mucinous vs. signet ring: NA.

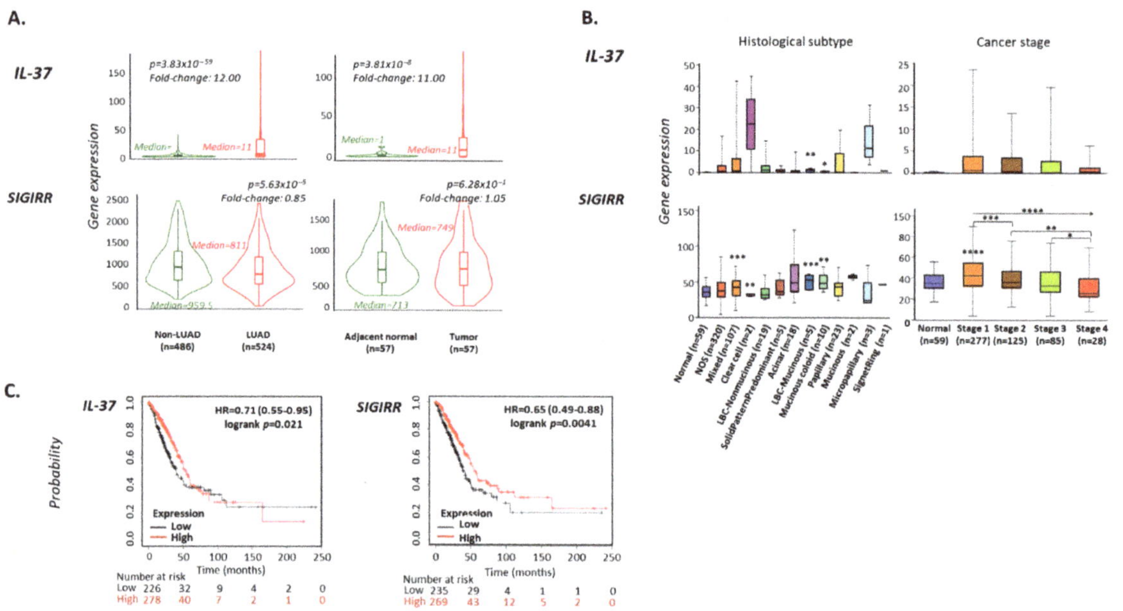

Figure 2. (**A**) Violin plots depicting the differential expression levels of *IL-37* and *SIGIRR* in lung tissues from LUAD (n = 486) vs. non-LUAD (n = 524) individuals or paired tumor vs. adjacent normal tissues from LUAD patients (n = 57 pairs). Median expression levels, Mann–Whitney *p*-values, and fold-changes between medians are reported. Data were obtained from www.tnmplot.com [32] (accessed on 3 September 2022). (**B**) Box and Tukey whiskers diagrams showing the differential distribution of *IL-37* and *SIGIRR* expression levels (FPKMs) as analyzed by RNA sequencing in LUAD samples of different histological type or stage. Data were obtained from http://ualcan.path.uab.edu/ (accessed on 20 September 2022). Asterisks designate statistically significant differences compared to normal samples; between groups (where accompanied by brackets) as analyzed by unpaired *t*-test with Welch's correction or statistically significant linear trend between group means and left-to-right order (where accompanied by an arrow) as analyzed by one-way ANOVA test; *: $p < 0.05$, **: $p < 0.01$, ***: $p < 0.001$, ****: $p < 0.0001$. (**C**) Kaplan–Meier plots depicting the probability of overall survival in months in LUAD patients exhibiting high (red) or low (black) expression levels of *IL-37*, *SIGIRR*. Hazard ratio (HR), logrank *p*-values, number of patients with either high or low gene expression categorized also in those who survived for 50, 100, 150, 200, and 250 months are reported. Graphs were exported from www.kmplot.com [34].

3.4. Increased IL-37 Expression Is a Favorable Prognostic Factor for Overall Survival in LUAD Patients

We further searched for possible correlations between *IL-37* expression levels and survival rates in LUAD patients utilizing the Kaplan–Meier Plotter tool [34]. Analysis of RNA sequencing data from 504 individuals with LUAD tumors revealed significant differences in overall survival (OS) time between patients with tumors expressing high (n = 226) vs. low *IL-37* levels (n = 278) (*log* rank $p = 0.021$). (Figure 2C). Specifically, high *IL-37* expression increases the probability of survival at 250 months by 29% (hazard ratio (HR) = 0.71. 95% CI = 0.53–0.95) compared to the low *IL-37*-expressing group. Further, median survival time for the *IL-37* high expression cohort was 54.07 months, whereas for the low expression cohort, it was 39.9 months. In the case of *SIGIRR*, its high expression increases the probability of survival at 250 months by 35% compared to low expression of the gene (HR = 0.65 (0.49–0.88), $p = 0.0041$; n = 235 and n = 269 for patients with high and low expression, respectively). The median survival time for the *SIGIRR* high expression cohort was 55.1 months, whereas for the low expression cohort, it was 40.3 months. We also checked the prognostic potential of the mean expression of the two genes, which was found to be weaker than the levels of each individual gene (HR = 0.74 (0.55–1), $p = 0.046$; n = 167 and n = 337 for patients with high and low mean expression, respectively).

Lastly, we analyzed the prognostic potential of *IL-37* and *SIGIRR* in individual group biopsies of different grade (1–4), stage (1–4), low or high mutation burden, and neoantigen load. The results revealed that both genes exhibit differential patterns and the ability to predict LUAD OS in patients bearing biopsies of distinct histopathological characteristics. Statistics of the analysis can be found in Supplementary Table S1. A low sample size, though, did not allow us to perform a combinatorial analysis of the parameters.

3.5. Correlation of IL-37 Expression with Cancer-Associated Genes in LUAD Tumors

The UALCAN portal [33] was used to explore possible linear association of *IL-37* mRNA levels with the expression of other genes in LUAD tumors. As shown in Figure 3A, there are 20 genes that are positively associated with *IL-37* expression, as revealed upon processing RNAseq data from TCGA biopsies and analysis with Pearson's correlation test. In details, *IL-37* levels ($\log_2$TPM + 1) were positively associated with those of: *PRODH* (enzyme proline dehydrogenase), *HINF1A* (hepatocyte nuclear factor 1-alpha), *DPP4* (dipeptidyl-peptidase 4), *DGCR5* (DiGeorge syndrome critical region gene 5), *DUSP6* (dual-specificity phosphatase 6), *NMNAT2* (nicotinamide nucleotide adenylyltransferase 2), *HLF* (hepatic leukemia factor), *ADORA1* (adenosine A1 receptor), *SHF* (Src homology 2 domain containing F), *MFSD4* (major facilitator superfamily domain containing 4A), *CXCL14* (C-X-C motif chemokine ligand 14), *ITGA2* (integrin subunit alpha 2), *DGCR9* (DiGeorge syndrome critical region gene 9), *CEACAM2* (carcinoembryonic antigen-related cell adhesion molecule 2), *PLAT* (plasminogen activator, tissue type), *PPP1R1B* (protein phosphatase 1 regulatory inhibitor subunit 1B), *STK39* (serine/threonine kinase 39), *MUC1* (mucin 1, cell surface-associated), *DPY19L1* (dpy-19 like C-mannosyltransferase 1), and *CDC42EP1* (CDC42 effector protein 1) (p for all < 0.0001) (Figure 3B). No gene was found to be negatively associated with *IL-37* in LUAD tumors.

Special interest was set on the exploration of possible associations of IL-37 with certain immune-checkpoint (IC) molecules. Since the UALCAN portal did not provide data on correlations for the non-significant associations, we explored the correlation analysis tool in the TNM plotter [32]. Spearman's r and p-values of linear associations with IL-37 were: $r = 0.05$ and $p = 0.265$ for PD-1 (or PDCD1; programmed cell death protein 1), $r = 0.13$, $p = 0.0035$ for PD-L1 (CD274), $r = 0.12$ and $p = 0.0065$ for CTLA-4 (cytotoxic T-lymphocyte-associated protein 4), $r = 0.1$ and $p = 0.023$ for VISTA (V-domain Ig suppressor of T cell activation or VSIR V-set immunoregulatory receptor), $r = -0.08$ and $p = 0.0892$ for LAG3 (lymphocyte activating 3), $r = -0.19$ and $p = 0.00001$ for TIM-3 (T cell membrane protein 3 or HAVCR2: hepatitis A virus cellular receptor 2).

STRING database [37] processing data from various depositories was utilized to pinpoint any potential cellular or molecular networks or biological processes that are regulated by the aforementioned set of positively correlated genes. "Regulation of leukocyte migration" and "regulation of response to external stimulus" were the two biological processes (Gene Ontology) enriched (false-discovery rate (FDR): 0.0032 for both; level of confidence for minimum required interaction score: 0.4). Since the activity of IL-37 lies mainly in the suppression of proinflammatory signaling factors such as mTOR, AKT, and PI3K, all involved in autophagy and key metabolic functions of immune and cancer cells [2], we further processed individual correlations with genes in these processes. In the aforementioned list, we pinpointed *HIF1A* as a key metabolic regulator, further implicated in key processes during cancer development and progression [38,39], and we also further explored for possible associations with OS of LUAD patients through the Kaplan–Meier Plotter tool [34]. No association was revealed when *HIF1A* was analyzed; however, the ratio of expression levels of *IL-37*-to-*HIF1A* was found to possess a prognostic potential in this cohort (HR = 0.67 (0.5–0.92), $p = 0.012$; low ratio samples n = 130, high expression samples n = 374).

Linear correlations with SIGIRR's expression levels were also explored. UALCAN portal [33] revealed 646 genes positively and 34 genes negatively associated with SIGIRR. The total list of the genes and the corresponding Pearson's r values can be found in Supplementary Table S2. Gene-set enrichment analysis of SIGIRR-related genes via the STRING database [37] did not reveal any cancer/LUAD-related entrance in any of the STRING-connected repositories.

3.6. IL-37 Protein Expression Correlates with the Grade of LUAD Tumor

Following gene expression analysis, investigation of the protein expression pattern through the Pathology tool of the Human Protein Atlas website [31] initially revealed that both IL-37 and SIGIRR are expressed in lung tumor biopsies (Figure 4A). IL-37 protein levels were similar between 111 LUAD and paired non-LUAD specimens (median z-value (range): 0.11 (−1.02–1.16) vs. −0.043 (−1.02–1.16), $p = 1.10 \times 10^{-1}$) (Table 3, Figure 4B). However, protein levels were associated with tumor grade: the lowest levels were observed in grade 2 biopsies and the highest in grade 3, whereas intermediate levels in grade 1 biopsies led to non-cancerous tissues. Regarding SIGIRR protein expression, this was modestly upregulated in tumor tissues compared to normal ones (−0.25 (−1.14–0.38) vs. 0 (−1.91–2.26), $p = 2.82 \times 10^{-2}$) and was associated with tumor grade, since SIGIRR levels exhibited a linear decrease following the grade 1-to-3 order ($r^2 = 0.1634$, $p = 0.0001$).

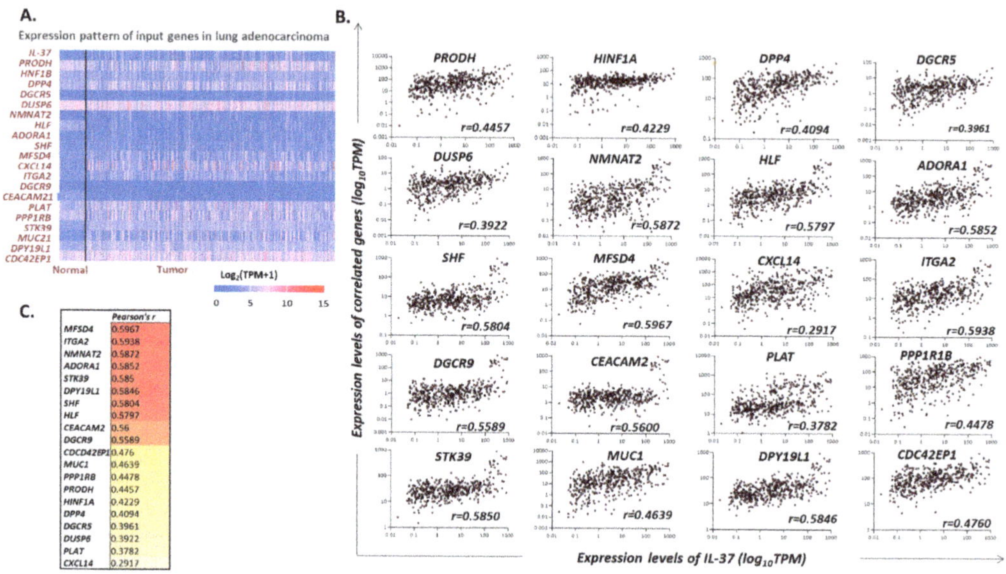

Figure 3. (**A**) Heatmap depicting the relative expression levels (log_2TPM + 1) of *IL-37* and twenty IL-37-correlated genes in normal and LUAD biopsies as analyzed by RNAseq. (**B**) Dot plot diagrams showing the correlation between *IL-37* and each of the twenty significantly positively correlated genes. No negative associations were detected. Pearson's *r* values are reported in each case. All *p*-values were <0.0001. Figures were exported from http://ualcan.path.uab.edu/ [33] (accessed on 10 September 2022). (**C**) Heatmap depicting the grade of association between the expression levels of each of the twenty *IL-37*-positively correlated genes and the levels of *IL-37*. Pearson's *r* values are reported.

Table 3. Differential distribution of IL-37 and SIGIRR protein levels in LUAD vs. normal tissues or among LUAD tumors of different grades. Number of patients in each group, the median levels of expression, and their range as obtained from UALCAN portal [33], as well as *p*-values of statistical differences are reported.

	IL-37			*SIGIRR*			
			Non-LUAD vs. LUAD				
	n	**Levels (Median; Range)**	***p*-Value**	**n**	**Levels (Median; Range)**	***p*-Value**	**Platform**
Normal	111	0.11 (−1.02–1.16)		111	−0.25 (−1.14–0.38)		UACLAN
Primary tumor	111	−0.043 (−1.02–1.16)	NS	111	0 (−1.91–2.26)	2.82×10^{-2}	UACLAN
			Grade				
Normal	111	0.11 (−1.02–1.16)		111	−0.25 (−1.14–0.38)		UACLAN
Grade 1	7	0.22 (0.21–0.24)	0.0313	7	0.93 (−0.23–1.35)	0.001	UACLAN
Grade 2	59	−0.28 (−0.75–1.07)	0.0351	59	0.35 (−1.55–2.26)	NS	UACLAN
Grade 3	39	0.66 (−0.68–1.90)	<0.0001	39	−0.48 (−1.57–1.26)	<0.0001	UACLAN
Linear trend Grade 1 → 3			NS			0.0001	

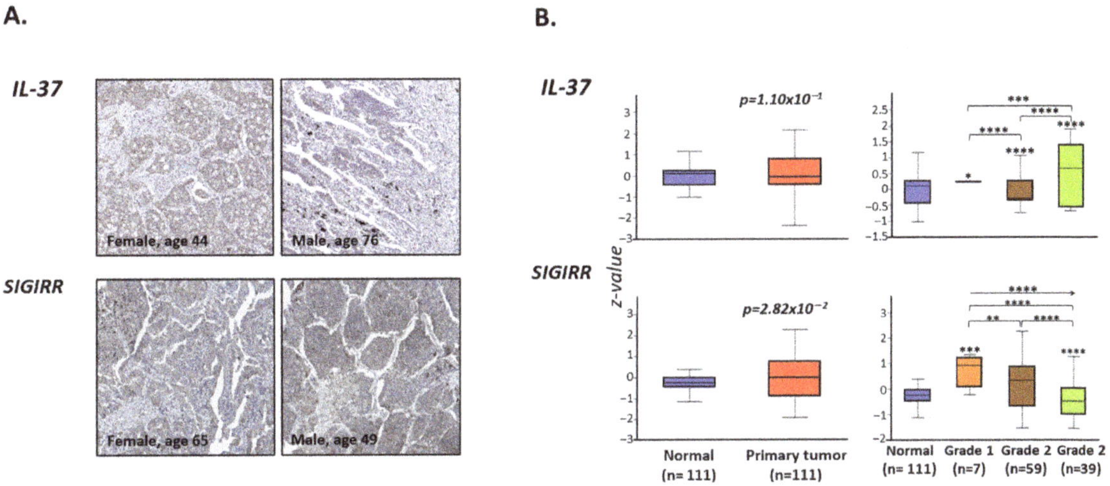

Figure 4. (**A**) Immunohistochemical staining of IL-37 and SIGIRR proteins in lung biopsies of LUAD patients. Pictures were obtained from www.proteinatlas.org [35] (accessed on 20 September 2022). (**B**) Box and Tukey whiskers diagrams depicting the differential protein expression levels of IL-37 and SIGIRR (z-values) in LUAD primary tumors (n = 111) vs. normal tissues (n = 111) and among LUAD tumors of different grade (1 to 3). Figures and graphs were exported from http://ualcan.path.uab.edu/ [33] (accessed on 20 September 2022). *: $p < 0.05$, **: $p < 0.01$, ***: $p < 0.001$, ****: $p < 0.0001$.

3.7. T-Lymphocytes and Macrophages of the Lung Express IL-37 and SIGIRR Genes

According to single-cell RNA sequencing (scRNAseq) data deposited and processed through the Single-Cell Atlas (of the Human Protein Atlas) [31], *IL-37* was found to be expressed by resident T lymphocytes and macrophages of the normal lung tissue (read counts normalized to transcripts per million protein coding genes (pTPM) = 2.1 and 1.2, respectively) (Supplementary Figure S1). Moreover, *SIGIRR* was found to be expressed by immune cell populations, including T cells, granulocytes, and macrophages, but also by other resident cell types including alveolar cell types 1 and 2, fibroblasts, club ciliated cells, and endothelial cells. The highest *SIGIRR* expression was detected in T cells (pTPM = 115.2) and the lowest in macrophages (pTPM = 23.6). It is also noteworthy that, based on the the scRNAseq analysis, IL-37 protein expression levels in normal human lung cell subsets are relatively lower compared to those of SIGIRR.

3.8. IL37 Gene Alterations Correlate with Differential Immune Cell Infiltration of the Lung Tumor

To investigate the effects of IL-37 gene mutations on immune cell infiltration in lung adenocarcinoma tumors, the "Mutation" module of the TIMER2.0 webserver was applied [36]. Our analysis indicates that tumors bearing non-synonymous, somatic mutations in the *IL-37* gene were characterized by significantly higher infiltration of CD4$^+$ T lymphocytes (CYBERSORT project; log_2fold-change = 2.105, Wilcoxon $p = 0.004$) and significantly lower infiltration of M2 macrophages (XCELL project; log_2fold-change = −2.709, Wilcoxon $p = 0.009$) and neutrophils (MCPCOUNTER project; log_2fold-change = −0.665, $p = 0.031$). The contribution of myeloid dendritic cells (mDCs) within tumor-association infiltration was of similar levels in patients with and without IL-37 non-synonymous mutations (log_2fold-change = 1.171, $p = 0.047$) (Figure 5A).

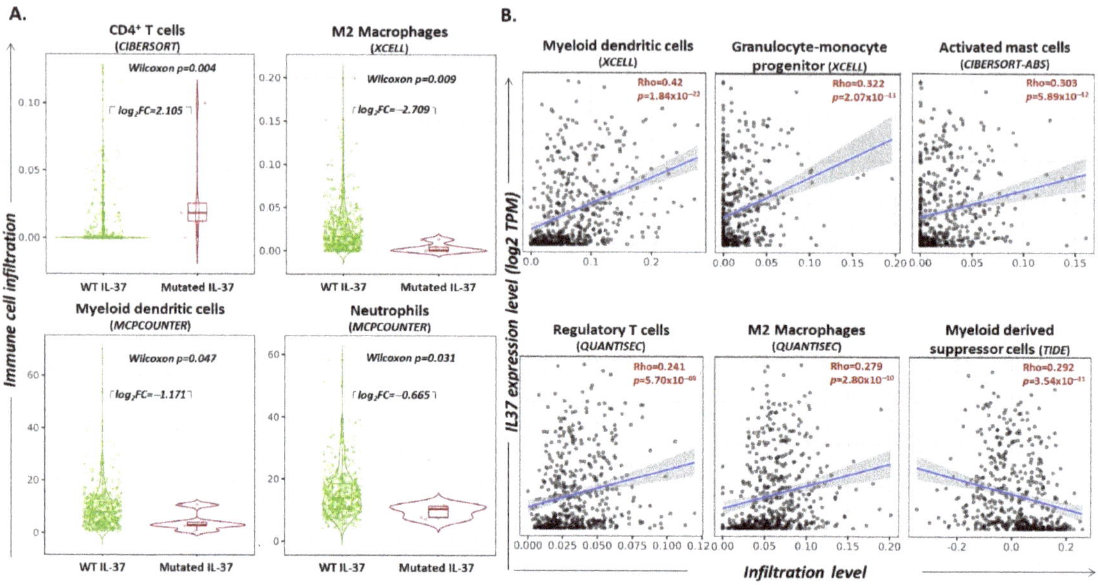

Figure 5. (**A**) Violin plots depicting the distribution of infiltrated T cells, M2 macrophages, myeloid dendritic cells, and neutrophils in LUAD tumors without versus with mutation on *IL-37*. Wilcoxon *p*-values and $\log_2$ (fold-changes, FC) are reported. (**B**) Scatter plot diagrams depicting the linear association between levels of *IL-37* gene expression ($\log_2$TPM; y-axis) and infiltration of the tumor by certain cell subsets (x-axis). Spearman's Rho and *p*-values are reported. Data were filtered for tumor purity. Graphs were exported from http://timer.cistrome.org/ [36] (accessed on 1 October 2022).

3.9. *IL37* Expression Levels Correlate with Infiltration Levels of Certain Immune Cell Subsets

Exploration through the "Gene" module of the TIMER2.0 webserver [36] revealed that *IL37* gene expression levels, as assessed in previous RNAseq experiments, were linearly correlated with certain immune cell populations infiltrating the LUAD tumor. More specifically, *IL37* expression levels ($\log_2$TPM) were positively associated with the infiltration rate of mDCs (XCELL project; Spearman's rho = 0.42, $p = 1.84 \times 10^{-22}$) progenitors of granulocytes-monocytes (GMPs) (XCELL project; Spearman's rho = 0.323, $p = 2.07 \times 10^{-13}$), activated mast cells (CIBERSORT-ABS project; Spearman's rho = 0.303, $p = 5.89 \times 10^{-12}$), Tregs (QUANTISEC project; Spearman's rho = 0.241, $p = 5.70 \times 10^{-8}$), and M2 macrophages (QUANTISEC project; Spearman's rho = 0.279, $p = 2.80 \times 10^{-10}$) and negatively associated with the infiltration rate of MDSCs (TIDE project; Spearman's rho = −0.292, $p = 3.54 \times 10^{-11}$) (Figure 5B); all correlations are summarized in Table 4.

Table 4. Associations between the levels of expression of *IL-37* and those of infiltration by certain immune cell types in LUAD tumors, as attested in 515 specimens. Spearman's *r* values are reported. For significant associations, *p*-values are also mentioned. Tumor purity filter was applied. Data were obtained from www.timer.cistrome.org, (accessed on 24 August 2022) [36].

	EPIC	TIMER	MCP-COUNTER	QUANTISEQ	CIBERSORT	CIBERSORT-ABS	XCELL	TIDE
$CD8^+$ T cell	−0.007; NS	0.032; NS	−0.029; NS	−0.002; NS	$-0.187; 3 \times 10^{-5}$	−0.051; NS	0.026; NS	NA
Naïve	NA	NA	NA	NA	NA	NA	−0.03	NA
Memory	NA	NA	NA	NA	NA	NA	0.069	NA
Central memory	NA	NA	NA	NA	NA	NA	−0.034; NS	NA
$CD4^+$ T cell	0.006; NS	0.06; NS		-0.168 *; 2×10^{-4}	NA	NA	0.088 *; NS	NA
Naïve	NA	NA	NA	NA	0.019; NS	0.019; NS	$0.126; 5 \times 10^{-3}$	NA
Memory	NA	NA	NA	NA	NA	NA	−0.019; NS	NA
Central memory	NA	NA	NA	NA	NA	NA	0.04; NS	NA
Effector memory	NA	NA	NA	NA	NA	NA	$0.384; 8 \times 10^{-20}$	NA
Memory activated	NA	NA	NA	NA	$-0.163; 3 \times 10^{-4}$	$-0.158; 4 \times 10^{-4}$	NA	NA
Memory resting	NA	NA	NA	NA	$0.208; 3 \times 10^{-6}$	$0.259; 5 \times 10^{-9}$	NA	NA

Table 4. *Cont.*

	EPIC	TIMER	MCP-COUNTER	QUANTISEQ	CIBERSORT	CIBERSORT-ABS	XCELL	TIDE
Th1	NA	NA	NA	NA	NA	NA	−0.103; 3×10^{-2}	NA
Th2	NA	NA	NA	NA	NA	NA	−0.323; 2×10^{-13}	NA
Tregs	NA	NA	NA	0.241; 6×10^{-8}	0.119; 6×10^{-3}	0.167; 2×10^{-4}	NA	NA
B cell	−0.017; NS	0.126; 5×10^{-3}	0.052; NS	−0.026; NS	NA	NA	0.066; NS	NA
Naïve	NA	NA	NA	NA	−0.103; 2×10^{-2}	−0.061; NS	0.047; NS	NA
Memory	0.26; NS	NA	NA	NA	0.202; 6×10^{-6}	0.204; 5×10^{-6}	−0.031; NS	NA
Neutrophil	NA	0.033; NS	0.145; 1×10^{-3}	0.311; 2×10^{-12}	−0.032; NS	−0.011; NS	−0.146; 1×10^{-3}	NA
Monocyte	NA	NA	−0.056; NS	−0.029; NS	0.207; 3×10^{-6}	0.226; 4×10^{-7}	0.26; 4×10^{-9}	NA
Macrophage	0.137; 2×10^{-3}	−0.01; NS	−0.056; NS	NA	−0.147; 1×10^{-3}	−0.113; 1×10^{-2}	0.12; 8×10^{-3}	NA
M1	NA	NA	NA	0.388; 4×10^{-19}	−0.109; 2×10^{-2}	−0.018; NS	−0.014; NS	NA
M2	NA	NA	NA	0.279; 3×10^{-10}	0.21; 3×10^{-6}	0.251; 2×10^{-8}	0.166; 2×10^{-4}	0.047; NS
Dendritic cell								
Myeloid	NA	NA	0.406; 5×10^{-21}	−0.141; NS	NA	NA	0.42; 2×10^{-22}	0.307; 4×10^{-12}
Myeloid activated	NA	NA	NA	NA	0.101; 2×10^{-2}	0.138; 2×10^{-3}	0.316; 7×10^{-13}	NA
Myeloid resting	NA	NA	NA	NA	NA	NA	NA	NA
Plasmacytoid	NA	NA	NA	NA	NA	NA	−0.124; 6×10^{-3}	NA
Natural killer	−0.086; NS	NA	−0.239; 8×10^{-8}	0.16; 7×10^{-4}	NA	NA	−0.018; NS	NA
Activated	NA	NA	NA	NA	−0.188; 3×10^{-5}	−0.064; NS	NA	NA
Resting	NA	NA	NA	NA	−0.078; NS	−0.077; NS	NA	NA
Mast cell	NA	NA	NA	NA	NA	NA	0.201; 7×10^{-6}	NA
Activated	NA	NA	NA	NA	0.272; 9×10^{-10}	0.303; 6×10^{-12}	NA	NA
Resting	NA	NA	NA	NA	−0.159; 4×10^{-4}	−0.138; 2×10^{-3}	NA	NA
Cancer-associated fibroblast	0.065; NS	NA	0.045; NS	NA	NA	NA	0.211; 2×10^{-6}	0.063; NS
Common lymphoid progenitor	NA	NA	NA	NA	NA	NA	−0.14; 2×10^{-3}	NA
Common myeloid progenitor	NA	NA	NA	NA	NA	NA	0.092; NS	NA
Endothelial cell	−0.007	NA	0.109; 2×10^{-3}	NA	NA	NA	−0.062; NS	NA
Eosinophil	NA	NA	NA	NA	0.067	NA	0.067; NS	0.005; NS
Granulocyte-lymphocyte progenitor	NA	NA	NA	NA	NA	NA	0.323; 2×10^{-13}	NA
Hematopoieitic stem cell	NA	NA	NA	NA	NA	NA	0.159; 4×10^{-4}	NA
follicular helper T cell	NA	NA	NA	NA	−0.187; NS	NA	NA	NA
$\gamma\ \delta$ T cell	NA	NA	NA	NA	−0.021; NS	−0.02; NS	0.025; NS	NA
NK T cell	NA	NA	NA	NA	NA	NA	0.43; NS	NA
MDSCs	NA	NA	NA	NA	NA	NA	NA	−0.292; 4×10^{-11}

* non-regulatory, NA: not available, NS: non-significant, Tregs: regulatory T cells, MDSCs: myeloid-derived suppressor cells.

3.10. IL-37 Signaling Shares Common Nodes with PD-1/PDL-1 and CTLA-4 Immune Checkpoint Pathways

Finally, analysis using the STRING database [37] processing data from the various depositories revealed that protein molecules involved in the IL-37 signaling pathway are also members of cancer-related pathways (Figure 6). Specifically, STAT3 is involved in "PD-L1 expression and PD-1 checkpoint pathway in cancer" (KEGG database) and "Cancer immunotherapy by PD-1 blockade" (WikiPathways), PTPN6 in "PD-L1 expression and PD-1 checkpoint pathway in cancer" (KEGG database)/"PD-1 signaling" (Reactome), and "Cancer immunotherapy by CTLA4 blockade" (WikiPathways), and PTPN11 is involved in all of the above. False discovery rates of the significance for each of the enhanced pathways are: 0.0015 for PD-L1 expression and PD-1 checkpoint pathway in cancer, 0.0029 for cancer immunotherapy by CTLA4 blockade, 0.0053 for cancer immunotherapy by PD-1 blockade, and 0.0081 for PD-1 signaling.

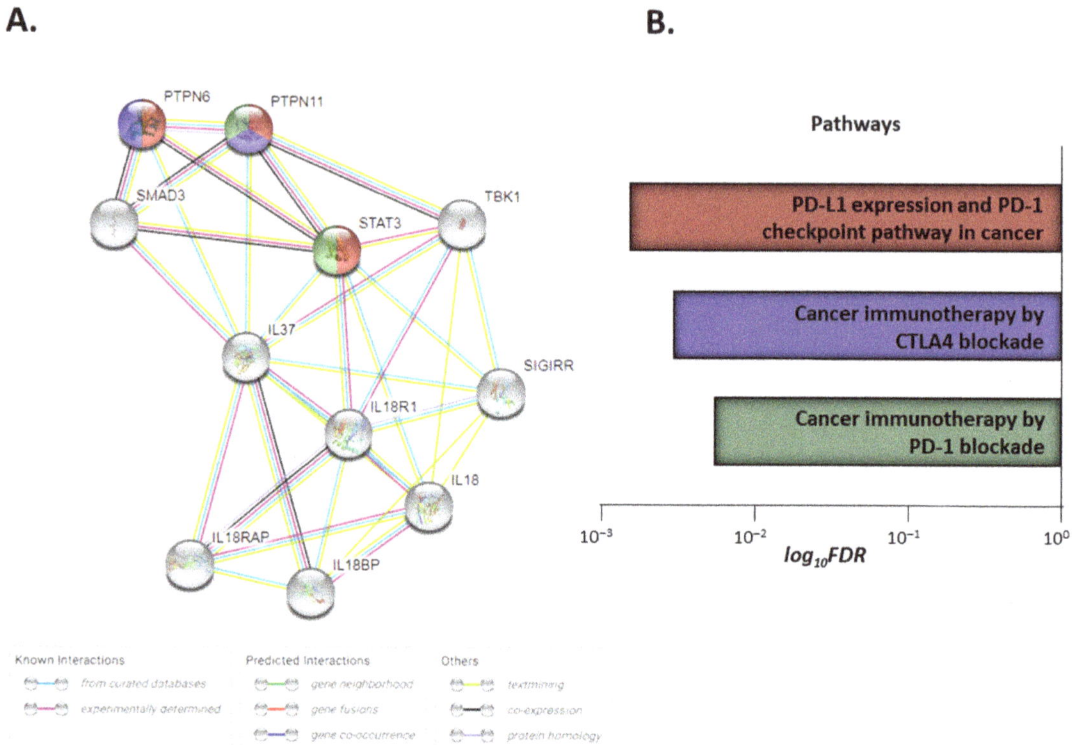

Figure 6. Network of interactions (exported by STRING portal [37]) (**A**) among proteins involved in IL-37 signaling and other proteins implicated in PD-1/PD-L1 and CTLA-4 immune-checkpoint pathways (**B**), as attested using the STRING (Search Tool for the Retrieval of Interacting Genes/Proteins). The above pathways share common nodes (each color corresponds to different pathway; nodes marked with more than one color belong to equal number of pathways). Edges represent protein–protein associations: known interactions, predicted interactions, or other associations. Level of confidence for minimum required interaction score was 0.4. Data were obtained from https://string-db.org [37] (accessed on 3 October 2022).

4. Discussion

Lung cancer is a leading cause of cancer death in both men and women, worldwide [40–42]; it is associated with almost three times the rate of deaths compared to prostate cancer in men and breast cancer in women. In 2022, 236.740 estimated new cases of lung cancer are expected to be diagnosed, and approximately 130.180 deaths are predicted to be recorded in the United States of America (USA) [43]. However, during the last 15 years, a steady decline in the incidence of new lung cancer diagnoses (2.8% in men and 1.4% in women) and deaths (50% and 67%, respectively) has been observed. This is probably attributed to recent advances in the management of patients, including the use of novel chemotherapeutic (cisplatin/pemetrexed or gemcitabine), molecular targeted (gefitinib), and/or immunotherapeutic (pembrolizumab, nivolumab, atezolizumab) agents, together with more personalized therapeutic approaches based on specific mutation patterns, such as those on *EGFR*, *ALK*, and *ROS-1* genes [44].

Lung adenocarcinoma (LUAD), which falls under the umbrella of non-small-cell lung cancer (NSCLC), represents about 40% of all lung cancer types [45], and even though there has been a significant decrease in its incidence and mortality rates, it remains the main cause of cancer death in the USA [45]. LUAD tumors that evolve primarily from mucosal glands are usually developed in the lung periphery but can also be found in scars or areas of chronic inflammation. The immune microenvironment exerts functions associated with the development and progression of the disease, as well as the response to therapy [40]. Alternatively activated MΦ (M2 type) and T cells, specifically resting memory $CD4^+$, are the predominant populations surrounding LUAD tumors [46]. Further, high immune cell infiltration rates correlate with a better prognosis compared to lower ones [47,48] and are characterized by an increased incidence of naïve B cells, plasma cells, follicular helper T cells, and classical

(M1 type) macrophages, as well as by a decreased prevalence of resting memory $CD4^+$ T cells, monocytes, and resting dendritic cells (DCs) [48].

Recently, Zuo et al. developed the immune-cell characteristic score (ICCS) model, which is suggested for the facilitation of LUAD prognosis [49]. This model assesses the infiltration rate of six immune cell populations in LUAD biopsies (B cells, immature DCs, eosinophils, mast cells, granzyme K expressing $CD8^+$ T cells, and Th2 cells) and independently predicts the overall survival (OS) of the patient. High infiltration of all the ICCS immune cell populations is associated with better prognosis, apart from that of Th2 cells, which indicate a poor outcome of the disease [49]. What is more, processing RNA-sequencing and clinical data from TCGA [50] and GEO databases [51] utilizing specific bioinformatics tools and algorithms has led to the construction of certain gene signatures with independent prognostic/predictive values for the survival of LUAD patients [47,48,52] or their response to immune-checkpoint inhibitors (ICIs) [53,54].

Focusing on the pivotal role of the tumor-immune microenvironment in LUAD development and progression, we aimed at the exploration of the expression patterns of *IL-37*, a novel cytokine with regulatory properties in LUAD tumors, using various bioinformatics tools and available databases. Based on our data, human LUAD tumors exhibit the highest gene-expression levels of the cytokine amongst various common human cancers, suggesting its possible pathophysiological involvement in this malignancy. Importantly, these are significantly increased in LUAD vs. non-LUAD tissues, suggesting a disease-specific involvement of IL-37 in the pathological lesion. Within LUAD, LBC-mucinous and mucinous-coloid histological subtypes are characterized by the highest *IL-37* levels; however, the small number of specimens assessed for each subtype does not allow for the extraction of safe conclusions regarding the histological-specific distribution of the cytokine.

It is of note that IL-37 expression exhibits a lung cancer stage-specific pattern: there is a trend of steady decline in its levels from earlier to later stages. This observation provides initial evidence for a possible prognostic value for *IL-37* expression patterns, but it also implies a potential pathogenetic role of this regulatory cytokine in the LUAD inflammatory lesion. However, there are clearly several unresolved questions regarding the molecular mechanisms underlying this differential expression profile, such as whether aberrant *IL-37* regulation is a cause or consequence of lung cancer development and progression. Therefore, it is of outmost importance to pinpoint the exact cellular source(s) of IL-37 that are responsible for the variability in its distribution (LUAD cells and/or immune cell subsets) before making any assumptions about the link between molecular/cellular pathogenetic mechanisms and the mRNA signature of *IL-37* in LUAD malignancies. Even though most of the current knowledge advocates towards a possible protective role of this cytokine in human malignancies, the enhanced expression in early-stage lung tumors, independently from the exact cellular source(s), may indicate a cellular response to promote anti-tumor immunity. Consequently, the observed gradual decrease in IL-37 expression during tumor progression and advanced lung cancer stages suggests that this anti-tumor effect could probably be attenuated. In contrast to the differential expression pattern of *IL-37* levels between LUAD and non-LUAD biopsies, this does not apply for its receptor *SIGIRR*. The receptor's gene expression levels between the groups are similar; nevertheless, within LUAD, the mixed, LBC-mucinous, and mucinous-coloid subtypes exhibit a tendency to express higher *SIGIRR* mRNA levels, possibly associated with the aforementioned enhanced *IL-37* levels in the same subtypes. What is of interest, though, is the sharp trend of decrease observed following the stage 1-to-stage 4 order, also matching with the corresponding *IL-37* signature.

Our results also reveal the prognostic potential of tumor-expressed *IL-37* and *SIGIRR* patterns. Patients whose biopsies exhibit high *IL-37* or *SIGIRR* mRNA levels have a better prognosis for OS compared to those with low levels. This, together with the stage-specific expression patterns described above, suggests that lower *IL-37* and *SIGIRR* gene expression could be associated with more advanced LUAD cases, of high grade/stage and/or metastatic status linked to poorer prognosis. Yet, *SIGIRR* seems to be an even stronger OS predictor, which is in accordance with the more significant, compared to *IL-37*, correlations with stage and metastasis status described herein.

IL-37 expression levels were found to be positively correlated with a 20-gene signature in LUAD biopsies. Most of these genes have an already established association with lung cancer and specifically with LUAD. Increased levels of *CXCL14* mRNA have been detected in LUAD biopsies with a micropapillary pattern [55], and smoking-induced *CXCL14* expression in the human airway epithelium has been implicated in chronic obstructive pulmonary disease (COPD)-mediated lung cancer development [56]. Similarly, *NMNAT2* expression was found to be increased in LUAD specimens and correlated negatively with OS of patients, whereas the DGUOK-NMNAT2-NAD+ axis was suggested as a potential therapeutic target for the disease [57]. Increased expression levels of *HLF*, shown to promote cell-cycle progression in various cancers [58], *ITGA2*, *MUC1*, and *DPY19L1* have also been proposed to confer prognostic value for the survival of patients suffering from LUAD malignancies [59–64]. Additionally, *DUSP6/MKP3* has been designated as a tumor suppressor phosphatase implicated in LUAD and cancer types, whereas several studies have revealed the clinical relevance of its expression patterns in lung cancer. In addition, various *DUSP6/MKP3*-associated SNPs have been linked with the response to chemoradiation therapy [65–69]. Similarly, *PPP1R1B* has been shown to interfere with the response to molecular targeted therapy in *EGFR*-mutated LUAD [70], and *SHP2* has been associated with *MET*-mutated NSCLC [71].

Moreover, additional *IL-37*-correlated genes have been described to be implicated in NSCLC pathogenesis and progression. *PLAT* has been reported to inhibit apoptosis in NSCLC cells, and its knockdown augments the therapeutic efficacy of gefitinib [72]. *PRODH* has been involved in NSCLC metastasis as shown, both in vitro and in vivo [73,74], and *ADORA1*, which is highly expressed in *EGFR*-mutant NSCLC biopsies [75], has been associated involved in tumor-immune evasion in NSCLC xenograft models [76]. *STK39* and the lncRNA *DGCR5* have been proposed as critical molecules for the regulation of the growth, migration, and invasion of NSCLC tumors [74,77–79]. *DGCR5* has been specifically implicated in the tumor progression of LUAD through the inhibition of hsa-mir-22-3p [80]. Regarding the remaining genes that exhibit an *IL-37*-correlated expression pattern, there is evidence for their association with the development of human cancers and for their potential to serve as disease biomarkers. This is the case with *HIF1A*, the most pivotal gene regulating metabolic pathways related to hypoxia, which is further implicated in proliferation, energy metabolism, invasion, and metastasis in a series of human cancers, and has been viewed as a highly promising therapeutic target [38,39]. Indeed, there is evidence that expression levels of *HIF1A* by tumor cells have a diagnostic and prognostic significance among different histological types of lung cancer [81,82]. Interestingly, in our study, the ratio of *IL-37*-to-*HIF1A* expression levels was found to have a favorable prognostic potential in LUAD patients. Further, *MFSD4* is considered to be a tumor-suppressor gene and a biomarker for hepatic metastasis in gastric cancer patients [83], as well as a diagnostic marker of esophageal carcinoma [83]. In addition, the lncRNA *DGCR9* has been reported as a potential tumor neoantigen [84], with a possible pathogenetic role in gastric cancer [85]. As for *CDC42EP1*, it was very recently described that certain gene mutations drive the development of parathyroid and oral tongue squamous cell carcinomas [86,87]. Lastly, *DPP-4* has been reported to possess a deleterious role and potential to be used as a biomarker in respiratory diseases, such as lung cancer, asthma, and chronic obstructive pulmonary disease (COPD) [88–90].

Regarding IL-37 protein expression, this was found to be similar for both proteins between LUAD and non-LUAD specimens. As with mRNA expression, its levels were associated with the tumor grade. It is essential to comment that the webportal used provided the opportunity to analyze the gene expression levels in regard to the stage of the tumor, as well as the protein levels with regard to its grade. Given the differences among these terms and scales, one can speculate that the significant trend of decrease in *SIGIRR* gene expression levels is pathophysiologically connected to the trend in protein levels, both associated with more severe disease. In the case of *IL-37*, things seem to be more complicated: biopsies of intermediate severity (Grade 2) express the highest gene expression but the lowest protein levels, respectively, compared to the rest of the subgroups in each classification. Further exploration for these seemingly opposite regressions needs to be performed.

By taking advantage of the TIMER2.0 webserver, we had the opportunity to search for possible correlations between *IL-37* gene alterations and infiltration rates of various immune cell types in LUAD. Interestingly, it was revealed that tumors with non-synonymous, somatic mutations of *IL-37* are characterized by a higher infiltration of $CD4^+$ T-cells and a lower infiltration of M2 MΦ and neutrophils. Although the differences are statistically significant, due to the low number of samples in the mutated arm (n = 5), the conclusions cannot be confidently evaluated. What is more, even though these mutations are expected to alter protein function, assumptions about the pathophysiological links of the above relationships, if not falsely positive, could be made upon determining their cell-specific distribution, as well as the activation status of the corresponding immunocytes. It is essential to confirm whether *IL-37* aberrations indeed affect the infiltrating incidence and/or function, especially of M2 MΦ and neutrophils, which dominate the myeloid-landscape of LUAD tumors and are of vital importance for their growth and metastatic potential [91].

Supplementing these observations, the *IL-37* gene expression pattern was found to be significantly associated with the infiltration rate of certain immune cell types. Myeloid dendritic cells (mDCs), which showed the most significant correlation, are known to support protective anti-tumor immunity in lung cancer, while being subjected to suppression mediated by cancer cells via different mechanisms [92,93]. In agreement, there was also a positive association between the percentage (%) of GMPs, which produce DCs and macrophages, and they have also been shown to be involved in LUAD immune cell infiltration [94]. Moreover, *IL-37* mRNA levels positively correlated with the percentage (%) of activated mast cells, which have been assigned as predictors of improved OS and PFS in NSCLC [49,95]. In contrast, *IL-37* expression was negatively correlated with the percentage (%) of infiltrating MDSCs, which are pivotal immunosuppressive partners and key targets for immunotherapy in lung cancer [96]. The above observations support the notion that IL-37 may exert tumor-protective immune functions. However, it should also be noted that our analyses revealed some, at first sight, conflicting evidence: *IL-37* expression levels correlated positively with the rate of infiltration by regulatory T cells, as well as M2 MΦ, that have been reported to exert pro-tumoral and anti-tumor immunity actions in LUAD [97,98]. The molecular and cellular networks responsible for these phenotypes need further investigation.

Finally, pathway analysis revealed that IL-37-signaling mediators, such as STAT3, are crucial partners of PD-1, PD-L1, and CTLA-4 pathways, providing hints for an interfering role of this cytokine in the immune-checkpoint blockade of anti-tumor immune responses.

Taken together, our data highlight the prognostic and diagnostic potential of *IL-37* mRNA levels in LUAD and provide evidence for its involvement in molecular networks and cellular distributions reported to play pivotal roles in LUAD tumorigenesis and progression. What is more, the interplay with the SIGIRR receptor, as well as its possible disturbance and/or independence, as possibly reflected by their differential expression pattern, points toward a crucial role in the differential disease phenotype, which could be further exploited as a potential immune-checkpoint therapeutic target. Our results are in agreement with previous studies supporting the implication of this cytokine and its receptor in the anti-tumor cytotoxic [24,99,100] and anti-angiogenic responses [101], as well as anti-invasion/metastatic processes in NSCLC [102,103]. Specifically, for LUAD, a study on patients' samples showed that the loss of or reduced IL-37 expression in the tumor correlates with metastasis development [23]. The protective effect of IL-37 has also been shown in other lung diseases such as idiopathic pulmonary fibrosis (IPF) [104].

However, the current study is limited in certain parameters, such as the fact that it is based on clinical samples analyzed through omics approaches. Therefore, it is essential that the resulted data are further validated in certain large patient cohorts of interest using specifically designed, targeted assays (including RT-qPCR). A detailed exploration of possible associations with clinical, histopathological, laboratory, and therapeutic parameters needs to be performed to empower the capacity of the *IL-37* to be conceivably used as a LUAD biomarker. For a better understanding of the pathogenetic implications of the cytokine, serial biopsies of LUAD specimens could be used to monitor its differential expression profile throughout the course of the disease and/or in association with certain treatment strategies. Further, peripheral blood vs. tumor samples should be comparably processed to explore the inflamed-tissue or peripheral LUAD specific distribution of *IL-37* as a demand of local or systemic immunoregulation, as previously described in other immune-mediated inflammatory disorders [105,106]. Moreover, it is important to further deepen our study and investigate the likely differential distribution of each of the five IL-37 isoforms [2], both in mRNA and protein levels. The analysis of the total mRNA and protein isoforms in -omics assays could hide specific patterns of certain variants, which also need to be further checked as to whether they exert similar or different functions in the LUAD microenvironment. The different isoforms bear different exons that have been implicated in extracellular and intranuclear activities of the cytokine, thus mediating different signaling regulations [2]. Additionally, and in combination with the above, it is of outmost importance to pinpoint the exact cellular source(s) of IL-37 and its receptor, to fully understand the effect of their aberrations in the intercellular responses within the LUAD microenvironment. Single-cell analyses supported their expression by lung tissue-resident T cells and MΦ, but relevant study should also be applied in LUAD tumor biopsies to explore the possible inducible expression of IL-37 isoforms and SIGIRR by malignant epithelial and/or tumor-infiltrating immune cell subsets. Towards that direction, our preliminary data support the fact that both IL-37 mRNA and protein are expressed by A549 human lung adenocarcinoma cells, as attested by a specifically developed RT-qPCR assay and flow cytometry (Supplementary Figure S2). SIGIRR expression was not detected in those cells by the aforementioned approaches.

Despite its limitations, the current study clearly supports the crucial involvement of IL-37 in LUAD pathogenesis and monitoring. Moreover, it highlights the plausible necessity for further investigation through mechanistic studies at the molecular and cellular level and validation in experimental models as well as in well-defined patient cohorts, in order to fully elucidate the exact role of this cytokine and to further exploit its potential for the improvement of LUAD patients' personalized management.

Supplementary Materials: The following are available online at https://www.mdpi.com/article/10.3390/biomedicines10123037/s1, Figure S1: UMAP plot (top) and bar chart (bottom) showing *IL-37* and *SIGIRR* expression levels (read counts normalized to transcripts per million protein coding genes, pTPM) in certain clusters of the lung. The numbers of cells in each cluster analyzed in single-cell RNA sequencing are included in parenthesis. Figures were exported from www.proteinatlas.org [35] (accessed on 20 August 2022). Figure S2: A. Indicative amplification and melting curve of RT-qPCR *IL-37* product in A549 human lung adenocarcinoma cells, B. Dot plot diagrams and related histograms depicting the expression of IL-37 protein by A549 cells as attested by flow cytometry. Table S1: OS prognostic potential of IL-37 and SIGIRR expression levels in LUAD patients bearing tumors of distinct histopathological characteristics. Table S2: Genes whose expression levels correlate with those of *SIGIRR* in LUAD tumors, as attested by the UALCAN portal (http://ualcan.path.uab.edu/) [33] (accessed on 20 September 2022). Pearson's *r* values for positive or negative associations are reported in each case.

Author Contributions: Study conception and supervision: M.-I.C.; Study design: M.-I.C. and P.P.; Acquisition of data: T.-C.K., P.C., P.B., M.A. and M.-I.C.; Statistical analysis: M.-I.C.; Drafting of the manuscript: T.-C.K., P.C. and P.B.; Writing of the manuscript: M.-I.C.; Critical revision of the manuscript: P.P., Y.J., D.X. and M.-I.C.; Response to reviewers: P.C. and M.-I.C. All authors have read and agreed to the published version of the manuscript.

Funding: This research received no funding.

Informed Consent Statement: Not applicable.

Data Availability Statement: Not applicable.

Conflicts of Interest: The authors declare no conflict of interest.

References

1. Su, Z.; Tao, X. Current Understanding of IL-37 in Human Health and Disease. *Front. Immunol.* **2021**, *12*, 696605. [CrossRef] [PubMed]
2. Cavalli, G.; Dinarello, C.A. Suppression of inflammation and acquired immunity by IL-37. *Immunol. Rev.* **2018**, *281*, 179–190. [CrossRef] [PubMed]
3. Sharaf, N.; Nicklin, M.J.; di Giovine, F.S. Long-range DNA interactions at the IL-1/IL-36/IL-37 gene cluster (2q13) are induced by activation of monocytes. *Cytokine* **2014**, *68*, 16–22. [CrossRef] [PubMed]
4. Bufler, P.; Gamboni-Robertson, F.; Azam, T.; Kim, S.H.; Dinarello, C.A. Interleukin-1 homologues IL-1F7b and IL-18 contain functional mRNA instability elements within the coding region responsive to lipopolysaccharide. *Biochem. J.* **2004**, *381 Pt 2*, 503–510. [CrossRef]
5. Sims, J.E.; Smith, D.E. The IL-1 family: Regulators of immunity. *Nat. Rev. Immunol.* **2010**, *10*, 89–102. [CrossRef]
6. Rudloff, I.; Cho, S.X.; Lao, J.C.; Ngo, D.; McKenzie, M.; Nold-Petry, C.A.; Nold, M.F. Monocytes and dendritic cells are the primary sources of interleukin 37 in human immune cells. *J. Leukoc. Biol.* **2017**, *101*, 901–911. [CrossRef]

7. Nold, M.F.; Nold-Petry, C.A.; Zepp, J.A.; Palmer, B.E.; Bufler, P.; Dinarello, C.A. IL-37 is a fundamental inhibitor of innate immunity. *Nat. Immunol.* **2010**, *11*, 1014–1022. [CrossRef]
8. Cavalli, G.; Justice, J.N.; Boyle, K.E.; D'Alessandro, A.; Eisenmesser, E.Z.; Herrera, J.J.; Hansen, K.C.; Nemkov, T.; Stienstra, R.; Garlanda, C.; et al. Interleukin 37 reverses the metabolic cost of inflammation, increases oxidative respiration, and improves exercise tolerance. *Proc. Natl. Acad. Sci. USA* **2017**, *114*, 2313–2318. [CrossRef]
9. Cavalli, G.; Koenders, M.; Kalabokis, V.; Kim, J.; Choon Tan, A.; Garlanda, C.; Mantovani, A.; Dagna, L.; Joosten, L.A.B.; Dinarello, C.A. Treating experimental arthritis with the innate immune inhibitor interleukin-37 reduces joint and systemic inflammation. *Rheumatology* **2017**, *56*, 2256. [CrossRef]
10. Moretti, S.; Bozza, S.; Oikonomou, V.; Renga, G.; Casagrande, A.; Iannitti, R.G.; Puccetti, M.; Garlanda, C.; Kim, S.; Li, S.; et al. IL-37 inhibits inflammasome activation and disease severity in murine aspergillosis. *PLoS Pathog.* **2014**, *10*, e1004462. [CrossRef]
11. Wald, D.; Qin, J.; Zhao, Z.; Qian, Y.; Naramura, M.; Tian, L.; Towne, J.; Sims, J.E.; Stark, G.R.; Li, X. SIGIRR, a negative regulator of Toll-like receptor-interleukin 1 receptor signaling. *Nat. Immunol.* **2003**, *4*, 920–927. [CrossRef] [PubMed]
12. Nold-Petry, C.A.; Lo, C.Y.; Rudloff, I.; Elgass, K.D.; Li, S.; Gantier, M.P.; Lotz-Havla, A.S.; Gersting, S.W.; Cho, S.X.; Lao, J.C.; et al. IL-37 requires the receptors IL-18Ralpha and IL-1R8 (SIGIRR) to carry out its multifaceted anti-inflammatory program upon innate signal transduction. *Nat. Immunol.* **2015**, *16*, 354–365. [CrossRef] [PubMed]
13. Bulau, A.M.; Nold, M.F.; Li, S.; Nold-Petry, C.A.; Fink, M.; Mansell, A.; Schwerd, T.; Hong, J.; Rubartelli, A.; Dinarello, C.A.; et al. Role of caspase-1 in nuclear translocation of IL-37, release of the cytokine, and IL-37 inhibition of innate immune responses. *Proc. Natl. Acad. Sci. USA* **2014**, *111*, 2650–2655. [CrossRef] [PubMed]
14. Grimsby, S.; Jaensson, H.; Dubrovska, A.; Lomnytska, M.; Hellman, U.; Souchelnytskyi, S. Proteomics-based identification of proteins interacting with Smad3: SREBP-2 forms a complex with Smad3 and inhibits its transcriptional activity. *FEBS Lett.* **2004**, *577*, 93–100. [CrossRef] [PubMed]
15. Sharma, S.; Kulk, N.; Nold, M.F.; Graf, R.; Kim, S.H.; Reinhardt, D.; Dinarello, C.A.; Bufler, P. The IL-1 family member 7b translocates to the nucleus and down-regulates proinflammatory cytokines. *J. Immunol.* **2008**, *180*, 5477–5482. [CrossRef]
16. Zhao, M.; Li, Y.; Guo, C.; Wang, L.; Chu, H.; Zhu, F.; Li, Y.; Wang, X.; Wang, Q.; Zhao, W.; et al. IL-37 isoform D downregulates pro-inflammatory cytokines expression in a Smad3-dependent manner. *Cell Death Dis.* **2018**, *9*, 582. [CrossRef]
17. Nascimento, C.; Ferreira, F. Tumor microenvironment of human breast cancer, and feline mammary carcinoma as a potential study model. *Biochim. Biophys. Acta Rev. Cancer.* **2021**, *1876*, 188587. [CrossRef]
18. Christodoulou, M.I.; Zaravinos, A. New Clinical Approaches and Emerging Evidence on Immune-Checkpoint Inhibitors as Anti-Cancer Therapeutics: CTLA-4 and PD-1 Pathways and Beyond. *Crit. Rev. Immunol.* **2019**, *39*, 379–408. [CrossRef]
19. Zhang, Z.; Zhang, J.; He, P.; Han, J.; Sun, C. Interleukin-37 suppresses hepatocellular carcinoma growth through inhibiting M2 polarization of tumor-associated macrophages. *Mol. Immunol.* **2020**, *122*, 13–20. [CrossRef]
20. Liu, Y.; Zhao, J.J.; Zhou, Z.Q.; Pan, Q.Z.; Zhu, Q.; Tang, Y.; Xia, J.C.; Weng, D.S. IL-37 induces anti-tumor immunity by indirectly promoting dendritic cell recruitment and activation in hepatocellular carcinoma. *Cancer Manag. Res.* **2019**, *11*, 6691–6702. [CrossRef]
21. Wang, Z.; Zeng, F.L.; Hu, Y.W.; Wang, X.Y.; Zhao, F.L.; Zhou, P.; Hu, J.; Xiao, Y.Y.; Hu, Z.L.; Guo, M.F.; et al. Interleukin-37 promotes colitis-associated carcinogenesis via SIGIRR-mediated cytotoxic T cells dysfunction. *Signal Transduct. Target. Ther.* **2022**, *7*, 19. [CrossRef] [PubMed]
22. Mei, Y.; Zhu, Y.; Teo, H.Y.; Liu, Y.; Song, Y.; Lim, H.Y.; Binte Hanafi, Z.; Angeli, V.; Liu, H. The indirect antiangiogenic effect of IL-37 in the tumor microenvironment. *J. Leukoc. Biol.* **2020**, *107*, 783–796. [CrossRef] [PubMed]
23. Li, Y.; Zhao, M.; Guo, C.; Chu, H.; Li, W.; Chen, X.; Wang, X.; Li, Y.; Jia, Y.; Koussatidjoa, S.; et al. Intracellular mature IL-37 suppresses tumor metastasis via inhibiting Rac1 activation. *Oncogene* **2018**, *37*, 1095–1106. [CrossRef] [PubMed]
24. Mu, X.; Zhao, Q.; Chen, W.; Zhao, Y.; Yan, Q.; Peng, R.; Zhu, J.; Yang, C.; Lan, K.; Gu, X.; et al. IL-37 Confers Anti-Tumor Activity by Regulation of m6A Methylation. *Front. Oncol.* **2020**, *10*, 526866. [CrossRef]
25. Zhu, B.; Luo, J.; Jiang, Y.; Yu, L.; Liu, M.; Fu, J. Prognostic significance of nomograms integrating IL-37 expression, neutrophil level, and MMR status in patients with colorectal cancer. *Cancer Med.* **2018**, *7*, 3682–3694. [CrossRef]
26. Haghshenas, M.R.; Hosseini, S.R.; Fattahi, M.J.; Malekzadeh, M.; Ariafar, A.; Ghaderi, A. Elevated IL-37 Serum Levels in Patients with Transitional Cell Carcinoma of Bladder. *Iran. J. Immunol.* **2021**, *18*, 346–353.
27. Osborne, D.G.; Domenico, J.; Luo, Y.; Reid, A.L.; Amato, C.; Zhai, Z.; Gao, D.; Ziman, M.; Dinarello, C.A.; Robinson, W.A.; et al. Interleukin-37 is highly expressed in regulatory T cells of melanoma patients and enhanced by melanoma cell secretome. *Mol. Carcinog.* **2019**, *58*, 1670–1679. [CrossRef]
28. Ding, L.; Zhao, X.; Zhu, N.; Zhao, M.; Hu, Q.; Ni, Y. The balance of serum IL-18/IL-37 levels is disrupted during the development of oral squamous cell carcinoma. *Surg. Oncol.* **2020**, *32*, 99–107. [CrossRef]
29. Wei, X.; Li, Y.; Zhang, G.; Wang, N.; Mi, M.; Xin, Y.; Jiang, H.; Sun, C. IL-37 Was Involved in Progress of Acute Myeloid Leukemia Through Regulating IL-6 Expression. *Cancer Manag. Res.* **2021**, *13*, 3393–3402. [CrossRef]
30. Farahani, N.; Mohagheghi, F.; Mosayebi, G.; Ghazavi, A.; Ganji, A. Reduced IL-37 gene expression and CD8 T lymphocytes in patients with metastatic breast cancer. *Breast Dis.* **2021**, *40*, 235–240. [CrossRef]
31. Uhlen, M.; Zhang, C.; Lee, S.; Sjostedt, E.; Fagerberg, L.; Bidkhori, G.; Benfeitas, R.; Arif, M.; Liu, Z.; Edfors, F.; et al. A pathology atlas of the human cancer transcriptome. *Science* **2017**, *357*, 2507. [CrossRef] [PubMed]

32. Bartha, A.; Gyorffy, B. TNMplot.com: A Web Tool for the Comparison of Gene Expression in Normal, Tumor and Metastatic Tissues. *Int. J. Mol. Sci.* **2021**, *22*, 2622. [CrossRef] [PubMed]
33. Chandrashekar, D.S.; Bashel, B.; Balasubramanya, S.A.H.; Creighton, C.J.; Ponce-Rodriguez, I.; Chakravarthi, B.; Varambally, S. UALCAN: A Portal for Facilitating Tumor Subgroup Gene Expression and Survival Analyses. *Neoplasia* **2017**, *19*, 649–658. [CrossRef] [PubMed]
34. Nagy, A.; Munkacsy, G.; Gyorffy, B. Pancancer survival analysis of cancer hallmark genes. *Sci. Rep.* **2021**, *11*, 6047. [CrossRef]
35. Szklarczyk, D.; Franceschini, A.; Wyder, S.; Forslund, K.; Heller, D.; Huerta-Cepas, J.; Simonovic, M.; Roth, A.; Santos, A.; Tsafou, K.P.; et al. STRING v10: Protein-protein interaction networks, integrated over the tree of life. *Nucleic Acids Res.* **2015**, *43*, D447–D452. [CrossRef] [PubMed]
36. Elzakra, N.; Kim, Y. HIF-1alpha Metabolic Pathways in Human Cancer. *Adv. Exp. Med. Biol.* **2021**, *1280*, 243–260. [PubMed]
37. Rashid, M.; Zadeh, L.R.; Baradaran, B.; Molavi, O.; Ghesmati, Z.; Sabzichi, M.; Ramezani, F. Up-down regulation of HIF-1alpha in cancer progression. *Gene* **2021**, *798*, 145796. [CrossRef]
38. Thul, P.J.; Akesson, L.; Wiking, M.; Mahdessian, D.; Geladaki, A.; Ait Blal, H.; Alm, T.; Asplund, A.; Bjork, L.; Breckels, L.M.; et al. A subcellular map of the human proteome. *Science* **2017**, *356*, eaal3321. [CrossRef]
39. Li, T.; Fu, J.; Zeng, Z.; Cohen, D.; Li, J.; Chen, Q.; Li, B.; Liu, X.S. TIMER2.0 for analysis of tumor-infiltrating immune cells. *Nucleic Acids Res.* **2020**, *48*, W509–W514. [CrossRef]
40. Barta, J.A.; Powell, C.A.; Wisnivesky, J.P. Global Epidemiology of Lung Cancer. *Ann. Glob. Health* **2019**, *85*, 8. [CrossRef]
41. Siegel, R.L.; Miller, K.D.; Fuchs, H.E.; Jemal, A. Cancer statistics, 2022. *CA Cancer J. Clin.* **2022**, *72*, 7–33. [CrossRef] [PubMed]
42. Sung, H.; Ferlay, J.; Siegel, R.L.; Laversanne, M.; Soerjomataram, I.; Jemal, A.; Bray, F. Global Cancer Statistics 2020: GLOBOCAN Estimates of Incidence and Mortality Worldwide for 36 Cancers in 185 Countries. *CA Cancer J. Clin.* **2021**, *71*, 209–249. [CrossRef] [PubMed]
43. Society, A.C. Cancer Facts and Figures 2022. Available online: https://www.cancer.org/research/cancer-facts-statistics/all-cancer-facts-figures/cancer-facts-figures-2022.html (accessed on 24 August 2022).
44. Jones, G.S.; Baldwin, D.R. Recent advances in the management of lung cancer. *Clin. Med.* **2018**, *18* (Suppl. 2), s41–s46. [CrossRef] [PubMed]
45. Myers, D.J.; Wallen, J.M. Lung adenocarcinoma. In *StatPearls*; StatPearls Publishing: Treasure Island, FL, USA, 2022.
46. Sun, B.; Zhao, H. A Study on Immune Cell Infiltration in Lung Adenocarcinoma. *Comb. Chem. High Throughput Screen* **2022**, *25*, 2082–2088. [PubMed]
47. Chen, G.; Dong, Z.; Wu, D.; Chen, Y. Profiles of immune infiltration in lung adenocarcinoma and their clinical significant: A gene-expression-based retrospective study. *J. Cell Biochem.* **2020**, *121*, 4431–4439. [CrossRef] [PubMed]
48. Zhao, M.; Li, M.; Chen, Z.; Bian, Y.; Zheng, Y.; Hu, Z.; Liang, J.; Huang, Y.; Yin, J.; Zhan, C.; et al. Identification of immune-related gene signature predicting survival in the tumor microenvironment of lung adenocarcinoma. *Immunogenetics* **2020**, *72*, 455–465. [CrossRef]
49. Zuo, S.; Wei, M.; Wang, S.; Dong, J.; Wei, J. Pan-Cancer Analysis of Immune Cell Infiltration Identifies a Prognostic Immune-Cell Characteristic Score (ICCS) in Lung Adenocarcinoma. *Front. Immunol.* **2020**, *11*, 1218. [CrossRef]
50. Insitute, N.C. The Cancer Genome Atlas. Available online: https://www.cancer.gov/about-nci/organization/ccg/research/structural-genomics/tcga (accessed on 24 August 2022).
51. Barrett, T.; Wilhite, S.E.; Ledoux, P.; Evangelista, C.; Kim, I.F.; Tomashevsky, M.; Marshall, K.A.; Phillippy, K.H.; Sherman, P.M.; Holko, M.; et al. NCBI GEO: Archive for functional genomics data sets—Update. *Nucleic Acids Res.* **2013**, *41*, D991–D995. [CrossRef]
52. Zhou, H.; Zheng, M.; Shi, M.; Wang, J.; Huang, Z.; Zhang, H.; Zhou, Y.; Shi, J. Characteristic of molecular subtypes in lung adenocarcinoma based on m6A RNA methylation modification and immune microenvironment. *BMC Cancer* **2021**, *21*, 938. [CrossRef]
53. Wu, C.; Rao, X.; Lin, W. Immune landscape and a promising immune prognostic model associated with TP53 in early-stage lung adenocarcinoma. *Cancer Med.* **2021**, *10*, 806–823. [CrossRef]
54. Zhao, Z.; He, B.; Cai, Q.; Zhang, P.; Peng, X.; Zhang, Y.; Xie, H.; Wang, X. Combination of tumor mutation burden and immune infiltrates for the prognosis of lung adenocarcinoma. *Int. Immunopharmacol.* **2021**, *98*, 107807. [CrossRef] [PubMed]
55. Sata, Y.; Nakajima, T.; Fukuyo, M.; Matsusaka, K.; Hata, A.; Morimoto, J.; Rahmutulla, B.; Ito, Y.; Suzuki, H.; Yoshino, I.; et al. High expression of CXCL14 is a biomarker of lung adenocarcinoma with micropapillary pattern. *Cancer Sci.* **2020**, *111*, 2588–2597. [CrossRef] [PubMed]
56. Shaykhiev, R.; Sackrowitz, R.; Fukui, T.; Zuo, W.L.; Chao, I.W.; Strulovici-Barel, Y.; Downey, R.J.; Crystal, R.G. Smoking-induced CXCL14 expression in the human airway epithelium links chronic obstructive pulmonary disease to lung cancer. *Am. J. Respir. Cell Mol. Biol.* **2013**, *49*, 418–425. [CrossRef] [PubMed]
57. Sang, L.; He, Y.J.; Kang, J.; Ye, H.; Bai, W.; Luo, X.D.; Sun, J. Mitochondrial Deoxyguanosine Kinase Regulates NAD(+) Biogenesis Independent of Mitochondria Complex I Activity. *Front. Oncol.* **2020**, *10*, 570656. [CrossRef]
58. Huang, J.; Zhang, Q.; Shen, J.; Chen, X.; Ma, S. Multi-omics analysis identifies potential mechanisms of AURKB in mediating poor outcome of lung adenocarcinoma. *Aging* **2021**, *13*, 5946–5966. [CrossRef]

59. Asada, K.; Kobayashi, K.; Joutard, S.; Tubaki, M.; Takahashi, S.; Takasawa, K.; Komatsu, M.; Kaneko, S.; Sese, J.; Hamamoto, R. Uncovering Prognosis-Related Genes and Pathways by Multi-Omics Analysis in Lung Cancer. *Biomolecules* **2020**, *10*, 524. [CrossRef]
60. Kato, T.; Ujiie, H.; Hatanaka, K.C.; Nange, A.; Okumura, A.; Tsubame, K.; Naruchi, K.; Sato, M.; Kaga, K.; Matsuno, Y.; et al. A novel Tn antigen epitope-recognizing antibody for MUC1 predicts clinical outcome in patients with primary lung adenocarcinoma. *Oncol. Lett.* **2021**, *21*, 202. [CrossRef]
61. Ren, J.; Yang, Y.; Li, C.; Xie, L.; Hu, R.; Qin, X.; Zhang, M. A Novel Prognostic Model of Early-Stage Lung Adenocarcinoma Integrating Methylation and Immune Biomarkers. *Front. Genet.* **2020**, *11*, 634634. [CrossRef]
62. Wang, Z.; Li, X.; Chen, H.; Han, L.; Ji, X.; Wang, Q.; Wei, L.; Miao, Y.; Wang, J.; Mao, J.; et al. Decreased HLF Expression Predicts Poor Survival in Lung Adenocarcinoma. *Med. Sci. Monit.* **2021**, *27*, e929333. [CrossRef]
63. Wu, Y.; Yang, L.; Zhang, L.; Zheng, X.; Xu, H.; Wang, K.; Weng, X. Identification of a Four-Gene Signature Associated with the Prognosis Prediction of Lung Adenocarcinoma Based on Integrated Bioinformatics Analysis. *Genes* **2022**, *13*, 238. [CrossRef]
64. Yang, D.; He, Y.; Wu, B.; Liu, R.; Wang, N.; Wang, T.; Luo, Y.; Li, Y.; Liu, Y. Predictions of the dysregulated competing endogenous RNA signature involved in the progression of human lung adenocarcinoma. *Cancer Biomark.* **2020**, *29*, 399–416. [CrossRef] [PubMed]
65. Ahmad, M.K.; Abdollah, N.A.; Shafie, N.H.; Yusof, N.M.; Razak, S.R.A. Dual-specificity phosphatase 6 (DUSP6): A review of its molecular characteristics and clinical relevance in cancer. *Cancer Biol. Med.* **2018**, *15*, 14–28. [PubMed]
66. Amelung, J.T.; Buhrens, R.; Beshay, M.; Reymond, M.A. Key genes in lung cancer translational research: A meta-analysis. *Pathobiology* **2010**, *77*, 53–63. [CrossRef] [PubMed]
67. Diaz-Garcia, C.V.; Agudo-Lopez, A.; Perez, C.; Prieto-Garcia, E.; Iglesias, L.; Ponce, S.; Rodriguez Garzotto, A.; Rodriguez-Peralto, J.L.; Cortes-Funes, H.; Lopez-Martin, J.A.; et al. Prognostic value of dual-specificity phosphatase 6 expression in non-small cell lung cancer. *Tumour. Biol.* **2015**, *36*, 1199–1206. [CrossRef]
68. Wang, T.L.; Ren, Y.W.; Wang, H.T.; Yu, H.; Zhao, Y.X. Association of Topoisomerase II (TOP2A) and Dual-Specificity Phosphatase 6 (DUSP6) Single Nucleotide Polymorphisms with Radiation Treatment Response and Prognosis of Lung Cancer in Han Chinese. *Med. Sci. Monit.* **2017**, *23*, 984–993. [CrossRef]
69. Wang, T.L.; Song, Y.Q.; Ren, Y.W.; Zhou, B.S.; Wang, H.T.; Bai, L.; Zhang, H.B.; Yu, H.; Zhao, Y.X. Dual Specificity Phosphatase 6 (DUSP6) Polymorphism Predicts Prognosis of Inoperable Non-Small Cell Lung Cancer after Chemoradiotherapy. *Clin. Lab.* **2016**, *62*, 301–310. [CrossRef]
70. Alam, S.K.; Zhang, Y.; Wang, L.; Zhu, Z.; Hernandez, C.E.; Zhou, Y.; Yang, N.; Lei, J.; Chen, X.; Zeng, L.; et al. DARPP-32 promotes ERBB3-mediated resistance to molecular targeted therapy in EGFR-mutated lung adenocarcinoma. *Oncogene* **2022**, *41*, 83–98. [CrossRef]
71. Pudelko, L.; Jaehrling, F.; Reusch, C.; Vitri, S.; Stroh, C.; Linde, N.; Sanderson, M.P.; Musch, D.; Lebrun, C.J.; Keil, M.; et al. SHP2 Inhibition Influences Therapeutic Response to Tepotinib in Tumors with MET Alterations. *iScience* **2020**, *23*, 101832. [CrossRef]
72. Yan, M.; Wang, W.; Zhou, J.; Chang, M.; Peng, W.; Zhang, G.; Li, J.; Li, H.; Bai, C. Knockdown of PLAT enhances the anticancer effect of gefitinib in non-small cell lung cancer. *J. Thorac. Dis.* **2020**, *12*, 712–723. [CrossRef]
73. Elia, I.; Broekaert, D.; Christen, S.; Boon, R.; Radaelli, E.; Orth, M.F.; Verfaillie, C.; Grunewald, T.G.P.; Fendt, S.M. Proline metabolism supports metastasis formation and could be inhibited to selectively target metastasizing cancer cells. *Nat. Commun.* **2017**, *8*, 15267. [CrossRef]
74. Liu, Y.; Mao, C.; Wang, M.; Liu, N.; Ouyang, L.; Liu, S.; Tang, H.; Cao, Y.; Liu, S.; Wang, X.; et al. Cancer progression is mediated by proline catabolism in non-small cell lung cancer. *Oncogene* **2020**, *39*, 2358–2376. [CrossRef] [PubMed]
75. Le, X.; Negrao, M.V.; Reuben, A.; Federico, L.; Diao, L.; McGrail, D.; Nilsson, M.; Robichaux, J.; Munoz, I.G.; Patel, S.; et al. Characterization of the Immune Landscape of EGFR-Mutant NSCLC Identifies CD73/Adenosine Pathway as a Potential Therapeutic Target. *J. Thorac. Oncol.* **2021**, *16*, 583–600. [CrossRef]
76. Liu, H.; Kuang, X.; Zhang, Y.; Ye, Y.; Li, J.; Liang, L.; Xie, Z.; Weng, L.; Guo, J.; Li, H.; et al. ADORA1 Inhibition Promotes Tumor Immune Evasion by Regulating the ATF3-PD-L1 Axis. *Cancer Cell* **2020**, *37*, 324–339 e8. [CrossRef] [PubMed]
77. Luo, J.; Zhu, H.; Jiang, H.; Cui, Y.; Wang, M.; Ni, X.; Ma, C. The effects of aberrant expression of LncRNA DGCR5/miR-873-5p/TUSC3 in lung cancer cell progression. *Cancer Med.* **2018**, *7*, 3331–3341. [CrossRef] [PubMed]
78. Kang, M.; Shi, J.; Li, B.; Luo, M.; Xu, S.; Liu, X. LncRNA DGCR5 regulates the non-small cell lung cancer cell growth, migration, and invasion through regulating miR-211-5p/EPHB6 axis. *Biofactors* **2019**, *45*, 788–794. [CrossRef]
79. Chen, E.G.; Zhang, J.S.; Xu, S.; Zhu, X.J.; Hu, H.H. Long non-coding RNA DGCR5 is involved in the regulation of proliferation, migration and invasion of lung cancer by targeting miR-1180. *Am. J. Cancer Res.* **2017**, *7*, 1463–1475. [PubMed]
80. Dong, H.X.; Wang, R.; Jin, X.Y.; Zeng, J.; Pan, J. LncRNA DGCR5 promotes lung adenocarcinoma (LUAD) progression via inhibiting hsa-mir-22-3p. *J. Cell Physiol.* **2018**, *233*, 4126–4136. [CrossRef]
81. Ren, W.; Mi, D.; Yang, K.; Cao, N.; Tian, J.; Li, Z.; Ma, B. The expression of hypoxia-inducible factor-1alpha and its clinical significance in lung cancer: A systematic review and meta-analysis. *Swiss Med. Wkly.* **2013**, *143*, w13855.
82. Yang, S.L.; Ren, Q.G.; Wen, L.; Hu, J.L. Clinicopathological and prognostic significance of hypoxia-inducible factor-1 alpha in lung cancer: A systematic review with meta-analysis. *J. Huazhong Univ. Sci. Technol. Med. Sci.* **2016**, *36*, 321–327. [CrossRef]
83. Xu, Y.; Wang, N.; Liu, R.; Lv, H.; Li, Z.; Zhang, F.; Gai, C.; Tian, Z. Epigenetic Study of Esophageal Carcinoma Based on Methylation, Gene Integration and Weighted Correlation Network Analysis. *OncoTargets Ther.* **2021**, *14*, 3133–3149. [CrossRef]

84. Xiang, R.; Ma, L.; Yang, M.; Zheng, Z.; Chen, X.; Jia, F.; Xie, F.; Zhou, Y.; Li, F.; Wu, K.; et al. Increased expression of peptides from non-coding genes in cancer proteomics datasets suggests potential tumor neoantigens. *Commun. Biol.* **2021**, *4*, 496. [CrossRef] [PubMed]
85. Ni, C.; Yang, P.; Guo, J.; Ye, M. Role of DiGeorge syndrome critical region gene 9, a long noncoding RNA, in gastric cancer. *OncoTargets Ther.* **2018**, *11*, 2259–2267. [CrossRef] [PubMed]
86. Hu, Y.; Zhang, X.; Wang, O.; Bi, Y.; Xing, X.; Cui, M.; Wang, M.; Tao, W.; Liao, Q.; Zhao, Y. The genomic profile of parathyroid carcinoma based on whole-genome sequencing. *Int. J. Cancer* **2020**, *147*, 2446–2457. [CrossRef] [PubMed]
87. Campbell, B.R.; Chen, Z.; Faden, D.L.; Agrawal, N.; Li, R.J.; Hanna, G.J.; Iyer, N.G.; Boot, A.; Rozen, S.G.; Vettore, A.L.; et al. The mutational landscape of early- and typical-onset oral tongue squamous cell carcinoma. *Cancer* **2021**, *127*, 544–553. [CrossRef]
88. Zou, H.; Zhu, N.; Li, S. The emerging role of dipeptidyl-peptidase-4 as a therapeutic target in lung disease. *Expert Opin. Ther. Targets* **2020**, *24*, 147–153. [CrossRef]
89. Ban, H.S.; Uto, Y.; Nakamura, H. Hypoxia-inducible factor (HIF) inhibitors: A patent survey (2016–2020). *Expert Opin. Ther. Pat.* **2021**, *31*, 387–397. [CrossRef]
90. Zhang, T.; Tong, X.; Zhang, S.; Wang, D.; Wang, L.; Wang, Q.; Fan, H. The Roles of Dipeptidyl Peptidase 4 (DPP4) and DPP4 Inhibitors in Different Lung Diseases: New Evidence. *Front. Pharmacol.* **2021**, *12*, 731453. [CrossRef]
91. Sangaletti, S.; Ferrara, R.; Tripodo, C.; Garassino, M.C.; Colombo, M.P. Myeloid cell heterogeneity in lung cancer: Implication for immunotherapy. *Cancer Immunol. Immunother.* **2021**, *70*, 2429–2438. [CrossRef]
92. Wang, J.B.; Huang, X.; Li, F.R. Impaired dendritic cell functions in lung cancer: A review of recent advances and future perspectives. *Cancer Commun.* **2019**, *39*, 43. [CrossRef]
93. Gardner, A.; de Mingo Pulido, A.; Ruffell, B. Dendritic Cells and Their Role in Immunotherapy. *Front. Immunol.* **2020**, *11*, 924. [CrossRef]
94. Kiss, M.; Caro, A.A.; Raes, G.; Laoui, D. Systemic Reprogramming of Monocytes in Cancer. *Front. Oncol.* **2020**, *10*, 1399. [CrossRef] [PubMed]
95. Leveque, E.; Rouch, A.; Syrykh, C.; Mazieres, J.; Brouchet, L.; Valitutti, S.; Espinosa, E.; Lafouresse, F. Phenotypic and Histological Distribution Analysis Identify Mast Cell Heterogeneity in Non-Small Cell Lung Cancer. *Cancers* **2022**, *14*, 1394. [CrossRef] [PubMed]
96. Sheida, F.; Razi, S.; Keshavarz-Fathi, M.; Rezaei, N. The role of myeloid-derived suppressor cells in lung cancer and targeted immunotherapies. *Expert Rev. Anticancer Ther.* **2022**, *22*, 65–81. [CrossRef] [PubMed]
97. Noyes, D.; Bag, A.; Oseni, S.; Semidey-Hurtado, J.; Cen, L.; Sarnaik, A.A.; Sondak, V.K.; Adeegbe, D. Tumor-associated Tregs obstruct antitumor immunity by promoting T cell dysfunction and restricting clonal diversity in tumor-infiltrating CD8+ T cells. *J. Immunother. Cancer* **2022**, *10*, e004605. [CrossRef] [PubMed]
98. Yuan, A.; Hsiao, Y.J.; Chen, H.Y.; Chen, H.W.; Ho, C.C.; Chen, Y.Y.; Liu, Y.C.; Hong, T.H.; Yu, S.L.; Chen, J.J.; et al. Opposite Effects of M1 and M2 Macrophage Subtypes on Lung Cancer Progression. *Sci. Rep.* **2015**, *5*, 14273. [CrossRef]
99. La Fleur, L.; Botling, J.; He, F.; Pelicano, C.; Zhou, C.; He, C.; Palano, G.; Mezheyeuski, A.; Micke, P.; Ravetch, J.V.; et al. Targeting MARCO and IL37R on Immunosuppressive Macrophages in Lung Cancer Blocks Regulatory T Cells and Supports Cytotoxic Lymphocyte Function. *Cancer Res.* **2021**, *81*, 956–967. [CrossRef]
100. Mei, Y.; Liu, H. IL-37: An anti-inflammatory cytokine with antitumor functions. *Cancer Rep.* **2019**, *2*, e1151. [CrossRef]
101. Ge, G.; Wang, A.; Yang, J.; Chen, Y.; Yang, J.; Li, Y.; Xue, Y. Interleukin-37 suppresses tumor growth through inhibition of angiogenesis in non-small cell lung cancer. *J. Exp. Clin. Cancer Res.* **2016**, *35*, 13. [CrossRef]
102. Chen, Y.H.; Zhou, B.Y.; Wu, X.J.; Xu, J.F.; Zhang, J.A.; Chen, Y.H.; Liang, S.S. [Corrigendum] CCL22 and IL37 inhibit the proliferation and epithelialmesenchymal transition process of NSCLC A549 cells. *Oncol. Rep.* **2021**, *45*, 786. [CrossRef]
103. Jiang, M.; Wang, Y.; Zhang, H.; Ji, Y.; Zhao, P.; Sun, R.; Zhang, C. IL-37 inhibits invasion and metastasis in non-small cell lung cancer by suppressing the IL-6/STAT3 signaling pathway. *Thorac. Cancer* **2018**, *9*, 621–629. [CrossRef]
104. Kim, M.S.; Baek, A.R.; Lee, J.H.; Jang, A.S.; Kim, D.J.; Chin, S.S.; Park, S.W. IL-37 Attenuates Lung Fibrosis by Inducing Autophagy and Regulating TGF-beta1 Production in Mice. *J. Immunol.* **2019**, *203*, 2265–2275. [CrossRef] [PubMed]
105. Christodoulou, M.I.; Kapsogeorgou, E.K.; Moutsopoulos, H.M. Characteristics of the minor salivary gland infiltrates in Sjogren's syndrome. *J. Autoimmun.* **2010**, *34*, 400–407. [CrossRef] [PubMed]
106. Christodoulou, M.I.; Kapsogeorgou, E.K.; Moutsopoulos, N.M.; Moutsopoulos, H.M. Foxp3+ T-regulatory cells in Sjogren's syndrome: Correlation with the grade of the autoimmune lesion and certain adverse prognostic factors. *Am. J. Pathol.* **2008**, *173*, 1389–1396. [CrossRef] [PubMed]

biomedicines

Review

Omics and Multi-Omics Analysis for the Early Identification and Improved Outcome of Patients with Psoriatic Arthritis

Robert Gurke [1,2,3,*], Annika Bendes [4], John Bowes [5,6], Michaela Koehm [1,2,7], Richard M. Twyman [8], Anne Barton [5,6], Dirk Elewaut [9], Carl Goodyear [10], Lisa Hahnefeld [1,2,3], Rainer Hillenbrand [11], Ewan Hunter [12], Mark Ibberson [13], Vassilios Ioannidis [13], Sabine Kugler [2,14], Rik J. Lories [15], Eduard Resch [1,2], Stefan Rüping [2,14], Klaus Scholich [1,2,3], Jochen M. Schwenk [4], James C. Waddington [16], Phil Whitfield [17], Gerd Geisslinger [1,2,3], Oliver FitzGerald [18], Frank Behrens [1,2,7], Stephen R. Pennington [16,18,*] and on behalf of the HIPPOCRATES Consortium

1 Fraunhofer Institute for Translational Medicine and Pharmacology ITMP, Theodor-Stern-Kai 7, 60596 Frankfurt am Main, Germany
2 Fraunhofer Cluster of Excellence Immune-Mediated Diseases CIMD, Theodor-Stern-Kai 7, 60596 Frankfurt am Main, Germany
3 Pharmazentrum Frankfurt/ZAFES, Institute of Clinical Pharmacology, Goethe University, Theodor-Stern-Kai 7, 60590 Frankfurt am Main, Germany
4 Science for Life Laboratory, School of Engineering Sciences in Chemistry, Biotechnology and Health, KTH Royal Institute of Technology, 171 65 Solna, Sweden
5 NIHR Manchester Biomedical Research Centre, Manchester University NHS Foundation Trust, Manchester Academic Health Science Centre, Manchester M13 9WU, UK
6 Centre for Genetics and Genomics Versus Arthritis, Centre for Musculoskeletal Research, Faculty of Biology, Medicine and Health, Manchester Academic Health Science Centre, The University of Manchester, Manchester M13 9PT, UK
7 Division of Rheumatology, University Hospital Frankfurt, Theodor-Stern-Kai 7, 60596 Frankfurt am Main, Germany
8 TRM Ltd., P.O. Box 493, Scarborough YO11 9FJ, UK
9 VIB-UGent Center for Inflammation Research, Ghent University, 9052 Ghent, Belgium
10 Institute of Infection, Immunity and Inflammation, University of Glasgow, Glasgow G12 8QQ, UK
11 Novartis Pharma AG, CH-4056 Basel, Switzerland
12 Oxford BioDynamics Limited, Oxford OX4 2JZ, UK
13 Vital-IT Group, SIB Swiss Institute of Bioinformatics, CH-1015 Lausanne, Switzerland
14 Fraunhofer IAIS, Institute for Intelligent Analysis and Information Systems, Schloss Birlinghoven 1, 53757 Sankt Augustin, Germany
15 Department of Development and Regeneration, KU Leuven, Skeletal Biology and Engineering Research Centre, P.O. Box 813 O&N, Herestraat 49, 3000 Leuven, Belgium
16 Atturos Ltd., c/o UCD Conway Institute, University College Dublin, D04 V1W8 Dublin, Ireland
17 Glasgow Polyomics, College of Medical, Veterinary and Life Sciences, Garscube Campus, University of Glasgow, Glasgow G61 1QH, UK
18 UCD Conway Institute, School of Medicine, University College Dublin, Belfield, D04 V1W8 Dublin, Ireland
* Correspondence: robert.gurke@itmp.fraunhofer.de (R.G.); stephen.pennington@ucd.ie (S.R.P.)

Citation: Gurke, R.; Bendes, A.; Bowes, J.; Koehm, M.; Twyman, R.M.; Barton, A.; Elewaut, D.; Goodyear, C.; Hahnefeld, L.; Hillenbrand, R.; et al. Omics and Multi-Omics Analysis for the Early Identification and Improved Outcome of Patients with Psoriatic Arthritis. *Biomedicines* **2022**, *10*, 2387. https://doi.org/10.3390/biomedicines10102387

Academic Editor: Marianna Christodoulou

Received: 11 August 2022
Accepted: 17 September 2022
Published: 24 September 2022

Abstract: The definitive diagnosis and early treatment of many immune-mediated inflammatory diseases (IMIDs) is hindered by variable and overlapping clinical manifestations. Psoriatic arthritis (PsA), which develops in ~30% of people with psoriasis, is a key example. This mixed-pattern IMID is apparent in entheseal and synovial musculoskeletal structures, but a definitive diagnosis often can only be made by clinical experts or when an extensive progressive disease state is apparent. As with other IMIDs, the detection of multimodal molecular biomarkers offers some hope for the early diagnosis of PsA and the initiation of effective management and treatment strategies. However, specific biomarkers are not yet available for PsA. The assessment of new markers by genomic and epigenomic profiling, or the analysis of blood and synovial fluid/tissue samples using proteomics, metabolomics and lipidomics, provides hope that complex molecular biomarker profiles could be developed to diagnose PsA. Importantly, the integration of these markers with high-throughput histology, imaging and standardized clinical assessment data provides an important opportunity to develop molecular profiles that could improve the diagnosis of PsA, predict its occurrence in cohorts of individuals with psoriasis, differentiate PsA from other IMIDs, and improve therapeutic responses. In this review, we consider the technologies that are currently deployed in the EU IMI2 project HIPPOCRATES to define biomarker profiles specific for PsA and discuss the advantages of combining multi-omics data to improve the outcome of PsA patients.

Keywords: psoriatic diseases; psoriatic arthritis; psoriasis; multi-omics; data integration

1. Introduction

1.1. Psoriasis and Psoriatic Arthritis

Psoriasis is a chronic, immune-mediated inflammatory disease (IMID) of the skin, which affects 0.91–8.5% of the population, varying by age, region and ethnicity [1]. The most common manifestation is plaque psoriasis (psoriasis vulgaris), which accounts for ~80% of cases and typically involves the formation of erythematous and scaly plaques on the head, ears, elbows and knees, as well as gluteal and umbilical areas. These skin changes are often highly conspicuous, and the resulting stigmatization can lead to psychosocial issues. There is also a high rate of comorbidities, including cardiovascular disease and obesity [2]. Approximately 30% of psoriasis patients go on to develop psoriatic arthritis (PsA) [3], a mixed-pattern IMID characterized by the inflammation of mainly entheseal and synovial musculoskeletal structures [4]. Predisposition to the development of PsA has a strong genetic basis [3] and correlates with the severity of psoriatic skin lesions, including nail involvement (pitting, cracking, separation or nail loss). However, in a minority of cases, the symptoms of PsA develop alongside psoriasis or even before it. Various environmental and lifestyle factors also increase the risk of PsA at the population level, including a high body mass index and smoking [5–7], although paradoxically, smoking is negatively associated with progression to PsA at the level of individual psoriasis patients [8]. There is also increasing evidence that dietary factors influence the risk of progressing to PsA [9,10]. PsA can lead to structural damage and loss of function of the joints due to bone erosion, new bone formation and cartilage loss [11]. It has diverse presentations including asymmetric oligo-articular forms of arthritis, polyarticular disease, dactylitis and spinal inflammation [12].

1.2. Current Diagnostic Practices and Disease Management Strategies

A diagnosis of psoriasis is usually based on the appearance of the skin [13]. Blood tests or other diagnostic procedures are generally unnecessary [14]. If clinical diagnosis is uncertain, psoriasis can be differentiated from visually similar conditions (such as certain forms of eczema) by skin biopsy, which will confirm epidermal thickening interdigitating with the dermis, changes to the stratum granulosum, the presence of nuclei in the superficial layer, and the presence of infiltrating T cells [15]. In contrast, there is no definitive diagnosis for PsA because the clinical manifestations overlap with other arthritic diseases, including rheumatoid arthritis (RA), osteoarthritis and inflammatory bowel disease (IBD)-associated arthritis. Current diagnostic practice is based on rheumatologic assessment involving physical examination, medical history, blood tests and imaging. More definitive diagnosis is generally dependent on the presence of inflammation and musculoskeletal damage, which makes early intervention much more challenging. The identification of early and specific biomarkers of PsA would facilitate immediate treatment with the most appropriate drugs, therefore offering a much better prognosis for PsA patients and even preventing disease progression in its early stages [16]. In addition to the need for early diagnosis so that treatment can improve patient outcomes, the management of chronic disease plays an important role with a focus on individualized and personalized treatment strategies. Even following the initiation of appropriate immunosuppressive therapy, up to ~40% of patients may not respond or experience adverse effects [17]. There is an urgent medical need for biomarkers that facilitate the early differentiation of PsA and allow the prediction and monitoring of therapeutic responses during the chronic disease stage, thus helping to normalize function and improve outcomes and quality of life.

1.3. The Promise of Omics and Multi-Omics Technology

Biomarkers that are distinct for specific groups of patients can be used for the early diagnosis of diseases because they often correspond to qualitative or even quantitative indicators of biological and pathological processes [18,19]. The genomics revolution in the 2000s identified a large panel of new genetic markers that are associated with particular disease phenotypes, but the potential of biomarkers expanded enormously as omics technology broadened to encompass the global analysis of DNA modifications (epigenomics), RNA (transcriptomics), proteins (proteomics) and metabolites (metabolomics). Furthermore, it is reasonable to differentiate between the analysis of polar metabolites and the analysis of lipids (lipidomics) because the physicochemical properties of these compounds are quite distinct and optimized methods for analyzing these groups are necessary. The advent of proteomics and metabolomics/lipidomics in particular has raised the possibility of using combinations of markers to differentiate between diseases or disease stages in a quantitative manner, which is not possible with genetic markers outside the field of oncology. As the corresponding technologies have become increasingly sophisticated, sensitive and automated, the cost of analysis has fallen and more ambitious studies are possible, including the correlation of multiple omics biomarker profiles across large groups of patients. This requires stringent quality control standards to be applied during sample collection, storage, preparation and analysis, including due attention to sample sizes and replicates, as well as appropriate randomization (Figure 1).

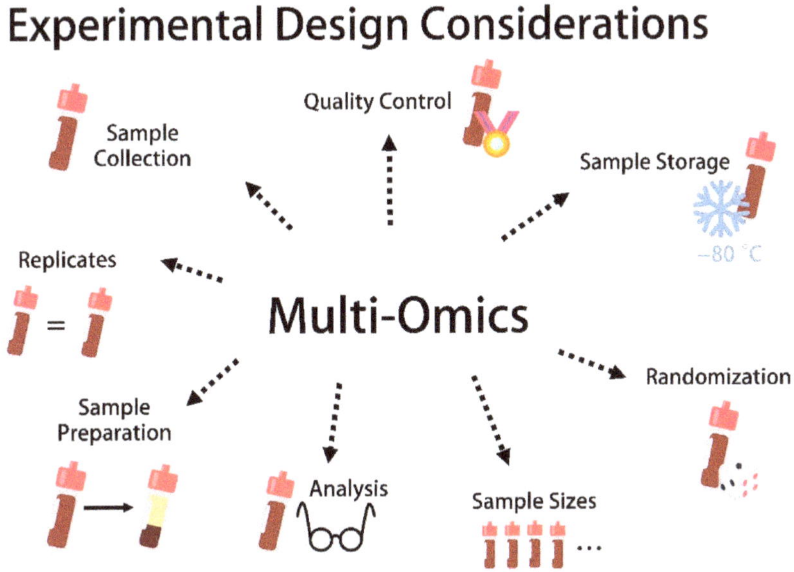

Figure 1. Experimental design considerations for the utilization of multi-omics data.

The EU-funded HIPPOCRATES project (https://hippocrates-imi.eu, accessed on 20 September 2022) is an ambitious collaboration that considers the potential of multiple molecular marker types across the spectrum of omics technology and seeks to combine them with conventional clinical diagnostic methods (imaging, medical records and physical examinations) for PsA. The value of omics technologies in the clinical care of PsA patients has been explored in a recent review article, including transcriptomics (which is not part of the HIPPOCRATES project) [20]. HIPPOCRATES aims to extend the concept by combining marker profiles for the differential diagnosis of psoriasis and PsA, as well as prognosis and the monitoring of treatment responses. In this review, we focus on the main objectives of the HIPPOCRATES project by considering the advantages and disadvantages of different omics technologies for the discovery of biomarkers for psoriasis and PsA, the potential of multi-omics approaches that combine different technologies to take advantage of synergies and how the diverse data formats may be combined and interrogated using advanced data evaluation tools (e.g., tools based on artificial intelligence) to identify patterns with diagnostic or prognostic value.

2. Genomics

2.1. Brief Overview of Relevant Genomics Technologies

Genomics is the branch of biology that deals with the analysis of genomes. In the context of psoriasis and PsA, genomics can be used to identify and characterize the genes, and more importantly the gene variants (alleles), that are associated with each disease. Many of the genes identified as associated with psoriasis have also been found to be associated with PsA when compared to population controls, highlighting their shared genetic basis. Susceptibility loci associated with PsA alone have also been identified, including several *HLA-B* alleles and *IL23R* [21,22]. The detection of pathological gene variants can be used to assist diagnosis and also to predict the age of onset, severity and likely symptoms of the disease. However, the multiple genes that distinguish between psoriasis and PsA may also be shared with other arthritic diseases, such as RA or ankylosing spondylitis.

The fundamental technology underlying the field of genomics is the genome-wide genotyping array, the contents of which are routinely enhanced by imputation, which provides the structure and sequence of key disease-associated genes and allows causative allelic variants to be identified. Genome-wide association studies (GWAS) and gene chip experiments have identified more than 20 additional loci outside the HLA system that are associated with PsA [23,24], some of which are exclusive (i.e., not also associated with psoriasis) [25]. The advent of next-generation sequencing platforms that are faster, cheaper and easier to automate than classic Sanger sequencing will enable researchers to amass a large body of sequence data from various patient cohorts, and this allows the comparison of patient groups to identify relevant alleles, in particular for rare variants not captured on genotyping arrays or by imputation.

2.2. Applications for Early Diagnosis, Prognosis and Treatment Monitoring

PsA is known to have a strong genetic component, which means that certain allelic variants are likely to be more prevalent among PsA patients than controls (or other disease cohorts). Because such genetic variation is present from conception, it should be possible to detect disease-causing alleles before the onset of symptoms and commence treatment as early as possible. Similarly, it should be possible to detect PsA-associated alleles in cohorts of psoriasis patients and thus identify those at the greatest risk of progression. Although many different alleles are associated with psoriasis, PsA or both, GWAS can be used to screen for large panels of variants in a single test, which is generally based on array hybridization or multiplex PCR [26,27]. The detection of one or more informative variants can therefore provide data to indicate causality. Other markers may be useful for the assessment of therapy, and to determine which subcomponents are heritable, and therefore more predictable [28]. Accordingly, prospective studies are needed in psoriasis patients, ideally recruited from primary care before disease-modifying therapy commences, to assess the ability of genetic variants to predict the onset of PsA.

2.3. Case Studies/Examples in Psoriasis and PsA

The primary genetic factors that distinguish PsA from psoriasis map to the *HLA-B* locus [29,30]. The alleles *HLA-B*39*, *HLA-B*07*, *HLA-B*38* and in particular *HLA-B*27* have been described as specific risk factors for PsA [31]. Although gene mapping is consistent across different studies, resolution to a precise allelic variant is conflicting when the reported index associations point to amino acid positions 45 or 97 (Table 1). Outside the HLA region, there is convincing evidence for a PsA-specific effect at the *IL23R* locus independent of the known psoriasis risk variant [32–34]. Other genes associated with PsA but not psoriasis include *KIR2D* [35], *IL4* and *KIF3A* [36], *B3GNT2* [37] and *PTPN22* [25].

Table 1. Genetic variants with evidence to support their ability to distinguish between PsA and cutaneous-only psoriasis.

Chromosome	Gene or Locus	Variant ID	Ref.
6	*HLA-B*	Amino acid position 45	[30]
6	*HLA-B*	Amino acid position 97	[29]
1	*IL23R*	rs12044149	[32–34]
5	*5q31 (IL4, KIF3A)*	rs715285	[36]
1	*PTPN22*	rs2476601	[25]
6	*TNFAIP3*	rs9321623	[34]
19	*KIR2D*		[35]
2	*B3GNT2*		[37]

3. Epigenomics

3.1. Brief Overview of Relevant Epigenomics Technologies

Epigenomics is the large-scale analysis of epigenetic phenomena, which include DNA methylation and histone modification as regulators of the 3D configuration of the genome, and the expression of small regulatory RNAs. Epigenetic mechanisms play a key role in the regulation of gene expression, and specific epigenetic markers can be associated with diseases such as psoriasis and PsA. Various technologies can be used to monitor genome-wide epigenetic phenomena, including chromatin immunoprecipitation (ChIP) followed by detection on microarrays (ChIP-chip) [38] or by sequencing (ChIP-Seq) [39], the detection of methylated DNA using bisulfite sequencing or (directly) by nanopore sequencing or SMRT sequencing [40], and enzyme-based chromatin accessibility assays [41]. The detection of chromosome conformation signatures (sequences that are likely to control the 3D structure of the genome) can also be used to pinpoint abnormal chromosome structures that are associated with diseases or responses to treatment. For example, the Oxford Biodynamics EpiSwitch platform is based on the testing of more than 10,000 samples in 30 disease indications, enabling the screening, evaluation, validation and monitoring of 3D genomic biomarkers [42].

3.2. Applications for Early Diagnosis, Prognosis and Treatment Monitoring

The EpiSwitch platform facilitates the discovery of stable and heriTable 3D genomic markers and the development of highly sensitive clinical assays based on non-invasive blood readouts. In the case of PsA, it can assist with a definitive diagnosis and prognosis in the context of comorbidities and overlapping symptoms, without resorting to biopsy. This technique has already delivered biomarkers that predict the response to methotrexate treatment in RA patients [43], that predict the response to immune checkpoint inhibitors in cancer [44], and that are prognostic for severe outcomes of COVID-19 based on individual patient immune health profiling [45]. The markers profiled by EpiSwitch technology are governed by all forms of genetic and epigenetic variation, and their combined influence has a major impact on the regulation of gene expression by controlling access to chromatin. Therefore, such markers are powerful high-level integrators of multi-omic signals [46]. In order to utilize the full potential of EpiSwitch, a representative cohort of whole blood samples with clinical annotations is required, representing extreme clinical outcomes. That spectrum will define the quality of the EpiSwitch biomarkers and their correlation with other modalities.

3.3. Case Studies/Examples in Psoriasis and PsA

Although chromosome conformation signatures for psoriasis and PsA are not yet available, the promise of the technique has been demonstrated in early RA patients commencing methotrexate treatment [43]. Using blood samples from responders, non-responders and healthy controls, a custom biomarker discovery array was refined to a five-marker chromosome conformation signature that could discriminate between responders and non-responders. Markers were validated using a blinded, independent cohort of 19 early RA patients (9 responders and 10 non-responders) and the corresponding loci were mapped to a RA-specific expression quantitative trait locus (eQTL). Finally, a five-marker chromosome conformation signature was found that could identify, at baseline, responders and non-responders to methotrexate. It consisted of binary chromosome conformations in the genomic regions of *IFNAR1*, *IL-21R*, *IL-23*, *CXCL13* and *IL-17A*. When tested on a cohort of 59 RA patients the marker provided a negative predictive value of 90% for methotrexate response. When tested on a blinded independent validation cohort of 19 early RA patients, the signature demonstrated a true negative response rate of 86%, and 90% sensitivity for the detection of non-responders. Only conformations in responders mapped to the RA-specific eQTL.

4. Proteomics

4.1. Brief Overview of Relevant Proteomics Technologies

Proteomics can be defined as the large-scale analysis of proteins. In the context of PsA, it has been applied mainly to identify biomarkers that can be detected in blood, synovial fluid or skin samples for the early diagnosis of PsA and its differentiation from psoriasis [47–49]. The proteome is much more complex and dynamic than the genome because there are an estimated ~20,000 protein-encoding genes in the human genome [50], but these give rise to multiple variants by alternative transcription, splicing and processing of RNA, post-translational modification and protein–protein interactions. About 10% of the human proteome lacks experimental evidence, and the combined effect of differential protein abundance, protein modifications, sequence variation and interactions further complicate the task of measuring all proteins in every sample [50].

The technologies used to interrogate the proteome can be broadly divided into untargeted methods that attempt to consider all proteins in a sample, and targeted methods that focus on specific proteins or classes of proteins. Mass spectrometry (MS) is a key platform in both approaches because it is a sensitive, high-throughput technology that is relatively easy to automate. Proteins are digested into peptides using a protease with known specificity such as trypsin, and the mass of each peptide, and its fragments generated inside a collision cell, is correlated with values in databases to achieve peptide and hence protein identification. Untargeted methods are based on the analysis of complex, uncharacterized peptide mixtures from multiple proteins. These are generally fractionated by liquid chromatography before injection into the mass spectrometer (LC-MS) and/or by multiple rounds of MS. In the latter case, data-dependent acquisition (DDA) involves the selection of specific peptides during the first round of MS for further fragmentation in subsequent rounds, whereas data-independent acquisition (DIA) involves the fragmentation and further analysis of all peptides from the first round [51]. Targeted methods involve the selection of one or a relatively small number of proteins from a sample for quantitative analysis [52]. Targeted analysis can be undertaken using MS-based methods as exemplified in the Atturos platform or methods that rely on highly specific affinity reagents. In the latter case, the production of high-quality data requires the use of validated binders (affinity reagents) that capture target proteins at low abundance [53]. Current affinity proteomics methods can detect more than 3000 proteins simultaneously by using different selectivity concepts, as well as the amplification capabilities of DNA-based readout methods. One relevant example is the Olink platform, a proximity extension assay that involves the recognition of proteins by antibodies linked to protein-specific DNA barcodes that can be amplified by qPCR or sequencing [54]. This may have a broader dynamic range and greater sensitivity than LC-MS and can simultaneously detect 3000 human proteins in plasma samples [55]. The use of slow off-rate DNA aptamers, provided by SomaLogic, has enabled large-scale studies of 10,000 donors targeting 4000 circulating proteins across human diseases [56].

4.2. Applications for Early Diagnosis, Prognosis and Treatment Monitoring

For the proteomic analysis of body fluids, particularly blood, further challenges arise due to the broad concentration range of different proteins, dynamic changes induced by disease processes, and analytical factors that influence protein detection [57]. Collectively, more than 4500 proteins have been detected in plasma samples by discovery-driven MS [58]. Many abundant plasma constituents are secreted by the liver, whereas other secreted proteins, such as inflammatory cytokines, are often elevated only transiently [59]. Accordingly, differences have been observed between individuals and between molecular profiles at longitudinal study time points [60]. When searching for protein biomarkers in healthy individuals, as well as psoriasis and PsA patients, the heterogeneity of signatures from circulating proteins should be expected.

Multiple candidate biomarkers of PsA have been reported in serum and plasma, in addition to a smaller number found in synovial fluid/tissue and skin biopsies [47,48]. Most of the biomarker candidates are proposed for the detection of PsA [61,62], differentiation between mild and severe forms [63,64], measuring disease activity [65], or predicting which psoriasis patients are likely to develop PsA [66]. However, others have been proposed to distinguish PsA from other arthritic diseases such as RA [49,67] or to monitor responses to therapy [68–71]. For example, the label-free MS analysis of synovial fluid from PsA patients revealed 12 candidate PsA markers including the injury marker MMP3, as well as the inflammatory proteins S100A9 and CRP [62]. A subsequent study using LC-MS identified periostin, which is related to cell-adhesion proteins, and the angiogenesis marker PGK1 [72]. More recently, a systematic search of five bibliographic databases for clinical, laboratory and genetic markers was used to determine the level of evidence for each marker and its association with concomitant/developing PsA [73]. These have been converted into proteomic biomarkers in Table 2. For the prediction of PsA in psoriasis patients, highly characterized cohorts of patients are needed with each disease, minimizing the proportion of undiagnosed subclinical PsA patients in the psoriasis group. Alternatively, longitudinal observation and sample collection in the psoriasis group may directly identify those progressing to PsA, allowing the retrospective analysis of early samples to look for predictive biomarkers.

4.3. Case Studies/Examples in Psoriasis and PsA

In a recent study, a set of 951 circulating proteins was analyzed in serum samples to interrogate possible differences between patients with PsA, psoriasis and healthy controls [74]. Sixty-eight differentially expressed proteins were identified when comparing PsA patients and healthy controls, but no differentially expressed proteins were identified when comparing PsA and psoriasis patients. This led the authors to propose a "shared serum proteomic signature" between psoriasis and PsA. However, the cohorts were very small and subclinical PsA in the psoriasis group could not be excluded. Indeed, no information was provided about patient inclusion/exclusion criteria or the criteria used for the differentiation of PsA from psoriasis, which is necessary in such studies. In conclusion, the authors recommended that future studies focus on skin and synovial tissue to find differences between PsA and psoriasis patients.

Table 2. Proteomic markers with evidence to support their ability to distinguish between PsA and cutaneous-only psoriasis. A gene-centric table of candidates was created by using the biomarkers listed by Mulder et al. [73]. The proteins and mRNAs were converted into gene-centric entries using the Human Protein Atlas portal (www.proteinatlas.org (accessed on 5 July 2022)), and were annotated for secretion location, tissue expression and biological functional based on the recent clustering of single-cell expression data [75].

Gene Name	Biomarker	UniProt ID	Category	Secretion	Tissue Expression	Biological Function
ADIPOQ	Adiponectin	Q15848	Lipid	Blood	Adipose tissue	ECM organization
APOA1	ApoA	P02467	Lipid	Blood	Liver	Metabolism
APOB	ApoB	P04114	Lipid	Blood	Liver	Metabolism
CMC2	C16ORF61	Q9NRP2	Skin	N/A	Non-specific	Mitochondria
COL2A1	C2C	P02458	Bone	ECM	Epididymis	Unknown function
CCL1	CCL1	P22362	mRNA	Blood	T cells	Adaptive immune response
CCL20	CCL20	P78556	mRNA	Blood	Smooth muscle tissue	Mixed function
CCL7	CCL7	P80098	mRNA	Blood	Neutrophils	Humoral immune response
CD5L	CD5L	O43866	Serum	Blood	Macrophages	Immune response
COMP	COMP	P49747	Bone	ECM	Skin	Epidermis development
C9	Complement C9	P02748	Serum	Blood	Liver	Hemostasis and lipid
COL2A1	CPII	P02458	Bone	ECM	Epididymis	Unknown function
CPN2	CPN2	P22792	Skin	Blood	Liver	Hemostasis
CRP	CRP	P02741	Inflammation	Blood	Liver	Hemostasis
COL1A1	CTX	P02452	Bone	ECM	Fibroblasts	ECM organization
CX3CL1	CX3CL1	P78423	mRNA	Blood	Adipose tissue	ECM organization
CXCL10	CXCL10	P02778	Cytokines	Blood	Immune cells	Immune response
CXCL12	CXCL12	P48061	Skin	Blood	Fibroblasts	ECM organization
CXCL2	CXCL2	P19875	mRNA	Blood	Liver	Metabolism
CXCL5	CXCL5	P42830	mRNA	Blood	Salivary gland	Salivary secretion
DKK1	DKK-1	O94907	Bone	Other	Adipose tissue	ECM organization
ESR1	ESR	P03372	Inflammation	N/A	Fibroblasts	ECM organization
FHL1	FHL1	Q13642	Skin	N/A	Striated muscle	Muscle contraction
GSN	Gelsolin	P06396	Serum	Blood	Fibroblasts	ECM organization
GPS1	GPS1	Q13098	Skin	N/A	Non-specific	Mitochondria
HAT1	HAT1	O14929	mRNA	N/A	Non-specific	Ribosome
IFI16	IFI16	Q16666	Serum	N/A	Immune cells	Immune response
IL12A	IL-12/23 p40	P29459	Cytokines	Blood	Brain and skin	Unknown function
IL12B	IL-12/23 p40	P29460	Cytokines	Blood	Non-specific	Cell cycle regulation
IL9	IL-12/23 p40	P15248	Cytokines	Blood	N/A	N/A
IL17A	IL-17	Q16552	Cell culture secretion	Blood	Immune cells	Immune response
IL17C	IL-17C	Q9P0M4	mRNA	Blood	Testis	DNA repair
IL17F	IL-17F	Q96PD4	mRNA	Blood	B cells	Humoral immune response
IL2	IL-2	P60568	Cell culture secretion	Blood	N/A	N/A
IL23	IL-23	Q9NPF7	Cytokines	Blood	B cells	Humoral immune response
IL23R	IL23R	Q5VWK5	Skin	N/A	Intestine	Brush border
IL3	IL-3	P08700	mRNA	Blood	N/A	N/A
IL33	IL-33	O95760	Cytokines	Blood	Fibroblasts	ECM organization
IL34	IL-34	Q6ZMJ4	Cytokines	Blood	Macrophages	Immune response
EBI3	IL-35	Q14213	Cytokines	Blood	Placenta	Pregnancy
IL12A	IL-35	P29459	Cytokines	Blood	Brain and skin	Unknown function
IL36A	IL-36a	Q9UHA7	Cytokines	Blood	Esophagus	Epithelial cell function
IL1F10	IL-38	Q8WWZ1	Cytokines	Blood	Skin	Cornification
IL6	IL-6	P05231	Cytokines, mRNA	Blood	Adipose tissue	ECM organization
CXCL8	IL-8	P10145	mRNA	Blood	Neutrophils	Humoral immune response
INS	Insulin	P01308	Lipid	Blood	Pancreas	Digestion
ISG20	ISG20	Q96AZ6	mRNA	N/A	Immune cells	Immune response
ITGB5	ITGB5	P18084	Serum	N/A	Adipose tissue	ECM organization
ITGB5	ITGB5	P18084	Skin	N/A	Adipose tissue	ECM organization
KRT17	K17	Q04695	Serum	N/A	Skin	Epidermis development
LEP	Leptin	P41159	Lipid	Blood	Adipose tissue	ECM organization
LGALS3BP	M2BP	Q08380	Serum	Blood	Stomach	Digestion
CSF1	M-CSF	P09603	Cytokines	Blood	Non-specific	Angiogenesis
MMP3	MMP3	P08254	Bone, mRNA	ECM	Salivary gland	Salivary secretion
MPO	MPO	P05164	Serum	Membrane	Neutrophils	Humoral immune response
NOTCH2NLA	NOTCH2NL	Q7Z3S9	mRNA	Blood	Testis	DNA repair
TNFRSF11B	OPG	O00300	Bone	Other	Kidney	Transmembrane transport
POSTN	POSTN	Q15063	Skin	ECM	Skin	Epidermis development
PTPA	PPP2R4	Q15257	Skin	N/A	Non-specific	Mitochondria
PRL	PRL	P01236	Serum	Blood	Pituitary gland	Hormone signaling
TNFSF11	RANKL	O14788	Bone	Blood	Immune cells	Immune response
SETD2	SETD2	Q9BYW2	mRNA	N/A	Non-specific	Transcription
IL2RA	sIL2R	P01589	Serum	N/A	Immune cells	Immune response
IL2RB	sIL2R	P14784	Serum	N/A	Immune cells	Immune response
IL2RG	sIL2R	P31785	Serum	N/A	T cells	Adaptive immune response
SNCA	SNCA	P37840	Skin	Membrane	Brain and bone marrow	Chromatin organization
SRP14	SRP14	P37108	Skin	N/A	Non-specific	Mitochondria
SRPX	SRPX	P78539	Skin	Unknown	Adipose tissue	ECM organization
STAT3	STAT3	P40763	mRNA	N/A	Non-specific	Mitochondria and proteasome
STAT6	STAT6	P42226	mRNA	N/A	Macrophages	Immune response

Table 2. *Cont.*

Gene Name	Biomarker	UniProt ID	Category	Secretion	Tissue Expression	Biological Function
STIP1	STIP1	P31948	Serum	N/A	Non-specific	Unknown function
SYK	SYK	P43405	mRNA	N/A	Non-specific	Transcription
TBX21	TBX21	Q9UL17	mRNA	N/A	Immune cells	Immune response
TNF	TNF-alpha	P01375	Cytokines	Blood	Neutrophils	Inflammatory response
VCP	VCP	P55072	Serum	N/A	Non-specific	Mitochondria
FLT4	VEGFR-3	P35916	Serum	Blood	Non-specific	Transcription
CHI3L1	YKL-40	P36222	Serum	Blood	Liver	Metabolism

5. Metabolomics

5.1. Brief Overview of Relevant Metabolomics Technologies

Metabolomics can be defined as the investigation of changes in the populations of endogenous and exogenous low-molecular-weight metabolites (<1500 Da), representing a shift from single metabolite monitoring to complex profiling and pattern recognition [76]. This is a considerable analytical challenge that involves the identification and quantification of a broad spectrum of molecules in biological matrices such as human plasma or urine, which contain hundreds or thousands of metabolites with diverse chemical and physical properties across a wide dynamic range of concentrations. The most widely used techniques include nuclear magnetic resonance (NMR) spectroscopy and mass spectrometry in combination with gas chromatography (GC-MS) or liquid chromatography (LC-MS). Advanced bioinformatics and statistical tools are used to maximize the recovery of information from the resulting metabolomic datasets.

5.2. Applications for Early Diagnosis, Prognosis and Treatment Monitoring

Low-molecular-weight metabolites are important indicators and even integrators of phenotypes, reflecting the biochemical activity of cells and tissues. Metabolomics recognizes that changes in cell function are most evident at the level of small-molecule metabolism and can provide a coherent view of the response of individuals to a variety of genetic and environmental influences [77]. The abnormal cellular processes associated with disease often disrupt the composition of low-molecular-weight metabolites. Perturbations in metabolite abundance and temporal profiles in readily accessible body fluids may provide an index of disease severity through the direct measurement of biochemical changes. As such, metabolomics has the potential to identify biomarkers of PsA that may improve diagnostic accuracy and predict disease progression as well as defining patient responses to specific therapeutic interventions. Similarly, metabolomics may offer additional insight into the metabolic pathways that drive the chronic, immune-mediated processes that are characteristic of PsA, opening routes to potential new drug targets.

5.3. Case Studies/Examples in Psoriasis and PsA

Researchers are increasingly using metabolomics for the clinical assessment of PsA [78–80]. Several studies have reported alterations in the metabolomes of PsA patients in comparison to healthy controls or individuals with related inflammatory diseases such as psoriasis or RA. The serum levels of various amino acids are modified in PsA patients relative to RA cohorts [81,82]. Changes in the levels of circulating glucuronic acid and α-ketoglutaric acid were detected among psoriasis patients with or without PsA [83] and a correlation was made between serum levels of the choline metabolite trimethylamine *N*-oxide (TMAO) and inflammation in PsA patients [84]. A more recent study used untargeted metabolomics to characterize the metabolic changes in the transition from psoriasis to PsA, revealing differences in the abundance of bile acids (particularly glycoursodeoxycholic acid sulfate) and butyrate to differentiate between psoriasis patients who did or did not progress to PsA [85].

Metabolite profiles in other matrices can also provide a window of opportunity to elucidate the metabolic changes in PsA. It was recently reported that α/β-turmerone, glycerol 1-hexadecanoate, dihydrosphingosine, pantothenic acid and glutamine may act as fecal biomarkers for PsA [86]. In addition, a metabolomic study focusing on urinary metabolites revealed lower levels of citrate, alanine, methylsuccinate and trigonelline in PsA patients compared to unaffected individuals [87]. Metabolomic approaches have also been used to evaluate PsA patient responses to anti-TNF therapy. For example, histamine, glutamine, phenylacetic acid, xanthine, xanthurenic acid and creatinine levels were elevated in urine samples from patients who responded to TNF antagonists, whereas ethanolamine, *p*-hydroxyphenylpyruvic acid and phosphocreatine levels were depleted [88].

6. Lipidomics

6.1. Brief Overview of Relevant Lipidomics Technologies

Recent technological improvements in LC-MS enable comprehensive lipidomic analysis in clinical studies, analyzing extensive sample sets for different lipids and lipid mediators. Depending on the specific lipids, targeted or untargeted LC-MS may be most appropriate. The untargeted approach is based on high-resolution mass spectrometry (HRMS) and can potentially examine the whole lipidome in a single run, but focuses on the more abundant lipids because the dynamic range is not sufficient to detect scarce molecules such as lipid mediators alongside abundant lipids such as triglycerides. Scarce lipids are analyzed using targeted approaches based on tandem mass spectrometry (MS/MS), which has greater sensitivity and selectivity. However, targeted methods cannot display the whole lipidome, so an approach combining untargeted and targeted methods is used for comprehensive lipidomics analysis, searching for lipids and lipid mediators relevant in the context of psoriatic diseases.

6.2. Applications for Early Diagnosis, Prognosis and Treatment Monitoring

Lipids and lipid mediators play a fundamental role in the immune system and changes in homeostatic status are closely related to IMIDs [89–93] such as RA [94], IBD [95–99] and psoriatic diseases [100–103]. Lipids are involved in many different processes and are also essential building blocks of membranes and key components in energy metabolism. Lipid mediators such as oxylipins and endocannabinoids, which are present at very low concentrations, are signaling molecules implicated in diverse physiological and pathological processes. Therefore, lipid profiles might be used as biomarkers for early diagnosis, prognosis of disease progression or the development of comorbidities, and to guide the selection of the most promising therapeutic approach.

Biomarker discovery in the field of lipidomics is challenging due to strict procedural requirements for sampling, sample preparation and analysis. This is partly due to the variable concentration of lipids in different biological matrices, the broad spectrum of isomeric compounds and the special procedures required to ensure lipid stability at all stages, including pre-analytical sample handling. On the other hand, lipidomics covers a field of up to several thousand different molecules and one of its key advantages is the close temporal linkage between these markers and individual clinical phenotypes or disease states [77,104].

6.3. Case Studies/Examples in Psoriasis and PsA

The close link between lipid profiles and IMIDs was recently demonstrated in PsA patients, where oxylipins [102,103,105,106], endocannabinoids [103,107], fatty acids [105–107] and phospholipids [106] were found to be potentially pathophysiologically relevant. A recent study also found that the level of inflammatory lipid mediators in psoriasis patients increased following a PsA diagnosis, particularly leukotriene B4 [85]. However, a comprehensive study is required to compare the lipid profiles of patients with psoriasis and PsA, and this will be the first step toward the identification of lipid biomarkers that improve the diagnosis and treatment of PsA.

7. Complementary Technologies—Multiple Sequential Immunohistochemistry

Several multiple immunohistochemistry systems have been developed that allow the staining of tissue sections with a theoretically unlimited number of antibodies. The technology makes use of directly labeled antibodies carrying a fluorophore or heavy metal ion. The antibodies are applied to the sample in an automated process, which includes a short incubation period, washing steps, imaging and signal removal. The latter involves either bleaching or chemical inactivation, and is followed by the addition of the next antibody [108,109]. This process can be repeated as often as necessary, and typically creates image stacks representing 30–50 antibodies. Recent developments include software that combines single-cell phenotyping and localized information about neighboring cells, facilitating a quantitative "tissue FACS analysis" (FACS = fluorescence-activated cell sorting) with the description of disease-specific immunological neighborhoods within inflamed tissues [108].

One of the key benefits of multiple immunohistochemistry systems in the context of PsA is single-cell phenotyping in different patient groups using 40–50 antibody probes in automated cycles. By detecting and quantifying a large panel of corresponding markers, it is possible to identify nearly all immune cells and their subtypes, and to characterize their cellular neighborhood to quantify and visualize cellular networks (information that is lost during FACS analysis). The comparison of samples from psoriasis and PsA cohorts can therefore identify differences between the patient groups and generate information about biomarkers and immune cell networks/interactions that may lead to new therapeutic options.

8. Data Management/Integration and Artificial Intelligence

To benefit from the wealth of methods used to mine multi-omics data, it is essential to align the data and verify their quality before integration. Data should be formatted according to international standards, including standard bioinformatics file formats (such as FASTA, FASTQ, SAM/BAM, VCF and GFF), and "minimum information" standards for omics experiments [110], including MIGS/MIMS for genomics [111] and MIAPE for proteomics [112]. The data must be checked to ensure they include the same annotation references (e.g., genome version, standard gene and protein names). This is challenging with lipidomics and metabolomics data where there are currently no widely accepted standards, although efforts are ongoing to establish equivalent minimum information standard such as MIAMET [113] as well as standards for lipidomics analysis [114,115]. Following de-identification, clinical data are standardized using the OMOP common data model (Observational Health Data Sciences and Informatics, OHDSI program, available at https://ohdsi.org/ (accessed on 20 June 2022) and aligned to standard dictionaries to ensure interoperability. Once formatted and standardized, data are stored and accessed via a secure data management infrastructure specifically designed to protect sensitive clinical and biomedical data [116].

Multiple processing steps should be considered to ensure data integrity, including missing value imputation, normalization, transformations, aggregation and batch effect correction [117–119]. Unsupervised multivariate analysis methods such as common dimensions [120] can be used to assess overall variability, trends and potential biases across multiple integrated layers of multi-omics and clinical data before further data exploration by supervised multivariate analysis methods such as OPLS-DA [121] or artificial intelligence-based approaches such as machine learning. A wide range of computational methods can be applied depending on the study design and research aim. In addition to classical statistical analysis, machine learning can be used to evaluate data in an unsupervised manner for preliminary exploration and dimensional reduction (e.g., clustering approaches such as DBSCAN or k-means algorithms, or dimensional reduction such as PCA or TSNE). Batch effects in dimensional reduction and clustering approaches can reveal outliers [118]. In a clinical setting, supervised machine learning often tackles classification problems rather than regression. Due to the high dimensionality of multi-omics data and the so-called "curse of dimensionality" (low number of subjects and high number of features), feature selection algorithms such as LASSO or ridge can be applied to enhance the results of supervised learning [118,122]. In addition to feature selection, class imbalances are common challenges in multi-omics analysis, but can be solved by sampling or cost-sensitive learning [122]. Commonly used algorithms such as random forests, support vector machines and the k-nearest neighbor algorithm can provide insight into the underlying structures of datasets and can be developed into powerful models for the support of clinical decision making [123–125]. In order to develop further hypotheses and integrate data with the literature, pathway analysis can embed the data in a broader context [126].

9. The Advantage of Multi-Omics Evaluation

As discussed above, several markers have been identified that commonly occur in PsA patients, but no single marker stands alone as a specific indicator of the disease. Even the most reliable markers are also present in other IMIDs, which therefore makes it difficult if not impossible to achieve a definitive diagnosis. In other fields, the lack of definitive qualitative markers has been addressed by (a) seeking quantitative markers, whose abundance rather than presence/absence correlates with a disease, and (b) profiles based on combinations of several markers that are more informative than single molecules, a strategy first applied to ovarian cancer [127]. Indeed, this approach has also been successful in RA, where the proteomic analysis of serum and synovial fluid has revealed the elevation of multiple biomarkers representing processes such as joint inflammation and injury (e.g., MMP3, IL-12, IL-15 and IL-18), cartilage integrity and bone or connective tissue degradation (e.g., MMP13 and neoepitopes of collagen) [128]. Considering that such panels of RA markers have been assembled based solely on proteomics data, it is clear that the combination of proteomics with orthogonal omics datasets plus more diverse data can increase the power of this approach exponentially, both for diagnosis/prognosis [129] and the monitoring of drug responses [130]. In one recent study, metabolomics and lipidomics analysis revealed that a combination of the bile acid conjugate glycoursodeoxycholic acid sulfate and lipid mediator leukotriene B4 provided a sensitive and specific predictor of progression from psoriasis to PsA [85]. However, adding new features will also require the careful evaluation of added value, both for discovery in basic research and translation to the clinic. Models that incorporate more markers may be more sensitive and specific, but the cost of acquiring the data in routine clinical practice may be prohibitive, although this may not always be the case [131]. A risk remains that expensive and large datasets merely report already known aspects, such as the effect of the body mass index or inflammation on disease progression. The field should also strive to identify causal markers rather than solely correlative observations without a direct biochemical connection to the phenotype. As the amount and complexity of the data increase, it becomes more difficult for humans to identify consistent patterns that correlate with certain diseases, but machine learning algorithms either supervised to assign samples to known categories or devising categories de novo using unsupervised analysis have the power to reveal these hidden patterns and then to apply the same approach when analyzing data from new patients, greatly improving the accuracy of the resulting predictions.

10. Conclusions and Outlook

The definitive diagnosis and early treatment of PsA are not yet possible because the clinical manifestations and associated biomarkers are not, on an individual basis, able to distinguish PsA from other IMIDs or predict those individuals with psoriasis who will progress to PsA. However, the combination of biomarker profiles based on data from multi-omics technologies and classical sources, such as imaging data and clinical evaluations, could provide the body of information required for early diagnosis and the initiation of effective treatment before symptoms emerge. Combinations of different types of biomarkers have proven effective in other fields, particularly oncology, but such biomarker profiles are often mainly based on a single method. The power of biomarker profiles may increase with the number of complementary modalities that can be tested simultaneously. For PsA, combining information from disease-associated alleles and chromatin structures, the levels of proteins, lipids and other metabolites, multimodal image analysis, histology and classical phenotyping will provide an important step forward. Ultimately, the use of multiple orthogonal technologies that embed machine learning will lead to the generation of unique molecular and clinical fingerprints that can be used for PsA diagnosis, prognosis and therapeutic monitoring. However, research on the identification of biomarker profiles/fingerprints using different omics technologies is still in the discovery phase with much work to be conducted to turn the anticipated results of these analyses into assays which are applicable in routine clinical settings. The HIPPOCRATES project is therefore strategically important because it combines expertise from all relevant fields with access to comprehensive cohorts, technologies and translational research experience. This will ultimately improve the quality of life for those living with PsA or at risk of developing PsA.

Author Contributions: All authors contributed to the conceptualization, writing, reviewing and editing of the manuscript. All authors have read and agreed to the published version of the manuscript.

Funding: HIPPOCRATES has received funding from the Innovative Medicines Initiative 2 Joint Undertaking (JU) under grant agreement no. 101007757. The JU receives support from the European Union's Horizon 2020 research and innovation program and EFPIA.

Institutional Review Board Statement: Not applicable.

Informed Consent Statement: Not applicable.

Data Availability Statement: Not applicable.

Acknowledgments: R.G., M.K., L.H., S.K., E.R., S.R., K.S., G.G. and F.B. acknowledge support from the Fraunhofer Cluster of Excellence Immune Mediated diseases (CIMD). A.B. is an NIHR Senior Investigator. A.B. and J.B. receive support from Versus Arthritis (grant ref. 21754) and the NIHR Manchester Biomedical Research Centre. The views expressed are those of the authors and not necessarily those of the NHS, National Institute for Health and Care Research or the Department of Health. The authors would like to thank Daniel Kratz and Samuel Rischke for support in creating the figures.

Conflicts of Interest: The authors, including Richard M. Twyman, who is the director of the company TRM, declare no conflict of interest.

References

1. Damiani, G.; Bragazzi, N.L.; Karimkhani Aksut, C.; Wu, D.; Alicandro, G.; McGonagle, D.; Guo, C.; Dellavalle, R.; Grada, A.; Wong, P.; et al. The Global, Regional, and National Burden of Psoriasis: Results and Insights from the Global Burden of Disease 2019 Study. *Front. Med.* **2021**, *8*, 743180. [CrossRef]
2. Armstrong, A.W.; Read, C. Pathophysiology, Clinical Presentation, and Treatment of Psoriasis: A Review. *JAMA* **2020**, *323*, 1945–1960. [CrossRef] [PubMed]
3. Ritchlin, C.T.; Colbert, R.A.; Gladman, D.D. Psoriatic Arthritis. *N. Engl. J. Med.* **2017**, *376*, 957–970. [CrossRef] [PubMed]
4. Simon, D.; Watad, A.; Rodrigues-Manica, S.; Perricone, C. Editorial: Early Origins of Psoriatic Arthritis. *Front. Med.* **2021**, *8*, 794229. [CrossRef]
5. Kumthekar, A.; Ogdie, A. Obesity and Psoriatic Arthritis: A Narrative Review. *Rheumatol. Ther.* **2020**, *7*, 447–456. [CrossRef]
6. Ogdie, A.; Harrison, R.W.; McLean, R.R.; Lin, T.-C.; Lebwohl, M.; Strober, B.E.; Zhuo, J.; Patel, V.; Mease, P.J. Prospective Cohort Study of Psoriatic Arthritis Risk in Patients With Psoriasis in a Real-World Psoriasis Registry. *J. Am. Acad. Dermatol.* 2022; *in press*. [CrossRef]
7. Gisondi, P.; Bellinato, F.; Maurelli, M.; Geat, D.; Zabotti, A.; McGonagle, D.; Girolomoni, G. Reducing the Risk of Developing Psoriatic Arthritis in Patients with Psoriasis. *Psoriasis* **2022**, *12*, 213–220. [CrossRef] [PubMed]
8. Pezzolo, E.; Naldi, L. The relationship between smoking, psoriasis and psoriatic arthritis. *Expert Rev. Clin. Immunol.* **2019**, *15*, 41–48. [CrossRef]
9. Owczarczyk-Saczonek, A.; Purzycka-Bohdan, D.; Nedoszytko, B.; Reich, A.; Szczerkowska-Dobosz, A.; Bartosińska, J.; Batycka-Baran, A.; Czajkowski, R.; Dobrucki, I.T.; Dobrucki, L.W.; et al. Pathogenesis of psoriasis in the "omic" era. Part III. Metabolic disorders, metabolomics, nutrigenomics in psoriasis. *Postepy Dermatol. Alergol.* **2020**, *37*, 452–467. [CrossRef]
10. Kocic, H.; Damiani, G.; Stamenkovic, B.; Tirant, M.; Jovic, A.; Tiodorovic, D.; Peris, K. Dietary compounds as potential modulators of microRNA expression in psoriasis. *Ther. Adv. Chronic Dis.* **2019**, *10*, 2040622319864805. [CrossRef]

11. Scher, J.U.; Ogdie, A.; Merola, J.F.; Ritchlin, C. Preventing psoriatic arthritis: Focusing on patients with psoriasis at increased risk of transition. *Nat. Rev. Rheumatol.* **2019**, *15*, 153–166. [CrossRef]
12. Ogdie, A.; Schwartzman, S.; Husni, M.E. Recognizing and managing comorbidities in psoriatic arthritis. *Curr. Opin. Rheumatol.* **2015**, *27*, 118–126. [CrossRef]
13. Weigle, N.; McBane, S. Psoriasis. *Am. Fam. Physician* **2013**, *87*, 626–633.
14. Raychaudhuri, S.K.; Maverakis, E.; Raychaudhuri, S.P. Diagnosis and classification of psoriasis. *Autoimmun. Rev.* **2014**, *13*, 490–495. [CrossRef]
15. Johnson, M.A.N.; Armstrong, A.W. Clinical and histologic diagnostic guidelines for psoriasis: A critical review. *Clin. Rev. Allergy Immunol.* **2013**, *44*, 166–172. [CrossRef]
16. Pennington, S.R.; FitzGerald, O. Early Origins of Psoriatic Arthritis: Clinical, Genetic and Molecular Biomarkers of Progression from Psoriasis to Psoriatic Arthritis. *Front. Med.* **2021**, *8*, 723944. [CrossRef]
17. Raychaudhuri, S.P.; Wilken, R.; Sukhov, A.C.; Raychaudhuri, S.K.; Maverakis, E. Management of psoriatic arthritis: Early diagnosis, monitoring of disease severity and cutting edge therapies. *J. Autoimmun.* **2017**, *76*, 21–37. [CrossRef]
18. Ghasemi, M.; Nabipour, I.; Omrani, A.; Alipour, Z.; Assadi, M. Precision medicine and molecular imaging: New targeted approaches toward cancer therapeutic and diagnosis. *Am. J. Nucl. Med. Mol. Imaging* **2016**, *6*, 310–327.
19. Vargas, A.J.; Harris, C.C. Biomarker development in the precision medicine era: Lung cancer as a case study. *Nat. Rev. Cancer* **2016**, *16*, 525–537. [CrossRef]
20. Grivas, A.; Fragoulis, G.; Garantziotis, P.; Banos, A.; Nikiphorou, E.; Boumpas, D. Unraveling the complexities of psoriatic arthritis by the use of -Omics and their relevance for clinical care. *Autoimmun. Rev.* **2021**, *20*, 102949. [CrossRef]
21. Caputo, V.; Strafella, C.; Termine, A.; Dattola, A.; Mazzilli, S.; Lanna, C.; Cosio, T.; Campione, E.; Novelli, G.; Giardina, E.; et al. Overview of the molecular determinants contributing to the expression of Psoriasis and Psoriatic Arthritis phenotypes. *J. Cell. Mol. Med.* **2020**, *24*, 13554–13563. [CrossRef]
22. Generali, E.; Scirè, C.A.; Favalli, E.G.; Selmi, C. Biomarkers in psoriatic arthritis: A systematic literature review. *Expert Rev. Clin. Immunol.* **2016**, *12*, 651–660. [CrossRef]
23. O'Rielly, D.D.; Jani, M.; Rahman, P.; Elder, J.T. The Genetics of Psoriasis and Psoriatic Arthritis. *J. Rheumatol. Suppl.* **2019**, *95*, 46–50. [CrossRef]
24. Shi, C.; Rattray, M.; Barton, A.; Bowes, J.; Orozco, G. Using functional genomics to advance the understanding of psoriatic arthritis. *Rheumatology* **2020**, *59*, 3137–3146. [CrossRef]
25. Bowes, J.; Loehr, S.; Budu-Aggrey, A.; Uebe, S.; Bruce, I.N.; Feletar, M.; Marzo-Ortega, H.; Helliwell, P.; Ryan, A.W.; Kane, D.; et al. PTPN22 is associated with susceptibility to psoriatic arthritis but not psoriasis: Evidence for a further PsA-specific risk locus. *Ann. Rheum. Dis.* **2015**, *74*, 1882–1885. [CrossRef]
26. Cortes, A.; Brown, M.A. Promise and pitfalls of the Immunochip. *Arthritis Res. Ther.* **2011**, *13*, 101. [CrossRef]
27. Budu-Aggrey, A.; Bowes, J.; Barton, A. Identifying a novel locus for psoriatic arthritis. *Rheumatology* **2016**, *55*, 25–32. [CrossRef]
28. Ramírez, J.; Fernández-Sueiro, J.L.; López-Mejías, R.; Montilla, C.; Arias, M.; Moll, C.; Alsina, M.; Sanmarti, R.; Lozano, F.; Cañete, J.D. FCGR2A/CD32A and FCGR3A/CD16A variants and EULAR response to tumor necrosis factor-α blockers in psoriatic arthritis: A longitudinal study with 6 months of followup. *J. Rheumatol.* **2012**, *39*, 1035–1041. [CrossRef]
29. Bowes, J.; Ashcroft, J.; Dand, N.; Jalali-Najafabadi, F.; Bellou, E.; Ho, P.; Marzo-Ortega, H.; Helliwell, P.S.; Feletar, M.; Ryan, A.W.; et al. Cross-phenotype association mapping of the MHC identifies genetic variants that differentiate psoriatic arthritis from psoriasis. *Ann. Rheum. Dis.* **2017**, *76*, 1774–1779. [CrossRef]
30. Okada, Y.; Han, B.; Tsoi, L.C.; Stuart, P.E.; Ellinghaus, E.; Tejasvi, T.; Chandran, V.; Pellett, F.; Pollock, R.; Bowcock, A.M.; et al. Fine mapping major histocompatibility complex associations in psoriasis and its clinical subtypes. *Am. J. Hum. Genet.* **2014**, *95*, 162–172. [CrossRef]
31. O'Rielly, D.D.; Rahman, P. Genetic, Epigenetic and Pharmacogenetic Aspects of Psoriasis and Psoriatic Arthritis. *Rheum. Dis. Clin. N. Am.* **2015**, *41*, 623–642. [CrossRef]
32. Bowes, J.; Budu-Aggrey, A.; Huffmeier, U.; Uebe, S.; Steel, K.; Hebert, H.L.; Wallace, C.; Massey, J.; Bruce, I.N.; Bluett, J.; et al. Dense genotyping of immune-related susceptibility loci reveals new insights into the genetics of psoriatic arthritis. *Nat. Commun.* **2015**, *6*, 6046. [CrossRef]
33. Budu-Aggrey, A.; Bowes, J.; Loehr, S.; Uebe, S.; Zervou, M.I.; Helliwell, P.; Ryan, A.W.; Kane, D.; Korendowych, E.; Giardina, E.; et al. Replication of a distinct psoriatic arthritis risk variant at the IL23R locus. *Ann. Rheum. Dis.* **2016**, *75*, 1417–1418. [CrossRef]
34. Stuart, P.E.; Nair, R.P.; Tsoi, L.C.; Tejasvi, T.; Das, S.; Kang, H.M.; Ellinghaus, E.; Chandran, V.; Callis-Duffin, K.; Ike, R.; et al. Genome-wide Association Analysis of Psoriatic Arthritis and Cutaneous Psoriasis Reveals Differences in Their Genetic Architecture. *Am. J. Hum. Genet.* **2015**, *97*, 816–836. [CrossRef]
35. Chandran, V.; Bull, S.B.; Pellett, F.J.; Ayearst, R.; Pollock, R.A.; Gladman, D.D. Killer-cell immunoglobulin-like receptor gene polymorphisms and susceptibility to psoriatic arthritis. *Rheumatology* **2014**, *53*, 233–239. [CrossRef]
36. Cascella, R.; Strafella, C.; Ragazzo, M.; Manzo, L.; Costanza, G.; Bowes, J.; Hüffmeier, U.; Potenza, S.; Sangiuolo, F.; Reis, A.; et al. KIF3A and IL-4 are disease-specific biomarkers for psoriatic arthritis susceptibility. *Oncotarget* **2017**, *8*, 95401–95411. [CrossRef]

37. Aterido, A.; Cañete, J.D.; Tornero, J.; Ferrándiz, C.; Pinto, J.A.; Gratacós, J.; Queiró, R.; Montilla, C.; Torre-Alonso, J.C.; Pérez-Venegas, J.J.; et al. Genetic variation at the glycosaminoglycan metabolism pathway contributes to the risk of psoriatic arthritis but not psoriasis. *Ann. Rheum. Dis.* **2019**, *78*, 355–364. [CrossRef]
38. Barski, A.; Cuddapah, S.; Cui, K.; Roh, T.-Y.; Schones, D.E.; Wang, Z.; Wei, G.; Chepelev, I.; Zhao, K. High-resolution profiling of histone methylations in the human genome. *Cell* **2007**, *129*, 823–837. [CrossRef]
39. Park, P.J. ChIP-seq: Advantages and challenges of a maturing technology. *Nat. Rev. Genet.* **2009**, *10*, 669–680. [CrossRef]
40. Schadt, E.E.; Banerjee, O.; Fang, G.; Feng, Z.; Wong, W.H.; Zhang, X.; Kislyuk, A.; Clark, T.A.; Luong, K.; Keren-Paz, A.; et al. Modeling kinetic rate variation in third generation DNA sequencing data to detect putative modifications to DNA bases. *Genome Res.* **2013**, *23*, 129–141. [CrossRef]
41. Zhang, Z.; Pugh, B.F. High-resolution genome-wide mapping of the primary structure of chromatin. *Cell* **2011**, *144*, 175–186. [CrossRef]
42. Merelli, I.; Tordini, F.; Drocco, M.; Aldinucci, M.; Liò, P.; Milanesi, L. Integrating multi-omic features exploiting Chromosome Conformation Capture data. *Front. Genet.* **2015**, *6*, 40. [CrossRef]
43. Carini, C.; Hunter, E.; Ramadass, A.S.; Green, J.; Akoulitchev, A.; McInnes, I.B.; Goodyear, C.S. Chromosome conformation signatures define predictive markers of inadequate response to methotrexate in early rheumatoid arthritis. *J. Transl. Med.* **2018**, *16*, 18. [CrossRef]
44. Hunter, E.; Dezfouli, M.; Koutsothanasi, C.; Wilson, A.; Santos, F.C.; Salter, M.; Westra, J.W.; Powell, R.; Dring, A.; Egan, B.; et al. Development and validation of blood-based predictive biomarkers for response to PD-(L)-1 checkpoint inhibitors: Evidence of a universal systemic core of 3D immunogenetic profiling across multiple oncological indications. *medRxiv* **2021**. [CrossRef]
45. Hunter, E.; Koutsothanasi, C.; Wilson, A.; Santos, F.C.; Salter, M.; Powell, R.; Dring, A.; Brajer, P.; Egan, B.; Westra, J.W.; et al. 3D genomic capture of regulatory immuno-genetic profiles in COVID-19 patients for prognosis of severe COVID disease outcome. *bioRxiv* **2021**. [CrossRef]
46. Tordini, F.; Aldinucci, M.; Milanesi, L.; Liò, P.; Merelli, I. The Genome Conformation as an Integrator of Multi-Omic Data: The Example of Damage Spreading in Cancer. *Front. Genet.* **2016**, *7*, 194. [CrossRef]
47. Mahendran, S.M.; Chandran, V. Exploring the Psoriatic Arthritis Proteome in Search of Novel Biomarkers. *Proteomes* **2018**, *6*, 5. [CrossRef]
48. Qi, F.; Tan, Y.; Yao, A.; Yang, X.; He, Y. Psoriasis to Psoriatic Arthritis: The Application of Proteomics Technologies. *Front. Med.* **2021**, *8*, 681172. [CrossRef]
49. Mc Ardle, A.; Kwasnik, A.; Szentpetery, A.; Hernandez, B.; Parnell, A.; de Jager, W.; de Roock, S.; FitzGerald, O.; Pennington, S.R. Identification and Evaluation of Serum Protein Biomarkers That Differentiate Psoriatic Arthritis From Rheumatoid Arthritis. *Arthritis Rheumatol.* **2022**, *74*, 81–91. [CrossRef]
50. Adhikari, S.; Nice, E.C.; Deutsch, E.W.; Lane, L.; Omenn, G.S.; Pennington, S.R.; Paik, Y.-K.; Overall, C.M.; Corrales, F.J.; Cristea, I.M.; et al. A high-stringency blueprint of the human proteome. *Nat. Commun.* **2020**, *11*, 5301. [CrossRef]
51. Bateman, N.W.; Goulding, S.P.; Shulman, N.J.; Gadok, A.K.; Szumlinski, K.K.; MacCoss, M.J.; Wu, C.C. Maximizing peptide identification events in proteomic workflows using data-dependent acquisition (DDA). *Mol. Cell. Proteom.* **2014**, *13*, 329–338. [CrossRef] [PubMed]
52. Ludwig, C.; Gillet, L.; Rosenberger, G.; Amon, S.; Collins, B.C.; Aebersold, R. Data-independent acquisition-based SWATH-MS for quantitative proteomics: A tutorial. *Mol. Syst. Biol.* **2018**, *14*, e8126. [CrossRef] [PubMed]
53. Suhre, K.; McCarthy, M.I.; Schwenk, J.M. Genetics meets proteomics: Perspectives for large population-based studies. *Nat. Rev. Genet.* **2021**, *22*, 19–37. [CrossRef]
54. Zhong, W.; Edfors, F.; Gummesson, A.; Bergström, G.; Fagerberg, L.; Uhlén, M. Next generation plasma proteome profiling to monitor health and disease. *Nat. Commun.* **2021**, *12*, 2493. [CrossRef]
55. Petrera, A.; von Toerne, C.; Behler, J.; Huth, C.; Thorand, B.; Hilgendorff, A.; Hauck, S.M. Multiplatform Approach for Plasma Proteomics: Complementarity of Olink Proximity Extension Assay Technology to Mass Spectrometry-Based Protein Profiling. *J. Proteome Res.* **2021**, *20*, 751–762. [CrossRef] [PubMed]
56. Pietzner, M.; Wheeler, E.; Carrasco-Zanini, J.; Cortes, A.; Koprulu, M.; Wörheide, M.A.; Oerton, E.; Cook, J.; Stewart, I.D.; Kerrison, N.D.; et al. Mapping the proteo-genomic convergence of human diseases. *Science* **2021**, *374*, eabj1541. [CrossRef] [PubMed]
57. Ignjatovic, V.; Geyer, P.E.; Palaniappan, K.K.; Chaaban, J.E.; Omenn, G.S.; Baker, M.S.; Deutsch, E.W.; Schwenk, J.M. Mass Spectrometry-Based Plasma Proteomics: Considerations from Sample Collection to Achieving Translational Data. *J. Proteome Res.* **2019**, *18*, 4085–4097. [CrossRef] [PubMed]
58. Deutsch, E.W.; Omenn, G.S.; Sun, Z.; Maes, M.; Pernemalm, M.; Palaniappan, K.K.; Letunica, N.; Vandenbrouck, Y.; Brun, V.; Tao, S.-C.; et al. Advances and Utility of the Human Plasma Proteome. *J. Proteome Res.* **2021**, *20*, 5241–5263. [CrossRef] [PubMed]
59. Uhlén, M.; Karlsson, M.J.; Hober, A.; Svensson, A.-S.; Scheffel, J.; Kotol, D.; Zhong, W.; Tebani, A.; Strandberg, L.; Edfors, F.; et al. The human secretome. *Sci. Signal.* **2019**, *12*, 609. [CrossRef]
60. Tebani, A.; Gummesson, A.; Zhong, W.; Koistinen, I.S.; Lakshmikanth, T.; Olsson, L.M.; Boulund, F.; Neiman, M.; Stenlund, H.; Hellström, C.; et al. Integration of molecular profiles in a longitudinal wellness profiling cohort. *Nat. Commun.* **2020**, *11*, 4487. [CrossRef]
61. Cretu, D.; Gao, L.; Liang, K.; Soosaipillai, A.; Diamandis, E.P.; Chandran, V. Differentiating Psoriatic Arthritis From Psoriasis Without Psoriatic Arthritis Using Novel Serum Biomarkers. *Arthritis Care Res.* **2018**, *70*, 454–461. [CrossRef]

62. Cretu, D.; Liang, K.; Saraon, P.; Batruch, I.; Diamandis, E.P.; Chandran, V. Quantitative tandem mass-spectrometry of skin tissue reveals putative psoriatic arthritis biomarkers. *Clin. Proteom.* **2015**, *12*, 1–8. [CrossRef]
63. Dalmády, S.; Kiss, M.; Képíró, L.; Kovács, L.; Sonkodi, G.; Kemény, L.; Gyulai, R. Higher levels of autoantibodies targeting mutated citrullinated vimentin in patients with psoriatic arthritis than in patients with psoriasis vulgaris. *Clin. Dev. Immunol.* **2013**, *2013*, 474028. [CrossRef]
64. Hansson, C.; Eriksson, C.; Alenius, G.-M. S-calprotectin (S100A8/S100A9): A potential marker of inflammation in patients with psoriatic arthritis. *J. Immunol. Res.* **2014**, *2014*, 696415. [CrossRef] [PubMed]
65. Alenius, G.-M.; Eriksson, C.; Rantapää Dahlqvist, S. Interleukin-6 and soluble interleukin-2 receptor alpha-markers of inflammation in patients with psoriatic arthritis? *Clin. Exp. Rheumatol.* **2009**, *27*, 120–123. [PubMed]
66. Abji, F.; Pollock, R.A.; Liang, K.; Chandran, V.; Gladman, D.D. Brief Report: CXCL10 Is a Possible Biomarker for the Development of Psoriatic Arthritis among Patients with Psoriasis. *Arthritis Rheumatol.* **2016**, *68*, 2911–2916. [CrossRef]
67. Grazio, S.; Razdorov, G.; Erjavec, I.; Grubisic, F.; Kusic, Z.; Punda, M.; Anticevic, D.; Vukicevic, S.; Grgurevic, L. Differential expression of proteins with heparin affinity in patients with rheumatoid and psoriatic arthritis: A preliminary study. *Clin. Exp. Rheumatol.* **2013**, *31*, 665–671. [PubMed]
68. Chandran, V.; Shen, H.; Pollock, R.A.; Pellett, F.J.; Carty, A.; Cook, R.J.; Gladman, D.D. Soluble biomarkers associated with response to treatment with tumor necrosis factor inhibitors in psoriatic arthritis. *J. Rheumatol.* **2013**, *40*, 866–871. [CrossRef]
69. Fiocco, U.; Sfriso, P.; Oliviero, F.; Roux-Lombard, P.; Scagliori, E.; Cozzi, L.; Lunardi, F.; Calabrese, F.; Vezzù, M.; Dainese, S.; et al. Synovial effusion and synovial fluid biomarkers in psoriatic arthritis to assess intraarticular tumor necrosis factor-α blockade in the knee joint. *Arthritis Res. Ther.* **2010**, *12*, R148. [CrossRef] [PubMed]
70. van Kuijk, A.W.R.; DeGroot, J.; Koeman, R.C.; Sakkee, N.; Baeten, D.L.; Gerlag, D.M.; Tak, P.P. Soluble biomarkers of cartilage and bone metabolism in early proof of concept trials in psoriatic arthritis: Effects of adalimumab versus placebo. *PLoS ONE* **2010**, *5*, e12556. [CrossRef]
71. Collins, E.S.; Butt, A.Q.; Gibson, D.S.; Dunn, M.J.; Fearon, U.; van Kuijk, A.W.; Gerlag, D.M.; Pontifex, E.; Veale, D.J.; Tak, P.P.; et al. A clinically based protein discovery strategy to identify potential biomarkers of response to anti-TNF-α treatment of psoriatic arthritis. *Proteom. Clin. Appl.* **2016**, *10*, 645–662. [CrossRef]
72. Mahendran, S.M.; Keystone, E.C.; Krawetz, R.J.; Liang, K.; Diamandis, E.P.; Chandran, V. Elucidating the endogenous synovial fluid proteome and peptidome of inflammatory arthritis using label-free mass spectrometry. *Clin. Proteom.* **2019**, *16*, 23. [CrossRef] [PubMed]
73. Mulder, M.L.M.; van Hal, T.W.; Wenink, M.H.; Koenen, H.J.P.M.; van den Hoogen, F.H.J.; de Jong, E.M.G.J.; van den Reek, J.M.P.A.; Vriezekolk, J.E. Clinical, laboratory, and genetic markers for the development or presence of psoriatic arthritis in psoriasis patients: A systematic review. *Arthritis Res. Ther.* **2021**, *23*, 168. [CrossRef]
74. Leijten, E.; Tao, W.; Pouw, J.; van Kempen, T.; Olde Nordkamp, M.; Balak, D.; Tekstra, J.; Muñoz-Elías, E.; DePrimo, S.; Drylewicz, J.; et al. Broad proteomic screen reveals shared serum proteomic signature in patients with psoriatic arthritis and psoriasis without arthritis. *Rheumatology* **2021**, *60*, 751–761. [CrossRef] [PubMed]
75. Karlsson, M.; Zhang, C.; Méar, L.; Zhong, W.; Digre, A.; Katona, B.; Sjöstedt, E.; Butler, L.; Odeberg, J.; Dusart, P.; et al. A single-cell type transcriptomics map of human tissues. *Sci. Adv.* **2021**, *7*, eabh2169. [CrossRef] [PubMed]
76. Wishart, D.S. Metabolomics for Investigating Physiological and Pathophysiological Processes. *Physiol. Rev.* **2019**, *99*, 1819–1875. [CrossRef] [PubMed]
77. López-López, Á.; López-Gonzálvez, Á.; Barker-Tejeda, T.C.; Barbas, C. A review of validated biomarkers obtained through metabolomics. *Expert Rev. Mol. Diagn.* **2018**, *18*, 557–575. [CrossRef]
78. Di, Y.; Afifi, L.; Jeon, C.; Trivedi, M.; Chang, H.W.; Lee, K.; Liao, W. The metabolomics of psoriatic disease. *Psoriasis* **2017**, *7*, 1–15. [CrossRef]
79. Koussiouris, J.; Looby, N.; Anderson, M.; Kulasingam, V.; Chandran, V. Metabolomics Studies in Psoriatic Disease: A Review. *Metabolites* **2021**, *11*, 375. [CrossRef] [PubMed]
80. Dorochow, E.; Köhm, M.; Hahnefeld, L.; Gurke, R. Metabolic Profiling in Rheumatoid Arthritis, Psoriatic Arthritis, and Psoriasis: Elucidating Pathogenesis, Improving Diagnosis, and Monitoring Disease Activity. *J. Pers. Med.* **2022**, *12*, 924. [CrossRef]
81. Madsen, R.K.; Lundstedt, T.; Gabrielsson, J.; Sennbro, C.-J.; Alenius, G.-M.; Moritz, T.; Rantapää-Dahlqvist, S.; Trygg, J. Diagnostic properties of metabolic perturbations in rheumatoid arthritis. *Arthritis Res. Ther.* **2011**, *13*, R19. [CrossRef] [PubMed]
82. Souto-Carneiro, M.; Tóth, L.; Behnisch, R.; Urbach, K.; Klika, K.D.; Carvalho, R.A.; Lorenz, H.-M. Differences in the serum metabolome and lipidome identify potential biomarkers for seronegative rheumatoid arthritis versus psoriatic arthritis. *Ann. Rheum. Dis.* **2020**, *79*, 499–506. [CrossRef]
83. Armstrong, A.W.; Wu, J.; Johnson, M.A.; Grapov, D.; Azizi, B.; Dhillon, J.; Fiehn, O. Metabolomics in psoriatic disease: Pilot study reveals metabolite differences in psoriasis and psoriatic arthritis. *F1000Research* **2014**, *3*, 248. [CrossRef]
84. Coras, R.; Kavanaugh, A.; Boyd, T.; Huynh, D.; Lagerborg, K.A.; Xu, Y.-J.; Rosenthal, S.B.; Jain, M.; Guma, M. Choline metabolite, trimethylamine N-oxide (TMAO), is associated with inflammation in psoriatic arthritis. *Clin. Exp. Rheumatol.* **2019**, *37*, 481–484.
85. Paine, A.; Brookes, P.S.; Bhattacharya, S.; Li, D.; La Garcia-Hernandez, M.D.L.; Tausk, F.; Ritchlin, C. Dysregulation of bile acids, lipids, and nucleotides in psoriatic arthritis revealed by unbiased profiling of serum metabolites. *Arthritis Rheumatol.* **2022**. [CrossRef]

86. Wang, N.; Yang, L.; Shang, L.; Liang, Z.; Wang, Y.; Feng, M.; Yu, S.; Li, X.; Gao, C.; Li, Z.; et al. Altered Fecal Metabolomics and Potential Biomarkers of Psoriatic Arthritis Differing From Rheumatoid Arthritis. *Front. Immunol.* **2022**, *13*, 812996. [CrossRef]
87. Alonso, A.; Julià, A.; Vinaixa, M.; Domènech, E.; Fernández-Nebro, A.; Cañete, J.D.; Ferrándiz, C.; Tornero, J.; Gisbert, J.P.; Nos, P.; et al. Urine metabolome profiling of immune-mediated inflammatory diseases. *BMC Med.* **2016**, *14*, 133. [CrossRef]
88. Kapoor, S.R.; Filer, A.; Fitzpatrick, M.A.; Fisher, B.A.; Taylor, P.C.; Buckley, C.D.; McInnes, I.B.; Raza, K.; Young, S.P. Metabolic profiling predicts response to anti-tumor necrosis factor α therapy in patients with rheumatoid arthritis. *Arthritis Rheum.* **2013**, *65*, 1448–1456. [CrossRef]
89. Maceyka, M.; Spiegel, S. Sphingolipid metabolites in inflammatory disease. *Nature* **2014**, *510*, 58–67. [CrossRef]
90. Dennis, E.A.; Norris, P.C. Eicosanoid storm in infection and inflammation. *Nat. Rev. Immunol.* **2015**, *15*, 511–523. [CrossRef]
91. Serhan, C.N. Discovery of specialized pro-resolving mediators marks the dawn of resolution physiology and pharmacology. *Mol. Aspects Med.* **2017**, *58*, 1–11. [CrossRef]
92. Zhang, C.; Wang, K.; Yang, L.; Liu, R.; Chu, Y.; Qin, X.; Yang, P.; Yu, H. Lipid metabolism in inflammation-related diseases. *Analyst* **2018**, *143*, 4526–4536. [CrossRef]
93. Lin, D.A.; Boyce, J.A. Lysophospholipids as Mediators of Immunity. *Adv. Immunol.* **2006**, *89*, 141–167. [CrossRef]
94. Brouwers, H.; von Hegedus, J.; Toes, R.; Kloppenburg, M.; Ioan-Facsinay, A. Lipid mediators of inflammation in rheumatoid arthritis and osteoarthritis. *Best Pract. Res. Clin. Rheumatol.* **2015**, *29*, 741–755. [CrossRef]
95. Fan, F.; Mundra, P.A.; Fang, L.; Galvin, A.; Moore, X.L.; Weir, J.M.; Wong, G.; White, D.A.; Chin-Dusting, J.; Sparrow, M.P.; et al. Lipidomic Profiling in Inflammatory Bowel Disease: Comparison Between Ulcerative Colitis and Crohn's Disease. *Inflamm. Bowel Dis.* **2015**, *21*, 1511–1518. [CrossRef]
96. Schwanke, R.C.; Marcon, R.; Bento, A.F.; Calixto, J.B. EPA- and DHA-derived resolvins' actions in inflammatory bowel disease. *Eur. J. Pharmacol.* **2016**, *785*, 156–164. [CrossRef]
97. Scoville, E.A.; Allaman, M.M.; Brown, C.T.; Motley, A.K.; Horst, S.N.; Williams, C.S.; Koyama, T.; Zhao, Z.; Adams, D.W.; Beaulieu, D.B.; et al. Alterations in Lipid, Amino Acid, and Energy Metabolism Distinguish Crohn's Disease from Ulcerative Colitis and Control Subjects by Serum Metabolomic Profiling. *Metabolomics* **2018**, *14*, 17. [CrossRef]
98. Grill, M.; Högenauer, C.; Blesl, A.; Haybaeck, J.; Golob-Schwarzl, N.; Ferreirós, N.; Thomas, D.; Gurke, R.; Trötzmüller, M.; Köfeler, H.C.; et al. Members of the endocannabinoid system are distinctly regulated in inflammatory bowel disease and colorectal cancer. *Sci. Rep.* **2019**, *9*, 2358. [CrossRef]
99. Bazarganipour, S.; Hausmann, J.; Oertel, S.; El-Hindi, K.; Brachtendorf, S.; Blumenstein, I.; Kubesch, A.; Sprinzl, K.; Birod, K.; Hahnefeld, L.; et al. The Lipid Status in Patients with Ulcerative Colitis: Sphingolipids are Disease-Dependent Regulated. *J. Clin. Med.* **2019**, *8*, 971. [CrossRef]
100. Zeng, C.; Wen, B.; Hou, G.; Lei, L.; Mei, Z.; Jia, X.; Chen, X.; Zhu, W.; Li, J.; Kuang, Y.; et al. Lipidomics profiling reveals the role of glycerophospholipid metabolism in psoriasis. *Gigascience* **2017**, *6*, 1–11. [CrossRef]
101. Sorokin, A.V.; Norris, P.C.; English, J.T.; Dey, A.K.; Chaturvedi, A.; Baumer, Y.; Silverman, J.; Playford, M.P.; Serhan, C.N.; Mehta, N.N. Identification of proresolving and inflammatory lipid mediators in human psoriasis. *J. Clin. Lipidol.* **2018**, *12*, 1047–1060. [CrossRef]
102. Coras, R.; Kavanaugh, A.; Boyd, T.; Huynh, Q.; Pedersen, B.; Armando, A.M.; Dahlberg-Wright, S.; Marsal, S.; Jain, M.; Paravar, T.; et al. Pro- and anti-inflammatory eicosanoids in psoriatic arthritis. *Metabolomics* **2019**, *15*, 65. [CrossRef]
103. Wójcik, P.; Biernacki, M.; Wroński, A.; Łuczaj, W.; Waeg, G.; Žarković, N.; Skrzydlewska, E. Altered Lipid Metabolism in Blood Mononuclear Cells of Psoriatic Patients Indicates Differential Changes in Psoriasis Vulgaris and Psoriatic Arthritis. *Int. J. Mol. Sci.* **2019**, *20*, 4249. [CrossRef] [PubMed]
104. Sethi, S.; Brietzke, E. Recent advances in lipidomics: Analytical and clinical perspectives. *Prostaglandins Other Lipid Mediat.* **2017**, *128–129*, 8–16. [CrossRef]
105. Looby, N.; Roszkowska, A.; Reyes-Garcés, N.; Yu, M.; Bączek, T.; Kulasingam, V.; Pawliszyn, J.; Chandran, V. Serum metabolic fingerprinting of psoriasis and psoriatic arthritis patients using solid-phase microextraction-liquid chromatography-high-resolution mass spectrometry. *Metabolomics* **2021**, *17*, 59. [CrossRef]
106. Ambrożewicz, E.; Wójcik, P.; Wroński, A.; Łuczaj, W.; Jastrząb, A.; Žarković, N.; Skrzydlewska, E. Pathophysiological Alterations of Redox Signaling and Endocannabinoid System in Granulocytes and Plasma of Psoriatic Patients. *Cells* **2018**, *7*, 159. [CrossRef]
107. Kishikawa, T.; Arase, N.; Tsuji, S.; Maeda, Y.; Nii, T.; Hirata, J.; Suzuki, K.; Yamamoto, K.; Masuda, T.; Ogawa, K.; et al. Large-scale plasma-metabolome analysis identifies potential biomarkers of psoriasis and its clinical subtypes. *J. Dermatol. Sci.* **2021**, *102*, 78–84. [CrossRef]
108. Hoch, T.; Schulz, D.; Eling, N.; Gómez, J.M.; Levesque, M.P.; Bodenmiller, B. Multiplexed imaging mass cytometry of the chemokine milieus in melanoma characterizes features of the response to immunotherapy. *Sci. Immunol.* **2022**, *7*, eabk1692. [CrossRef]
109. Schubert, W.; Bonnekoh, B.; Pommer, A.J.; Philipsen, L.; Böckelmann, R.; Malykh, Y.; Gollnick, H.; Friedenberger, M.; Bode, M.; Dress, A.W.M. Analyzing proteome topology and function by automated multidimensional fluorescence microscopy. *Nat. Biotechnol.* **2006**, *24*, 1270–1278. [CrossRef]
110. Schriml, L.M. A decade of GigaScience: 10 years of the evolving genomic and biomedical standards landscape. *Gigascience* **2022**, *11*, giac047. [CrossRef]

111. Field, D.; Amaral-Zettler, L.; Cochrane, G.; Cole, J.R.; Dawyndt, P.; Garrity, G.M.; Gilbert, J.; Glöckner, F.O.; Hirschman, L.; Karsch-Mizrachi, I.; et al. The Genomic Standards Consortium. *PLoS Biol.* **2011**, *9*, e1001088. [CrossRef] [PubMed]
112. Taylor, C.F.; Paton, N.W.; Lilley, K.S.; Binz, P.-A.; Julian, R.K.; Jones, A.R.; Zhu, W.; Apweiler, R.; Aebersold, R.; Deutsch, E.W.; et al. The minimum information about a proteomics experiment (MIAPE). *Nat. Biotechnol.* **2007**, *25*, 887–893. [CrossRef]
113. Fiehn, O.; Kristal, B.; van Ommen, B.; Sumner, L.W.; Sansone, S.-A.; Taylor, C.; Hardy, N.; Kaddurah-Daouk, R. Establishing reporting standards for metabolomic and metabonomic studies: A call for participation. *Omics A J. Integr. Biol.* **2006**, *10*, 158–163. [CrossRef] [PubMed]
114. Köfeler, H.C.; Ahrends, R.; Baker, E.S.; Ekroos, K.; Han, X.; Hoffmann, N.; Holčapek, M.; Wenk, M.R.; Liebisch, G. Recommendations for good practice in MS-based lipidomics. *J. Lipid Res.* **2021**, *62*, 100138. [CrossRef] [PubMed]
115. Burla, B.; Arita, M.; Arita, M.; Bendt, A.K.; Cazenave-Gassiot, A.; Dennis, E.A.; Ekroos, K.; Han, X.; Ikeda, K.; Liebisch, G.; et al. MS-based lipidomics of human blood plasma: A community-initiated position paper to develop accepted guidelines. *J. Lipid Res.* **2018**, *59*, 2001–2017. [CrossRef] [PubMed]
116. Coman Schmid, D.; Crameri, K.; Oesterle, S.; Rinn, B.; Sengstag, T.; Stockinger, H. SPHN—The BioMedIT Network: A Secure IT Platform for Research with Sensitive Human Data. *Stud. Health Technol. Inform.* **2020**, *270*, 1170–1174. [CrossRef]
117. Wehrens, R.; Hageman, J.A.; van Eeuwijk, F.; Kooke, R.; Flood, P.J.; Wijnker, E.; Keurentjes, J.J.B.; Lommen, A.; van Eekelen, H.D.L.M.; Hall, R.D.; et al. Improved batch correction in untargeted MS-based metabolomics. *Metabolomics* **2016**, *12*, 88. [CrossRef]
118. Lancaster, S.M.; Sanghi, A.; Wu, S.; Snyder, M.P. A Customizable Analysis Flow in Integrative Multi-Omics. *Biomolecules* **2020**, *10*, 1606. [CrossRef]
119. Kohl, M.; Megger, D.A.; Trippler, M.; Meckel, H.; Ahrens, M.; Bracht, T.; Weber, F.; Hoffmann, A.-C.; Baba, H.A.; Sitek, B.; et al. A practical data processing workflow for multi-OMICS projects. *Biochim. Biophys. Acta* **2014**, *1844*, 52–62. [CrossRef] [PubMed]
120. Jouan-Rimbaud Bouveresse, D.; Pinto, R.C.; Schmidtke, L.M.; Locquet, N.; Rutledge, D.N. Identification of significant factors by an extension of ANOVA–PCA based on multi-block analysis. *Chemom. Intell. Lab. Syst.* **2011**, *106*, 173–182. [CrossRef]
121. Boccard, J.; Rutledge, D.N. A consensus orthogonal partial least squares discriminant analysis (OPLS-DA) strategy for multiblock Omics data fusion. *Anal. Chim. Acta* **2013**, *769*, 30–39. [CrossRef]
122. Picard, M.; Scott-Boyer, M.-P.; Bodein, A.; Périn, O.; Droit, A. Integration strategies of multi-omics data for machine learning analysis. *Comput. Struct. Biotechnol. J.* **2021**, *19*, 3735–3746. [CrossRef] [PubMed]
123. Kingsmore, K.M.; Puglisi, C.E.; Grammer, A.C.; Lipsky, P.E. An introduction to machine learning and analysis of its use in rheumatic diseases. *Nat. Rev. Rheumatol.* **2021**, *17*, 710–730. [CrossRef] [PubMed]
124. Xing, L.; Yu, L.; Zhou, N.; Zhang, Z.; Pu, Y.; Wu, J.; Shu, H. Exploration of biomarkers of psoriasis through combined multiomics analysis. *Res. Sq.* 2022; *Preprint*. [CrossRef]
125. Stafford, I.S.; Kellermann, M.; Mossotto, E.; Beattie, R.M.; MacArthur, B.D.; Ennis, S. A systematic review of the applications of artificial intelligence and machine learning in autoimmune diseases. *NPJ Digit. Med.* **2020**, *3*, 30. [CrossRef] [PubMed]
126. Wörheide, M.A.; Krumsiek, J.; Kastenmüller, G.; Arnold, M. Multi-omics integration in biomedical research—A metabolomics-centric review. *Anal. Chim. Acta* **2021**, *1141*, 144–162. [CrossRef]
127. Petricoin, E.F.; Ardekani, A.M.; Hitt, B.A.; Levine, P.J.; Fusaro, V.A.; Steinberg, S.M.; Mills, G.B.; Simone, C.; Fishman, D.A.; Kohn, E.C.; et al. Use of proteomic patterns in serum to identify ovarian cancer. *Lancet* **2002**, *359*, 572–577. [CrossRef]
128. Park, Y.-J.; Chung, M.K.; Hwang, D.; Kim, W.-U. Proteomics in Rheumatoid Arthritis Research. *Immune Netw.* **2015**, *15*, 177–185. [CrossRef] [PubMed]
129. Tsuchiya, H.; Ota, M.; Fujio, K. Multiomics landscape of synovial fibroblasts in rheumatoid arthritis. *Inflamm. Regen.* **2021**, *41*, 7. [CrossRef]
130. Tasaki, S.; Suzuki, K.; Kassai, Y.; Takeshita, M.; Murota, A.; Kondo, Y.; Ando, T.; Nakayama, Y.; Okuzono, Y.; Takiguchi, M.; et al. Multi-omics monitoring of drug response in rheumatoid arthritis in pursuit of molecular remission. *Nat. Commun.* **2018**, *9*, 2755. [CrossRef] [PubMed]
131. Karlsson, A.A.; Hao, S.; Jauhiainen, A.; Elfström, K.M.; Egevad, L.; Nordström, T.; Heintz, E.; Clements, M.S. The cost-effectiveness of prostate cancer screening using the Stockholm3 test. *PLoS ONE* **2021**, *16*, e0246674. [CrossRef]

biomedicines

MDPI

Review

Rebooting Regulatory T Cell and Dendritic Cell Function in Immune-Mediated Inflammatory Diseases: Biomarker and Therapy Discovery under a Multi-Omics Lens

Dimitra Kerdidani [1,2,†], Nikos E. Papaioannou [1,2,†], Evangelia Nakou [1,2,†] and Themis Alissafi [1,2,*]

1 Immune Regulation Laboratory, Center of Basic Research, Biomedical Research Foundation Academy of Athens, 11527 Athens, Greece
2 Laboratory of Biology, Medical School, National and Kapodistrian University of Athens, 12462 Athens, Greece
* Correspondence: talissafi@bioacademy.gr
† These authors contributed equally to this work.

Abstract: Immune-mediated inflammatory diseases (IMIDs) are a group of autoimmune and chronic inflammatory disorders with constantly increasing prevalence in the modern world. The vast majority of IMIDs develop as a consequence of complex mechanisms dependent on genetic, epigenetic, molecular, cellular, and environmental elements, that lead to defects in immune regulatory guardians of tolerance, such as dendritic (DCs) and regulatory T (Tregs) cells. As a result of this dysfunction, immune tolerance collapses and pathogenesis emerges. Deeper understanding of such disease driving mechanisms remains a major challenge for the prevention of inflammatory disorders. The recent renaissance in high throughput technologies has enabled the increase in the amount of data collected through multiple omics layers, while additionally narrowing the resolution down to the single cell level. In light of the aforementioned, this review focuses on DCs and Tregs and discusses how multi-omics approaches can be harnessed to create robust cell-based IMID biomarkers in hope of leading to more efficient and patient-tailored therapeutic interventions.

Keywords: immune-mediated inflammatory disorders; autoimmune diseases; immune regulation; dendritic cells; regulatory T cells; omics; therapeutic targeting; biomarkers; chronic inflammation

Citation: Kerdidani, D.; Papaioannou, N.E.; Nakou, E.; Alissafi, T. Rebooting Regulatory T Cell and Dendritic Cell Function in Immune-Mediated Inflammatory Diseases: Biomarker and Therapy Discovery under a Multi-Omics Lens. *Biomedicines* **2022**, *10*, 2140. https://doi.org/10.3390/biomedicines10092140

Academic Editor: Marianna Christodoulou

Received: 27 July 2022
Accepted: 29 August 2022
Published: 31 August 2022

1. Introduction

Immune-mediated inflammatory diseases (IMIDs) are a diverse group of incurable clinical disorders that constitute a unique conceptual and medical challenge for the scientific community. Under the umbrella of the broad term IMIDs, many autoimmune as well as chronic inflammatory diseases, such as rheumatoid arthritis (RA), inflammatory bowel disease (IBD), systemic lupus erythematosus (SLE), type 1 diabetes (T1D), cutaneous inflammatory disorders (including psoriasis and atopic dermatitis (AD)), asthma and autoimmune neurological diseases such as multiple sclerosis (MS), can be incorporated. IMIDs develop as a consequence of complex mechanisms that depend on genetic, epigenetic, molecular, cellular, and environmental elements and result in defects in immune regulatory checkpoints of tolerance [1,2]. This breakdown of self-tolerance leads to the aberrant activation of lymphocytes against otherwise harmless self or foreign antigens causing chronic unrestrained inflammation that destroys self-organs and tissues.

Two key checkpoints of self-tolerance and decision-makers of the type and magnitude of the immune response are dendritic (DC) and regulatory T (Tregs) cells. On the one side, DCs, by up-taking environmental cues, self or foreign antigens and translating them into signals for the proper initiation of the immune response, constitute the sensors of the immune system and the link between innate and adaptive immunity [3,4]. On the other side are Tregs, that respond to signals of DCs, regulating and restraining exacerbated inflammation, thus comprising the brakes of the immune response [5,6]. During IMIDs, both cell types have been reported to be dysregulated, with altered frequencies in the periphery of patients, overt activation, and certain degrees of imbalance in their phenotype and function [7–10], thus leading to the breakdown of self-tolerance. Although the previous two decades have been transformative for the understanding of the mechanisms that govern immune dysregulation in IMIDs, effective and highly targeted treatments have proven to be elusive. Evidently, IMIDs remain a major burden on health systems around the world, accounting annually for several billion EUR in medical costs and lost income. Deciphering in depth the cellular and molecular mechanisms that contribute to the breakdown of immune tolerance is thus an important goal, with the prospect that this knowledge will pave the way to new clinical advances in the treatment of IMIDs.

The recent breakthrough in advanced multi-omics technologies provides the essential tools to ease the massive and in-depth understanding of the mechanisms driving immune dysfunction in IMIDs. Indeed, bulk and single-cell omics, multi-parameter flow and mass cytometry, next-generation spatial omics, and systems biology are among the current approaches expected to be applied in daily clinical practice for the upgrade of patients' management and quality of life. Here, we focus on Tregs and DCs, the two fundamental gatekeepers of the immune tolerance and discuss how recent advances in the field of IMIDs, illuminated by the dawn of omics technologies, can be harnessed to create robust cell-based biomarkers and patient-tailored therapeutic interventions.

2. Regulatory T Cells as Multifaceted Orchestrators of Immune Responses

Tregs are an important immune system component, critical for maintaining homeostasis and immunological self-tolerance [11,12]. Tregs exert their suppressive functions either by cell-to-cell contact or secretion of cytokines. More specifically, Tregs can effectively suppress immune responses via (a) secretion of anti-inflammatory cytokines such as interleukin-10 (IL-10), IL-35, and TGF-β [13–16], (b) granzyme and perforin-mediated cytolysis [17–19], (c) expression of nucleotide-metabolizing enzymes such as CD39 and CD73 [20–22], (d) competing with effector T cells for IL-2, an essential T cell survival cytokine [23,24], and (e) dampening the maturation/antigen-presenting capacity of dendritic cells [25–27].

Both human and murine Tregs are phenotypically distinguishable by the expression of the transcription factor Foxp3 and the IL-2 receptor alpha chain (IL-2Rα, CD25). However, since CD25 can also be highly expressed in other subsets of activated $CD4^+$ T cells in humans, the absence of the IL-7 receptor alpha chain (IL-7Rα, CD127) is complementarily used to identify human Tregs [28]. Expression of the master regulator Foxp3 is a cardinal feature of Tregs, fundamental for their development and suppressive function [6,29]. Therefore, loss-of-function mutations of the *FOXP3* gene in humans lead to the development of a severe autoimmune disease termed immune dysregulation, polyendocrinopathy, enteropathy, and X-linked (IPEX) syndrome [30,31]. Miyara et al. classified $CD4^+Foxp3^+$ Tregs from the peripheral blood of healthy individuals into three main fractions: Fr.I naïve Tregs ($CD45RA^+/CD25^{low}$), Fr. II effector Tregs ($CD45RA^-/CD25^{high}$) and Fr. III not Tregs ($CD45RA^-/CD25^{low}$) [32]. This classification, which is based on surface markers, nicely correlates to Tregs' epigenetic profile and suppressive function with Fr. I and II being suppressive resting or activated Tregs, respectively, and Fr. III being non-suppressive and cytokine secreting non-Tregs [8].

Tregs are either generated in the thymus (thymic-derived Tregs, tTreg) or in the periphery through conversion of $CD4^+Foxp3^-$ T conventional cells following antigenic stimulation in the presence of TGF-β and IL-2 (induced Treg, iTreg) [33,34]. Whereas Treg cells were traditionally considered a terminally differentiated population, it is now well accepted that they acquire plasticity that allows them to adapt to the cues of the microenvironment [8]. By acquiring expression of specific lineage T cell-transcription factors, such as T-bet, GATA-3, IRF-4, STAT-3, RORγt, Bcl-6, and chemokine receptors, Tregs can skew to Th1, Th2, Th17, or T follicular helper cell-like phenotypes [35–41]. These functional adaptability is context and tissue-dependent [8].

Th1-like Tregs circulate in the blood of patients with autoimmune diseases [9]. Except for T-bet, they also upregulate CCR5, CXCR3, and secrete IFN-γ, while displaying reduced suppressive capacity when compared to Tregs. IFN-γ secretion has been shown in vitro to depend on PI3K/AKT/FoxO signaling [9,42]. Respectively, Th2-like Tregs upregulate GATA-3 and IRF-4 and secrete IL-4 and IL-13 [9]. In the setting of IMIDs, Th2-like Tregs have been found in tissues rather than the periphery [43]. Th17-like Tregs upregulate the transcription factor RORγt and secrete IL-17A. Although it is yet unclear whether they are a stable subcluster of Tregs or a transitory stage of Tregs to Th17 cells, Th17-like Tregs are found in steady-state in the gastrointestinal tract, where they have a protective role [40,44], but also in the synovium of arthritic patients and in psoriatic lesions where they contribute to disease pathogenesis [9,45–47].

Adding up to their heterogeneity, Tregs possess also a certain degree of instability. Unstable Tregs, named ex-Tregs, produce inflammatory cytokines, downregulate Foxp3 expression, and concomitantly lose their suppressive function [46,48,49]. Post-translational modifications of the Foxp3 protein, namely acetylation, phosphorylation, and ubiquitination of specific residues also contribute to Tregs' instability and plasticity as they may lead to Foxp3 protein stabilization or proteasomal degradation [50–52]. Tregs' instability seems to have a key role in the pathogenesis of autoimmune diseases [49]. However, the extent to which both Treg plasticity and instability contribute to the pathogenesis of IMIDs and whether the modulation of Tregs' state can be proven therapeutically relevant is under investigation.

Lastly, major advances in the field have uncovered Tregs that reside in non-lymphoid structures and contribute to tissue homeostasis rather than immune surveillance [53]. Tissue-resident Tregs have been identified in several tissues including adipose tissue, skin, lung and gastrointestinal tract where they become epigenetically adapted to microenvironment's specific cues [54]. Thus, the transcriptomic profile of tissue-resident Tregs varies significantly with that of their lymphoid tissue counterparts, as well as among different tissues. Several markers have been identified that distinguish tissue-Treg precursors that reside in lymphoid organs prior to their transport to homing tissues, such as PPARγ^{low}, TCF1^{low}, ID3^{low}, and NFIL3^{+} [55]. Nevertheless, tissue-resident Treg biology remains largely unexplored and constitutes a fruitful field of research.

Tregs are instrumental in preventing IMIDs and preserving immune homeostasis. In fact, most autoimmune diseases bear numerical or functional alterations in their Treg cell compartment. For example, in T1D, the activated Tregs (CD4^{+}CD45RA^{-}Foxp3^{high}) in peripheral blood of patients are increased in numbers and functionally impaired akin to a pro-inflammatory phenotype [56–58]. In RA patients, although the frequencies of Tregs (CD4^{+}CD25^{+}CD127^{-}) in the periphery are either similar or lower compared to healthy controls [32,59,60], Tregs in the synovial fluid are increased in numbers and less suppressive [32,61]. Individuals suffering from relapsing-remitting MS, in most of the studies, have decreased numbers of CD4^{+}CD25^{+} Tregs and increased frequencies of Th1-like (CD4^{+}CD25^{high}CD45RA^{-}CD127^{-}Foxp3^{+}) Tregs in their blood [32,59,60,62,63]. The latter have been shown to express the pro-inflammatory cytokine IFN-γ and have reduced suppressive function when co-cultured with effector T cells in vitro [61]. In line with the perturbed function and frequency of Tregs noted in various autoimmune diseases, Tregs (CD4^{+}CD25^{high}) numbers are also decreased in the peripheral blood of SLE patients and demonstrate reduced suppressive capacity relative to healthy controls [32,60,64]. Although human studies that investigate Treg frequencies and function in various autoimmune diseases suffer from discrepancies due to a lack of consistency in Treg definition markers, they nevertheless reveal the significance of Tregs for immune homeostasis [59].

Studies of Tregs in IMIDs derive mostly from data acquisition of flow cytometry and ex vivo assays, thus lacking collective and high-throughput insight. Recent technological advances have established multi-omics platforms in the field of research and offer holistic approaches to data acquisition that are unbiased and hypothesis-driven independent.

Herein, we review Treg-specific multi-omics approaches that have been applied in IMIDs research up to date.

2.1. Transcriptomic Studies Paving the Way for Illuminating Tregs' Functional Profiles and Subsets in IMIDs

The study of bulk mRNA transcripts within a biological sample, termed transcriptomics, has now become a standard approach for investigating molecular mechanisms that underlie steady-state and pathogenic conditions, as transcriptional profiling of cells is able to reveal gene function and gene structure [65]. By moving onward to the single-cell era it became apparent that transcriptomics at the single-cell level have reshaped modern research and have uncovered cellular differences and the heterogeneity of biological samples that have been masked by bulk RNA-sequencing (RNA-seq). Mostly bulk, and to a lesser extent single-cell RNA-sequencing (scRNA-seq), have been applied thus far in studying Tregs in the context of IMIDs.

A recent study in our lab interrogated the transcriptomic profile of Tregs from the peripheral blood of individuals suffering from MS, RA and SLE [60]. RNA-seq analysis revealed a plethora of deregulated transcripts when compared to healthy controls. Tregs were predominately altered in metabolic pathways related to oxidative stress, mitochondrial dysfunction, cell death and DNA damage response. Interestingly, this signature was consistent across all autoimmune disease settings studied [60].

As mentioned above, Tregs are able to adapt to specific microenvironments, thus conditions such as excessive inflammation imprint onto Treg profile. It has been reported by two independent studies that in juvenile idiopathic arthritis (JIA), Tregs obtained from inflamed joints have a specific effector profile [66,67]. Both studies compared, among others, the transcriptome of Tregs from synovial fluid to those of peripheral blood of individuals with JIA. Differential gene expression analysis revealed that Tregs in the synovial fluid express a Th1 transcriptomic signature that is characterized by the expression of transcription factor *TBX21* (T-bet), chemokine receptor *CXCR3*, and IL-12 receptor β2 (*IL12RB2*). IFN-γ was also found upregulated in one of the studies [66], nevertheless, when Tregs were stimulated ex vivo they failed to produce this cytokine [67]. Despite high expression of Th1-related proteins, Tregs preserved their suppressive features as shown by the maintenance of a robust Treg-associated transcriptional program [66] and functional assays [66,67].

Julé et al. employed scRNA-seq on Tregs sorted from synovial fluid of individuals experiencing JIA. Among five Treg clusters identified in this study, cluster 1 matched the expression profile of Th1-like Tregs while preserving the Treg transcriptomic signature, thus confirming the uncovering of a stable effector Treg population that maintains Treg-specific demethylation patterns and suppressive capacity, as identified by bulk RNA-seq. The newly identified and highly suppressive population of Th1-like Tregs, which was unveiled through the prism of transcriptomics, could constitute an attractive target with important therapeutic benefits for individuals with JIA. Four additional Tregs subpopulations were identified that spanned from the classical and highly activated HLA-DR$^+$ Tregs that robustly express Treg signature genes to the CD161$^+$ and IFN-induced Tregs that share some genes with effector T cell clusters [66].

Recently, the transcriptome of Tregs in the blood of individuals experiencing autoimmune polyendocrine syndrome type I (APS-1) versus healthy controls has been interrogated [68]. Whereas only subtle changes were observed between disease and healthy groups, the G Protein-Coupled Receptor 15 (*GPR15*) gene was found significantly downregulated, while the Fatty Acid Synthase (*FASN*) gene was upregulated in APS-1 Tregs [68]. Given that individuals with APS-1 suffer from gastrointestinal manifestations and GPR15 is a homing receptor for the gut, it was speculated that GPR15 downregulation might be indicative of a defective influx of Tregs in the gut [68]. In addition, an increase in FASN, important for fatty acid synthesis, is suggestive of metabolic reprogramming of APS-1

Tregs [68]. However, data were strictly descriptive and deprived of functional evidence, thus results must be considered cautiously.

Despite their dominant role in immunosuppression, so far only a very limited number of studies have focused on the single-cell analysis of tissue-specific Treg cells in IMIDs. In one of them, scRNA-seq was used to characterize Treg cells isolated from the peripheral blood and synovial fluid of two individuals with ankylosing spondylitis (AS) [69]. Analysis revealed ten specialized Treg clusters, present in both tissues, with unique gene expression signatures. Among them, a $CD8^+$ Treg subset expressing cytotoxic markers such as granzyme B and granulysin was significantly enriched in the synovial fluid of individuals with AS, whereas a Th17-like $RORC^+KLRB1^+$ Treg subset characterized by IL-10 and LAG-3 expression was significantly enriched in the blood of AS patients. Despite the small size of samples, these two clusters were also identified in the peripheral blood and synovial fluid of individuals with psoriatic arthritis, another type of spondyloarthritis (SpA) [69]. Total synovial fluid Tregs were characterized by the upregulation of activation and inhibitory markers, as well as TNF and interferon response genes, and they were clonally expanded suggesting tissue-specific adaptation. Thus, targeting these unique characteristics of joint-specific Treg subsets could have promising applications for the amelioration of SpA.

Immune-related adverse events (irAEs) are an atypical IMID that is worth mentioning. The impressive success of immune checkpoint therapies in the treatment of various types of cancer is often overshadowed by irAEs that arise due to excessive activation of the immune system. Previous studies in our lab applying RNA-seq have demonstrated that Tregs from the peripheral blood of individuals developing irAEs bear a pro-inflammatory profile accompanied by enrichment in the apoptotic and metabolic pathways [70]. Moreover, irAEs-Treg signature is shared across different types of cancer and resembles Treg traits of individuals with autoimmune diseases [70]. Unraveling phenotypic switches of Tregs that drive or precondition the development of irAEs is of utmost importance for the prevention of toxicities that often accompany cancer immunotherapies.

Although not an IMID per se, graft-versus-host disease (GVHD) manifests as an autoimmune disease, and transcriptomic approaches have been employed to dissect Treg complexity in patients receiving hematopoietic stem cell transplantation [71]. Specifically, single-cell transcriptomic analysis was performed in Tregs of the peripheral blood and bone marrow of healthy donors and patients after hematopoietic stem cell transplantation that were either experiencing GVHD or not. The analysis resolved nine clusters both in the peripheral blood and bone marrow of individuals that included naïve ($CCR7^{hi}$), activated ($HLA\text{-}DR^{hi}$), $LIMS1^{hi}$, effector ($Foxp3^{hi}$), and proliferative ($MKI67^{hi}$) Tregs. Functional evaluation revealed $MKI67^{hi}$ and $Foxp3^{hi}$ clusters as highly suppressive, followed by $HLA\text{-}DR^{hi}$ and $LIMS1^{hi}$ clusters. Pseudotime trajectory analysis uncovered the transition among clusters according to which naïve Tregs followed two distinct differentiation pathways towards either $Foxp3^{hi}$ Tregs (Path 1) or $MKI67^{hi}$ Tregs (Path2). Whereas similar clusters, spanning from naïve to activated/effector Tregs, were identified in all groups, effector Tregs clusters in individuals developing GVHD displayed downregulation of suppression and migration pathways as well as a senescence-like signature compared with non-GVHD patients [71]. Although the latter can be attributed to the age gap between GVHD and non-GVHD patients, Treg interrogation on a single-cell level offered a greater understanding of Treg features upon GVHD.

Regarding organ-specific immune-mediated diseases, the role of cell-based omics technologies, and particularly the advances in single-cell TCR$\alpha\beta$ sequencing, is of primary importance to illuminate the antigen specificities of the pathogenic cells that mediate tissue damage, or of the regulatory cells that suppress the former in the inflammatory niche. Such knowledge will be decisive during the design of more efficient and targeted therapeutic approaches such as autoantigen-specific TCR engineering.

One such case is T1D in which Treg cells have already been exploited in therapies, with early phase clinical trials of ex vivo expanded polyclonal Treg cells showing promising results [72,73]. However, since polyclonal Tregs are not antigen-specific, the approach utilized in these clinical trials could potentially lead to systemic unwanted immunosuppression. Interestingly, preclinical studies using the non-obese diabetic (NOD) murine model for T1D revealed that relatively small numbers of antigen-specific Treg cells, either pancreatic lymph node-derived or genetically engineered, and not polyclonal Treg cells, could prevent and even reverse T1D, pointing to therapies utilizing diabetogenic TCR-expressing Treg cells [74,75]. Still, most antigen-specific Treg cells are tissue-resident and only a small portion of them circulates in the bloodstream, rendering them difficult to isolate and characterize in humans. Additionally, so far, the attempts to create tailored Tregs utilize recombinant TCRs from Teff cells [76,77]. Due to these challenges, up to now the identification of the exact TCR sequences specific for dominant diabetogenic epitopes in Treg cells has been restricted only to NOD mice. To this end, Spence et al. employed TCR repertoire profiling and TCR$\alpha\beta$ scRNA-seq to determine the specificity of Treg cells in the islets of Langerhans. Treg clonotypes were found to be expanded and the least diverse in inflamed islets compared to other lymphoid organs, while some of their TCRs were specific for islet-derived antigens including insulin B:9–23 and proinsulin, implying tissue-specific antigen-driven expansion of Treg clonotypes [78]. Their transcriptomic observation was further confirmed utilizing insulin B:9-23 tetramers able to detect increased insulin-specific Treg clones in the islets of NOD mice. Moreover, the adoptive transfer of total Treg cells from the islets, but not of Tregs from lymphoid organs, in NOD.CD28$^{-/-}$ mice could lead to disease rescue, further supporting the suitability of engineered Treg cells expressing insulin-specific TCRs as a promising strategy for suppressing autoimmune reactions against beta cells.

In JIA, TCR repertoire assessment on a single-cell level revealed that the Th1-like Tregs identified in the joints of individuals with JIA are bone fide Tregs, as their clonotypic composition was similar to that of other Treg clusters and not to effector T cells [66]. Another study has identified a subpopulation of activated Tregs (HLA-DR^{+}) in the blood of JIA and RA patients that has been negatively correlated to response to therapy [79]. In JIA, the so-called inflammation associated (ia) Tregs expand during inflammation and decrease when the disease is inactive. It is important to note that iaTregs also expand when children have poor responsiveness to therapy. TCR-seq revealed antigenic stimulation and shared clonotypes between these iaTregs and Tregs from the synovium [79]. This observation confirmed the fact that HLA-DR^{+} Tregs recirculate between the synovium and blood, which could only be hypothesized up to then by the expression of tissue-homing receptors [79]. Migrating to blood-synovial Tregs could offer easy, non-invasive access to arthritis-associated clonotypes and at the same time could be exploited to monitor response to therapy [79].

2.2. *Unraveling the Epigenetic Mechanisms Governing Tregs Links Molecular Traits to Pathogenicity*

Marking the epigenetic changes across many genes is another available multi-omic tool termed epigenomics. Gene expression is driven by promoters, enhancers, insulators, etc. Epigenetic regulation of enhancers via histone modifications, which reveals gene regulation, has been used in IMIDs research [80]. Epigenomic approaches often act conjointly with transcriptomics to uncover context-specific gene regulation, as changes noted at the mRNA level are sought to be reflected also at the epigenetic level [67]. ChIP-seq was performed to profile histone modification marks that indicate transcriptionally active enhancers (acetylation of lysine 27 on histone H3 (H3K27ac) and monomethylation of lysine 4 on histone (H3K4me1)) using Tregs obtained from the synovial fluid versus peripheral blood of individuals with JIA. The study validated the Th1-like profile of synovial fluid-Tregs that was observed from RNA-seq data [66,67]. Specifically, ChIP-seq identified super-enhancers of genes that were found upregulated in mRNA levels such as *TBX21* and *IL12RB2* as well as super-enhancers associated with putative Treg markers, indicating that Tregs in the inflammatory environment of arthritic joints are adapted to a Th1-related profile while maintaining Treg specific features [67]. The same study uncovered vitamin D receptor (VDR) as one of the top predicted regulators of Treg differentiation in the arthritic joints, marking it as an attractive therapeutic target. Ex vivo stimulation with vitamin D3 skewed Tregs towards an effector Treg profile [67].

Similar epigenetic profiling was performed in peripheral blood-Tregs in individuals with T1D versus healthy controls [81]. ChIP-seq and subsequent sophisticated in silico analysis revealed that (a) T1D-Tregs have fewer active enhancers compared to healthy Tregs, many of which regulate genes implicated in T1D pathogenesis, and (b) certain single nucleotide polymorphisms (SNPs) in enhancer regions disrupt the binding of key transcription factors that regulate transcriptome changes in T1D-Tregs [81]. Similar studies that translate, via multi-omics approaches, non-coding genetic variants to functional/pathological states of Tregs are needed for the prediction and understanding of IMIDs.

ChIP-seq along with ATAC-seq that determines chromatin accessibility and RNA-seq have also been used to highlight Tregs' contribution to the development of IMIDs at large [82–86]. Epigenetic profiling of Tregs from the peripheral blood of healthy individuals revealed that autoimmune disease-associated SNPs are enriched in hypomethylated regions of naïve Tregs that control transcription and epigenetic changes, hence Treg function [86]. A recent study further supports the functional relevance of SNPs by showing that immune disease variants reside in chromosomal loci involved in Treg cell activation and IL-2 signaling [83]. In general, genetic variants associated with immune diseases are found enriched in regulatory regions of Tregs [82,84,85]. Multi-omics approaches combined with genome-wide association studies (GWAS) pave the way for the understanding of Tregs' contribution to IMIDs and the discovery of new Treg-specific therapeutic targets.

2.3. Proteomic Studies Shed Light on Distinct Treg Subsets with Opposing Functions

Proteomic analyses have helped us elucidate the mechanisms of inflammation-mediated pathology. They have also long been considered a valuable platform for the identification of autoimmune disease biomarkers for diagnostic and prognostic purposes in accessible biological fluids. In the new multi-omics era, approaches combining cell-type-based proteomics with transcriptomics could foster the characterization of disease-specific Treg subtypes which may serve as biomarkers for disease initiation or progression. However, up to date, only one study has focused on the proteomic profiling of Tregs in the context of IMIDs.

In this study, Weerakoon et al. employed proteomics in sorted Tregs ($CD4^{+}CD25^{high}CD127^{-}$) and conventional $CD4^{+}$ T cells ($CD4^{+}CD25^{-}$) from the peripheral blood of IBD patients. Their analysis pinpoints the absence or presence of integrin CD49f as a marker that distinguishes conventional T cells from Tregs. However, among Tregs, CD49f expression was also variable, and could separate two Treg subsets with distinct functions in the peripheral blood of IBD patients. $CD49f^{-}$ Tregs show increased suppressive ability and expression of inhibitory receptors, whereas $CD49f^{high}$ Tregs possess a proinflammatory phenotype and they are increased in the blood of IBD patients with active disease. They also suggest that the ratio $CD49f^{high}/CD49f^{-}$ Tregs may constitute a useful predictor of disease activity, but this result should be validated in larger cohorts of patients [87]. Still, it is beyond doubt that more studies in the field of Treg proteomics in IMIDs are needed in order to disentangle the protein profile of these cells and identify novel Treg-specific markers and potential therapeutic targets. Furthermore, following the road paved by single-cell transcriptomics, newly developed single-cell proteomic platforms have the potential to uncover additional layers of Tregs' complexity in the setting of IMIDs.

2.4. Microbiome-16S-Sequencing at the Crossroads between Tregs and Microbiota, Leading the Way to Microbiota-Related Therapeutic Interventions

Commensal microbes colonize barrier sites where they are essential for immune homeostasis predominantly by modulating the generation of Treg cells. The advancements in 16S rRNA and metagenomics sequencing technologies have shed light on the composition and function of the human microbiome, as well as its direct role in modulating immune responses through its components or metabolites [88]. Increasing evidence suggests that gut dysbiosis is implicated in many IMIDs including SLE [89,90], RA [91,92], IBD [93,94],

T1D [95], Grave's disease [96] and MS [97,98], and it is characterized by a reduction in small-chain fatty acid (SCFA)-producing species. Given the importance of Treg cells in establishing immune tolerance to self-antigens and commensal microbes, researchers' attention is now shifted towards Treg–microbiota interactions in autoimmune disorders, which may underpin the decreased numbers and/or dysfunction of Tregs in these conditions. Specifically, it has been shown that in mice, the SCFA butyrate promotes the induction of Treg cells, whereas treatment of naïve T cells with butyrate-enhanced histone 3 acetylation in the promoter and conserved non-coding sequence regions of the *FOXP3* locus leads to differentiation into Treg cells [99]. These unique effects of butyrate on Treg cells could provide protection from diabetes in NOD mice fed with a diet that generates large amounts of butyrate after colonic fermentation [100].

Another study showed that microbial species found in fecal samples of SLE patients induced a pro-inflammatory immune phenotype characterized by lymphocyte activation and Th17 differentiation. Interestingly, supplementation of SLE stool samples with Treg-inducing bacteria could restore Treg/Th17/Th1 imbalance [90]. Furthermore, long-term propionic acid supplementation in MS patients could reduce the annual relapse rate and ameliorate disease progression by increasing Treg cell numbers and suppressive function [98]. Thus, further exploring the crosstalk between Tregs and microbiota by integrating information from different high-throughput technologies (single-cell, metabolomics) will facilitate the development of therapeutic interventions that restore immunological tolerance through manipulation of the microbiome.

Key observations by studies employing transcriptomic, proteomic and epigenomic approaches have provided insight into Treg cells' function and contribution to the pathogenesis of numerous IMIDs (Figure 1). In the single-cell era, multi-omics approaches are indispensable for understanding the perplexing mechanisms that underlie Treg cell biology in IMIDs, the elucidation of which can lead to specific and effective therapeutic regimes.

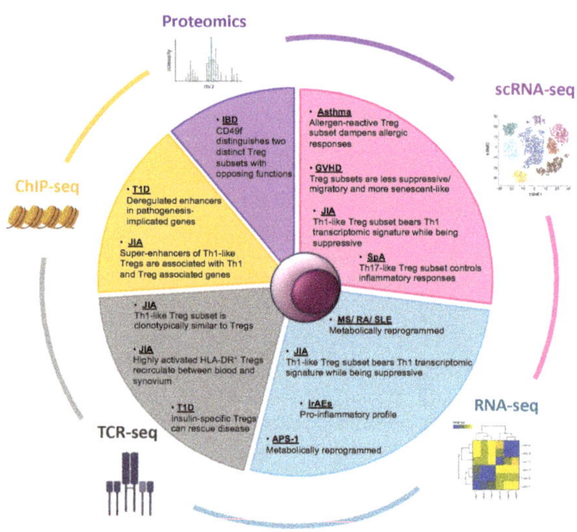

Figure 1. Multi-omics approaches utilized in IMIDs research, focusing on regulatory T cells. The Pie chart depicts omics technologies that have been used to study the contribution of regulatory T cells in the pathology of IMIDs. Predominantly RNA-seq but also proteomic and epigenomic technologies have revealed Treg profiles that are suppressive, pro-inflammatory, or metabolically reprogrammed, as well as distinct Treg subsets across various diseases. scRNA-seq, single cell RNA-seq; ChIP-seq, Chromatin Immunoprecipitation sequencing; TCR-seq, T Cell Receptor sequencing; GVHD, Graft Versus Host Disease; JIA, Juvenile Rheumatoid Arthritis; SpA, Spondyloarthritis; MS, Multiple Sclerosis; RA, Rheumatoid Arthritis; SLE, Systemic Lupus Erythematosus; irAEs, immune related Adverse Events; APS-1, Autoimmune Polyendocrine Syndrome Type I; T1D, Type I Diabetes; IBD, Inflammatory Bowel Disease.

3. Dendritic Cells as Multifaceted Orchestrators of Immune Responses

Ever since their initial discovery by Steinmann and Cohn [101], DCs have grown from simply being viewed as highly motile stellate cells to being recognized as an essential connective link between the innate and adaptive arm of immunity in mammals. DCs constantly sample their microenvironment by engulfing self or non-self antigenic molecules. By possessing a large array of surface and intracellular receptors, they integrate the context in which these molecules are met and thus whether they are associated with invading pathogens, damaged cells or constitute innocuous antigens. After antigen processing, DCs present peptides to T cells, thereby activating them in an antigen-specific way. Most importantly, the induced T cell activation is polarized accordingly, through the production of cytokines and provision of specific costimulatory signals in order to ensure either sufficient protection against the pathogen met, or establishment and maintenance of tolerance against self and innocuous antigens [102–104]. This has earned them the title of orchestrators of immune responses.

The multifaceted role of DCs in immune responses is a derivative of their heterogeneity. Notably, the DC term functions as an umbrella that encloses several cell subsets, each possessing distinct developmental requirements, phenotype and functional properties [102,105]. While DCs have initially been studied more extensively in mice, with the help of multi-omics approaches, recent publications have elegantly dissected the human DC compartment, elucidating in parallel a high interspecies conservation of their development, phenotype and function [105–107]. Among DCs, two main distinct lineages can be distinguished, namely conventional DCs (cDCs) and plasmacytoid DCs (pDCs).

In both mice and humans, pDCs have a prominent role in anti-viral defense due to their ability to secrete copious amounts of type I interferons (IFN) in response to virally derived nucleic acids [108]. The efficiency of pDCs in antigen presentation and T cell activation is still not clearly defined due to controversial findings between different experimental settings [109–112]. While their exact developmental trajectory has also been a highly debated topic in recent years, [113–115] the consensus is that their differentiation is dependent on the transcription factor E2-2 in both species [107,108]. On the contrary, their major defining phenotypic markers seem to be not so well-conserved. Despite MHC-II/HLA-DR expression being a common trait, murine pDCs are characterized as B220$^+$, SiglecH$^+$, CD317$^+$, Ly6C$^+$, CD11c^{intermediate},while in humans characteristic pDC markers are CD123, CD303, CD304, combined with a lack of CD11c and CD5 expression [107,108,116].

cDCs excel in the activation of adaptive immune responses by presenting antigens to T cells [105]. They are subsequently divided into cDC1 and cDC2 and exhibit a remarkable division of labor when it comes to their role in immune responses [105]. Both cDC subsets are characterized by the expression of CD11c and MHC-II/HLA-DR but are distinct in dependence on transcription factors and the expression of other surface markers. Continuous and high expression of the transcription factors IRF8 and BATF3 is a prerequisite for maintaining the developmental and functional program of both human and murine cDC1 [106,117–119]. Genetic approaches have additionally elucidated the role of ID2 [120] and NFIL3 [121,122] in mouse cDC1 development, however, their implication in humans has yet to be determined. In terms of their phenotype, murine cDC1 can be reliably identified across tissues by the expression of XCR-1, CLEC9A, CD24 and CD205 [105]. Moreover, CD8α and CD103 are used as cDC1 characteristic markers in lymphoid and non-lymphoid tissues, respectively, despite the latter also being expressed in an intestinal cDC2 population [105]. In addition to XCR-1 and CLEC9A, human cDC1 in both blood and non-lymphoid tissues have characteristic expression of CD141 and CADM1 [107,116]. Functionally, cDC1 play a dominant role in inducing cytotoxic CD8$^+$ T and Th1 polarized CD4$^+$ T cell responses against intracellular pathogens, such as viruses and bacteria, but also participate majorly in antitumor immunity [105]. They do so via producing ample amounts of IL-12 that activates T cells both directly and indirectly by promoting a Th1-favorable cytokine milieu from bystander cells [119,123,124]. Added to the above, their remarkable potential as CD8$^+$ T cell activators is extended by their ability to cross-present extracellular antigens on MHC-I molecules [119,123,124]. In contrast to the pro-inflammatory role described above, especially in mice, the high potential of cDC1 to induce peripheral regulatory T cells has also been proposed [125,126].

In contrast to pDCs and cDC1 subtypes, the phenotype and developmental requirements of cDC2 between humans and mice seem to overlap the least. In mice, studies have identified transcription factors IRF-4, ZEB2, KLF4 and RELB as central mediators of cDC2 development [102,105,122] as well as pathways with more tissue-specific context such as NOTCH and retinoic acid signaling [127]. While human cDC2 distinctively expresses IRF-4, its role in their development is not yet elucidated. Characteristic murine cDC2 surface markers include CD11b, CD172a, CD4 and CLEC4A4 [105] of which only CD172a is a common defining marker with their human counterparts. The latter are additionally identified by their expression of CD1c, FcεR1α and CLEC10A [107,116]. Functionally, human and murine cDC2 align and are believed to be more efficient in inducing CD4$^+$ T cell activation and polarization towards Tfh, Th2 or Th17 effector responses, crucial for T cell-dependent antibody production by B cells, defense against multicellular pathogens such as helminths or extracellular bacteria and fungi, respectively [128–135]. Their CD4$^+$ T cell activation pattern also extends to regulatory directions via the induction of Tregs both in the thymus and in peripheral tissues [136,137]. Remarkably, cDC2 have been found to exhibit the highest intra-subset diversity compared to pDCs and cDC1. This heterogeneity, despite being ever-growing, has been studied in detail in mice [138–140], however, it has only recently been appreciated in humans.

In addition to pDCs and cDCs, a new cell subset termed transitional DCs has quite recently been identified in both humans and mice [141]. As implied by their name, these cells are placed in between the two aforementioned populations in the DC spectrum and have been described to possess shared pDC and cDC properties. Nonetheless, their exact function is yet to be defined and needs to be investigated further.

Many studies focusing on DCs, and especially in humans, use peripheral blood monocytes as a source to generate them in vitro. While not ontogenetically related to pDCs and cDCs, these monocyte-derived DCs (moDCs) have been used extensively due to the existence of established protocols for their generation, the enhanced availability of monocytes in the peripheral blood and their implementation in clinical practice [107,142]. Similar protocols exist in mice, however, bone marrow rather than peripheral blood is the selected source to generate such cells [107]. Data suggest that mature in vitro-differentiated moDCs likely align with monocyte-derived cells arising under inflammatory conditions in vivo [107,143]. The latter cells are characterized by the expression of CD11c, MHC-II/HLA-DR, CD14, CD64, CD11b, CCR2, CD209 and CD206 in mice and humans with Ly6C positivity being an extra distinctive phenotypic trait of the murine cells [143]. In vivo generated moDCs have a profound pro-inflammatory potential and functional specialization, primarily connected with direct anti-microbial effector function, evident by the fact that there were first described in mice infected with *L. monocytogenes* [144]. Their T cell activation potential in most cases does not match that of cDCs, however, it is not redundant for the clearance of some pathogens requiring strong Th1 immunity [145]. As expected, their pro-inflammatory role can function as a double-edged sword, since these cells have been postulated to enhance many IMIDs manifestations [107,143].

3.1. Elucidating the Role of Dendritic Cells in IMIDs Utilizing Multi-Omics Approaches

Given their role in maintaining the balance between protective immune responses and self-tolerance, DCs play a critical part in IMID manifestations in which this balance is by default perturbed. Their detailed role has been extensively reviewed elsewhere [7], and in brief entails the dysregulation of one or more of the following functional properties: (a) perturbation in the pattern of secreted cytokines, quantitatively and qualitatively, that promote pro-inflammatory responses from other innate and adaptive immune system cells; (b) enhanced antigen presentation of primarily self-antigens; and (c) altered distribution in terms of both frequency and spatial arrangement, often related to differences in their migratory capacity, that affects especially the inflamed tissues but also peripheral blood. Here, we aim to report cases in which the role of DCs in IMIDs has been refined or enriched by the advent of recent omics approaches (Figure 2).

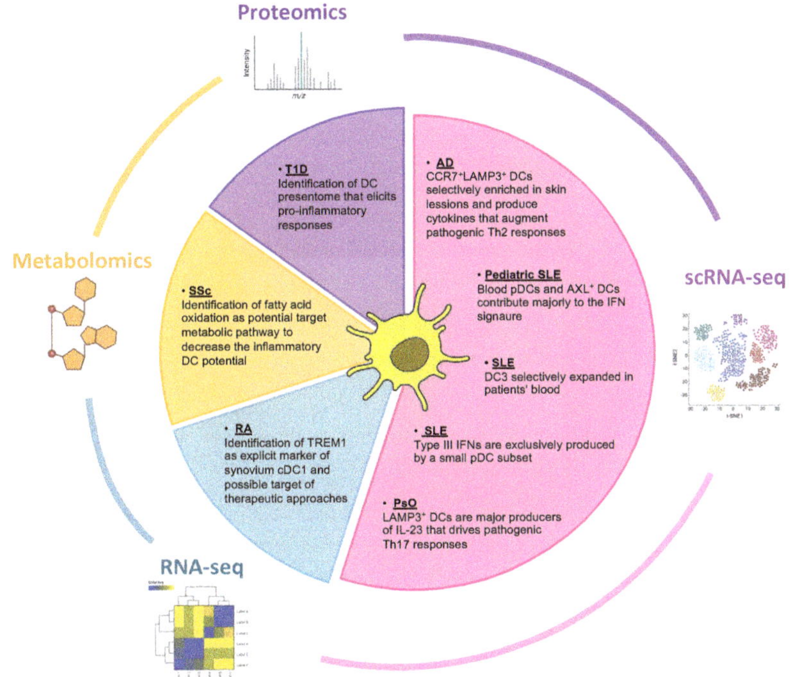

Figure 2. Multi-omics approaches utilized in IMIDs research focusing on dendritic cells. The Pie chart depicts omics technologies that have been used to study the contribution of dendritic cells in the pathology of IMIDs. Mainly scRNA-seq but also proteomic and metabolomic studies have highlighted dendritic cell subsets and inflammatory signatures that drive pathogenic responses in the disease spectrum of IMIDs. scRNA-seq, single cell RNA-sequencing; RA, Rheumatoid Arthritis; SLE, Systemic Lupus Erythematosus; T1D, Type I Diabetes; AD, Atopic Dermatitis; PsO, Psoriasis; Ssc, Systemic sclerosis.

3.2. Bulk and Single-Cell RNA Sequencing Have Expanded the Portfolio of DC Subsets and Illuminated Their Role in IMIDs Perturbations

One remarkable advantage of multi-omics approaches is their potential for single-cell resolution. This was made apparent especially for human cDC2, as recent studies identified novel subsets within the CD1c^{+} cDC2 population using scRNA-seq coupled with index sorting [146,147]. The subdivision of these new subpopulations, namely DC2 (CD5$^{+/-}$CD163^{-}CD14^{-}) and DC3 (CD5^{-}CD163^{+}CD14$^{+/-}$), based on their immunophenotype was also found to be accompanied by functional differences [147,148]. In the context of IMIDs, CD163^{+} DC3s were found to be expanded in the blood of SLE patients and presented a highly activated phenotype compared to healthy controls. Interestingly, their frequency in blood was highly correlated to clinical scores. Secretome analysis showed that, among cDC2 subsets, DC3s uniquely produced many pro-inflammatory mediators when activated by the serum of SLE patients [147]. Given the above, it would be intriguing to investigate the performance of these cells as disease biomarkers and establish whether manipulating their function could ameliorate disease progression. Additionally, their role in other IMIDs such as RA and Psoriatic Arthritis (PsA) warrants further investigation due to their increased potential for induction of IL-17A producing T cells [147].

DC3s were also found selectively expanded, among cDC2, as assessed by scRNA-seq in pediatric SLE (cSLE) patients' peripheral blood mononuclear cells (PBMCs), compared to age-matched healthy individuals [149]. Interestingly, overtaking even DC3s, the majorly expanded cDC cluster resembled the AXL$^+$ DCs first identified by Villani et al. [146]. Additionally, although pDCs were found decreased as a total population in cSLE samples, further analysis revealed four distinct subclusters, one of which was profoundly expanded in SLE compared to healthy individuals. Notably, the defining markers of this expanded pDCs subcluster consisted primarily of interferon-induced genes, accompanied by genes connected to transcription factors (e.g., *STAT1*, *IRF7*) and antigen presentation (e.g., *CD74*, *HLA-DRA*, *CTSB*) [149]. The latter could point towards a yet unexplored role of these cells in propagating the IMID by activation of autoreactive T cells. In line with their initial placing on the verge between cDCs and pDCs, AXL$^+$ DCs together with the expanded pDC subcluster were found to be among the PBMC clusters contributing the most to the SLE IFN signature. The above study proceeded a step further by aligning side by side the pediatric samples to corresponding samples from adults, highlighting age as another contributor to the fluctuation of disease-specific subclusters. Keeping up with the pDC and SLE field, Hjorton et al. investigated the cellular source of type III IFNs, a cytokine group whose contribution to the SLE IFN signature and disease progression remains poorly studied. To this end, they isolated pDCs from healthy donors, used a stimulation mix containing RNA immunocomplexes (used widely as IFN inducers in these cells) plus IFN2ab and IL-3, and subjected them to scRNA-seq [150]. Unexpectedly, they found that only a small population of single-cell sequenced pDCs contributed mostly to the total detected transcripts of both IFN III and I. Compared to the non-IFN III-producing cells, the identified small pDC cluster was also characterized by higher mRNA levels of genes connected to immune activation such as *TNF*, *CD40*, *CD83* and *IL12A*. While not explored by the authors, it would be interesting to speculate as to whether their identified pDC subcluster aligns with the expanded one mentioned by Belaid et al. [149], as in both cases its frequency among healthy donor pDCs was minimal. Thus, the importance of single-cell resolution in identifying the disease and age-relevant cell populations was once more signified.

In most cases, peripheral blood has been used as a mirror to study DC properties in IMIDs, however, analyses from inflamed tissues are equally or even more important as suggested by the expected effect of tissue microenvironments in DC transcriptional and functional signatures [151,152]. This site-specific analysis has been bolstered by recent omics advances, since their high throughput performance automatically decreases the required cell numbers to conduct meaningful experiments. As an example, Caravan and colleagues, studied the impact of the synovial microenvironment in cDCs from RA patients [153,154]. Using multiparameter flow cytometry and RNA sequencing, they found not only that cDCs were enriched in the synovial tissue of RA patients, compared to the blood of the same individuals as well as that of healthy controls, but that they also exhibited a highly activated phenotype as assessed by expression of costimulatory molecules [153,154]. Regarding the CD1c$^+$ cDC2, the synovial microenvironment was shown to induce metabolic alterations, polarizing them to a more glycolytic phenotype while a more detailed analysis was performed for CD141$^+$ cDC1. For the latter, the hypoxic synovium was shown to specifically induce the expression of TREM-1 as part of a site and disease-specific signature. Interestingly, in vitro crosslinking of TREM-1 in cDC1 isolated from synovial tissue could induce their activated phenotype in parallel to an increased ability to induce pro-inflammatory cytokine production from heterologous and autologous T cells [153]. Additionally, supernatants from these cDC1-T cell co-cultures could activate synovial fibroblasts to produce an array of soluble mediators consistent with the acquisition of an invasive phenotype. The authors concluded that the discovered synovium-specific signatures could be harnessed in order to design novel therapeutic and cDC targeted strategies, with TREM-1 being a frontline example.

Omics analysis targeted to the inflamed tissue is additionally essential for another IMID, namely AD. Two recent studies have attempted to interrogate the immune and non-immune skin compartments of patients with AD and healthy controls using scRNA-seq [155,156]. Both studies found that DC populations were expanded in the pathogenic samples in relation to healthy skin, with cDC2 probably being the more over-represented population due to their characteristic expression pattern of surface markers. Alongside the "typical" cDC populations, another smaller cluster expressing *CCR7* and *LAMP3* was identified. Despite their small numbers, these cells exhibited some very interesting traits such as clear characteristics of mature and migratory behavior and selective enrichment in the lesional skin of AD patients combined with their almost complete absence from healthy samples [155,156]. He et al. also found that these $LAMP3^{+}CCR7^{+}$ DCs robustly expressed type 2 chemokines such as *CCL17* and *CCL22*. These data were nicely corroborated by the fact that T cell populations with Th2 and Th22 polarization states were additionally enriched in the AD skin samples, opening the possibility that DCs are the major innate immune cell to attract these pathogenic T cells in the site of inflammation. Notably, these type 2 chemokines have already been used as reliable biomarkers to measure disease progression and response to therapy [157], however, the source cells were not clearly defined. Moreover, Rojahn et al. reported that added to type 2 chemokines, myeloid cells including DCs, produced amphiregulin in the lesional skin that can activate keratinocytes and thus worsen the clinical manifestations of AD [156]. Collectively, the above studies could be the starting point of further investigations on whether DCs are a major source of the above soluble factors and if so, implement their targeting as better therapeutic interventions and/or evaluation of them as more accurate biomarkers.

On the same page and similarly focusing on IMIDs with skin-related pathological manifestations, Kim and colleagues [158] interrogated the immune compartment of skin biopsies from patients with psoriasis as compared to healthy volunteers. To cope with the inherent issues introduced by enzymatic digestion of the skin as well as with the low leukocyte frequencies in that tissue, they implemented a novel approach by profiling with scRNA-seq the cells naturally emigrating from skin biopsies over the course of 48 h. In line with the studies above, they found DCs to be majorly expanded in the samples of psoriatic patients with a reported increase in their numbers over three-fold compared to healthy skin [158]. Interestingly, they identified DCs with both a mature and semi-mature phenotype. Semi-mature DCs, in both sample groups, were found to express genes encoding for IL-10 and CD141. While the authors did not elaborate further on this, the description could fit skin resident cDC1 and at the same time render these cells as potential targets of tolerance, re-establishing therapeutic approaches. Mature DCs, on the other hand, had higher expression of genes related to antigen presentation machinery and costimulatory signals, a signature that was further reinforced in psoriasis samples. A defining marker of these mature DCs was *LAMP3*, highlighting, in conjunction with the above studies in AD, that the same DC populations can have disease-promoting roles in a broad spectrum of IMIDs. Additionally, in psoriatic samples mature DCs expressed considerably more *IL-23A*, a cytokine related to the establishment of a pathogenic Th17 profile, while at the same time had markedly less expression of *KYNU*, an enzyme participating in the kynurenine pathway known for its immunomodulatory role. Going a step further, by using a computational algorithm to simulate cell-to-cell communication events, the authors were able to show that the increased *IL-23A* production by mature DCs in psoriatic skin would signal in *IL-17F* producing Th17 cells, shown to express the highest amount of the cognate receptor. Interestingly, these $IL\text{-}17F^{+}$ cells were the largest subset of IL-17 producing T cells in psoriasis samples, therefore suggesting that their expansion and pathogenic function is a derivative of the pro-inflammatory secretory behavior of mature DCs.

3.3. Contribution of Proteomics in the Identification of DC-Presented Epitopes in IMIDs

As mentioned above, a prominent DC function lies in the processing and presentation of autoantigens to autoreactive T cells. However, an up-to-the-point question concerns whether specific epitopes dominate and are preferentially presented, even in cases that a particulate cell population is targeted. In the example of T1D, a recent study aimed to delineate the naturally processed and presented epitopes by DCs as derived from pancreatic beta cells [159]. The authors isolated peripheral blood CD14$^+$ monocytes from healthy donors and cultured them in vitro with GM-CSF and IL-4 in order to generate moDCs. Then, these moDCs were pulsed in vitro with various pancreatic islet autoantigens and then their "presentome" was analyzed (the eluted epitopes presented in the surface HLA-DR molecules) with mass spectrometry. Their experimental set up also held augmented clinical relevance as they selectively used moDCs from individuals possessing the alleles HLA-DR3 and HLA-DR4, associated with high-risk of disease emergence. Among their findings was the addition of new epitopes to those already characterized for some peptide autoantigens as well as the discovery of some derived from pancreatic islet peptides for which epitope generation had not been previously reported. Interestingly, they were able to show that not all the discovered epitopes induce a pro-inflammatory reaction, evident by the response magnitude and its induced IL-10 or IFN-α signature upon incubation of PBMCs from T1D patients with them. Importantly, the most immunodominant epitopes were generated by moDCs when compared to B cells, another immune system cell with antigen presentation capacity, in an HLA-DR allele independent manner. Thus, such approaches could help increase the efficacy of peptide-based tolerogenic immunotherapies as well as their patient-specific tailoring by using epitopes preferentially presented by each HLA haplotype in moDCs.

3.4. Bridging the Metabolic Profile and the Function of DCs in IMIDs: An Emerging Field of Research

The relevance of metabolomics in IMIDs is constantly increasing, however, its targeted application in DCs is only lately gaining attention. Towards that point, a recent study aimed to identify metabolic pathways exhibiting similar dysregulation in the circulation and DCs of patients with systemic sclerosis [160]. To this end, they first performed metabolite analysis using plasma of patients and healthy individuals and discovered evidence of imbalanced fatty acid and carnitine levels in systemic sclerosis samples. In line with this, they also found increased levels of L-carnitine in moDCs, derived from GM-CSF and IL-4 cultures of peripheral blood monocytes from systemic sclerosis patients, after their stimulation with TLR agonists. As a continuation of their observations, the authors tested the effect of etoposide, a carnitine transporter inhibitor widely used for cancer treatment, on the activation of patient-derived moDCs after TLR stimulation and showed secretion of reduced levels of pro-inflammatory cytokines such as IL-6 in its presence. As carnitine transports fatty acids into the mitochondria in order to facilitate their oxidation, the above observations suggest that targeted suppression of fatty acid oxidation in DCs could be helpful in decreasing the inflammation related to the particular IMID.

4. Conclusions, Challenges and Open Questions

Multi-omics data can play a crucial role in clinical practice in the near future, for predicting disease susceptibility, disease severity and treatment response or identifying new therapeutic targets for IMIDs. However, we are still in the dawn of this exciting new era. Building large-scale patient cohorts with high-quality clinical data consisting of patient demographics, disease response and multiple layers of omics data, as well as refined analytic approaches to handle these data, would contribute to a better understanding of mechanisms governing IMIDs biology and accelerate precision medicine.

Certain barriers need to be considered and overcome towards the vision of biomarker discovery and targeted new therapies for IMIDs. First of all, so far, high-throughput analysis is mostly restricted to total PBMCs of patients, with data extracted from diverse immune cell types being very limited. To analyze the genome, which is regarded as a stable feature for each individual, an easily accessible tissue, such as blood and analysis of whole PBMCs is broadly acceptable. However, many other types of omics, such as transcriptome, proteome and metabolome, vary between diverse immune cell types and tissues. Due to the high degree of complexity of the immune system, selective targeting of specific immune cell populations dictating the complex immune responses during IMIDs, such as DCs and Tregs, allows a deeper understanding of the mechanisms driving disease pathogenesis, with the prospect of identifying more precise therapeutic targets avoiding broad immunosuppression. Additional multi-omics data extracted from the analysis of Tregs and DCs specifically are needed to elucidate the degree of dysfunction rendering these cells pathogenic for IMIDs.

Secondly, the few existing studies utilizing diverse omics approaches to analyze DCs and Tregs in IMIDs are restricted to information extrapolated from a single omic level. A single omic data layer characterizes a specific biological process from one aspect. However, biological processes are based on interactions among genes, proteomes, metabolites, etc., and are regulated by epigenetic modifications. Single biomolecules or signaling pathways cannot fully explain biological mechanisms or functions. To acquire a comprehensive picture of the intrinsic molecular mechanisms driving disease pathogenesis, a systematic collection of multi-omics data is required. This increasing availability of multi-omic platforms and layers poses new challenges in data analysis. Integration and common visualization of multi-omics data are fundamental in comprehending connections across diverse molecular layers and in fully utilizing the multi-omics resources available to make breakthroughs in biomarker and therapy discovery. Artificial intelligence (AI) and machine learning (ML) approaches are techniques further required to identify and uncover clinically relevant biomarkers and biological processes that can be targeted for therapy. To achieve this vision, the interdisciplinary collaboration of biologists, computer scientists, mathematicians, and physicians is indispensable for the task of precision medicine, that holds the promise of clinically meaningful benefits for the individual patient with IMIDs.

Thirdly, the tissue (or source) where the immune cells to be studied are located is another critical aspect to be considered. Indeed, due to sample accessibility, fewer studies have been performed on tissues rather than blood. Taking into account: (a) the diversity of IMIDs, each manifesting into different tissue of the body; and (b) the recently appreciated residency of DCs and Tregs in non-lymphoid tissues such as skin, adipose tissue, lung, bone marrow, etc., with the ability to control local inflammatory responses and to express diverse transcriptional programs compared to peripheral blood or lymphoid organs, additional multi-omics studies investigating the role and function of distinct immune cells in diverse tissues are required, in order to acquire a more holistic view of the complexity of the mechanisms governing the development of IMIDs.

Finally, taking into consideration the dynamics, rapid responses and spatial particularity of the immune system, temporal and spatial omics studies will be meaningful in providing insights into the dynamic process dictating the manifestation of IMIDs. For example, the process of antigen uptake, presentation, immunological synapsis and cell-to-cell contact in the interplay of DCs and Tregs is highly dynamic and depends on the spatial position of immune cells, stroma and other non-immune counterparts. Despite its importance in the function of an immune response and immune-mediated diseases, our current knowledge is only basic, which calls for more extensive research.

Author Contributions: D.K., N.E.P., E.N. and T.A. wrote the article and revised the manuscript. All authors have read and agreed to the published version of the manuscript.

Funding: This work was funded by the European Research Council (ERC) under the European Union's Horizon 2020 research and innovation program (agreement no. 947975) and the Hellenic Foundation for Research and Innovation (H.F.R.I.), under the "2nd Call for H.F.R.I. Research Projects to support Post-Doctoral Researchers" (Project Number: 166).

Acknowledgments: The authors would like to thank Pegy Melissa for providing technical and administrative assistance and Dora Togia for financial and administrative management. Figures were partly created using BioRender.com.

Conflicts of Interest: The authors declare no conflict of interest.

References

1. Wahren-Herlenius, M.; Dörner, T. Immunopathogenic Mechanisms of Systemic Autoimmune Disease. *Lancet* **2013**, *382*, 819–831. [CrossRef]
2. Bluestone, J.A.; Anderson, M. Tolerance in the Age of Immunotherapy. *N. Engl. J. Med.* **2020**, *383*, 1156–1166. [CrossRef] [PubMed]
3. Steinman, R.M.; Hemmi, H. Dendritic Cells: Translating Innate to Adaptive Immunity. *Curr. Top. Microbiol. Immunol.* **2006**, *311*, 17–58. [CrossRef]
4. Mellman, I.; Steinman, R.M. Dendritic Cells Specialized and Regulated Antigen Processing Machines. *Cell* **2001**, *106*, 255–258. [CrossRef]
5. Sakaguchi, S.; Yamaguchi, T.; Nomura, T.; Ono, M. Regulatory T Cells and Immune Tolerance. *Cell* **2008**, *133*, 775–787. [CrossRef] [PubMed]
6. Fontenot, J.D.; Gavin, M.A.; Rudensky, A.Y. Foxp3 Programs the Development and Function of $CD4^+CD25^+$ Regulatory T Cells. *Nat. Immunol.* **2003**, *4*, 330–336. [CrossRef] [PubMed]
7. Coutant, F.; Miossec, P. Altered Dendritic Cell Functions in Autoimmune Diseases: Distinct and Overlapping Profiles. *Nat. Rev. Rheumatol.* **2016**, *12*, 703–715. [CrossRef] [PubMed]
8. Sakaguchi, S.; Mikami, N.; Wing, J.B.; Tanaka, A.; Ichiyama, K.; Ohkura, N. Regulatory T Cells and Human Disease. *Annu. Rev. Immunol.* **2020**, *38*, 541–566. [CrossRef]
9. Dominguez-Villar, M.; Hafler, D.A. Regulatory T Cells in Autoimmune Disease. *Nat. Immunol.* **2018**, *19*, 665–673. [CrossRef]
10. Hatzioannou, A.; Boumpas, A.; Papadopoulou, M.; Papafragkos, I.; Varveri, A.; Alissafi, T.; Verginis, P. Regulatory T Cells in Autoimmunity and Cancer: A Duplicitous Lifestyle. *Front. Immunol.* **2021**, *12*, 731947. [CrossRef]
11. Lahl, K.; Loddenkemper, C.; Drouin, C.; Freyer, J.; Arnason, J.; Eberl, G.; Hamann, A.; Wagner, H.; Huehn, J.; Sparwasser, T. Selective Depletion of $Foxp3^+$ Regulatory T Cells Induces a Scurfy-like Disease. *J. Exp. Med.* **2007**, *204*, 57–63. [CrossRef] [PubMed]
12. Wing, K.; Sakaguchi, S. Regulatory T Cells Exert Checks and Balances on Self Tolerance and Autoimmunity. *Nat. Immunol.* **2010**, *11*, 7–13. [CrossRef] [PubMed]
13. Annacker, O.; Asseman, C.; Read, S.; Powrie, F. Interleukin-10 in the Regulation of T Cell-Induced Colitis. *J. Autoimmun.* **2003**, *20*, 277–279. [CrossRef]
14. Collison, L.W.; Workman, C.J.; Kuo, T.T.; Boyd, K.; Wang, Y.; Vignali, K.M.; Cross, R.; Sehy, D.; Blumberg, R.S.; Vignali, D.A. The Inhibitory Cytokine IL-35 Contributes to Regulatory T-Cell Function. *Nature* **2007**, *450*, 566–569. [CrossRef] [PubMed]
15. Green, E.A.; Gorelik, L.; McGregor, C.M.; Tran, E.H.; Flavell, R.A. $CD4^+CD25^+$ T Regulatory Cells Control Anti-Islet $CD8^+$ T Cells through TGF-β–TGF-β Receptor Interactions in Type 1 Diabetes. *Proc. Natl. Acad. Sci. USA* **2003**, *100*, 10878–10883. [CrossRef] [PubMed]
16. Nakamura, K.; Kitani, A.; Strober, W. Cell Contact–Dependent Immunosuppression by $Cd4^+Cd25^+$Regulatory T Cells Is Mediated by Cell Surface–Bound Transforming Growth Factor β. *J. Exp. Med.* **2001**, *194*, 629–644. [CrossRef]
17. Cao, X.; Cai, S.F.; Fehniger, T.A.; Song, J.; Collins, L.I.; Piwnica-Worms, D.R.; Ley, T.J. Granzyme B and Perforin Are Important for Regulatory T Cell-Mediated Suppression of Tumor Clearance. *Immunity* **2007**, *27*, 635–646. [CrossRef]
18. Grossman, W.J.; Verbsky, J.W.; Tollefsen, B.L.; Kemper, C.; Atkinson, J.P.; Ley, T.J. Differential Expression of Granzymes A and B in Human Cytotoxic Lymphocyte Subsets and T Regulatory Cells. *Blood* **2004**, *104*, 2840–2848. [CrossRef]
19. Zhao, D.M.; Thornton, A.M.; DiPaolo, R.J.; Shevach, E.M. Activated $CD4^+CD25^+$ T Cells Selectively Kill B Lymphocytes. *Blood* **2006**, *107*, 3925–3932. [CrossRef]
20. Borsellino, G.; Kleinewietfeld, M.; Mitri, D.D.; Sternjak, A.; Diamantini, A.; Giometto, R.; Höpner, S.; Centonze, D.; Bernardi, G.; Dell'Acqua, M.L.; et al. Expression of Ectonucleotidase CD39 by $Foxp3^+$ Treg Cells: Hydrolysis of Extracellular ATP and Immune Suppression. *Blood* **2007**, *110*, 1225–1232. [CrossRef]
21. Deaglio, S.; Dwyer, K.M.; Gao, W.; Friedman, D.; Usheva, A.; Erat, A.; Chen, J.-F.; Enjyoji, K.; Linden, J.; Oukka, M.; et al. Adenosine Generation Catalyzed by CD39 and CD73 Expressed on Regulatory T Cells Mediates Immune Suppression. *J. Exp. Med.* **2007**, *204*, 1257–1265. [CrossRef]
22. Kobie, J.J.; Shah, P.R.; Yang, L.; Rebhahn, J.A.; Fowell, D.J.; Mosmann, T.R. T Regulatory and Primed Uncommitted CD4 T Cells Express CD73, Which Suppresses Effector CD4 T Cells by Converting 5′-Adenosine Monophosphate to Adenosine. *J. Immunol.* **2006**, *177*, 6780–6786. [CrossRef] [PubMed]
23. Thornton, A.M.; Shevach, E.M. $CD4^+CD25^+$ Immunoregulatory T Cells Suppress Polyclonal T Cell Activation In Vitro by Inhibiting Interleukin 2 Production. *J. Exp. Med.* **1998**, *188*, 287–296. [CrossRef] [PubMed]
24. De la Rosa, M.; Rutz, S.; Dorninger, H.; Scheffold, A. Interleukin-2 Is Essential for $CD4^+CD25^+$ Regulatory T Cell Function. *Eur. J. Immunol.* **2004**, *34*, 2480–2488. [CrossRef] [PubMed]
25. Cederbom, L.; Hall, H.; Ivars, F. $CD4^+CD25^+$ Regulatory T Cells Down-regulate Co-stimulatory Molecules on Antigen-presenting Cells. *Eur. J. Immunol.* **2000**, *30*, 1538–1543. [CrossRef]
26. Serra, P.; Amrani, A.; Yamanouchi, J.; Han, B.; Thiessen, S.; Utsugi, T.; Verdaguer, J.; Santamaria, P. CD40 Ligation Releases Immature Dendritic Cells from the Control of Regulatory $CD4^+CD25^+$ T Cells. *Immunity* **2003**, *19*, 877–889. [CrossRef]
27. Misra, N.; Bayry, J.; Lacroix-Desmazes, S.; Kazatchkine, M.D.; Kaveri, S.V. Cutting Edge: Human $CD4^+CD25^+$ T Cells Restrain the Maturation and Antigen-Presenting Function of Dendritic Cells. *J. Immunol.* **2004**, *172*, 4676–4680. [CrossRef]

28. Liu, W.; Putnam, A.L.; Xu-yu, Z.; Szot, G.L.; Lee, M.R.; Zhu, S.; Gottlieb, P.A.; Kapranov, P.; Gingeras, T.R.; de St. Groth, B.F.; et al. CD127 Expression Inversely Correlates with FoxP3 and Suppressive Function of Human CD4+ T Reg Cells. *J. Exp. Med.* **2006**, *203*, 1701–1711. [CrossRef]
29. Hori, S.; Nomura, T.; Sakaguchi, S. Control of Regulatory T Cell Development by the Transcription Factor Foxp3. *Science* **2003**, *299*, 1057–1061. [CrossRef]
30. Bennett, C.L.; Christie, J.; Ramsdell, F.; Brunkow, M.E.; Ferguson, P.J.; Whitesell, L.; Kelly, T.E.; Saulsbury, F.T.; Chance, P.F.; Ochs, H.D. The Immune Dysregulation, Polyendocrinopathy, Enteropathy, X-Linked Syndrome (IPEX) Is Caused by Mutations of FOXP3. *Nat. Genet.* **2001**, *27*, 20–21. [CrossRef] [PubMed]
31. Wildin, R.S.; Ramsdell, F.; Peake, J.; Faravelli, F.; Casanova, J.L.; Buist, N.; Levy-Lahad, E.; Mazzella, M.; Goulet, O.; Perroni, L.; et al. X-Linked Neonatal Diabetes Mellitus, Enteropathy and Endocrinopathy Syndrome Is the Human Equivalent of Mouse Scurfy. *Nat. Genet.* **2001**, *27*, 18–20. [CrossRef] [PubMed]
32. Miyara, M.; Gorochov, G.; Ehrenstein, M.; Musset, L.; Sakaguchi, S.; Amoura, Z. Human FoxP3+ Regulatory T Cells in Systemic Autoimmune Diseases. *Autoimmun. Rev.* **2011**, *10*, 744–755. [CrossRef] [PubMed]
33. Josefowicz, S.Z.; Lu, L.F.; Rudensky, A.Y. Regulatory T Cells: Mechanisms of Differentiation and Function. *Annu. Rev. Immunol.* **2012**, *30*, 531–564. [CrossRef] [PubMed]
34. Kanamori, M.; Nakatsukasa, H.; Okada, M.; Lu, Q.; Yoshimura, A. Induced Regulatory T Cells: Their Development, Stability, and Applications. *Trends Immunol.* **2016**, *37*, 803–811. [CrossRef]
35. Chaudhry, A.; Rudra, D.; Treuting, P.; Samstein, R.M.; Liang, Y.; Kas, A.; Rudensky, A.Y. CD4+ Regulatory T Cells Control TH17 Responses in a Stat3-Dependent Manner. *Science* **2009**, *326*, 986–991. [CrossRef]
36. Chung, Y.; Tanaka, S.; Chu, F.; Nurieva, R.I.; Martinez, G.J.; Rawal, S.; Wang, Y.H.; Lim, H.; Reynolds, J.M.; Zhou, X.H.; et al. Follicular Regulatory T Cells Expressing Foxp3 and Bcl-6 Suppress Germinal Center Reactions. *Nat. Med.* **2011**, *17*, 983–988. [CrossRef]
37. Cretney, E.; Xin, A.; Shi, W.; Minnich, M.; Masson, F.; Miasari, M.; Belz, G.T.; Smyth, G.K.; Busslinger, M.; Nutt, S.L.; et al. The Transcription Factors Blimp-1 and IRF4 Jointly Control the Differentiation and Function of Effector Regulatory T Cells. *Nat. Immunol.* **2011**, *12*, 304–311. [CrossRef]
38. Levine, A.G.; Mendoza, A.; Hemmers, S.; Moltedo, B.; Niec, R.E.; Schizas, M.; Hoyos, B.E.; Putintseva, E.V.; Chaudhry, A.; Dikiy, S.; et al. Stability and Function of Regulatory T Cells Expressing the Transcription Factor T-Bet. *Nature* **2017**, *546*, 421–425. [CrossRef]
39. Ohnmacht, C.; Park, J.H.; Cording, S.; Wing, J.B.; Atarashi, K.; Obata, Y.; Gaboriau-Routhiau, V.; Marques, R.; Dulauroy, S.; Fedoseeva, M.; et al. The Microbiota Regulates Type 2 Immunity through RORγt+ T Cells. *Science* **2015**, *349*, 989–993. [CrossRef]
40. Sefik, E.; Geva-Zatorsky, N.; Oh, S.; Konnikova, L.; Zemmour, D.; McGuire, A.M.; Burzyn, D.; Ortiz-Lopez, A.; Lobera, M.; Yang, J.; et al. Individual Intestinal Symbionts Induce a Distinct Population of RORγ+ Regulatory T Cells. *Science* **2015**, *349*, 993–997. [CrossRef]
41. Wohlfert, E.A.; Grainger, J.R.; Bouladoux, N.; Konkel, J.E.; Oldenhove, G.; Ribeiro, C.H.; Hall, J.A.; Yagi, R.; Naik, S.; Bhairavabhotla, R.; et al. GATA3 Controls Foxp3+ Regulatory T Cell Fate during Inflammation in Mice. *J. Clin. Investig.* **2011**, *121*, 4503–4515. [CrossRef] [PubMed]
42. Kitz, A.; de Marcken, M.; Gautron, A.; Mitrovic, M.; Hafler, D.A.; Dominguez-Villar, M. AKT Isoforms Modulate Th1-like Treg Generation and Function in Human Autoimmune Disease. *EMBO Rep.* **2016**, *17*, 1169–1183. [CrossRef] [PubMed]
43. Halim, L.; Romano, M.; McGregor, R.; Correa, I.; Pavlidis, P.; Grageda, N.; Hoong, S.-J.; Yuksel, M.; Jassem, W.; Hannen, R.F.; et al. An Atlas of Human Regulatory T Helper-like Cells Reveals Features of Th2-like Tregs That Support a Tumorigenic Environment. *Cell Rep.* **2017**, *20*, 757–770. [CrossRef] [PubMed]
44. Yang, B.-H.; Hagemann, S.; Mamareli, P.; Lauer, U.; Hoffmann, U.; Beckstette, M.; Föhse, L.; Prinz, I.; Pezoldt, J.; Suerbaum, S.; et al. Foxp3+ T Cells Expressing RORγt Represent a Stable Regulatory T-Cell Effector Lineage with Enhanced Suppressive Capacity during Intestinal Inflammation. *Mucosal Immunol.* **2016**, *9*, 444–457. [CrossRef]
45. Bovenschen, H.J.; van de Kerkhof, P.C.; van Erp, P.E.; Woestenenk, R.; Joosten, I.; Koenen, H.J.P.M. Foxp3+ Regulatory T Cells of Psoriasis Patients Easily Differentiate into IL-17A-Producing Cells and Are Found in Lesional Skin. *J. Investig. Dermatol.* **2011**, *131*, 1853–1860. [CrossRef]
46. Komatsu, N.; Okamoto, K.; Sawa, S.; Nakashima, T.; Oh-hora, M.; Kodama, T.; Tanaka, S.; Bluestone, J.A.; Takayanagi, H. Pathogenic Conversion of Foxp3+ T Cells into TH17 Cells in Autoimmune Arthritis. *Nat. Med.* **2014**, *20*, 62–68. [CrossRef]
47. Jung, M.K.; Kwak, J.-E.; Shin, E.-C. IL-17A-Producing Foxp3+ Regulatory T Cells and Human Diseases. *Immune Netw.* **2017**, *17*, 276–286. [CrossRef]
48. Bailey-Bucktrout, S.L.; Martinez-Llordella, M.; Zhou, X.; Anthony, B.; Rosenthal, W.; Luche, H.; Fehling, H.J.; Bluestone, J.A. Self-Antigen-Driven Activation Induces Instability of Regulatory T Cells during an Inflammatory Autoimmune Response. *Immunity* **2013**, *39*, 949–962. [CrossRef]
49. Zhou, X.; Bailey-Bucktrout, S.; Jeker, L.T.; Bluestone, J.A. Plasticity of CD4+ FoxP3+ T Cells. *Curr. Opin. Immunol.* **2009**, *21*, 281–285. [CrossRef]
50. Van Loosdregt, J.; Vercoulen, Y.; Guichelaar, T.; Gent, Y.Y.; Beekman, J.M.; van Beekum, O.; Brenkman, A.B.; Hijnen, D.J.; Mutis, T.; Kalkhoven, E.; et al. Regulation of Treg Functionality by Acetylation-Mediated Foxp3 Protein Stabilization. *Blood* **2010**, *115*, 965–974. [CrossRef]

51. Deng, G.; Nagai, Y.; Xiao, Y.; Li, Z.; Dai, S.; Ohtani, T.; Banham, A.; Li, B.; Wu, S.-L.; Hancock, W.; et al. Pim-2 Kinase Influences Regulatory T Cell Function and Stability by Mediating Foxp3 Protein N-Terminal Phosphorylation. *J. Biol. Chem.* **2015**, *290*, 20211–20220. [CrossRef] [PubMed]
52. Barbi, J.; Pardoll, D.M.; Pan, F. Ubiquitin-Dependent Regulation of Foxp3 and Treg Function. *Immunol. Rev.* **2015**, *266*, 27–45. [CrossRef] [PubMed]
53. Panduro, M.; Benoist, C.; Mathis, D. Tissue Tregs. *Annu. Rev. Immunol.* **2016**, *34*, 609–633. [CrossRef] [PubMed]
54. Delacher, M.; Imbusch, C.D.; Weichenhan, D.; Breiling, A.; Hotz-Wagenblatt, A.; Träger, U.; Hofer, A.-C.; Kägebein, D.; Wang, Q.; Frauhammer, F.; et al. Genome-Wide DNA Methylation Landscape Defines Specialization of Regulatory T Cells in Tissues. *Nat. Immunol.* **2017**, *18*, 1160–1172. [CrossRef]
55. Muñoz-Rojas, A.R.; Mathis, D. Tissue Regulatory T Cells: Regulatory Chameleons. *Nat. Rev. Immunol.* **2021**, *21*, 597–611. [CrossRef]
56. Brusko, T.M.; Wasserfall, C.H.; Clare-Salzler, M.J.; Schatz, D.A.; Atkinson, M.A. Functional Defects and the Influence of Age on the Frequency of $CD4^{+}CD25^{+}$ T-Cells in Type 1 Diabetes. *Diabetes* **2005**, *54*, 1407–1414. [CrossRef]
57. Haseda, F.; Imagawa, A.; Murase-Mishiba, Y.; Terasaki, J.; Hanafusa, T. $CD4^{+}CD45RA-$FoxP3high Activated Regulatory T Cells Are Functionally Impaired and Related to Residual Insulin-secreting Capacity in Patients with Type 1 Diabetes. *Clin. Exp. Immunol.* **2013**, *173*, 207–216. [CrossRef]
58. Lindley, S.; Dayan, C.M.; Bishop, A.; Roep, B.O.; Peakman, M.; Tree, T.I.M. Defective Suppressor Function in $CD4^{+}CD25^{+}$ T-Cells From Patients With Type 1 Diabetes. *Diabetes* **2005**, *54*, 92–99. [CrossRef]
59. Göschl, L.; Scheinecker, C.; Bonelli, M. Treg Cells in Autoimmunity: From Identification to Treg-Based Therapies. *Semin. Immunopathol.* **2019**, *41*, 301–314. [CrossRef]
60. Alissafi, T.; Kalafati, L.; Lazari, M.; Filia, A.; Kloukina, I.; Manifava, M.; Lim, J.-H.; Alexaki, V.I.; Ktistakis, N.T.; Doskas, T.; et al. Mitochondrial Oxidative Damage Underlies Regulatory T Cell Defects in Autoimmunity. *Cell Metab.* **2020**, *32*, 591–604.e7. [CrossRef]
61. Van Roon, J.A.; Hartgring, S.A.; van der Wurff-Jacobs, K.M.; Bijlsma, J.W.; Lafeber, F.P. of $CD25^{+}Foxp3^{+}$ T Cells That Lack the IL-7 Receptor Are Increased Intra-Articularly and Have Impaired Suppressive Function in RA Patients. *Rheumatology* **2010**, *49*, 2084–2089. [CrossRef]
62. Dominguez-Villar, M.; Baecher-Allan, C.M.; Hafler, D.A. Identification of T Helper Type 1–like, $Foxp3^{+}$ Regulatory T Cells in Human Autoimmune Disease. *Nat. Med.* **2011**, *17*, 673–675. [CrossRef] [PubMed]
63. Viglietta, V.; Baecher-Allan, C.; Weiner, H.L.; Hafler, D.A. Loss of Functional Suppression by $CD4^{+}CD25^{+}$ Regulatory T Cells in Patients with Multiple Sclerosis. *J. Exp. Med.* **2004**, *199*, 971–979. [CrossRef]
64. Bonelli, M.; Savitskaya, A.; von Dalwigk, K.; Steiner, C.W.; Aletaha, D.; Smolen, J.S.; Scheinecker, C. Quantitative and Qualitative Deficiencies of Regulatory T Cells in Patients with Systemic Lupus Erythematosus (SLE). *Int. Immunol.* **2008**, *20*, 861–868. [CrossRef] [PubMed]
65. Dong, Z.; Chen, Y. Transcriptomics: Advances and Approaches. *Sci. China Life Sci.* **2013**, *56*, 960–967. [CrossRef]
66. Julé, A.M.; Hoyt, K.J.; Wei, K.; Gutierrez-Arcelus, M.; Taylor, M.L.; Ng, J.; Lederer, J.A.; Case, S.M.; Chang, M.H.; Cohen, E.M.; et al. Th1 Polarization Defines the Synovial Fluid T Cell Compartment in Oligoarticular Juvenile Idiopathic Arthritis. *JCI Insight* **2021**, *6*, e149185. [CrossRef]
67. Mijnheer, G.; Lutter, L.; Mokry, M.; van der Wal, M.; Scholman, R.; Fleskens, V.; Pandit, A.; Tao, W.; Wekking, M.; Vervoort, S.; et al. Conserved Human Effector Treg Cell Transcriptomic and Epigenetic Signature in Arthritic Joint Inflammation. *Nat. Commun.* **2021**, *12*, 2710. [CrossRef]
68. Berger, A.H.; Bratland, E.; Sjøgren, T.; Heimli, M.; Tyssedal, T.; Bruserud, Ø.; Johansson, S.; Husebye, E.S.; Oftedal, B.E.; Wolff, A.S.B. Transcriptional Changes in Regulatory T Cells From Patients With Autoimmune Polyendocrine Syndrome Type 1 Suggest Functional Impairment of Lipid Metabolism and Gut Homing. *Front. Immunol.* **2021**, *12*, 722860. [CrossRef]
69. Simone, D.; Penkava, F.; Ridley, A.; Sansom, S.; Al-Mossawi, M.H.; Bowness, P. Single Cell Analysis of Spondyloarthritis Regulatory T Cells Identifies Distinct Synovial Gene Expression Patterns and Clonal Fates. *Commun. Biol.* **2021**, *4*, 1395. [CrossRef] [PubMed]
70. Grigoriou, M.; Banos, A.; Hatzioannou, A.; Kloetgen, A.; Kouzis, P.; Aggouraki, D.; Zakopoulou, R.; Bamias, G.; Kassi, E.; Mavroudis, D.; et al. Regulatory T-Cell Transcriptomic Reprogramming Characterizes Adverse Events by Checkpoint Inhibitors in Solid Tumors. *Cancer Immunol. Res.* **2021**, *9*, 726–734. [CrossRef]
71. Luo, Y.; Xu, C.; Wang, B.; Niu, Q.; Su, X.; Bai, Y.; Zhu, S.; Zhao, C.; Sun, Y.; Wang, J.; et al. Single-Cell Transcriptomic Analysis Reveals Disparate Effector Differentiation Pathways in Human Treg Compartment. *Nat. Commun.* **2021**, *12*, 3913. [CrossRef] [PubMed]
72. Marek-Trzonkowska, N.; Myśliwiec, M.; Dobyszuk, A.; Grabowska, M.; Techmańska, I.; Juścińska, J.; Wujtewicz, M.A.; Witkowski, P.; Młynarski, W.; Balcerska, A.; et al. Administration of $CD4^{+}$CD25highCD127− Regulatory T Cells Preserves β-Cell Function in Type 1 Diabetes in Children. *Diabetes Care* **2012**, *35*, 1817–1820. [CrossRef] [PubMed]
73. Bluestone, J.A.; Buckner, J.H.; Fitch, M.; Gitelman, S.E.; Gupta, S.; Hellerstein, M.K.; Herold, K.C.; Lares, A.; Lee, M.R.; Li, K.; et al. Type 1 Diabetes Immunotherapy Using Polyclonal Regulatory T Cells. *Sci. Transl. Med.* **2015**, *7*, 315ra189. [CrossRef] [PubMed]
74. Green, E.A.; Choi, Y.; Flavell, R.A. Pancreatic Lymph Node-Derived $CD4^{+}CD25^{+}$ Treg Cells Highly Potent Regulators of Diabetes That Require TRANCE-RANK Signals. *Immunity* **2002**, *16*, 183–191. [CrossRef]

75. Tang, Q.; Henriksen, K.J.; Bi, M.; Finger, E.B.; Szot, G.; Ye, J.; Masteller, E.L.; McDevitt, H.; Bonyhadi, M.; Bluestone, J.A. In Vitro–Expanded Antigen-Specific Regulatory T Cells Suppress Autoimmune Diabetes. *J. Exp. Med.* **2004**, *199*, 1455–1465. [CrossRef] [PubMed]
76. Kim, Y.C.; Zhang, A.-H.; Su, Y.; Rieder, S.A.; Rossi, R.J.; Ettinger, R.A.; Pratt, K.P.; Shevach, E.M.; Scott, D.W. Engineered Antigen-Specific Human Regulatory T Cells: Immunosuppression of FVIII-Specific T- and B-Cell Responses. *Blood* **2015**, *125*, 1107–1115. [CrossRef]
77. Kim, Y.C.; Zhang, A.-H.; Yoon, J.; Culp, W.E.; Lees, J.R.; Wucherpfennig, K.W.; Scott, D.W. Engineered MBP-Specific Human Tregs Ameliorate MOG-Induced EAE through IL-2-Triggered Inhibition of Effector T Cells. *J. Autoimmun.* **2018**, *92*, 77–86. [CrossRef]
78. Spence, A.; Purtha, W.; Tam, J.; Dong, S.; Kim, Y.; Ju, C.-H.; Sterling, T.; Nakayama, M.; Robinson, W.H.; Bluestone, J.A.; et al. Revealing the Specificity of Regulatory T Cells in Murine Autoimmune Diabetes. *Proc. Natl. Acad. Sci. USA* **2018**, *115*, 5265–5270. [CrossRef]
79. Rossetti, M.; Spreafico, R.; Consolaro, A.; Leong, J.Y.; Chua, C.; Massa, M.; Saidin, S.; Magni-Manzoni, S.; Arkachaisri, T.; Wallace, C.A.; et al. TCR Repertoire Sequencing Identifies Synovial Treg Cell Clonotypes in the Bloodstream during Active Inflammation in Human Arthritis. *Ann. Rheum. Dis.* **2017**, *76*, 435–441. [CrossRef]
80. Heintzman, N.D.; Hon, G.C.; Hawkins, R.D.; Kheradpour, P.; Stark, A.; Harp, L.F.; Ye, Z.; Lee, L.K.; Stuart, R.K.; Ching, C.W.; et al. Histone Modifications at Human Enhancers Reflect Global Cell-Type-Specific Gene Expression. *Nature* **2009**, *459*, 108–112. [CrossRef]
81. Gao, P.; Uzun, Y.; He, B.; Salamati, S.E.; Coffey, J.; Tsalikian, E.; Tan, K. Risk Variants Disrupting Enhancers of TH1 and TREG Cells in Type 1 Diabetes. *Proc. Natl. Acad. Sci. USA* **2019**, *116*, 7581–7590. [CrossRef] [PubMed]
82. Amariuta, T.; Luo, Y.; Gazal, S.; Davenport, E.E.; van deGeijn, B.; Ishigaki, K.; Westra, H.-J.; Teslovich, N.; Okada, Y.; Yamamoto, K.; et al. IMPACT: Genomic Annotation of Cell-State-Specific Regulatory Elements Inferred from the Epigenome of Bound Transcription Factors. *Am. J. Hum. Genet.* **2019**, *104*, 879–895. [CrossRef] [PubMed]
83. Bossini-Castillo, L.; Glinos, D.A.; Kunowska, N.; Golda, G.; Lamikanra, A.A.; Spitzer, M.; Soskic, B.; Cano-Gamez, E.; Smyth, D.J.; Cattermole, C.; et al. Immune Disease Variants Modulate Gene Expression in Regulatory $CD4^+$ T Cells. *Cell Genom.* **2022**, *2*, 100117. [CrossRef]
84. Trynka, G.; Sandor, C.; Han, B.; Xu, H.; Stranger, B.E.; Liu, X.S.; Raychaudhuri, S. Chromatin Marks Identify Critical Cell Types for Fine Mapping Complex Trait Variants. *Nat. Genet.* **2013**, *45*, 124–130. [CrossRef] [PubMed]
85. Huang, H.; Fang, M.; Jostins, L.; Umićević Mirkov, M.; Boucher, G.; Anderson, C.A.; Andersen, V.; Cleynen, I.; Cortes, A.; Crins, F.; et al. Fine-Mapping Inflammatory Bowel Disease Loci to Single-Variant Resolution. *Nature* **2017**, *547*, 173–178. [CrossRef] [PubMed]
86. Ohkura, N.; Yasumizu, Y.; Kitagawa, Y.; Tanaka, A.; Nakamura, Y.; Motooka, D.; Nakamura, S.; Okada, Y.; Sakaguchi, S. Regulatory T Cell-Specific Epigenomic Region Variants Are a Key Determinant of Susceptibility to Common Autoimmune Diseases. *Immunity* **2020**, *52*, 1119–1132.e4. [CrossRef]
87. Weerakoon, H.; Straube, J.; Lineburg, K.; Cooper, L.; Lane, S.; Smith, C.; Alabbas, S.; Begun, J.; Miles, J.J.; Hill, M.M.; et al. Expression of CD49f Defines Subsets of Human Regulatory T Cells with Divergent Transcriptional Landscape and Function That Correlate with Ulcerative Colitis Disease Activity. *Clin. Transl. Immunol.* **2021**, *10*, e1334. [CrossRef]
88. Chen, B.; Sun, L.; Zhang, X. Integration of Microbiome and Epigenome to Decipher the Pathogenesis of Autoimmune Diseases. *J. Autoimmun.* **2017**, *83*, 31–42. [CrossRef]
89. Hevia, A.; Milani, C.; López, P.; Cuervo, A.; Arboleya, S.; Duranti, S.; Turroni, F.; González, S.; Suárez, A.; Gueimonde, M.; et al. Intestinal Dysbiosis Associated with Systemic Lupus Erythematosus. *Mbio* **2014**, *5*, e01548-14. [CrossRef]
90. López, P.; de Paz, B.; Rodríguez-Carrio, J.; Hevia, A.; Sánchez, B.; Margolles, A.; Suárez, A. Th17 Responses and Natural IgM Antibodies Are Related to Gut Microbiota Composition in Systemic Lupus Erythematosus Patients. *Sci. Rep.* **2016**, *6*, 24072. [CrossRef]
91. Liu, X.; Zou, Q.; Zeng, B.; Fang, Y.; Wei, H. Analysis of Fecal Lactobacillus Community Structure in Patients with Early Rheumatoid Arthritis. *Curr. Microbiol.* **2013**, *67*, 170–176. [CrossRef] [PubMed]
92. Maeda, Y.; Kurakawa, T.; Umemoto, E.; Motooka, D.; Ito, Y.; Gotoh, K.; Hirota, K.; Matsushita, M.; Furuta, Y.; Narazaki, M.; et al. Dysbiosis Contributes to Arthritis Development via Activation of Autoreactive T Cells in the Intestine. *Arthritis Rheumatol.* **2016**, *68*, 2646–2661. [CrossRef] [PubMed]
93. Manichanh, C.; Rigottier-Gois, L.; Bonnaud, E.; Gloux, K.; Pelletier, E.; Frangeul, L.; Nalin, R.; Jarrin, C.; Chardon, P.; Marteau, P.; et al. Reduced Diversity of Faecal Microbiota in Crohn's Disease Revealed by a Metagenomic Approach. *Gut* **2006**, *55*, 205. [CrossRef] [PubMed]
94. Frank, D.N.; St. Amand, A.L.; Feldman, R.A.; Boedeker, E.C.; Harpaz, N.; Pace, N.R. Molecular-Phylogenetic Characterization of Microbial Community Imbalances in Human Inflammatory Bowel Diseases. *Proc. Natl. Acad. Sci. USA* **2007**, *104*, 13780–13785. [CrossRef] [PubMed]
95. Murri, M.; Leiva, I.; Gomez-Zumaquero, J.M.; Tinahones, F.J.; Cardona, F.; Soriguer, F.; Queipo-Ortuño, M.I. Gut Microbiota in Children with Type 1 Diabetes Differs from That in Healthy Children: A Case-Control Study. *BMC Med.* **2013**, *11*, 46. [CrossRef]
96. Su, X.; Yin, X.; Liu, Y.; Yan, X.; Zhang, S.; Wang, X.; Lin, Z.; Zhou, X.; Gao, J.; Wang, Z.; et al. Gut Dysbiosis Contributes to the Imbalance of Treg and Th17 Cells in Graves' Disease Patients by Propionic Acid. *J. Clin. Endocrinol. Metab.* **2020**, *105*, 3526–3547. [CrossRef]

97. Miyake, S.; Kim, S.; Suda, W.; Oshima, K.; Nakamura, M.; Matsuoka, T.; Chihara, N.; Tomita, A.; Sato, W.; Kim, S.-W.; et al. Dysbiosis in the Gut Microbiota of Patients with Multiple Sclerosis, with a Striking Depletion of Species Belonging to Clostridia XIVa and IV Clusters. *PLoS ONE* **2015**, *10*, e0137429. [CrossRef]
98. Duscha, A.; Gisevius, B.; Hirschberg, S.; Yissachar, N.; Stangl, G.I.; Eilers, E.; Bader, V.; Haase, S.; Kaisler, J.; David, C.; et al. Propionic Acid Shapes the Multiple Sclerosis Disease Course by an Immunomodulatory Mechanism. *Cell* **2020**, *180*, 1067–1080.e16. [CrossRef]
99. Furusawa, Y.; Obata, Y.; Fukuda, S.; Endo, T.A.; Nakato, G.; Takahashi, D.; Nakanishi, Y.; Uetake, C.; Kato, K.; Kato, T.; et al. Commensal Microbe-Derived Butyrate Induces the Differentiation of Colonic Regulatory T Cells. *Nature* **2013**, *504*, 446–450. [CrossRef]
100. Mariño, E.; Richards, J.L.; McLeod, K.H.; Stanley, D.; Yap, Y.A.; Knight, J.; McKenzie, C.; Kranich, J.; Oliveira, A.C.; Rossello, F.J.; et al. Gut Microbial Metabolites Limit the Frequency of Autoimmune T Cells and Protect against Type 1 Diabetes. *Nat. Immunol.* **2017**, *18*, 552–562. [CrossRef]
101. Steinman, R.M.; Cohn, Z.A. Identification of a Novel Cell Type in Peripheral Lymphoid Organs of Mice. I. Morphology, Quantitation, Tissue Distribution. *J. Exp. Med.* **1973**, *137*, 1142–1162. [CrossRef] [PubMed]
102. Durai, V.; Murphy, K.M. Functions of Murine Dendritic Cells. *Immunity* **2016**, *45*, 719–736. [CrossRef] [PubMed]
103. Iberg, C.A.; Jones, A.; Hawiger, D. Dendritic Cells As Inducers of Peripheral Tolerance. *Trends Immunol.* **2017**, *38*, 793–804. [CrossRef] [PubMed]
104. Hilligan, K.L.; Ronchese, F. Antigen Presentation by Dendritic Cells and Their Instruction of CD4$^+$ T Helper Cell Responses. *Cell. Mol. Immunol.* **2020**, *17*, 587–599. [CrossRef]
105. Cabeza-Cabrerizo, M.; Cardoso, A.; Minutti, C.M.; da Costa, M.P.; Reis e Sousa, C. Dendritic Cells Revisited. *Annu. Rev. Immunol.* **2021**, *39*, 131–166. [CrossRef] [PubMed]
106. Guilliams, M.; Dutertre, C.-A.; Scott, C.L.; McGovern, N.; Sichien, D.; Chakarov, S.; Gassen, S.V.; Chen, J.; Poidinger, M.; Prijck, S.D.; et al. Unsupervised High-Dimensional Analysis Aligns Dendritic Cells across Tissues and Species. *Immunity* **2016**, *45*, 669–684. [CrossRef] [PubMed]
107. Patente, T.A.; Pinho, M.P.; Oliveira, A.A.; Evangelista, G.C.M.; Bergami-Santos, P.C.; Barbuto, J.A.M. Human Dendritic Cells: Their Heterogeneity and Clinical Application Potential in Cancer Immunotherapy. *Front. Immunol.* **2019**, *9*, 3176. [CrossRef]
108. Reizis, B. Plasmacytoid Dendritic Cells: Development, Regulation, and Function. *Immunity* **2019**, *50*, 37–50. [CrossRef]
109. See, P.; Dutertre, C.-A.; Chen, J.; Günther, P.; McGovern, N.; Irac, S.E.; Gunawan, M.; Beyer, M.; Händler, K.; Duan, K.; et al. Mapping the Human DC Lineage through the Integration of High-Dimensional Techniques. *Science* **2017**, *356*, eaag3009. [CrossRef]
110. Matsui, T.; Connolly, J.E.; Michnevitz, M.; Chaussabel, D.; Yu, C.I.; Glaser, C.; Tindle, S.; Pypaert, M.; Freitas, H.; Piqueras, B.; et al. CD2 Distinguishes Two Subsets of Human Plasmacytoid Dendritic Cells with Distinct Phenotype and Functions. *J. Immunol.* **2009**, *182*, 6815. [CrossRef]
111. Villadangos, J.A.; Young, L. Antigen-Presentation Properties of Plasmacytoid Dendritic Cells. *Immunity* **2008**, *29*, 352–361. [CrossRef] [PubMed]
112. Hoeffel, G.; Ripoche, A.C.; Matheoud, D.; Nascimbeni, M.; Escriou, N.; Lebon, P.; Heshmati, F.; Guillet, J.G.; Gannagé, M.; Caillat-Zucman, S.; et al. Antigen Crosspresentation by Human Plasmacytoid Dendritic Cells. *Immunity* **2007**, *27*, 481–492. [CrossRef]
113. Rodrigues, P.F.; Alberti-Servera, L.; Eremin, A.; Grajales-Reyes, G.E.; Ivanek, R.; Tussiwand, R. Distinct Progenitor Lineages Contribute to the Heterogeneity of Plasmacytoid Dendritic Cells. *Nat. Immunol.* **2018**, *19*, 711–722. [CrossRef] [PubMed]
114. Dress, R.J.; Dutertre, C.-A.; Giladi, A.; Schlitzer, A.; Low, I.; Shadan, N.B.; Tay, A.; Lum, J.; Kairi, M.F.B.M.; Hwang, Y.Y.; et al. Plasmacytoid Dendritic Cells Develop from Ly6D$^+$ Lymphoid Progenitors Distinct from the Myeloid Lineage. *Nat. Immunol.* **2019**, *20*, 852–864. [CrossRef] [PubMed]
115. Feng, J.; Pucella, J.N.; Jang, G.; Alcántara-Hernández, M.; Upadhaya, S.; Adams, N.M.; Khodadadi-Jamayran, A.; Lau, C.M.; Stoeckius, M.; Hao, S.; et al. Clonal Lineage Tracing Reveals Shared Origin of Conventional and Plasmacytoid Dendritic Cells. *Immunity* **2022**, *55*, 405–422.e11. [CrossRef]
116. Mair, F.; Liechti, T. Comprehensive Phenotyping of Human Dendritic Cells and Monocytes. *Cytom. Part A* **2021**, *99*, 231–242. [CrossRef] [PubMed]
117. Poulin, L.F.; Reyal, Y.; Uronen-Hansson, H.; Schraml, B.U.; Sancho, D.; Murphy, K.M.; Hakansson, U.K.; Moita, L.F.; Agace, W.W.; Bonnet, D.; et al. DNGR-1 Is a Specific and Universal Marker of Mouse and Human Batf3-Dependent Dendritic Cells in Lymphoid and Nonlymphoid Tissues. *Blood* **2012**, *119*, 6052–6062. [CrossRef]
118. Sichien, D.; Scott, C.L.; Martens, L.; Vanderkerken, M.; Gassen, S.V.; Plantinga, M.; Joeris, T.; Prijck, S.D.; Vanhoutte, L.; Vanheerswynghels, M.; et al. IRF8 Transcription Factor Controls Survival and Function of Terminally Differentiated Conventional and Plasmacytoid Dendritic Cells, Respectively. *Immunity* **2016**, *45*, 626–640. [CrossRef]
119. Hildner, K.; Edelson, B.T.; Purtha, W.E.; Diamond, M.; Matsushita, H.; Kohyama, M.; Calderon, B.; Schraml, B.U.; Unanue, E.R.; Diamond, M.S.; et al. Batf3 Deficiency Reveals a Critical Role for CD8α^+ Dendritic Cells in Cytotoxic T Cell Immunity. *Science* **2008**, *322*, 1097–1100. [CrossRef]
120. Hacker, C.; Kirsch, R.D.; Ju, X.S.; Hieronymus, T.; Gust, T.C.; Kuhl, C.; Jorgas, T.; Kurz, S.M.; Rose-John, S.; Yokota, Y.; et al. Transcriptional Profiling Identifies Id2 Function in Dendritic Cell Development. *Nat. Immunol.* **2003**, *4*, 380–386. [CrossRef]

121. Kashiwada, M.; Pham, N.L.; Pewe, L.L.; Harty, J.T.; Rothman, P.B. NFIL3/E4BP4 Is a Key Transcription Factor for CD8α+ Dendritic Cell Development. *Blood* **2011**, *117*, 6193–6197. [CrossRef] [PubMed]
122. Liu, T.-T.; Kim, S.; Desai, P.; Kim, D.-H.; Huang, X.; Ferris, S.T.; Wu, R.; Ou, F.; Egawa, T.; Dyken, S.J.V.; et al. Ablation of cDC2 Development by Triple Mutations within the Zeb2 Enhancer. *Nature* **2022**, *607*, 142–148. [CrossRef] [PubMed]
123. Reis e Sousa, C.; Hieny, S.; Scharton-Kersten, T.; Jankovic, D.; Charest, H.; Germain, R.N.; Sher, A. In Vivo Microbial Stimulation Induces Rapid CD40 Ligand-Independent Production of Interleukin 12 by Dendritic Cells and Their Redistribution to T Cell Areas. *J. Exp. Med.* **1997**, *186*, 1819–1829. [CrossRef] [PubMed]
124. Mashayekhi, M.; Sandau, M.M.; Dunay, I.R.; Frickel, E.M.; Khan, A.; Goldszmid, R.S.; Sher, A.; Ploegh, H.L.; Murphy, T.L.; Sibley, L.D.; et al. CD8α+ Dendritic Cells Are the Critical Source of Interleukin-12 That Controls Acute Infection by Toxoplasma Gondii Tachyzoites. *Immunity* **2011**, *35*, 249–259. [CrossRef]
125. Idoyaga, J.; Fiorese, C.; Zbytnuik, L.; Lubkin, A.; Miller, J.; Malissen, B.; Mucida, D.; Merad, M.; Steinman, R.M. Specialized Role of Migratory Dendritic Cells in Peripheral Tolerance Induction. *J. Clin. Investig.* **2013**, *123*, 844–854. [CrossRef]
126. Tabansky, I.; Keskin, D.B.; Watts, D.; Petzold, C.; Funaro, M.; Sands, W.; Wright, P.; Yunis, E.J.; Najjar, S.; Diamond, B.; et al. Targeting DEC-205-DCIR2+ Dendritic Cells Promotes Immunological Tolerance in Proteolipid Protein-Induced Experimental Autoimmune Encephalomyelitis. *Mol. Med.* **2018**, *24*, 17. [CrossRef]
127. Bosteels, C.; Scott, C.L. Transcriptional Regulation of DC Fate Specification. *Mol. Immunol.* **2020**, *121*, 38–46. [CrossRef]
128. Dudziak, D.; Kamphorst, A.O.; Heidkamp, G.F.; Buchholz, V.R.; Trumpfheller, C.; Yamazaki, S.; Cheong, C.; Liu, K.; Lee, H.W.; Park, C.G.; et al. Differential Antigen Processing by Dendritic Cell Subsets in Vivo. *Science* **2007**, *315*, 107–111. [CrossRef]
129. Shin, C.; Han, J.-A.; Choi, B.; Cho, Y.-K.; Do, Y.; Ryu, S. Intrinsic Features of the CD8α− Dendritic Cell Subset in Inducing Functional T Follicular Helper Cells. *Immunol. Lett.* **2016**, *172*, 21–28. [CrossRef]
130. Persson, E.K.; Uronen-Hansson, H.; Semmrich, M.; Rivollier, A.; Hägerbrand, K.; Marsal, J.; Gudjonsson, S.; Håkansson, U.; Reizis, B.; Kotarsky, K.; et al. IRF4 Transcription-Factor-Dependent CD103+CD11b+ Dendritic Cells Drive Mucosal T Helper 17 Cell Differentiation. *Immunity* **2013**, *38*, 958–969. [CrossRef]
131. Schlitzer, A.; McGovern, N.; Teo, P.; Zelante, T.; Atarashi, K.; Low, D.; Ho, A.W.S.; See, P.; Shin, A.; Wasan, P.S.; et al. IRF4 Transcription Factor-Dependent CD11b+ Dendritic Cells in Human and Mouse Control Mucosal IL-17 Cytokine Responses. *Immunity* **2013**, *38*, 970–983. [CrossRef] [PubMed]
132. Tussiwand, R.; Everts, B.; Grajales-Reyes, G.E.; Kretzer, N.M.; Iwata, A.; Bagaitkar, J.; Wu, X.; Wong, R.; Anderson, D.A.; Murphy, T.L.; et al. Klf4 Expression in Conventional Dendritic Cells Is Required for T Helper 2 Cell Responses. *Immunity* **2015**, *42*, 916–928. [CrossRef] [PubMed]
133. Rojas, I.M.L.; Mok, W.-H.; Pearson, F.E.; Minoda, Y.; Kenna, T.J.; Barnard, R.T.; Radford, K.J. Human Blood CD1c+ Dendritic Cells Promote Th1 and Th17 Effector Function in Memory CD4+ T Cells. *Front. Immunol.* **2017**, *8*, 971. [CrossRef] [PubMed]
134. Binnewies, M.; Mujal, A.M.; Pollack, J.L.; Combes, A.J.; Hardison, E.A.; Barry, K.C.; Tsui, J.; Ruhland, M.K.; Kersten, K.; Abushawish, M.A.; et al. Unleashing Type-2 Dendritic Cells to Drive Protective Antitumor CD4+ T Cell Immunity. *Cell* **2019**, *177*, 556–571.e16. [CrossRef]
135. Mayer, J.U.; Hilligan, K.L.; Chandler, J.S.; Eccles, D.A.; Old, S.I.; Domingues, R.G.; Yang, J.; Webb, G.R.; Munoz-Erazo, L.; Hyde, E.J.; et al. Homeostatic IL-13 in Healthy Skin Directs Dendritic Cell Differentiation to Promote TH2 and Inhibit TH17 Cell Polarization. *Nat. Immunol.* **2021**, *22*, 1538–1550. [CrossRef]
136. Bamboat, Z.M.; Stableford, J.A.; Plitas, G.; Burt, B.M.; Nguyen, H.M.; Welles, A.P.; Gonen, M.; Young, J.W.; DeMatteo, R.P. Human Liver Dendritic Cells Promote T Cell Hyporesponsiveness. *J. Immunol.* **2009**, *182*, 1901. [CrossRef]
137. Hu, Z.; Li, Y.; Nieuwenhuijze, A.V.; Selden, H.J.; Jarrett, A.M.; Sorace, A.G.; Yankeelov, T.E.; Liston, A.; Ehrlich, L.I.R. CCR7 Modulates the Generation of Thymic Regulatory T Cells by Altering the Composition of the Thymic Dendritic Cell Compartment. *Cell Rep.* **2017**, *21*, 168–180. [CrossRef]
138. Brown, C.C.; Gudjonson, H.; Pritykin, Y.; Deep, D.; Lavallée, V.-P.; Mendoza, A.; Fromme, R.; Mazutis, L.; Ariyan, C.; Leslie, C.; et al. Transcriptional Basis of Mouse and Human Dendritic Cell Heterogeneity. *Cell* **2019**, *179*, 846–863.e24. [CrossRef]
139. Lewis, K.L.; Caton, M.L.; Bogunovic, M.; Greter, M.; Grajkowska, L.T.; Ng, D.; Klinakis, A.; Charo, I.F.; Jung, S.; Gommerman, J.L.; et al. Notch2 Receptor Signaling Controls Functional Differentiation of Dendritic Cells in the Spleen and Intestine. *Immunity* **2011**, *35*, 780–791. [CrossRef]
140. Bosteels, C.; Neyt, K.; Vanheerswynghels, M.; Helden, M.J.V.; Sichien, D.; Debeuf, N.; Prijck, S.D.; Bosteels, V.; Vandamme, N.; Martens, L.; et al. Inflammatory Type 2 cDCs Acquire Features of cDC1s and Macrophages to Orchestrate Immunity to Respiratory Virus Infection. *Immunity* **2020**, *52*, 1039–1056.e9. [CrossRef]
141. Leylek, R.; Alcántara-Hernández, M.; Lanzar, Z.; Lüdtke, A.; Perez, O.A.; Reizis, B.; Idoyaga, J. Integrated Cross-Species Analysis Identifies a Conserved Transitional Dendritic Cell Population. *Cell Rep.* **2019**, *29*, 3736–3750.e8. [CrossRef] [PubMed]
142. Passeri, L.; Marta, F.; Bassi, V.; Gregori, S. Tolerogenic Dendritic Cell-Based Approaches in Autoimmunity. *Int. J. Mol. Sci.* **2021**, *22*, 8415. [CrossRef] [PubMed]
143. Schlitzer, A.; McGovern, N.; Ginhoux, F. Dendritic Cells and Monocyte-Derived Cells: Two Complementary and Integrated Functional Systems. *Semin. Cell Dev. Biol.* **2015**, *41*, 9–22. [CrossRef] [PubMed]
144. Serbina, N.V.; Salazar-Mather, T.P.; Biron, C.A.; Kuziel, W.A.; Pamer, E.G. TNF/iNOS-Producing Dendritic Cells Mediate Innate Immune Defense against Bacterial Infection. *Immunity* **2003**, *19*, 59–70. [CrossRef]

145. León, B.; López-Bravo, M.; Ardavín, C. Monocyte-Derived Dendritic Cells Formed at the Infection Site Control the Induction of Protective T Helper 1 Responses against Leishmania. *Immunity* **2007**, *26*, 519–531. [CrossRef]
146. Villani, A.-C.; Satija, R.; Reynolds, G.; Sarkizova, S.; Shekhar, K.; Fletcher, J.; Griesbeck, M.; Butler, A.; Zheng, S.; Lazo, S.; et al. Single-Cell RNA-Seq Reveals New Types of Human Blood Dendritic Cells, Monocytes, and Progenitors. *Science* **2017**, *356*, eaah4573. [CrossRef]
147. Dutertre, C.-A.; Becht, E.; Irac, S.E.; Khalilnezhad, A.; Narang, V.; Khalilnezhad, S.; Ng, P.Y.; van den Hoogen, L.L.; Leong, J.Y.; Lee, B.; et al. Single-Cell Analysis of Human Mononuclear Phagocytes Reveals Subset-Defining Markers and Identifies Circulating Inflammatory Dendritic Cells. *Immunity* **2019**, *51*, 573–589.e8. [CrossRef]
148. Bourdely, P.; Anselmi, G.; Vaivode, K.; Ramos, R.N.; Missolo-Koussou, Y.; Hidalgo, S.; Tosselo, J.; Nuñez, N.; Richer, W.; Vincent-Salomon, A.; et al. Transcriptional and Functional Analysis of $CD1c^+$ Human Dendritic Cells Identifies a $CD163^+$ Subset Priming $CD8^+CD103^+$ T Cells. *Immunity* **2020**, *53*, 335–352.e8. [CrossRef]
149. Nehar-Belaid, D.; Hong, S.; Marches, R.; Chen, G.; Bolisetty, M.; Baisch, J.; Walters, L.; Punaro, M.; Rossi, R.J.; Chung, C.-H.; et al. Mapping Systemic Lupus Erythematosus Heterogeneity at the Single-Cell Level. *Nat. Immunol.* **2020**, *21*, 1094–1106. [CrossRef]
150. Hjorton, K.; Hagberg, N.; Pucholt, P.; Eloranta, M.-L.; Rönnblom, L. The Regulation and Pharmacological Modulation of Immune Complex Induced Type III IFN Production by Plasmacytoid Dendritic Cells. *Arthritis Res. Ther.* **2020**, *22*, 130. [CrossRef]
151. Liu, K.; Victora, G.D.; Schwickert, T.A.; Guermonprez, P.; Meredith, M.M.; Yao, K.; Chu, F.-F.; Randolph, G.J.; Rudensky, A.Y.; Nussenzweig, M. In Vivo Analysis of Dendritic Cell Development and Homeostasis. *Science* **2009**, *324*, 392–397. [CrossRef] [PubMed]
152. Granot, T.; Senda, T.; Carpenter, D.J.; Matsuoka, N.; Weiner, J.; Gordon, C.L.; Miron, M.; Kumar, B.V.; Griesemer, A.; Ho, S.-H.; et al. Dendritic Cells Display Subset and Tissue-Specific Maturation Dynamics over Human Life. *Immunity* **2017**, *46*, 504–515. [CrossRef] [PubMed]
153. Canavan, M.; Walsh, A.M.; Bhargava, V.; Wade, S.M.; McGarry, T.; Marzaioli, V.; Moran, B.; Biniecka, M.; Convery, H.; Wade, S.; et al. Enriched $Cd141^+$ DCs in the Joint Are Transcriptionally Distinct, Activated, and Contribute to Joint Pathogenesis. *JCI Insight* **2018**, *3*, e95228. [CrossRef] [PubMed]
154. Canavan, M.; Marzaioli, V.; McGarry, T.; Bhargava, V.; Nagpal, S.; Veale, D.J.; Fearon, U. Rheumatoid Arthritis Synovial Microenvironment Induces Metabolic and Functional Adaptations in Dendritic Cells. *Clin. Exp. Immunol.* **2020**, *202*, 226–238. [CrossRef]
155. He, H.; Suryawanshi, H.; Morozov, P.; Gay-Mimbrera, J.; Duca, E.D.; Kim, H.J.; Kameyama, N.; Estrada, Y.; Der, E.; Krueger, J.G.; et al. Single-Cell Transcriptome Analysis of Human Skin Identifies Novel Fibroblast Subpopulation and Enrichment of Immune Subsets in Atopic Dermatitis. *J. Allergy Clin. Immun.* **2020**, *145*, 1615–1628. [CrossRef]
156. Rojahn, T.B.; Vorstandlechner, V.; Krausgruber, T.; Bauer, W.M.; Alkon, N.; Bangert, C.; Thaler, F.M.; Sadeghyar, F.; Fortelny, N.; Gernedl, V.; et al. Single-Cell Transcriptomics Combined with Interstitial Fluid Proteomics Defines Cell Type–Specific Immune Regulation in Atopic Dermatitis. *J. Allergy Clin. Immun.* **2020**, *146*, 1056–1069. [CrossRef]
157. Renert-Yuval, Y.; Thyssen, J.P.; Bissonnette, R.; Bieber, T.; Kabashima, K.; Hijnen, D.; Guttman-Yassky, E. Biomarkers in Atopic Dermatitis—A Review on Behalf of the International Eczema Council. *J. Allergy Clin. Immun.* **2021**, *147*, 1174–1190.e1. [CrossRef]
158. Kim, J.; Lee, J.; Kim, H.J.; Kameyama, N.; Nazarian, R.; Der, E.; Cohen, S.; Guttman-Yassky, E.; Putterman, C.; Krueger, J.G. Single-Cell Transcriptomics Applied to Emigrating Cells from Psoriasis Elucidate Pathogenic versus Regulatory Immune Cell Subsets. *J. Allergy Clin. Immun.* **2021**, *148*, 1281–1292. [CrossRef]
159. Van Lummel, M.; van Veelen, P.A.; de Ru, A.H.; Janssen, G.M.C.; Pool, J.; Laban, S.; Joosten, A.M.; Nikolic, T.; Drijfhout, J.W.; Mearin, M.L.; et al. Dendritic Cells Guide Islet Autoimmunity through a Restricted and Uniquely Processed Peptidome Presented by High-Risk HLA-DR. *J. Immunol.* **2016**, *196*, 3253–3263. [CrossRef]
160. Ottria, A.; Hoekstra, A.T.; Zimmermann, M.; van der Kroef, M.; Vazirpanah, N.; Cossu, M.; Chouri, E.; Rossato, M.; Beretta, L.; Tieland, R.G.; et al. Fatty Acid and Carnitine Metabolism Are Dysregulated in Systemic Sclerosis Patients. *Front. Immunol.* **2020**, *11*, 822. [CrossRef]

biomedicines

Review

Contribution of the Environment, Epigenetic Mechanisms and Non-Coding RNAs in Psoriasis

Charalabos Antonatos [1], Katerina Grafanaki [2], Paschalia Asmenoudi [1], Panagiotis Xiropotamos [1], Paraskevi Nani [1], Georgios K. Georgakilas [1,3], Sophia Georgiou [2] and Yiannis Vasilopoulos [1,*]

1 Laboratory of Genetics, Section of Genetics, Cell Biology and Development, Department of Biology, University of Patras, 26504 Patras, Greece
2 Department of Dermatology, School of Medicine, University Hospital of Patras, University of Patras, 26504 Patras, Greece
3 Laboratory of Hygiene and Epidemiology, Department of Clinical and Laboratory Research, Faculty of Medicine, University of Thessaly, 38334 Volos, Greece
* Correspondence: iovasilop@upatras.gr

Abstract: Despite the increasing research and clinical interest in the predisposition of psoriasis, a chronic inflammatory skin disease, the multitude of genetic and environmental factors involved in its pathogenesis remain unclear. This complexity is further exacerbated by the several cell types that are implicated in Psoriasis's progression, including keratinocytes, melanocytes and various immune cell types. The observed interactions between the genetic substrate and the environment lead to epigenetic alterations that directly or indirectly affect gene expression. Changes in DNA methylation and histone modifications that alter DNA-binding site accessibility, as well as non-coding RNAs implicated in the post-transcriptional regulation, are mechanisms of gene transcriptional activity modification and therefore affect the pathways involved in the pathogenesis of Psoriasis. In this review, we summarize the research conducted on the environmental factors contributing to the disease onset, epigenetic modifications and non-coding RNAs exhibiting deregulation in Psoriasis, and we further categorize them based on the under-study cell types. We also assess the recent literature considering therapeutic applications targeting molecules that compromise the epigenome, as a way to suppress the inflammatory cutaneous cascade.

Keywords: epigenetics; psoriasis; methylation; histone; ncRNAs

Citation: Antonatos, C.; Grafanaki, K.; Asmenoudi, P.; Xiropotamos, P.; Nani, P.; Georgakilas, G.K.; Georgiou, S.; Vasilopoulos, Y. Contribution of the Environment, Epigenetic Mechanisms and Non-Coding RNAs in Psoriasis. *Biomedicines* **2022**, *10*, 1934. https://doi.org/10.3390/biomedicines10081934

Academic Editor: Marianna Christodoulou

Received: 7 July 2022
Accepted: 8 August 2022
Published: 9 August 2022

Publisher's Note: MDPI stays neutral with regard to jurisdictional claims in published maps and institutional affiliations.

1. Introduction

Psoriasis (PsO) is a chronic, inflammatory skin disease with its prevalence ranging from 1.83 to 5.32% in central European adults [1], with similar frequencies observed in white individuals in the United States [2]. PsO's manifestation lies in the epidermal keratinocytes (KCs), where the perturbation of inflammatory and cell-cycle-related pathways leads to their uncontrolled proliferation, aberrant differentiation and the development of distinctive, erythematous plaques on the skin surface [3]. Significant progress has been accomplished in characterizing the mechanisms involved in the pathogenesis of PsO, which are related to the activation of immune cell types and the maintenance of the chronic inflammation through the production by KCs of numerous signaling and chemotactic molecules [3,4]. Specifically, antimicrobial peptides (AMPs) produced by KCs and melanocyte auto-antigens, such as ADAMTSL5, in response to cell damage and altered microbial environment stimulate Toll-like Receptor (TLR) 9 and 8 signaling pathways in the plasmacytoid dendritic cells (pDCs) and myeloid dendritic cells (mDCs), respectively [5,6]. The pDCs are activated and type I IFN is produced, inducing both the maturation of mDCs and the secretion of proinflammatory cytokines, such as IL-12, IL-23 and tumor necrosis factor-alpha (TNF-α), and consequently the expansion of T helper (Th) cells. [7]. Both interleukins modulate the differentiation and proliferation of Th1 and Th17 cell subtypes [8], while TNFα is capable of enhancing the mitotic rhythm of KCs through the stimulation of cutaneous fibroblasts, production of the Keratinocyte Growth Factor (KGF) [9,10], as well as fostering leukocyte migration and T regulatory (Treg) cells suppression [11]. Specifically, the diverse role of TNFα in both the facilitation of the leukocyte migration in the cutaneous inflammation and the stimulation of KCs for the production of inflammatory cytokines has been exploited by the development of anti-TNFα drugs, displaying a high remission rate amongst PsO patients [3]. TNFα is secreted by the majority of the implicated cell types, including IFNγ-producing Th1 cells, which stimulate chemokine synthesis by the KCs [4] (Figure 1). The inflammatory cascade is amplified through the dysregulated IL-23/Th17 axis that plays a central role in the pathogenesis of the disease via the secretion of IL-17 and IL-22 [12]; IL-23 exhibits a lesional-specific increased expression profile, in contrast to other Th17-dependent inflammatory diseases, highlighting the accumulated cutaneous levels of Th17 as well as the role of IL-23 in the Th17 polarization [9]. The direct influence of IL-17 on the inflamed KCs via their increased proliferative activity is enhanced due to its secretion from the majority of the diverse cell types implicated in the disease pathogenesis, with, nevertheless, the IL-23-dependent IL-17 secretion through the activation of Th17 being widely demonstrated as a core pathogenetic mechanism and utilized as a therapeutic approach [12]. The inflammatory milieu is preserved by the abundant production of chemokines, AMPs and proinflammatory cytokines by the KCs [9] (Figure 1).

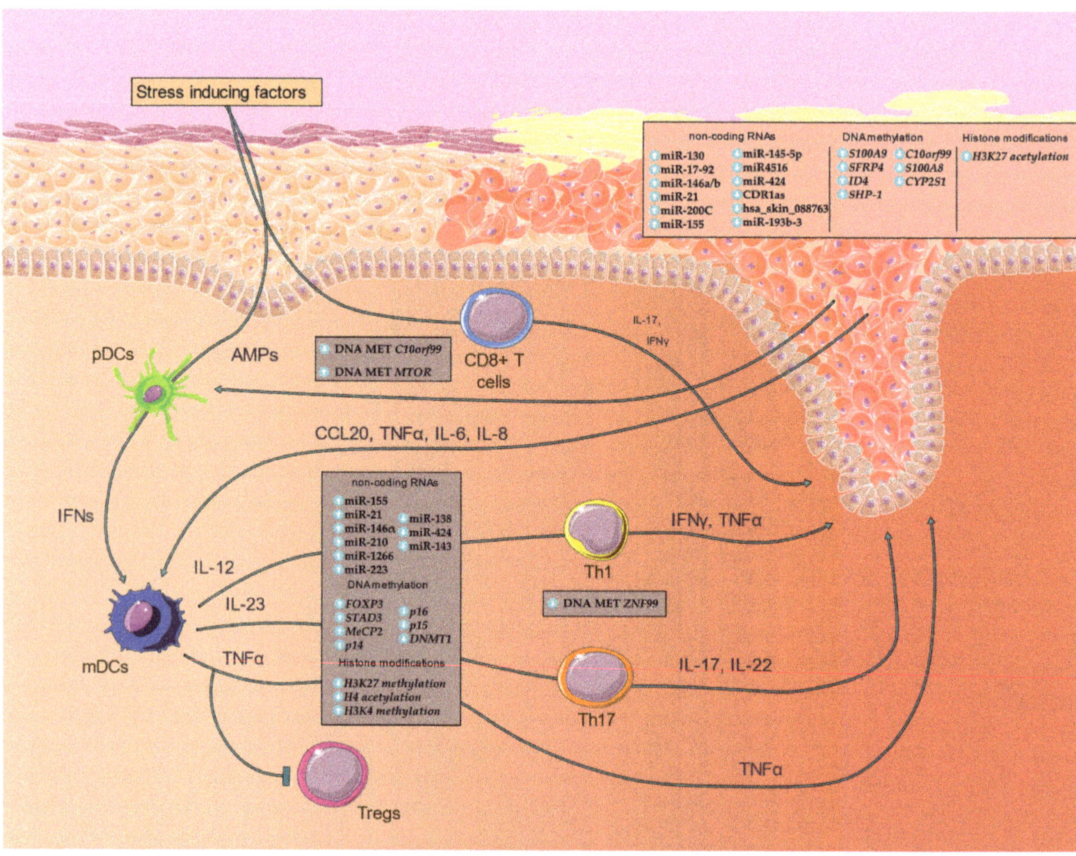

Figure 1. Overview of the inflammatory cascade observed in PsO, as well as key deregulated epigenetic factors and ncRNAs during the disease progression.

Disruption of such core mechanisms that regulate the immune response and cell proliferation is mediated through a multi-layered interaction between genetic and environmental factors. The largest genome-wide association analysis, conducted in 2017 by Tsoi et al., has uncovered 63 associated genetic loci, mapped in genes that participate in the inflammatory cascade occurring in PsO, such as the adaptive immune response and differentiation of lymphocytes [13]. Nevertheless, the complexity of PsO considering the disease's development and progression is also attributed to environmental factors that aggravate the existing genetic predisposition. The continuous infiltration of the epidermal barricade from immunogenic stimuli, as well as smoking [14], diet [15] and sun exposure, [16] significantly alter the epigenomic profile of the diverse collection of cell types involved in the pathogenesis of PsO. Epigenetic alterations are characterized as reversible, chemical modifications in the structure of DNA without affecting the genomic sequence, thus modifying gene expression [17]. DNA methylation and post-translational histone modifications contribute to transcriptional activity, whereas post-transcriptional regulation is performed by non-coding RNA molecules (ncRNAs) [18] (Figure 1). While epigenetic changes can normally be utilized as a tool to control gene expression throughout the developmental stages of a cell type, multiple studies have associated aberrant epigenetic changes with the pathogenesis of cancer [19] and cardiovascular and autoimmune diseases [20]. This review focuses on the research conducted in characterizing the contribution of environmental factors, epigenetic factors and ncRNAs in PsO onset, in the context of KCs and the various implicated immune cell types, as well as their potential clinical relevance as biomarkers and therapeutic targeting.

2. Microbiome and Environmental Risk Factors

Since PsO is a multifactorial disease, genetic and environmental factors affect its onset and progression. The microbiome [21,22] is responsible for triggering adaptive and innate immune responses that extensively affect numerous immunomodulatory mechanisms (Figure 1). The connection between PsO and bacterial infections was established decades ago by providing evidence that streptococcal infection [23] can lead to PsO, and PsO patients can be distinguished from healthy individuals based on differences in their skin and gut microbiome [24]. Lifestyle factors such as diet, smoking and alcohol intake can further alter the gut microbiota composition, while metabolites derived from the latter can influence epigenetic modifying enzymes (Figure 1).

2.1. Skin and Gut Microbiome

Bacterial and fungal populations on the skin differ between healthy and psoriatic patients; *Streptococcus*, *Staphylococcus* and *Malassezia* [25] are increased and *Propionibacterium* and *Corynebacterium* are decreased [22]. It is suggested that *Streptococcus'* M protein, which is highly homologous to type I keratins, can induce the expression of superantigens by T cells, further targeting KCs and causing chronic inflammation and proliferation of KCs [23]. Such superantigens are also produced by the gram-positive *Staphylococcus aureus* through the secretion of pyrogenic exotoxins, leading to severe cutaneous inflammation. HLA-DR, expressed by KCs, bind these superantigens along with secreted TNFα leading to inflammatory cascades [26]. Voluminous fungal populations of *Malassezia* spp. [27] can disrupt the epidermal barrier by producing lipases and phospholipases, attracting polymorphonuclear leukocytes and causing local skin sensitization. Production of propionate and radical oxygenase by *Propionibacterium* [26] reduces the oxidative stress levels and prevents skin inflammation. Additionally, it can modulate Th17 cells to maintain immune homeostasis. Decreased populations of *Corynebacterium* are associated with the onset and exacerbation of PsO [28], as *Corynebacterium* possesses anti-inflammatory abilities by negatively regulating interferon signaling in pDCs [28,29].

The gut microbiome is associated with skin diseases via the intestinal barrier, inflammatory mediators, and metabolites. In general, PsO patients appear to have inadequate intestinal flora [30,31], which is characterized by a reduced population of *Bacteroides* and abundant *Actinobacteria* and *Firmicutes*. *Bacteroides* possess anti-inflammatory capabilities through the production of polysaccharide-A, which is able to activate Tregs, stimulate anti-inflammatory pathways (i.e., IL-10), and thus inhibit the maintenance of inflammation. *Firmicutes* and *Bacteroides* can decrease the level of short-chain fatty acids (SCFAs) that cause inflammation and increase the vulnerability of the intestinal barrier [32]. A disrupted barrier enables microbiota dysbiosis via the circulatory system inducing both local and systemic immune responses.

2.2. Lifestyle

It is widely accepted that the interaction of environmental and genetic factors via epigenetic modifications contributes to the onset of a wide spectrum of diseases, including PsO (Figure 1). Food and nutrient intake can lead to alterations in the composition of the gut microbiome, allowing differential growth [33] of certain populations that are associated with PsO. In parallel with microbiota alterations, several nutrients such as sulphoraphane, curcumin [34] and omega-3 polyunsaturated fatty acids [35] can induce DNA methylation in leukocytes [36] and histone modifications by activating the epigenetic related enzymes DNMTs, HDAC and HAT. Dietary habits display a further added risk in developing PsO via the increased obesity prevalence amongst PsO cases, as described in numerous studies [37–39]. Meta-analysis of the leptin levels in patients with PsO confirmed the increased levels of the pro-inflammatory adipokine, a hormone that inhibits hunger and autoregulation of T cells, despite the increased between-study heterogeneity [40,41]. The implication of the adipose tissue in the inflammation through the secretion of both adipokines and the classical pro-inflammatory cytokines, such as TNFα and IL-6, establishes abdominal obesity as a risk factor for PsO, nevertheless without fully clarifying the exact causal mechanisms [42]. Smoking and alcohol can additionally enhance such psoriatic signals through a variety of mechanisms implicated in immunological disorders (i.e., KC hyperproliferation) due to the overexpression [43] of a5 integrin, cyclin D1, KGF receptor and pro-inflammatory cytokines. The epigenetic effect of tobacco is based on the induction of CpG island [44] methylation, decreased HDAC activity, increased histone methylation [45] levels and altered expression of non-coding RNAs [46,47].

Psychological factors as well as mood disorders, such as stress and depression, appear to play an important role in the onset and exacerbation of PsO. Stress is implicated in PsO pathogenesis through immune regulation and abnormal T cell activation. Actually, patients with PsO have lower cortisol levels [48] when stressed. Moreover, cortisol in addition to its anti-inflammatory effects, induces epigenetic changes such as DNA methylation [49], histone modifications [50], and may affect the expression of ncRNAs [51]. The psychological burden stimulates the secretion of pro-inflammatory cytokines [52], including TNFα and IL-6, further strengthening the correlation between depression and inflammatory disorders. Specifically, PsO patients undergo a significant social stigmatization due to the presence of the psoriatic plaques, leading to increased risk of social anxiety and depression [53].

UV radiation, especially UVB, is used to treat psoriatic plaques, although in some cases exposure to low light UVA may trigger photosensitivity of the skin and cause inflammation by enabling the local infiltration of neutrophiles and lymphocytes [54]. The above therapeutic mechanism of the UVB radiation is highlighted via the elevated serum levels of 25(OH) Vitamin D (the serum marker of vitamin D) in PsO patients undergoing UVB phototherapy [55]; Vitamin D binds to the Vitamin D receptor (VDR) exhibiting an immunomodulatory activity by decreasing IL-17 and IFNγ levels on Peripheral Blood Mononuclear Cells (PBMCs) [56], while Vitamin D deficiency displays a perturbated differentiation and increased proliferation of KCs [57]. Epigenetic modifications related to the Vitamin D show an anti-inflammatory and anti-proliferative profile [58], further strengthening the role of Vitamin D as an anti-inflammatory mediator.

3. DNA Methylation

DNA methylation is a well-studied epigenetic alteration mechanism mainly occurring in CpG islands localized on gene promoters. The addition of a methyl group into the cytosine's C5 position, forming a 5-methylcytosine (5mC), can significantly reduce the accessibility of Transcription Factor (TF) and RNA polymerase binding sites on the DNA helix, thus repressing the transcriptional activity. DNA methylation is catalyzed through the DNA methyltransferase enzymes (DNMTs), which consist of DNMT1, DNMT3a, DNMT3b and DNMT3L. DNMT1 prefers hemimethylated DNA and is characterized as a "maintenance DNMT" due to its repairing activity, while DNMT3L induces the de novo methyl group transfer activities catalyzed by both DNMT3a and DNMT3b [59,60]. Demethylation of 5mC is performed either from the ten-eleven translocation (TET) enzymes with the formation of the intermediate 5 hydroxyl-methylated cytosine (5hmC), or by the deamination of 5mC and the utilization of the base excision repair (BER) pathway [60].

3.1. DNA Methylation in KCs

The role of KCs in PsO pathogenesis includes both the formation of the psoriatic plaques due to their increased proliferation and the maintenance of inflammation through their contribution in the inflammatory milieu and the production of multiple proinflammatory cytokines [61]. Thus, epigenetic modifications affecting the transcriptional activity of genes involved in these pathways may contribute to PsO pathogenesis.

The pro-inflammatory Ca^{2+} binding proteins S100A8 and S100A9 are members of the S100 family. The heterodimeric protein S100A8/A9 is released actively at the time of inflammation, modulating its progression by stimulating leukocyte recruitment and inducing cytokine secretion [62]. These molecules are highly up-regulated in KCs and leukocytes of psoriatic skin and their expression is induced by IL-10 in differentiated human dendritic cells [63]. The methylation status of their genes' promoter has been characterized by multiple whole-genome methylation analyses (Figure 1).

Roberson et al. were the first to observe global alterations of CpG methylation in skin from PsO patients compared to skin from healthy volunteers. They identified 674 hypermethylated and 444 hypomethylated CpG sites, which were mainly localized on gene promoters, unveiling significant correlation between methylation and the expression of nearby genes such as *C10orf99*, *OAS2* and *KYNU* [64] (Figure 1). Furthermore, the methylation patterns were shown to be reversible to a non-psoriatic state after the administration of anti-TNFα therapy for one month. In the study by Zhang et al., whole-genome DNA methylation of psoriatic and non-psoriatic skin samples showed more hypermethylated regions (15,684) than hypomethylated (11,084). *PDCD5* and *TIMP2*, which induce KC proliferation, were hypermethylated and hypomethylated, respectively, exhibiting reversed expression levels [65]. Chandra et al. reported that 25% of differentially methylated CpGs were located at characterized PsO susceptibility (PSORS) loci, including PSORS2, PSORS4, PSORS6 and PSORS7, encoding several genes such as *S100A9*, *SELENBP1*, *CARD14*, *KAZN* and *PTPN22* with an inverse correlation between methylation and expression (Figure 1). Differentially methylated genes associated with histopathological aspects were also found, including *AIF1*, *FFAR2* and *TREM1*, which are implicated in neutrophil and leukocyte chemotactic events [66]. Hou et al. detected 96 hypermethylated genes, including MAPK signaling-related genes, such as *CACNA2D3* and *SRF*, and 234 hypomethylated genes participating in the increased angiogenesis of psoriatic lesion, namely, *NRP2*, *VEGF*, and *VASH1* [67]. In PsO skin, Zhou et al. discovered nine differentially methylated sites near metabolism-related genes, including *CYP2S1*, *ECE1*, *EIF2C2*, *MAN1C1*, and *DLGAP4*, whose methylation was negatively correlated with their expression. In the intergenic area surrounding *CYP21*, considerably low methylation has been observed [68].

There is ample evidence in the literature linking the methylation status and expression level of genes that are candidates for PsO pathogenesis. Bisulfite sequencing in skin lesions revealed that hypermethylation of $p14^{ARF}$ promoter resulted in its downregulation [69]. The low expression of *SFRP4* in psoriatic skin, which is involved in the Wnt pathway and

KC's hyperproliferation, is correlated with its promoter's hypermethylation [70] (Figure 1). The Wnt pathway plays a key role in PsO by regulating the proliferation and differentiation of KCs [71]. Additional evidence exists regarding the non-malignant effect of specialized promoters' methylation status in a human malady. The promoter of the SHP-1 isoform was found to be hypomethylated in psoriatic lesions, indicating that the methylation of *SHP-1*'s promoter in PsO might be related to the STAT3's binding affinity, due to the upregulation of the latter in the lesional skin [72] (Figure 1). The same group also studied ID4, a protein that participates in cell proliferation, differentiation and apoptosis, as well as in tumorigenesis (cholangiocarcinoma, breast cancer, lymphoma). *ID4* promoter hypermethylation promoter was linked with parakeratosis, which refers to deficient development of KCs, and skin-related cellular differentiation in PsO cases [73] (Figure 1). The promoter of the $p16^{INK4a}$ gene, which is involved in hyperproliferative skin diseases, was found to be hypermethylated in psoriatic skin [74]. Sheng et al. investigated the hypomethylation of *CYP2S1* and further identified the hypomethylation of two extra loci within the *CYP2S1* region, leading to its upregulation in psoriatic tissues [75] (Figure 1). Members of the growth arrest and DNA damage-inducible gene family, such as *GADD45a* and *GADD45b*, exhibit low expression in psoriatic lesional skin [76]. Specifically the expression *GADD45a*, which has a demethylase activity, was found to be positively correlated with IFN-γ and TNFα expression. Its depletion leads to hypermethylation of *UCHL1* promoter in PsO cases [77]. The expression of *WIF1*, an inhibitor of Wnt signaling, is linked to its promoter's demethylation, which is a result of *DNMT1* silencing, while its hypermethylation is a consequence of *DNMT1* overexpression. Indirubin, a traditional medicine utilized for treating various inflammatory diseases, inhibits the expression of *DNMT1* and the methylation of *WIF1* promoter, as well as the expression of Wnt-pathway core genes, such as *FZD2*, *FZD5*, and β-catenin [78].

3.2. DNA Methylation in Immune Cells

Despite the predominant role of KCs in the pathogenesis of cutaneous diseases, PsO, as an autoimmune disorder, is driven by the substantial activation of multiple immune cell types, which further stimulate KCs' proliferation through the secretion of pro-inflammatory cytokines. Moreover, PsO is considered to be a T-cell-mediated autoimmune disease, since the role of Th cells, as well as their secretome, has been extensively studied and targeted therapeutically [79]. Immune cells can be isolated either from lesional skin or PBMCs. The latter is a non-invasive approach and has therefore been established as the standard method for studying PsO-related immune cells.

PBMCs from psoriatic patients have been shown to exhibit aberrant DNA methylation, for example, hypermethylation of $p14^{ARF}$, *MBD2*, *MeCP2* and hypomethylation of *DNMT1* [69] (Figure 1). Genome-wide DNA methylation profiling of $CD4^+$ T cells unveiled the significantly hypermethylated promoters of immune-related X chromosome genes, such as *SLITRK4*, *EMD*, *ZIC3*, *CXorf40A*, *HDAC6*, *IKBKG*, *SH3KBP1*, *OTUD5*, *NDUFA1*, *WNK3* and *MSL3* [80]. Recent studies uncovered the hypermethylated profile of genes that are implicated in the TGFβ pathway, including *SNX25*, *STAD3* and *BRG1*, from whole blood samples of monozygotic twins [81] (Figure 1). Additional comparison of $CD8^+$ T cells in monozygotic twins from psoriatic samples and healthy controls identified 110 hypermethylated and 224 hypomethylated loci. DNA methylation analyses of $CD8^+$ T cells between PsO, psoriatic arthritis cases and healthy controls revealed numerous differences, indicating that DNA methylation screening in these cell subtypes could act as a potential diagnostic biomarker [82]. Furthermore, genome-wide DNA methylation profiling from peripheral whole blood displayed that *FOXP3* is hypermethylated, leading to reduced T_{reg} levels in patients with PsO [83] (Figure 1).

Bisulfite-sequencing on a targeted gene panel revealed the low methylation levels of *p15* and *p21* promoters in hematopoietic stem cells; *p15*, *p21* and *p16* exhibit similar methylation patterns and play a well-established role in controlling cell cycle [84,85] (Figure 1). Research conducted in $CD4^+$ T cells of monozygotic twins with PsO showed the promoter hypomethylation of transcription-regulator *ZNF99* gene. In CD8+ T cells associated with PsO, hypomethylation of the serine/threonine MAST3 and MTOR kinases, and hypermethylation of the *PM20D1* peptidase gene was shown [86] (Figure 1).

4. Histone Modifications

The post-translational histone modification (PTM) process is an important mechanism of gene expression regulation since these proteins directly participate in DNA organization and accessibility. Briefly, nucleosomes are the fundamental subunit of chromatin and are composed of histone octamers. Each histone (H2A, H2B, H3 and H4) is represented twice in the nucleosome structure that forms a binding scaffold for 147 DNA base pairs [87]. PTMs usually occur in the overhanging N-terminal tails of histones, resulting in either enhanced transcriptional activity through nucleosome unwinding and euchromatin formation, or strengthened DNA-histone interactions that form heterochromatin and suppress gene expression. Several types of histone modifications, implicated in both gene silencing and enhanced transcription, have been described, including adenylation, methylation, phosphorylation, ADP ribosylation and sumoylation among others [88].

4.1. Histone Modifications in KCs

Epigenetic regulation of KCs is an essential part of chronic skin inflammation. Recent research demonstrated that decreased H3K9 dimethylation leads to increased IL-23 expression in KCs; H3K9me2 levels play a key role in regulating basal and TNF-induced IL-23A expression [89]. H3K27me3 and *EZH2*, a histone methyl-transferase enzyme, were significantly enriched in cutaneous biopsies from individuals with PsO when compared to healthy controls. *EZH2* is implicated in cell proliferation and tumorigenesis [90] and thus its transcriptional silencing affects the proliferation and differentiation of KCs.

Sirtuin (SIRT) is a family of (NAD^+)-dependent deacetylases, involved in cell apoptosis, gene transcription, tumor development, autoimmune inflammation and epigenetic modification processes. Specifically, *SIRT1* regulates inflammation-associated signaling pathways [91,92]. Hwang et al. showed that *HDAC-1* is overexpressed and *SIRT1* displays a decreased expression in skin biopsies of patients with PsO [93]. GLS1-mediated glutaminolysis induces proliferation of KCs in PsO and promotes Th17 and $\gamma\delta$ T17 cell differentiation through the acetylation of H3 on Il17a promoter [94]. The first whole-genome study for histone modifications showed that H3K27 is hyperacetylated in 60% of the overexpressed gene promoters in cutaneous lesions and binding sites of overexpressed TFs in lesional skin, such as *GRHL* [95] (Figure 1). *WT1*, a TF implicated in cell proliferation and apoptosis, is highly expressed in psoriatic skin lesions and has two binding sites in the IL-1β gene promoter. IL-1β is produced by KCs, with its gene promoter showing considerably high histone acetylation levels that positively correlate with histone acetyltransferases p300 (P300) expression [96]. H3K27 hyperacetylation of *RPL22* promoter in PsO lesional skin leads to overexpression of *RPL22*, which is linked to CyclinD1 upregulation, inducing KC proliferation. *RLP22* also prevents KC apoptosis and is involved in CD4+ T cell chemotaxis [97].

4.2. Histone Modifications in Immune Cells

PBMCs from PsO patients exhibit higher H3K4 methylation levels when compared to healthy individuals. Responders and non-responders in biological therapies tend to have different H3K27 and H3K4 methylation profiles [98] (Figure 1). Regulation of Th17 cell differentiation can be achieved by TCR-induced H3K27 demethylase Jmjd3, which is overexpressed and decreases H3K27me3 levels. Jmjd3 controls chromatin accessibility of numerous Th17-related loci, such as Th17-specific gene promoters, and induces Th17 cell differentiation by decreasing H3K27me3 enrichment [99]. Zhang et al. unveiled decreased H4 acetylation in psoriatic PBMCs, downregulation of *P300*, *CBP* and *SIRT1* as well as increased HDAC1 levels [100] (Figure 1).

5. Non-Coding RNAs

Non-coding RNAs (ncRNAs) are RNA molecules that are not translated into functional proteins. They are typically grouped into distinct families according to their size and function. Most ncRNA families play a key role in directly or indirectly affecting gene expression at the transcriptional and post-transcriptional level. The microRNA (miRNA) family consists of small (~18–23 nucleotides), single-stranded RNA molecules that are loaded on and guide the RNA-induced silencing complex (RISC) to mediate mRNA degradation and/or translation suppression [101]. Circular RNAs (cirRNAs) form a continuous loop through their linkage on the 5′ and 3′ termini, establishing them as stable, exonuclease-proof RNA molecules with numerous roles in transcriptional and translational regulation [102]. Long non-coding RNAs (lncRNAs) are defined as longer than 200 nucleotides transcripts with an emerging role in the pathogenesis of autoimmune diseases [91]. LncRNAs and circRNAs can regulate gene expression by participating in processes that alter chromatic conformation, forming triplexes with DNA as well as interfering with transcription enzymes [103].

Due to the abundance of the ncRNAs implicated and studied in the context of PsO, we conducted an exhaustive literature search regarding the differential expression of ncRNAs in cutaneous biopsies, serum levels as well as PBMCs. We filtered the screened studies according to the importance of the ncRNAs under study, evaluated through their statistically significant differential expression in contrast to healthy controls as well as their identified, direct or indirect target genes. Non-coding RNAs screened by more than one study were also identified as important regulators of the psoriatic transcriptome.

5.1. MiRNAs in KCs

The immortalized nontumorigenic human epidermal (HaCaT) cell line and KCs have been widely used in PsO studies as they are easy to isolate and provide reliable results regarding the transcriptome and proteome profile of the disease [104]. Fibronectin 1 (*FN1*) and integrin subunit α9 (*ITGA9*) signaling pathways are both implicated in cell motility and direct targets of miR-4516, which was found to be downregulated in cutaneous PsO biopsies (Figure 1). Specifically, Chowdhari et al. found significant overexpression of both *FN1 and ITGA9* as well as STAT3 in PsO, which could be partly responsible for the KCs activated state in lesional skin since it induces proliferation and terminal differentiation [105]. MiR-424 has been also investigated as a potential biomarker in PsO given its regulatory role in signaling pathways that orchestrate differentiation and cell cycle regulation in KCs [106] (Figure 1). The decreased miR-424 expression has been associated with the overexpression of *MEK1* and *CCNE1*, members of the metabolic pathways responsible for the abnormal KC proliferation observed in the clinical manifestation of the disease; however, the exact regulatory mechanism has yet to be defined. Another miRNA whose down-regulation has been associated with PsO is miR-145, a molecule that has been widely studied in immune-mediated inflammatory diseases due to its inhibitory role in cell proliferation and immune responses [107] (Figure 1). It has been observed that miR-145 downregulation promotes proliferation and chemokine expression in the lesional skin, as it directly targets *MLK3*, which in turn regulates STAT3 and NF-κB TFs. Increased expression of miR-21 has been associated with the epidermal downregulation of *TIMP-3*, leading to the activation of *TACE*, which subsequently induces TNFα overexpression and a psoriasis-like phenotype [108] (Figure 1).

MiR-200c, a miRNA involved in apoptosis and senescence of KCs, was found to be up-regulated in lesional skin compared to non-lesional skin biopsies and healthy controls [109] (Figure 1). MiR-200c is also known to directly repress *SIRT1*, which has a key role in oxidative stress and the regulation of skin inflammation, as well as *eNOS* and *FOXO1*, which have a significant role in regulating the function and preservation of endothelial cells. MiR-200c expression also shows a positive correlation, without being a direct regulator of molecules involved in inflammation such as *IL-6* and *COX-2*, and plaque destabilization such as *MMP-1* and *MMP-9*. Table 1 presents further deregulated miRNAs in KCs from PsO patients.

Table 1. Deregulated ncRNAs in PsO in keratinocytes.

Author, Year	ncRNA	Expression	Direct Targets (Indirectly Affected Genes)	Cell Type
Yan et al., 2018 [107]	miR-145-5p	Down	*MLK3, (NF-κB, STAT3, AKT, GSK, MAPK, MTOR)*	HEKs
Chowdhari et al., 2017 [105]	miR-4516	Down	*FN1, ITG9, STAT3*	HaCaT
Ichihara et al., 2011 [106]	miR-424	Down	*(MEK1, CCNE1)*	HEKs
Huang et al., 2021 [110]	miR-193b-3	Down	*ERBB4, (STAT3, NF-κB, IL-6, CXCL1, CCL20, BD-2)*	HaCaT
Pan et al., 2019 [111]	miR-125b	Down	*BRD4, (JAG1)*	HaCaT
Xu et al., 2011 [112]	miR-125b	Down	*FGFR2*	HEKs
Yu et al., 2017 [113]	miR-194	Down	*GRHL2, SOX5*	HEKs
Rongna et al., 2018 [114]	miR-876-5p	Down	*ANGPT1*	HaCaT
Zheng et al., 2019 [115]	miR-181b-5p	Down	*AKT3*	HEKs
Zheng et al., 2019 [115]	miR-125b-5	Down	*AKT3*	HEKs
Jiang et al., 2017 [116]	miR-486-3p	Down	*K17*	HaCaT
Zhao et al., 2022 [117]	miR-214-3p	Down	*FOXM1*	HaCaT
Yan et al., 2015 [118]	miR-31	Up	*PPP6C*	HEKs
Guinea-Viniegra et al., 2014 [108]	miR-21	Up	*TIMP3*	HEKs
Zhang et al., 2018 [119]	miR-17-92	Up	*CDKN2B, SOCS1*	HEKs
Zhang et al., 2020 [120]	miR-142-3p	Up	*SEMA3A*	HaCaT
Xu et al., 2017 [121]	miR-155	Up	(PTEN, PIP3, AKT, BAX, Bcl-2)	HaCaT
Wang et al., 2019 [122]	miR-223	Up	*PTEN*	HaCaT
Sonkoly et al., 2007 [123]	miR-203	Up	*SOCS3, (STAT3)*	HEKs
Zibert et al., 2010 [124]	miR-221, miR-222	Up	*(TIMP3)*	HEKs

Abbreviations: ncRNA, non-coding RNA; miR, microRNA; HEK, human embryonic kidney cell line; HaCaT, human epidermal keratinocyte cell line.

5.2. MiRNAs in Immune Cells and Serum

Fu et al. showed that miR-138, whose expression levels regulate the balance between Th1 and Th2 cells by targeting RUNX3, was found downregulated in PBMCs of Pso patients [125] (Figure 1). RUNX3 is an important TF regulating cell proliferation and apoptosis, and its increased expression in PsO classifies it as a key gene for PsO susceptibility. MiR-143 downregulation exhibits a significant correlation with PsO severity; specifically, patients with stable disease stages showed higher miR-143 expression levels, while patients in progressive stages had lower expression levels [126] (Figure 1). *BCL2*, which is thought to be responsible for shortening the lifetime of cortical cells, has been proven a direct target of miR-143 [126]. Additionally, García-Rodríguez et al. showed that the upregulation of miR-155 in PsO plasma samples is an important regulator of *SOCS1*, a susceptibility locus of PsO [127]. The miRNA-155/SOCS1 pathway is targeted in macrophages by Vitamin D or Vitamin D Receptor (VDR) signaling to reduce the inflammatory response [128] (Figure 1). The upregulation of miR-210 in $CD4^+$ T cells has also been found to significantly correlate with PsO onset and progression, mainly by targeting *FOXP3* [129] (Figure 1). *FOXP3* displays a central role in the development and diverse functionality of Treg cells, as it appears to facilitate their differentiation through genetic programming [129], thus establishing it as an important contributor to the pathogenesis of the disease. In normal $CD4^+$ T cells, overexpression of miR-210 can indirectly, contribute to the expression of inflammatory cytokines such as IFN-γ and IL-17 while suppressing other cytokines such as IL-10 and TGF-β, which are secreted by Tregs. Table 2 presents further deregulated miRNAs in PBMCs and plasma from PsO patients.

Table 2. Deregulated ncRNAs in PsO in immune cells and plasma.

Author, Year	ncRNA	Expression	Direct Targets (Indirectly Affected Genes)	Cell Type
		Immune cells		
Zheng et al., 2017 [126]	miR-143	Down	*BCL2*	PBMCs
Fu et al., 2015 [125]	miR-138	Down	*RUNX3*	CD4⁺ T cells
Garcıa–Rodrıguez et al., 2016 [128]	miR-146a	Up	*TRAF6, IRAK1, (NF-κB, IL-6, TNFα)*	PBMCs
Garcıa–Rodrıguez et al., 2016 [128]	miR-21	Up	*PDCD4*	PBMCs
Garcıa–Rodrıguez et al., 2016 [128]	miR-155	Up	*SOCS1*	PBMCs
		Serum levels		
Duan et al., 2019 [130]	miR-126	Down	-	Plasma
Ichihara et al., 2011 [106]	miR-424	Down	*(MEK1, CCNE1)*	Plasma
Zhao et al., 2014 [129]	miR-210	Up	*FOXP3, (IFN-γ, IL-17, IL-10, TGF-β)*	Plasma
Magenta et al., 2019 [109]	miR-200c	Up	*SIRT1, eNOS, FOXO1*	Plasma
Wang et al., 2017 [131]	miR-200a	Up	-	Plasma
Borska et al., 2017 [132]	miR-31	Up	-	Plasma
García-Rodríguez et al., 2014 [133]	miR-33	Up	*ABCA1*	Plasma
Guo et al., 2013 [134]	miR-369-3p	Up	-	Plasma

Abbreviations: ncRNA, non-coding RNA; miR, microRNA; PBMCs, peripheral blood mononuclear cells.

5.3. LncRNAs in PsO

Despite the established regulatory role of lncRNAs, there is still limited evidence regarding their participation in PsO pathogenesis. The maternally expressed gene 3 (*MEG3*), a downregulated lncRNA in HaCaT cells, has an identified miR-21 binding site, thus acting as a sponge or decoy for miR-21. It is postulated that *MEG3* participates in the regulation of PsO KCs proliferation and apoptosis as well as the expression of *CASP8* through its interplay with miR-21 [135]. By utilizing a luciferase reporter assay, it was confirmed that MSX2P1, a lncRNA overexpressed in HaCaT and KCs, is a direct target of miR-6731, thus negatively affecting its function on other RNAs. Further research in IL-22-stimulated KCs revealed MSX2P1's indirect role in increasing the protein levels of S100A7, IL-23, NF-κB, TNFα, IL-12β, HLA-C, and CCHCR [136]. Psoriasis-susceptibility-related RNA Gene Induced by Stress (PRINS) [137] and GAS5 [138] are lncRNAs displaying an under- and over-expression pattern, respectively, in the serum levels of patients with PsO. Specifically, plasma levels of PRINS transcripts were found to be down-regulated in patients with PsO, similar to its direct target *G1P3* and interacting partner *NPM*, while PRINS' miRNA targets that function as decoys, consisting of miR-124, miR-203a, miR-129, miR-146a and miR-9, were overexpressed, thus indicating a possible lncRNA–miRNA–mRNA axis of diagnostic value.

While the role of circRNAs in PsO progression remains obscure, CDR1as, a ciRNA significantly downregulated in cutaneous PsO biopsies, has been associated with numerous genes that are involved in the pathogenesis of the disease, such as *EGR3, GATA6, GATA3* and *FOXN3* [139]. However, the direct regulatory mechanism remains unclear (Figure 1). Xiaoxin Liu et al. also showed that circRNA hsa_skin_088763 was down-regulated in lesional skin compared to normal controls. It is postulated that this circRNA is indirectly associated with several PsO-related genes such as *GATA6, SIK2, IL17RD, EGR3, FAS, LRIG1*, and *PPARGC1A*, due to their shared regulatory ncRNAs (Figure 1), thus characterizing it as a competing endogenous RNA (ceRNA). A comprehensive list of all deregulated lncRNAs is presented in Tables 3 and 4, stratified based on the cell type under study.

Table 3. Deregulated lncRNAs in psoriasis in keratinocytes. In this table, miRNAs, which are targets of lncRNAs that function as decoys, are highlighted in bold.

Author, Year	Non-Coding RNA	Expression	Target Genes (Indirect Targets)	Cell Type
Liu et al., 2021 [139]	CDR1as	Down	(*EGR3, GATA6, GATA3, FOXN3*)	HEKs
Liu et al., 2021 [139]	hsa_skin_088763	Down	(*GATA6, SIK2, IL17RD, EGR3, FAS, LRIG1, PPARGC1A*)	HEKs
Jia et al., 2019 [135]	MEG3	Down	**miR-21**	HaCaT
Liu et al., 2021 [139]	hsa_skin_05227	Down	*GATA6*	HEKs
Yazıcı et al., 2021 [140]	7SL-RNA	Down	-	HEKs
Moldovan et al., 2020 [141]	circEXOC6B	Down	-	HEKs
Moldovan et al., 2020 [142]	circSLC8A	Down	-	HEKs
Moldovan et al., 2020 [142]	circRHOBTB	Down	-	HEKs
Qiao et al., 2018 [136]	MSX2P1	Up	**miR-6731**, (*S100A7*)	HaCaT, NHEK
Gao et al., 2018 [142]	MIR31HG	Up	-	HaCaT
Li et al., 2017 [143]	lncRNA-H19	Up	**miR-130b**, (*DSG1*)	HEKs
Szegedi et al., 2010 [144]	PRINS	Up	-	HeLa
Cai et al., 2019 [145]	PRANCR	Up	-	HEKs
Moldovan et al., 2020 [141]	cTNFRSF21,	Up	-	HEKs
Moldovan et al., 2020 [141]	circDOCK1	Up	-	HEKs
Moldovan et al., 2020 [141]	circARAP2	Up	-	HEKs
Moldovan et al., 2020 [141]	circDDX21	Up	-	HEKs
Moldovan et al., 2020 [141]	circZRANB1	Up	-	HEKs

Abbreviations: circRNA, circular RNA; lncRNA, long non-coding RNA; HEKs, human embryonic kidney cell line; HaCaT, human epidermal keratinocyte cell line; NHEK, normal human epidermal keratinocyte.

Table 4. Deregulated lncRNAs in psoriasis in serum. In this table, miRNAs, which are targets of lncRNAs that function as decoys, are highlighted in bold.

Author, Year	Non-Coding RNA	Expression	Target Genes (Indirect Targets)	Cell Type
Abdallah et al., 2022 [137]	PRINS	Down	**miR-124**, **miR-203a**, **miR-129**, **miR-146a**, **miR-9**, *GIP3*	Plasma
Shehata et al., 2021 [138]	GAS5	Up	-	Plasma

6. Therapeutic Approaches Targeting the Epigenetic Mechanisms

In recent years, biologic drugs targeting TNF, IL-23 and IL-17 and small-molecule drugs such as phosphodiesterase-4 (apremilast) and Janus kinase (JAK) inhibitors have been effective in plaque psoriasis clinical management. The predominant role of epigenetic modifications in the pathogenesis of complex diseases, as analyzed in the framework of PsO, establishes the therapeutic interventions targeting the epigenome, a promising clinical field. Numerous inhibitors of molecules that participate in the epigenetic reprogramming, including DNMTs and HDAC, have been extensively utilized in clinical trials. Recently FDA-approved agents, combined with cytotoxic chemotherapies, show promising results despite their limited implementation, mostly on hematologic malignancies [146]. Such repurposing approaches, already applicable in cancer, might prove beneficial in PsO considering the diverse cell subtypes that are involved in its pathogenesis.

Reservatol, a polyphenol with anti-inflammatory properties, was shown to stimulate the expression of *SIRT1* leading HaCaT cells to death [147]. Recently, trichostatin A (TSA), a class I and II HDAC inhibitor (HDACi), significantly decreased KC's proliferative phenotype, both in vitro and in vivo [148]. These results are in accordance with previous studies examining the effect of TSA on human Tregs and the prevention of their differentiation into IL-17A producing cells through the overexpression of *FOXP3* [149,150]. Another example of an epigenetic-driven therapeutic intervention is peroxisome proliferator-activated receptor gamma (PPARγ) and/or alpha (PPARα) antagonists, which inhibit *AQP3* expression in KCs, while agonists induce the differentiation of the latter, establishing them as a topical treatment for cutaneous diseases such as PsO. AQP3 is a water channel protein that regulates multiple aspects of KCs [151,152], with its expression induced by HDACs, particularly HDAC3, via acetylated transcription factors such as the family of p53 and PPARs.

MiRNA-mediated gene expression regulation is another approach in the anti-PsO therapeutic arsenal. MiRNAs exhibit a diverse role through their interaction with numerous transcripts and can therefore affect multiple pathways implicated in PsO. Imiquimod-induced psoriasis-like murine models were treated with a miR-210 antisense molecule-containing topical gel. Based on the predominant role of miR-210 in the regulation of multiple genes expressed in $CD4^+$ T cells, as analyzed before, cellular markers of cell proliferation were significantly decreased in KCs, while the imbalance of $CD4^+$ T cells was reversed to a non-pathological state [153]. Another topical-application approach referred to the usage of quaternized starch (Q-starch) as a miRNA-197 delivery system in lesional-xenotransplantated mice, alongside ultrasound for increased cutaneous permeability. The Q-starch/miRNA-197 complex was able to alleviate the psoriatic symptoms through the targeting of IL-22RA1 and IL-17RA transcripts, nevertheless without a homogeneous effect along the transplanted psoriatic skin sample [154]. Locked nucleid acid (LNA) *anti-miR-21* oligonucleotides were also assessed for their efficacy as a therapeutic approach in a double-knockout PsO murine model, displaying an important amelioration in the histopathological symptoms of the disease [108], further highlighting the role of miR-21 overexpression in PsO. Additionally, Qiao et al. inhibited miR-6731 in IL-22-stimulated HaCaT cells, showing an increased proliferative activity of the IL-22-induced KCs, while protein expression levels of therapeutic targets of PsO, including TNFα, IL-23, HLA-C as well as inflammatory molecules such as NF-κB and the PSORS1 locus were significantly overexpressed. These results imply a protective role of miR-21 in the pathogenesis of PsO [135].

7. Discussion

The etiopathology of PsO lies in the complex interactions between genetic, immunological and environmental factors, including but not limited to the imbalance in the gut and skin microbiome, as well as lifestyle and stress-inducing factors. The effect of genome–environment interactions can extend to the epigenome with direct and indirect modulation of DNA methylation and histone modifications in PsO-related loci. These cascading effects are further amplified by the function of ncRNAs, which in numerous ways regulate the expression of PsO-associated genes and pathways. However, epigenetic modifications and deregulation of ncRNA molecules occur in the spectrum of both KCs and immune cells, while the polygenicity of PsO aggravates candidate-gene approaches, thus obscuring the characterization of the exact mechanisms that alter the disease predisposition in individuals. Furthermore, skin tissue and immune cells consist of abundant and heterogeneous cellular populations, where each cell type exhibits a distinct epigenomic and, thus, transcriptomic profile. This diversity dramatically increases the complexity of research conducted in the field. The secretome of implicated cell types is affected by the distinct epigenetic and regulatory profiles of each sub-population, which is triggered by stress-inducing factors and the disease progression (Figure 1). Nevertheless, with the advent of next-generation sequencing, modern flow cytometry techniques and genome-wide analyses, the epigenetic reprogramming that participates in homeostasis disruption and the cutaneous inflammatory cascade is gradually elucidated, leading to the clarification of PsO pathogenesis.

Even though the epigenetic modifications cannot be utilized as clinical biomarkers for the disease progression, shedding light upon the molecular mechanisms governing their development and maintenance can potentially uncover novel therapeutic targets associated with the induced epigenetic changes. In contrast, ncRNAs and especially miRNAs have a rich history of being used as disease biomarkers and targets of therapeutic intervention, despite the difficulties imposed by the off-target effects due to the wide spectrum of pathways affected by miRNAs.

Future studies should focus on developing methods for reversing DNA methylation in the context of PsO therapeutic interventions, as a way to understand the molecular discrepancies between responders and non-responders to therapy. Additionally, the catalogue of ncRNAs implicated in PsO should be significantly enriched to provide an extensive view of the targeted genes set and an accurate description of the vast interactome that underlies

PsO. The gene regulatory networks that will emerge from the combination of epigenetic and ncRNA meta-analyses, as well as the genetic predisposition to PsO, will facilitate the development of a new generation of highly precise therapeutic approaches with minimum adverse effects and maximum impact.

Author Contributions: Conceptualization, C.A. and Y.V.; resources, C.A., P.A., P.X. and P.N.; data curation, C.A., P.A., P.X., P.N., G.K.G. and Y.V.; writing—original draft preparation, C.A., P.A., P.X., P.N. and G.K.G.; writing—review and editing, C.A., K.G., P.A., P.X., P.N., G.K.G., S.G. and Y.V.; visualization, C.A, P.X. and Y.V.; supervision, Y.V. All authors have read and agreed to the published version of the manuscript.

Funding: The publication of this article has been financed by the Research Committee of the University of Patras.

Institutional Review Board Statement: Not applicable.

Informed Consent Statement: Not applicable.

Data Availability Statement: The datasets generated or analyzed during the current study are available from the corresponding author on reasonable request.

Conflicts of Interest: The authors declare no conflict of interest.

References

1. Parisi, R.; Iskandar, I.Y.K.; Kontopantelis, E.; Augustin, M.; Griffiths, C.E.M.; Ashcroft, D.M. National, Regional, and Worldwide Epidemiology of Psoriasis: Systematic Analysis and Modelling Study. *BMJ* **2020**, *369*, m1590. [CrossRef] [PubMed]
2. Armstrong, A.W.; Mehta, M.D.; Schupp, C.W.; Gondo, G.C.; Bell, S.J.; Griffiths, C.E.M. Psoriasis Prevalence in Adults in the United States. *JAMA Dermatol.* **2021**, *157*, 940. [CrossRef] [PubMed]
3. Rendon, A.; Schäkel, K. Psoriasis Pathogenesis and Treatment. *Int. J. Mol. Sci.* **2019**, *20*, 1475. [CrossRef] [PubMed]
4. Lowes, M.A.; Suárez-Fariñas, M.; Krueger, J.G. Immunology of Psoriasis. *Annu. Rev. Immunol.* **2014**, *32*, 227–255. [CrossRef]
5. Chang, H.-W.; Yan, D.; Singh, R.; Liu, J.; Lu, X.; Ucmak, D.; Lee, K.; Afifi, L.; Fadrosh, D.; Leech, J.; et al. Alteration of the Cutaneous Microbiome in Psoriasis and Potential Role in Th17 Polarization. *Microbiome* **2018**, *6*, 154. [CrossRef]
6. Fuentes-Duculan, J.; Bonifacio, K.; Kunjravia, N.; Cueto, I.; Li, X.; Garcet, S.; Krueger, J.G. Autoantigens ADAMTSL5 and LL-37 Are Significantly Upregulated in Active Psoriasis and Associated with Dendritic Cells and Macrophages. *J. Investig. Dermatol.* **2017**, *137*, S1. [CrossRef]
7. Hänsel, A.; Günther, C.; Ingwersen, J.; Starke, J.; Schmitz, M.; Bachmann, M.; Meurer, M.; Rieber, E.P.; Schäkel, K. Human Slan (6-Sulfo LacNAc) Dendritic Cells Are Inflammatory Dermal Dendritic Cells in Psoriasis and Drive Strong T 17/T 1 T-Cell Responses. *J. Allergy Clin. Immunol.* **2011**, *127*, 787–794. [CrossRef]
8. Fitch, E.; Harper, E.; Skorcheva, I.; Kurtz, S.E.; Blauvelt, A. Pathophysiology of Psoriasis: Recent Advances on IL-23 and Th17 Cytokines. *Curr. Rheumatol. Rep.* **2007**, *9*, 461–467. [CrossRef]
9. Di Cesare, A.; Di Meglio, P.; Nestle, F.O. The IL-23/Th17 Axis in the Immunopathogenesis of Psoriasis. *J. Investig. Dermatol.* **2009**, *129*, 1339–1350. [CrossRef]
10. Banno, T.; Gazel, A.; Blumenberg, M. Effects of Tumor Necrosis Factor-α (TNFα) in Epidermal Keratinocytes Revealed Using Global Transcriptional Profiling. *J. Biol. Chem.* **2004**, *279*, 32633–32642. [CrossRef]
11. Chen, X.; Hamano, R.; Subleski, J.J.; Hurwitz, A.A.; Howard, O.M.Z.; Oppenheim, J.J. Expression of Costimulatory TNFR2 Induces Resistance of CD4$^+$ FoxP3$^-$ Conventional T Cells to Suppression by CD4$^+$ FoxP3$^+$ Regulatory T Cells. *J. Immunol.* **2010**, *185*, 174–182. [CrossRef] [PubMed]
12. Schön, M.P.; Erpenbeck, L. The Interleukin-23/Interleukin-17 Axis Links Adaptive and Innate Immunity in Psoriasis. *Front. Immunol.* **2018**, *9*, 1323. [CrossRef] [PubMed]
13. Tsoi, L.C.; Stuart, P.E.; Tian, C.; Gudjonsson, J.E.; Das, S.; Zawistowski, M.; Ellinghaus, E.; Barker, J.N.; Chandran, V.; Dand, N.; et al. Large Scale Meta-Analysis Characterizes Genetic Architecture for Common Psoriasis Associated Variants. *Nat. Commun.* **2017**, *8*, 15382. [CrossRef]
14. Naldi, L. Psoriasis and Smoking: Links and Risks. *PTT* **2016**, *6*, 65. [CrossRef]
15. Wolters, M. Diet and Psoriasis: Experimental Data and Clinical Evidence. *Br. J. Dermatol.* **2005**, *153*, 706–714. [CrossRef]
16. Søyland, E.; Heier, I.; Rodríguez-Gallego, C.; Mollnes, T.E.; Johansen, F.-E.; Holven, K.B.; Halvorsen, B.; Aukrust, P.; Jahnsen, F.L.; de la Rosa Carrillo, D.; et al. Sun Exposure Induces Rapid Immunological Changes in Skin and Peripheral Blood in Patients with Psoriasis: Immunological Changes in Sun-Exposed Patients with Psoriasis. *Br. J. Dermatol.* **2011**, *164*, 344–355. [CrossRef] [PubMed]
17. Schuebel, K.; Gitik, M.; Domschke, K.; Goldman, D. Making Sense of Epigenetics. *Int. J. Neuropsychopharmacol.* **2016**, *19*, pyw058. [CrossRef]

18. Zhang, J.; Zhang, Y.-Z.; Jiang, J.; Duan, C.-G. The Crosstalk Between Epigenetic Mechanisms and Alternative RNA Processing Regulation. *Front. Genet.* **2020**, *11*, 998. [CrossRef]
19. Sharma, S.; Kelly, T.K.; Jones, P.A. Epigenetics in Cancer. *Carcinogenesis* **2010**, *31*, 27–36. [CrossRef]
20. Mazzone, R.; Zwergel, C.; Artico, M.; Taurone, S.; Ralli, M.; Greco, A.; Mai, A. The Emerging Role of Epigenetics in Human Autoimmune Disorders. *Clin. Epigenet.* **2019**, *11*, 34. [CrossRef]
21. Hsu, D.K.; Fung, M.A.; Chen, H.-L. Role of Skin and Gut Microbiota in the Pathogenesis of Psoriasis, an Inflammatory Skin Disease. *Med. Microecol.* **2020**, *4*, 100016. [CrossRef]
22. Roszkiewicz, M.; Dopytalska, K.; Szymańska, E.; Jakimiuk, A.; Walecka, I. Environmental Risk Factors and Epigenetic Alternations in Psoriasis. *Ann. Agric. Environ. Med.* **2020**, *27*, 335–342. [CrossRef] [PubMed]
23. Leung, D.Y.; Travers, J.B.; Giorno, R.; Norris, D.A.; Skinner, R.; Aelion, J.; Kazemi, L.V.; Kim, M.H.; Trumble, A.E.; Kotb, M. Evidence for a Streptococcal Superantigen-Driven Process in Acute Guttate Psoriasis. *J. Clin. Investig.* **1995**, *96*, 2106–2112. [CrossRef]
24. Benhadou, F.; Mintoff, D.; Schnebert, B.; Thio, H. Psoriasis and Microbiota: A Systematic Review. *Diseases* **2018**, *6*, 47. [CrossRef]
25. Paulino, L.C.; Tseng, C.-H.; Blaser, M.J. Analysis of Malassezia Microbiota in Healthy Superficial Human Skin and in Psoriatic Lesions by Multiplex Real-Time PCR: Malassezia Microbiota in Health and Psoriasis by QPCR. *FEMS Yeast Res.* **2008**, *8*, 460–471. [CrossRef] [PubMed]
26. Allhorn, M.; Arve, S.; Brüggemann, H.; Lood, R. A Novel Enzyme with Antioxidant Capacity Produced by the Ubiquitous Skin Colonizer Propionibacterium Acnes. *Sci. Rep.* **2016**, *6*, 36412. [CrossRef] [PubMed]
27. Baroni, A.; Orlando, M.; Donnarumma, G.; Farro, P.; Iovene, M.R.; Tufano, M.A.; Buommino, E. Toll-like Receptor 2 (TLR2) Mediates Intracellular Signalling in Human Keratinocytes in Response to Malassezia Furfur. *Arch. Derm. Res.* **2006**, *297*, 280–288. [CrossRef]
28. Fyhrquist, N.; Muirhead, G.; Prast-Nielsen, S.; Jeanmougin, M.; Olah, P.; Skoog, T.; Jules-Clement, G.; Feld, M.; Barrientos-Somarribas, M.; Sinkko, H.; et al. Microbe-Host Interplay in Atopic Dermatitis and Psoriasis. *Nat. Commun.* **2019**, *10*, 4703. [CrossRef]
29. Reizis, B.; Colonna, M.; Trinchieri, G.; Barrat, F.; Gilliet, M. Plasmacytoid Dendritic Cells: One-Trick Ponies or Workhorses of the Immune System? *Nat. Rev. Immunol.* **2011**, *11*, 558–565. [CrossRef]
30. Hidalgo-Cantabrana, C.; Gómez, J.; Delgado, S.; Requena-López, S.; Queiro-Silva, R.; Margolles, A.; Coto, E.; Sánchez, B.; Coto-Segura, P. Gut Microbiota Dysbiosis in a Cohort of Patients with Psoriasis. *Br. J. Dermatol.* **2019**, *181*, 1287–1295. [CrossRef]
31. Scher, J.U.; Ubeda, C.; Artacho, A.; Attur, M.; Isaac, S.; Reddy, S.M.; Marmon, S.; Neimann, A.; Brusca, S.; Patel, T.; et al. Decreased Bacterial Diversity Characterizes the Altered Gut Microbiota in Patients With Psoriatic Arthritis, Resembling Dysbiosis in Inflammatory Bowel Disease: Gut Microbiota in PsA. *Arthritis Rheumatol.* **2015**, *67*, 128–139. [CrossRef] [PubMed]
32. Shapiro, H.; Thaiss, C.A.; Levy, M.; Elinav, E. The Cross Talk between Microbiota and the Immune System: Metabolites Take Center Stage. *Curr. Opin. Immunol.* **2014**, *30*, 54–62. [CrossRef] [PubMed]
33. Bhat, M.I.; Kapila, R. Dietary Metabolites Derived from Gut Microbiota: Critical Modulators of Epigenetic Changes in Mammals. *Nutr. Rev.* **2017**, *75*, 374–389. [CrossRef] [PubMed]
34. Remely, M.; Lovrecic, L.; de la Garza, A.L.; Migliore, L.; Peterlin, B.; Milagro, F.I.; Martinez, A.J.; Haslberger, A.G. Therapeutic Perspectives of Epigenetically Active Nutrients: Therapeutic Epigenetically Active Nutrients. *Br. J. Pharm.* **2015**, *172*, 2756–2768. [CrossRef] [PubMed]
35. Fujii, R.; Yamada, H.; Munetsuna, E.; Yamazaki, M.; Mizuno, G.; Ando, Y.; Maeda, K.; Tsuboi, Y.; Ohashi, K.; Ishikawa, H.; et al. Dietary Fish and ω-3 Polyunsaturated Fatty Acids Are Associated with Leukocyte ABCA1 DNA Methylation Levels. *Nutrition* **2021**, *81*, 110951. [CrossRef]
36. Karimi, M.; Vedin, I.; Freund Levi, Y.; Basun, H.; Faxén Irving, G.; Eriksdotter, M.; Wahlund, L.-O.; Schultzberg, M.; Hjorth, E.; Cederholm, T.; et al. DHA-Rich n–3 Fatty Acid Supplementation Decreases DNA Methylation in Blood Leukocytes: The OmegAD Study. *Am. J. Clin. Nutr.* **2017**, *106*, 1157–1165. [CrossRef]
37. Naldi, L.; Chatenoud, L.; Linder, D.; Belloni Fortina, A.; Peserico, A.; Virgili, A.R.; Bruni, P.L.; Ingordo, V.; Scocco, G.L.; Solaroli, C.; et al. Cigarette Smoking, Body Mass Index, and Stressful Life Events as Risk Factors for Psoriasis: Results from an Italian Case–Control Study. *J. Investig. Dermatol.* **2005**, *125*, 61–67. [CrossRef]
38. Setty, A.R.; Curhan, G.; Choi, H.K. Obesity, Waist Circumference, Weight Change, and the Risk of Psoriasis in Women. *ARCH Intern. Med.* **2007**, *167*, 6. [CrossRef]
39. Snekvik, I.; Smith, C.H.; Nilsen, T.I.L.; Langan, S.M.; Modalsli, E.H.; Romundstad, P.R.; Saunes, M. Obesity, Waist Circumference, Weight Change, and Risk of Incident Psoriasis: Prospective Data from the HUNT Study. *J. Investig. Dermatol.* **2017**, *137*, 2484–2490. [CrossRef]
40. Zhu, K.-J.; Zhang, C.; Li, M.; Zhu, C.-Y.; Shi, G.; Fan, Y.-M. Leptin Levels in Patients with Psoriasis: A Meta-Analysis. *Clin. Exp. Dermatol.* **2013**, *38*, 478–483. [CrossRef]
41. Kong, Y.; Zhang, S.; Wu, R.; Su, X.; Peng, D.; Zhao, M.; Su, Y. New Insights into Different Adipokines in Linking the Pathophysiology of Obesity and Psoriasis. *Lipids Health Dis.* **2019**, *18*, 171. [CrossRef] [PubMed]
42. Kunz, M.; Simon, J.C.; Saalbach, A. Psoriasis: Obesity and Fatty Acids. *Front. Immunol.* **2019**, *10*, 1807. [CrossRef] [PubMed]
43. Farkas, Á.; Kemény, L. Alcohol, Liver, Systemic Inflammation and Skin: A Focus on Patients with Psoriasis. *Ski. Pharm. Physiol.* **2013**, *26*, 119–126. [CrossRef] [PubMed]

44. Joehanes, R.; Just, A.C.; Marioni, R.E.; Pilling, L.C.; Reynolds, L.M.; Mandaviya, P.R.; Guan, W.; Xu, T.; Elks, C.E.; Aslibekyan, S.; et al. Epigenetic Signatures of Cigarette Smoking. *Circ. Cardiovasc. Genet.* **2016**, *9*, 436–447. [CrossRef]
45. Sundar, I.K.; Nevid, M.Z.; Friedman, A.E.; Rahman, I. Cigarette Smoke Induces Distinct Histone Modifications in Lung Cells: Implications for the Pathogenesis of COPD and Lung Cancer. *J. Proteome Res.* **2014**, *13*, 982–996. [CrossRef]
46. Izzotti, A.; Larghero, P.; Longobardi, M.; Cartiglia, C.; Camoirano, A.; Steele, V.E.; De Flora, S. Dose-Responsiveness and Persistence of MicroRNA Expression Alterations Induced by Cigarette Smoke in Mouse Lung. *Mutat. Res./Fundam. Mol. Mech. Mutagenesis* **2011**, *717*, 9–16. [CrossRef]
47. Bi, H.; Zhou, J.; Wu, D.; Gao, W.; Li, L.; Yu, L.; Liu, F.; Huang, M.; Adcock, I.M.; Barnes, P.J.; et al. Microarray Analysis of Long Non-Coding RNAs in COPD Lung Tissue. *Inflamm. Res.* **2015**, *64*, 119–126. [CrossRef]
48. Evers, A.W.M.; Verhoeven, E.W.M.; Kraaimaat, F.W.; De Jong, E.M.G.J.; De Brouwer, S.J.M.; Schalkwijk, J.; Sweep, F.C.G.J.; Van De Kerkhof, P.C.M. How Stress Gets under the Skin: Cortisol and Stress Reactivity in Psoriasis: Stress Reactivity in Psoriasis. *Br. J. Dermatol.* **2010**, *163*, 986–991. [CrossRef]
49. Thomassin, H. Glucocorticoid-Induced DNA Demethylation and Gene Memory during Development. *EMBO J.* **2001**, *20*, 1974–1983. [CrossRef]
50. Dowlati, Y.; Herrmann, N.; Swardfager, W.; Liu, H.; Sham, L.; Reim, E.K.; Lanctôt, K.L. A Meta-Analysis of Cytokines in Major Depression. *Biol. Psychiatry* **2010**, *67*, 446–457. [CrossRef]
51. Zill, J.M.; Dirmaier, J.; Augustin, M.; Dwinger, S.; Christalle, E.; Härter, M.; Mrowietz, U. Psychosocial Distress of Patients with Psoriasis: Protocol for an Assessment of Care Needs and the Development of a Supportive Intervention. *JMIR Res. Protoc.* **2018**, *7*, e22. [CrossRef] [PubMed]
52. Vockley, C.M.; D'Ippolito, A.M.; McDowell, I.C.; Majoros, W.H.; Safi, A.; Song, L.; Crawford, G.E.; Reddy, T.E. Direct GR Binding Sites Potentiate Clusters of TF Binding across the Human Genome. *Cell* **2016**, *166*, 1269–1281.e19. [CrossRef] [PubMed]
53. Dwivedi, Y.; Roy, B.; Lugli, G.; Rizavi, H.; Zhang, H.; Smalheiser, N.R. Chronic Corticosterone-Mediated Dysregulation of MicroRNA Network in Prefrontal Cortex of Rats: Relevance to Depression Pathophysiology. *Transl. Psychiatry* **2015**, *5*, e682. [CrossRef] [PubMed]
54. Rutter, K.J.; Watson, R.E.B.; Cotterell, L.F.; Brenn, T.; Griffiths, C.E.M.; Rhodes, L.E. Severely Photosensitive Psoriasis: A Phenotypically Defined Patient Subset. *J. Investig. Dermatol.* **2009**, *129*, 2861–2867. [CrossRef] [PubMed]
55. Cicarma, E.; Mørk, C.; Porojnicu, A.C.; Juzeniene, A.; Tam, T.T.T.; Dahlback, A.; Moan, J. Influence of Narrowband UVB Phototherapy on Vitamin D and Folate Status: Narrowband UVB, Vitamin D and Folate. *Exp. Dermatol.* **2009**, *19*, e67–e72. [CrossRef] [PubMed]
56. Sloka, S.; Silva, C.; Wang, J.; Yong, V.W. Predominance of Th2 Polarization by Vitamin D through a STAT6-Dependent Mechanism. *J. Neuroinflamm.* **2011**, *8*, 56. [CrossRef] [PubMed]
57. Xie, Z.; Komuves, L.; Yu, Q.-C.; Elalieh, H.; Ng, D.C.; Leary, C.; Chang, S.; Crumrine, D.; Bikle, D.D.; Yoshizawa, T.; et al. Lack of the Vitamin D Receptor Is Associated with Reduced Epidermal Differentiation and Hair Follicle Growth. *J. Investig. Dermatol.* **2002**, *118*, 11–16. [CrossRef]
58. Fetahu, I.S.; Höbaus, J.; KÃ¡llay, E. Vitamin D and the Epigenome. *Front. Physiol.* **2014**, *5*. [CrossRef]
59. Zhang, W.; Xu, J. DNA Methyltransferases and Their Roles in Tumorigenesis. *Biomark. Res.* **2017**, *5*, 1. [CrossRef]
60. Moore, L.D.; Le, T.; Fan, G. DNA Methylation and Its Basic Function. *Neuropsychopharmacol* **2013**, *38*, 23–38. [CrossRef]
61. Zhou, X.; Chen, Y.; Cui, L.; Shi, Y.; Guo, C. Advances in the Pathogenesis of Psoriasis: From Keratinocyte Perspective. *Cell Death Dis.* **2022**, *13*, 81. [CrossRef] [PubMed]
62. Wang, S.; Song, R.; Wang, Z.; Jing, Z.; Wang, S.; Ma, J. S100A8/A9 in Inflammation. *Front. Immunol.* **2018**, *9*, 1298. [CrossRef]
63. Christmann, C.; Zenker, S.; Martens, L.; Hübner, J.; Loser, K.; Vogl, T.; Roth, J. Interleukin 17 Promotes Expression of Alarmins S100A8 and S100A9 During the Inflammatory Response of Keratinocytes. *Front. Immunol.* **2021**, *11*, 599947. [CrossRef]
64. Roberson, E.D.O.; Liu, Y.; Ryan, C.; Joyce, C.E.; Duan, S.; Cao, L.; Martin, A.; Liao, W.; Menter, A.; Bowcock, A.M. A Subset of Methylated CpG Sites Differentiate Psoriatic from Normal Skin. *J. Investig. Dermatol.* **2012**, *132*, 583–592. [CrossRef] [PubMed]
65. Zhang, P.; Zhao, M.; Liang, G.; Yin, G.; Huang, D.; Su, F.; Zhai, H.; Wang, L.; Su, Y.; Lu, Q. Whole-Genome DNA Methylation in Skin Lesions from Patients with Psoriasis Vulgaris. *J. Autoimmun.* **2013**, *41*, 17–24. [CrossRef] [PubMed]
66. Chandra, A.; Senapati, S.; Roy, S.; Chatterjee, G.; Chatterjee, R. Epigenome-Wide DNA Methylation Regulates Cardinal Pathological Features of Psoriasis. *Clin. Epigenet.* **2018**, *10*, 108. [CrossRef] [PubMed]
67. Hou, R.; Yin, G.; An, P.; Wang, C.; Liu, R.; Yang, Y.; Yan, X.; Li, J.; Li, X.; Zhang, K. DNA Methylation of Dermal MSCs in Psoriasis: Identification of Epigenetically Dysregulated Genes. *J. Dermatol. Sci.* **2013**, *72*, 103–109. [CrossRef]
68. Zhou, F.; Wang, W.; Shen, C.; Li, H.; Zuo, X.; Zheng, X.; Yue, M.; Zhang, C.; Yu, L.; Chen, M.; et al. Epigenome-Wide Association Analysis Identified Nine Skin DNA Methylation Loci for Psoriasis. *J. Investig. Dermatol.* **2016**, *136*, 779–787. [CrossRef]
69. Zhang, P.; Su, Y.; Chen, H.; Zhao, M.; Lu, Q. Abnormal DNA Methylation in Skin Lesions and PBMCs of Patients with Psoriasis Vulgaris. *J. Dermatol. Sci.* **2010**, *60*, 40–42. [CrossRef]
70. Bai, J.; Liu, Z.; Xu, Z.; Ke, F.; Zhang, L.; Zhu, H.; Lou, F.; Wang, H.; Fei, Y.; Shi, Y.-L.; et al. Epigenetic Downregulation of SFRP4 Contributes to Epidermal Hyperplasia in Psoriasis. *J. Immunol.* **2015**, *194*, 4185–4198. [CrossRef]
71. Tian, F.; Mauro, T.M.; Li, Z. The Pathological Role of Wnt5a in Psoriasis and Psoriatic Arthritis. *J. Cell Mol. Med.* **2019**, *23*, 5876–5883. [CrossRef] [PubMed]

72. Ruchusatsawat, K.; Wongpiyabovorn, J.; Shuangshoti, S.; Hirankarn, N.; Mutirangura, A. SHP-1 Promoter 2 Methylation in Normal Epithelial Tissues and Demethylation in Psoriasis. *J. Mol. Med.* **2006**, *84*, 175–182. [CrossRef] [PubMed]
73. Ruchusatsawat, K.; Wongpiyabovorn, J.; Protjaroen, P.; Chaipipat, M.; Shuangshoti, S.; Thorner, P.S.; Mutirangura, A. Parakeratosis in Skin Is Associated with Loss of Inhibitor of Differentiation 4 via Promoter Methylation. *Hum. Pathol.* **2011**, *42*, 1878–1887. [CrossRef] [PubMed]
74. Chen, M.; Chen, Z.-Q.; Cui, P.-G.; Yao, X.; Li, Y.-M.; Li, A.-S.; Gong, J.-Q.; Cao, Y.-H. The Methylation Pattern of P16 ^{INK4a} Gene Promoter in Psoriatic Epidermis and Its Clinical Significance. *Br. J. Derm.* **2008**, *158*, 987–993. [CrossRef]
75. Sheng, Y.; Wen, L.; Zheng, X.; Li, M.; Wang, D.; Chen, S.; Li, R.; Tang, L.; Zhou, F. *CYP2S1* Might Regulate Proliferation and Immune Response of Keratinocyte in Psoriasis. *Epigenetics* **2021**, *16*, 618–628. [CrossRef]
76. Maeda, T.; Hanna, A.N.; Sim, A.B.; Chua, P.P.; Chong, M.T.; Tron, V.A. GADD45 Regulates G2/M Arrest, DNA Repair, and Cell Death in Keratinocytes Following Ultraviolet Exposure. *J. Investig. Dermatol.* **2002**, *119*, 22–26. [CrossRef]
77. Rodríguez-Jiménez, P.; Fernández-Messina, L.; Ovejero-Benito, M.C.; Chicharro, P.; Vera-Tomé, P.; Vara, A.; Cibrian, D.; Martínez-Fleta, P.; Jiménez-Fernández, M.; Sánchez-García, I.; et al. Growth Arrest and DNA Damage-Inducible Proteins (GADD45) in Psoriasis. *Sci. Rep.* **2021**, *11*, 14579. [CrossRef]
78. Liu, S.G.; Luo, G.P.; Qu, Y.B.; Chen, Y.F. Indirubin Inhibits Wnt/β-Catenin Signal Pathway via Promoter Demethylation of WIF-1. *BMC Complement Med.* **2020**, *20*, 250. [CrossRef]
79. Cai, Y.; Fleming, C.; Yan, J. New Insights of T Cells in the Pathogenesis of Psoriasis. *Cell Mol. Immunol.* **2012**, *9*, 302–309. [CrossRef]
80. Han, J.; Park, S.-G.; Bae, J.-B.; Choi, J.; Lyu, J.-M.; Park, S.H.; Kim, H.S.; Kim, Y.-J.; Kim, S.; Kim, T.-Y. The Characteristics of Genome-Wide DNA Methylation in Naïve CD4+ T Cells of Patients with Psoriasis or Atopic Dermatitis. *Biochem. Biophys. Res. Commun.* **2012**, *422*, 157–163. [CrossRef]
81. Vecellio, M.; Paraboschi, E.M.; Ceribelli, A.; Isailovic, N.; Motta, F.; Cardamone, G.; Robusto, M.; Asselta, R.; Brescianini, S.; Sacrini, F.; et al. DNA Methylation Signature in Monozygotic Twins Discordant for Psoriatic Disease. *Front. Cell Dev. Biol.* **2021**, *9*, 778677. [CrossRef] [PubMed]
82. Charras, A.; Garau, J.; Hofmann, S.R.; Carlsson, E.; Cereda, C.; Russ, S.; Abraham, S.; Hedrich, C.M. DNA Methylation Patterns in CD8+ T Cells Discern Psoriasis From Psoriatic Arthritis and Correlate With Cutaneous Disease Activity. *Front. Cell Dev. Biol.* **2021**, *9*, 746145. [CrossRef] [PubMed]
83. Ngalamika, O.; Liang, G.; Zhao, M.; Yu, X.; Yang, Y.; Yin, H.; Liu, Y.; Yung, S.; Chan, T.M.; Lu, Q. Peripheral Whole Blood FOXP3 TSDR Methylation: A Potential Marker in Severity Assessment of Autoimmune Diseases and Chronic Infections. *Immunol. Investig.* **2015**, *44*, 126–136. [CrossRef] [PubMed]
84. Zhang, K.; Zhang, R.; Li, X.; Yin, G.; Niu, X. Promoter Methylation Status of P15 and P21 Genes in HPP-CFCs of Bone Marrow of Patients with Psoriasis. *Eur. J. Dermatol.* **2009**, *19*, 141–146. [CrossRef]
85. Zhang, K.; Zhang, R.; Li, X.; Yin, G.; Niu, X.; Hou, R. The MRNA Expression and Promoter Methylation Status of the P16 Gene in Colony-Forming Cells with High Proliferative Potential in Patients with Psoriasis. *Clin. Exp. Derm.* **2007**, *32*, 702–708. [CrossRef]
86. Li, X.; Zhao, X.; Xing, J.; Li, J.; He, F.; Hou, R.; Wang, Q.; Yin, G.; Li, X.; Zhang, K. Different Epigenome Regulation and Transcriptome Expression of CD4 $^+$ and CD8 $^+$ T Cells from Monozygotic Twins Discordant for Psoriasis. *Australas. J. Derm.* **2020**, *61*. [CrossRef]
87. MacAlpine, D.M.; Almouzni, G. Chromatin and DNA Replication. *Cold Spring Harb. Perspect. Biol.* **2013**, *5*, a010207. [CrossRef]
88. Bannister, A.J.; Kouzarides, T. Regulation of Chromatin by Histone Modifications. *Cell Res.* **2011**, *21*, 381–395. [CrossRef]
89. Li, H.; Yao, Q.; Mariscal, A.G.; Wu, X.; Hülse, J.; Pedersen, E.; Helin, K.; Waisman, A.; Vinkel, C.; Thomsen, S.F.; et al. Epigenetic Control of IL-23 Expression in Keratinocytes Is Important for Chronic Skin Inflammation. *Nat. Commun.* **2018**, *9*, 1420. [CrossRef]
90. Zhang, T.; Yang, L.; Ke, Y.; Lei, J.; Shen, S.; Shao, S.; Zhang, C.; Zhu, Z.; Dang, E.; Wang, G. EZH2-Dependent Epigenetic Modulation of Histone H3 Lysine-27 Contributes to Psoriasis by Promoting Keratinocyte Proliferation. *Cell Death Dis.* **2020**, *11*, 826. [CrossRef]
91. Busch, F.; Mobasheri, A.; Shayan, P.; Stahlmann, R.; Shakibaei, M. Sirt-1 Is Required for the Inhibition of Apoptosis and Inflammatory Responses in Human Tenocytes. *J. Biol. Chem.* **2012**, *287*, 25770–25781. [CrossRef] [PubMed]
92. McCormick, R.; McDonagh, B.; Goljanek-Whysall, K. MicroRNA-SIRT-1 Interactions: Key Regulators of Adult Skeletal Muscle Homeostasis?: Perspectives. *J. Physiol.* **2017**, *595*, 3253–3254. [CrossRef]
93. Hwang, Y.-J.; Na, J.-I.; Byun, S.-Y.; Kwon, S.-H.; Yang, S.-H.; Lee, H.-S.; Choi, H.-R.; Cho, S.; Youn, S.W.; Park, K.-C. Histone Deacetylase 1 and Sirtuin 1 Expression in Psoriatic Skin: A Comparison between Guttate and Plaque Psoriasis. *Life* **2020**, *10*, 157. [CrossRef] [PubMed]
94. Xia, X.; Cao, G.; Sun, G.; Zhu, L.; Tian, Y.; Song, Y.; Guo, C.; Wang, X.; Zhong, J.; Zhou, W.; et al. GLS1-Mediated Glutaminolysis Unbridled by MALT1 Protease Promotes Psoriasis Pathogenesis. *J. Clin. Investig.* **2020**, *130*, 5180–5196. [CrossRef] [PubMed]
95. Masalha, M.; Ben-Dov, I.Z.; Ram, O.; Meningher, T.; Jacob-Hirsch, J.; Kassem, R.; Sidi, Y.; Avni, D. H3K27Ac Modification and Gene Expression in Psoriasis. *J. Dermatol. Sci.* **2021**, *103*, 93–100. [CrossRef]
96. Liao, Y.; Su, Y.; Wu, R.; Zhang, P.; Feng, C. Overexpression of Wilms Tumor 1 Promotes IL-1β Expression by Upregulating Histone Acetylation in Keratinocytes. *Int. Immunopharmacol.* **2021**, *96*, 107793. [CrossRef]
97. Zeng, J.; Zhang, Y.; Zhang, H.; Zhang, Y.; Gao, L.; Tong, X.; Xie, Y.; Hu, Q.; Chen, C.; Ding, S.; et al. RPL22 Overexpression Promotes Psoriasis-Like Lesion by Inducing Keratinocytes Abnormal Biological Behavior. *Front. Immunol.* **2021**, *12*, 699900. [CrossRef]

98. Ovejero-Benito, M.C.; Reolid, A.; Sánchez-Jiménez, P.; Saiz-Rodríguez, M.; Muñoz-Aceituno, E.; Llamas-Velasco, M.; Martín-Vilchez, S.; Cabaleiro, T.; Román, M.; Ochoa, D.; et al. Histone Modifications Associated with Biological Drug Response in Moderate-to-severe Psoriasis. *Exp. Derm.* **2018**, *27*, 1361–1371. [CrossRef]
99. Liu, Z.; Cao, W.; Xu, L.; Chen, X.; Zhan, Y.; Yang, Q.; Liu, S.; Chen, P.; Jiang, Y.; Sun, X.; et al. The Histone H3 Lysine-27 Demethylase Jmjd3 Plays a Critical Role in Specific Regulation of Th17 Cell Differentiation. *J. Mol. Cell Biol.* **2015**, *7*, 505–516. [CrossRef]
100. Zhang, P.; Su, Y.; Zhao, M.; Huang, W.; Lu, Q. Abnormal Histone Modifications in PBMCs from Patients with Psoriasis Vulgaris. *Eur. J. Dermatol.* **2011**, *21*, 552–557. [CrossRef]
101. O'Brien, J.; Hayder, H.; Zayed, Y.; Peng, C. Overview of MicroRNA Biogenesis, Mechanisms of Actions, and Circulation. *Front. Endocrinol.* **2018**, *9*, 402. [CrossRef]
102. Meng, S.; Zhou, H.; Feng, Z.; Xu, Z.; Tang, Y.; Li, P.; Wu, M. CircRNA: Functions and Properties of a Novel Potential Biomarker for Cancer. *Mol. Cancer* **2017**, *16*, 94. [CrossRef] [PubMed]
103. Wu, G.-C.; Pan, H.-F.; Leng, R.-X.; Wang, D.-G.; Li, X.-P.; Li, X.-M.; Ye, D.-Q. Emerging Role of Long Noncoding RNAs in Autoimmune Diseases. *Autoimmun. Rev.* **2015**, *14*, 798–805. [CrossRef] [PubMed]
104. Jiang, B.-W.; Zhang, W.-J.; Wang, Y.; Tan, L.-P.; Bao, Y.-L.; Song, Z.-B.; Yu, C.-L.; Wang, S.-Y.; Liu, L.; Li, Y.-X. Convallatoxin Induces HaCaT Cell Necroptosis and Ameliorates Skin Lesions in Psoriasis-like Mouse Models. *Biomed. Pharmacother.* **2020**, *121*, 109615. [CrossRef] [PubMed]
105. Chowdhari, S.; Sardana, K.; Saini, N. MiR-4516, a MicroRNA Downregulated in Psoriasis Inhibits Keratinocyte Motility by Targeting Fibronectin/Integrin A9 Signaling. *Biochim. Et Biophys. Acta (BBA)—Mol. Basis Dis.* **2017**, *1863*, 3142–3152. [CrossRef]
106. Ichihara, A.; Jinnin, M.; Yamane, K.; Fujisawa, A.; Sakai, K.; Masuguchi, S.; Fukushima, S.; Maruo, K.; Ihn, H. MicroRNA-Mediated Keratinocyte Hyperproliferation in Psoriasis Vulgaris: MicroRNA-Mediated Keratinocyte Hyperproliferation in Psoriasis. *Br. J. Dermatol.* **2011**, *165*, 1003–1010. [CrossRef] [PubMed]
107. Yan, J.J.; Qiao, M.; Li, R.H.; Zhao, X.T.; Wang, X.Y.; Sun, Q. Downregulation of MiR-145-5p Contributes to Hyperproliferation of Keratinocytes and Skin Inflammation in Psoriasis. *Br. J. Dermatol.* **2019**, *180*, 365–372. [CrossRef]
108. Guinea-Viniegra, J.; Jiménez, M.; Schonthaler, H.B.; Navarro, R.; Delgado, Y.; José Concha-Garzón, M.; Tschachler, E.; Obad, S.; Daudén, E.; Wagner, E.F. Targeting *MiR-21* to Treat Psoriasis. *Sci. Transl. Med.* **2014**, *6*. [CrossRef]
109. Magenta, A.; D'Agostino, M.; Sileno, S.; Di Vito, L.; Uras, C.; Abeni, D.; Martino, F.; Barillà, F.; Madonna, S.; Albanesi, C.; et al. The Oxidative Stress-Induced MiR-200c Is Upregulated in Psoriasis and Correlates with Disease Severity and Determinants of Cardiovascular Risk. *Oxidative Med. Cell. Longev.* **2019**, *2019*, 8061901. [CrossRef]
110. Huang, C.; Zhong, W.; Ren, X.; Huang, X.; Li, Z.; Chen, C.; Jiang, B.; Chen, Z.; Jian, X.; Yang, L.; et al. MiR-193b-3p–ERBB4 Axis Regulates Psoriasis Pathogenesis via Modulating Cellular Proliferation and Inflammatory-Mediator Production of Keratinocytes. *Cell Death Dis.* **2021**, *12*, 963. [CrossRef]
111. Pan, M.; Huang, Y.; Zhu, X.; Lin, X.; Luo, D. MiR-125b-mediated Regulation of Cell Proliferation through the Jagged-1/Notch Signaling Pathway by Inhibiting BRD4 Expression in Psoriasis. *Mol. Med. Rep.* **2019**. [CrossRef] [PubMed]
112. Xu, N.; Brodin, P.; Wei, T.; Meisgen, F.; Eidsmo, L.; Nagy, N.; Kemeny, L.; Ståhle, M.; Sonkoly, E.; Pivarcsi, A. MiR-125b, a MicroRNA Downregulated in Psoriasis, Modulates Keratinocyte Proliferation by Targeting FGFR2. *J. Investig. Dermatol.* **2011**, *131*, 1521–1529. [CrossRef] [PubMed]
113. Yu, X.; An, J.; Hua, Y.; Li, Z.; Yan, N.; Fan, W.; Su, C. MicroRNA-194 Regulates Keratinocyte Proliferation and Differentiation by Targeting Grainyhead-like 2 in Psoriasis. *Pathol.—Res. Pract.* **2017**, *213*, 89–97. [CrossRef] [PubMed]
114. Rongna, A.; Yu, P.; Hao, S.; Li, Y. MiR-876-5p Suppresses Cell Proliferation by Targeting Angiopoietin-1 in the Psoriasis. *Biomed. Pharmacother.* **2018**, *103*, 1163–1169. [CrossRef]
115. Zheng, Y.; Cai, B.; Li, X.; Li, D.; Yin, G. MiR-125b-5p and MiR-181b-5p Inhibit Keratinocyte Proliferation in Skin by Targeting Akt3. *Eur. J. Pharmacol.* **2019**, *862*, 172659. [CrossRef]
116. Jiang, M.; Sun, Z.; Dang, E.; Li, B.; Fang, H.; Li, J.; Gao, L.; Zhang, K.; Wang, G. TGFβ/SMAD/MicroRNA-486-3p Signaling Axis Mediates Keratin 17 Expression and Keratinocyte Hyperproliferation in Psoriasis. *J. Investig. Dermatol.* **2017**, *137*, 2177–2186. [CrossRef]
117. Zhao, J.; Wang, F.; Tian, Q.; Dong, J.; Chen, L.; Hu, R. Involvement of MiR-214-3p/FOXM1 Axis During the Progression of Psoriasis. *Inflammation* **2022**, *45*, 267–278. [CrossRef]
118. Yan, S.; Xu, Z.; Lou, F.; Zhang, L.; Ke, F.; Bai, J.; Liu, Z.; Liu, J.; Wang, H.; Zhu, H.; et al. NF-KB-Induced MicroRNA-31 Promotes Epidermal Hyperplasia by Repressing Protein Phosphatase 6 in Psoriasis. *Nat. Commun.* **2015**, *6*, 7652. [CrossRef]
119. Zhang, W.; Yi, X.; An, Y.; Guo, S.; Li, S.; Song, P.; Chang, Y.; Zhang, S.; Gao, T.; Wang, G.; et al. MicroRNA-17-92 Cluster Promotes the Proliferation and the Chemokine Production of Keratinocytes: Implication for the Pathogenesis of Psoriasis. *Cell Death Dis.* **2018**, *9*, 567. [CrossRef]
120. Zhang, D.; Wang, Y.; Xia, Y.; Huo, J.; Zhang, Y.; Yang, P.; Zhang, Y.; Wang, X. Repression of MiR-142–3p Alleviates Psoriasis-like Inflammation by Repressing Proliferation and Promoting Apoptosis of Keratinocytes via Targeting Sema3A. *Mol. Cell. Probes* **2020**, *52*, 101573. [CrossRef]
121. Xu, L.; Leng, H.; Shi, X.; Ji, J.; Fu, J.; Leng, H. MiR-155 Promotes Cell Proliferation and Inhibits Apoptosis by PTEN Signaling Pathway in the Psoriasis. *Biomed. Pharmacother.* **2017**, *90*, 524–530. [CrossRef] [PubMed]

122. Wang, R.; Wang, F.; Cao, H.; Yang, J. MiR-223 Regulates Proliferation and Apoptosis of IL-22-Stimulated HaCat Human Keratinocyte Cell Lines via the PTEN/Akt Pathway. *Life Sci.* **2019**, *230*, 28–34. [CrossRef] [PubMed]
123. Sonkoly, E.; Wei, T.; Janson, P.C.J.; Sääf, A.; Lundeberg, L.; Tengvall-Linder, M.; Norstedt, G.; Alenius, H.; Homey, B.; Scheynius, A.; et al. MicroRNAs: Novel Regulators Involved in the Pathogenesis of Psoriasis? *PLoS ONE* **2007**, *2*, e610. [CrossRef] [PubMed]
124. Zibert, J.R.; Løvendorf, M.B.; Litman, T.; Olsen, J.; Kaczkowski, B.; Skov, L. MicroRNAs and Potential Target Interactions in Psoriasis. *J. Dermatol. Sci.* **2010**, *58*, 177–185. [CrossRef]
125. Fu, D.; Yu, W.; Li, M.; Wang, H.; Liu, D.; Song, X.; Li, Z.; Tian, Z. MicroRNA-138 Regulates the Balance of Th1/Th2 via Targeting RUNX3 in Psoriasis. *Immunol. Lett.* **2015**, *166*, 55–62. [CrossRef]
126. Zheng, Y.-Z.; Chen, C.-F.; Jia, L.-Y.; Yu, T.-G.; Sun, J.; Wang, X.-Y. Correlation between MicroRNA-143 in Peripheral Blood Mononuclear Cells and Disease Severity in Patients with Psoriasis Vulgaris. *Oncotarget* **2017**, *8*, 51288–51295. [CrossRef]
127. Collaborative Association Study of Psoriasis (CASP); Genetic Analysis of Psoriasis Consortium; Psoriasis Association Genetics Extension; Wellcome Trust Case Control Consortium 2; Tsoi, L.C.; Spain, S.L.; Knight, J.; Ellinghaus, E.; Stuart, P.E.; Capon, F.; et al. Identification of 15 New Psoriasis Susceptibility Loci Highlights the Role of Innate Immunity. *Nat. Genet.* **2012**, *44*, 1341–1348. [CrossRef]
128. García-Rodríguez, S.; Arias-Santiago, S.; Blasco-Morente, G.; Orgaz-Molina, J.; Rosal-Vela, A.; Navarro, P.; Magro-Checa, C.; Martínez-López, A.; Ruiz, J.-C.; Raya, E.; et al. Increased Expression of MicroRNA-155 in Peripheral Blood Mononuclear Cells from Psoriasis Patients Is Related to Disease Activity. *J. Eur. Acad. Derm. Venereol.* **2017**, *31*, 312–322. [CrossRef]
129. Zhao, M.; Wang, L.; Liang, G.; Zhang, P.; Deng, X.; Tang, Q.; Zhai, H.; Chang, C.C.; Su, Y.; Lu, Q. Up-Regulation of MicroRNA-210 Induces Immune Dysfunction via Targeting FOXP3 in CD4+ T Cells of Psoriasis Vulgaris. *Clin. Immunol.* **2014**, *150*, 22–30. [CrossRef]
130. Duan, Y.; Zou, J.; Mao, J.; Guo, D.; Wu, M.; Xu, N.; Zhou, J.; Zhang, Y.; Guo, W.; Jin, W. Plasma MiR-126 Expression Correlates with Risk and Severity of Psoriasis and Its High Level at Baseline Predicts Worse Response to Tripterygium Wilfordii Hook F in Combination with Acitretin. *Biomed. Pharmacother.* **2019**, *115*, 108761. [CrossRef]
131. Wang, X.-Y.; Chen, X.-Y.; Li, J.; Zhang, H.-Y.; Liu, J.; Sun, L.-D. MiR-200a Expression in CD4+ T Cells Correlates with the Expression of Th17/Treg Cells and Relevant Cytokines in Psoriasis Vulgaris: A Case Control Study. *Biomed. Pharmacother.* **2017**, *93*, 1158–1164. [CrossRef] [PubMed]
132. Borska, L.; Andrys, C.; Chmelarova, M.; Kovarikova, H.; Krejsek, J.; Hamakova, K.; Beranek, M.; Palicka, V.; Kremlacek, J.; Borsky, P.; et al. Roles of MiR-31 and Endothelin-1 in Psoriasis Vulgaris: Pathophysiological Functions and Potential Biomarkers. *Physiol. Res.* **2017**, *66*, 987–992. [CrossRef] [PubMed]
133. García-Rodríguez, S.; Arias-Santiago, S.; Orgaz-Molina, J.; Magro-Checa, C.; Valenzuela, I.; Navarro, P.; Naranjo-Sintes, R.; Sancho, J.; Zubiaur, M. Abnormal Levels of Expression of Plasma MicroRNA-33 in Patients With Psoriasis. *Actas Dermo-Sifiliográficas (Engl. Ed.)* **2014**, *105*, 497–503. [CrossRef]
134. Guo, S.; Zhang, W.; Wei, C.; Wang, L.; Zhu, G.; Shi, Q.; Li, S.; Ge, R.; Li, K.; Gao, L.; et al. Serum and Skin Levels of MiR-369-3p in Patients with Psoriasis and Their Correlation with Disease Severity. *Eur. J. Dermatol.* **2013**, *23*, 608–613. [CrossRef] [PubMed]
135. Jia, H.-Y.; Zhang, K.; Lu, W.-J.; Xu, G.-W.; Zhang, J.-F.; Tang, Z.-L. LncRNA MEG3 Influences the Proliferation and Apoptosis of Psoriasis Epidermal Cells by Targeting MiR-21/Caspase-8. *BMC Mol. Cell Biol.* **2019**, *20*, 46. [CrossRef]
136. Qiao, M.; Li, R.; Zhao, X.; Yan, J.; Sun, Q. Up-Regulated LncRNA-MSX2P1 Promotes the Growth of IL-22-Stimulated Keratinocytes by Inhibiting MiR-6731-5p and Activating S100A7. *Exp. Cell Res.* **2018**, *363*, 243–254. [CrossRef]
137. Abdallah, H.Y.; Tawfik, N.Z.; Soliman, N.H.; Eldeen, L.A.T. The LncRNA PRINS-MiRNA-MRNA Axis Gene Expression Profile as a Circulating Biomarker Panel in Psoriasis. *Mol. Diagn.* **2022**. [CrossRef]
138. Ahmed Shehata, W.; Maraee, A.; Abd El Monem Ellaithy, M.; Tayel, N.; Abo-Ghazala, A.; Mohammed El-Hefnawy, S. Circulating Long Noncoding RNA Growth Arrest-specific Transcript 5 as a Diagnostic Marker and Indicator of Degree of Severity in Plaque Psoriasis. *Int. J. Derm.* **2021**, *60*, 973–979. [CrossRef]
139. Liu, X.; Frost, J.; Bowcock, A.; Zhang, W. Canonical and Interior Circular RNAs Function as Competing Endogenous RNAs in Psoriatic Skin. *Int. J. Mol. Sci.* **2021**, *22*, 5182. [CrossRef]
140. Yazici, S.; Guner, R.; Akyol, M.; Tuzemen Bayyurt, E.; Arslan, S. Research on Hotair and 7SL-RNA Gene Expression Levels in Psoriasis Vulgaris. *Indian J. Derm.* **2021**, *66*, 704. [CrossRef]
141. Moldovan, L.I.; Tsoi, L.C.; Ranjitha, U.; Hager, H.; Weidinger, S.; Gudjonsson, J.E.; Kjems, J.; Kristensen, L.S. Characterization of Circular RNA Transcriptomes in Psoriasis and Atopic Dermatitis Reveals Disease-specific Expression Profiles. *Exp. Derm.* **2021**, *30*, 1187–1196. [CrossRef] [PubMed]
142. Gao, J.; Chen, F.; Hua, M.; Guo, J.; Nong, Y.; Tang, Q.; Zhong, F.; Qin, L. Knockdown of LncRNA MIR31HG Inhibits Cell Proliferation in Human HaCaT Keratinocytes. *Biol. Res.* **2018**, *51*, 30. [CrossRef] [PubMed]
143. Li, C.; Li, H.; Huang, L.; Kong, Y.; Chen, F.; Liang, J.; Yu, H.; Yao, Z. H19 LncRNA Regulates Keratinocyte Differentiation by Targeting MiR-130b-3p. *Cell Death Dis.* **2017**, *8*, e3174. [CrossRef] [PubMed]
144. Szegedi, K.; Sonkoly, E.; Nagy, N.; Németh, I.B.; Bata-Csörgő, Z.; Kemény, L.; Dobozy, A.; Széll, M. The Anti-Apoptotic Protein G1P3 Is Overexpressed in Psoriasis and Regulated by the Non-Coding RNA, PRINS. *Exp. Dermatol.* **2010**, *19*, 269–278. [CrossRef]
145. Cai, P.; Otten, A.B.C.; Cheng, B.; Ishii, M.A.; Zhang, W.; Huang, B.; Qu, K.; Sun, B.K. A Genome-Wide Long Noncoding RNA CRISPRi Screen Identifies *PRANCR* as a Novel Regulator of Epidermal Homeostasis. *Genome Res.* **2020**, *30*, 22–34. [CrossRef] [PubMed]

146. Jin, N.; George, T.L.; Otterson, G.A.; Verschraegen, C.; Wen, H.; Carbone, D.; Herman, J.; Bertino, E.M.; He, K. Advances in Epigenetic Therapeutics with Focus on Solid Tumors. *Clin. Epigenet.* **2021**, *13*, 83. [CrossRef]
147. Lee, J.-H.; Kim, J.-S.; Park, S.-Y.; Lee, Y.-J. Resveratrol Induces Human Keratinocyte Damage via the Activation of Class III Histone Deacetylase, Sirt1. *Oncol. Rep.* **2016**, *35*, 524–529. [CrossRef]
148. Samuelov, L.; Bochner, R.; Magal, L.; Malovitski, K.; Sagiv, N.; Nousbeck, J.; Keren, A.; Fuchs-Telem, D.; Sarig, O.; Gilhar, A.; et al. Vorinostat, a Histone Deacetylase Inhibitor, as a Potential Novel Treatment for Psoriasis. *Exp. Dermatol.* **2022**, *31*, 567–576. [CrossRef]
149. Jorn Bovenschen, H.; van de Kerkhof, P.C.; van Erp, P.E.; Woestenenk, R.; Joosten, I.; Koenen, H.J.P.M. Foxp3+ Regulatory T Cells of Psoriasis Patients Easily Differentiate into IL-17A-Producing Cells and Are Found in Lesional Skin. *J. Investig. Dermatol.* **2011**, *131*, 1853–1860. [CrossRef]
150. Tao, R.; de Zoeten, E.F.; Özkaynak, E.; Chen, C.; Wang, L.; Porrett, P.M.; Li, B.; Turka, L.A.; Olson, E.N.; Greene, M.I.; et al. Deacetylase Inhibition Promotes the Generation and Function of Regulatory T Cells. *Nat. Med.* **2007**, *13*, 1299–1307. [CrossRef] [PubMed]
151. Yang, R.; Chowdhury, S.; Choudhary, V.; Chen, X.; Bollag, W.B. Keratinocyte Aquaporin-3 Expression Induced by Histone Deacetylase Inhibitors Is Mediated in Part by Peroxisome Proliferator-activated Receptors (PPARs). *Exp. Derm.* **2020**, *29*, 380–386. [CrossRef] [PubMed]
152. Bollag, W.B.; Aitkens, L.; White, J.; Hyndman, K.A. Aquaporin-3 in the Epidermis: More than Skin Deep. *Am. J. Physiol.-Cell Physiol.* **2020**, *318*, C1144–C1153. [CrossRef] [PubMed]
153. Feng, H.; Wu, R.; Zhang, S.; Kong, Y.; Liu, Z.; Wu, H.; Wang, H.; Su, Y.; Zhao, M.; Lu, Q. Topical Administration of Nanocarrier MiRNA-210 Antisense Ameliorates Imiquimod-induced Psoriasis-like Dermatitis in Mice. *J. Dermatol.* **2020**, *47*, 147–154. [CrossRef]
154. Lifshiz Zimon, R.; Lerman, G.; Elharrar, E.; Meningher, T.; Barzilai, A.; Masalha, M.; Chintakunta, R.; Hollander, E.; Goldbart, R.; Traitel, T.; et al. Ultrasound Targeting of Q-Starch/MiR-197 Complexes for Topical Treatment of Psoriasis. *J. Control. Release* **2018**, *284*, 103–111. [CrossRef] [PubMed]

biomedicines

MDPI

Review

Multi-Omic Biomarkers for Patient Stratification in Sjogren's Syndrome—A Review of the Literature

Lucia Martin-Gutierrez [1], Robert Wilson [2], Madhura Castelino [2], Elizabeth C. Jury [1] and Coziana Ciurtin [3,*]

1 Centre for Rheumatology Research, Division of Medicine, University College London, London WC1E 6JF, UK; l.martin-gutierrez@ucl.ac.uk (L.M.-G.); e.jury@ucl.ac.uk (E.C.J.)
2 Department of Rheumatology, University College London Hospitals NHS Trust, London NW1 2PG, UK; robert.wilson24@nhs.net (R.W.); madhura.castelino@nhs.net (M.C.)
3 Centre for Adolescent Rheumatology Versus Arthritis, Division of Medicine, University College London, London WC1E 6JF, UK
* Correspondence: c.ciurtin@ucl.ac.uk

Abstract: Sjögren's syndrome (SS) is a heterogeneous autoimmune rheumatic disease (ARD) characterised by dryness due to the chronic lymphocytic infiltration of the exocrine glands. Patients can also present other extra glandular manifestations, such as arthritis, anaemia and fatigue or various types of organ involvement. Due to its heterogenicity, along with the lack of effective treatments, the diagnosis and management of this disease is challenging. The objective of this review is to summarize recent multi-omic publications aiming to identify biomarkers in tears, saliva and peripheral blood from SS patients that could be relevant for their better stratification aiming at improved treatment selection and hopefully better outcomes. We highlight the relevance of pro-inflammatory cytokines and interferon (IFN) as biomarkers identified in higher concentrations in serum, saliva and tears. Transcriptomic studies confirmed the upregulation of IFN and interleukin signalling in patients with SS, whereas immunophenotyping studies have shown dysregulation in the immune cell population frequencies, specifically CD4^{+}and C8^{+}T activated cells, and their correlations with clinical parameters, such as disease activity scores. Lastly, we discussed emerging findings derived from different omic technologies which can provide integrated knowledge about SS pathogenesis and facilitate personalised medicine approaches leading to better patient outcomes in the future.

Keywords: Sjogren's syndrome; patient stratification; clinical relevance; multi-omics

Citation: Martin-Gutierrez, L.; Wilson, R.; Castelino, M.; Jury, E.C.; Ciurtin, C. Multi-Omic Biomarkers for Patient Stratification in Sjogren's Syndrome—A Review of the Literature. *Biomedicines* **2022**, *10*, 1773. https://doi.org/10.3390/biomedicines10081773

Academic Editor: Marianna Christodoulou

Received: 22 June 2022
Accepted: 20 July 2022
Published: 22 July 2022

1. Introduction

Sjögren's syndrome (SS) is an autoimmune rheumatic disease (ARD) characterised by a chronic inflammatory process associated with lymphocytic infiltrate affecting the exocrine glands. The disease has significant heterogeneity in clinical presentation according to age at disease onset, type of organ involvement, as well as serological features and response to therapy [1,2]. When the disease occurs on its own, it is called primary SS (pSS), while when it accompanies other autoimmune conditions, it is defined as secondary SS (sSS). Various classification criteria have been used to define pSS and exclude mimicking pathology, with the most recent ones being the data and consensus-driven American College of Rheumatology/European League Against Rheumatism Classification Criteria proposed in 2016 [3].

There are currently no universally accepted classification criteria for sSS and some experts argue that making a distinction between pSS and sSS is not adequate anymore, as both phenotypes represent the same disease [4]. Moreover, the classification criteria validated in adults have minimal utility in SS with childhood-onset (defined as disease onset before the age of 18 years), as the disease presentation in children and young people, although rare, is different [5]. This, in addition to the lack of validated classification criteria for childhood-onset SS, further limits the research opportunities for younger people affected by this disease [6].

The disease manifestations vary among patients; some have predominant exocrine glandular involvement leading to dryness, which is the hallmark symptom of the disease. Glandular involvement manifests as dry mouth (xerostomia), dry eyes (xerophthalmia), dry skin (xerosis cutis), as well as vaginal dryness, dry cough, pancreatic dysfunction, salivary gland inflammation/enlargement (e.g., parotitis), etc., while patients with extra-glandular involvement can experience frequent musculoskeletal, haematological, and rarely hepatic, renal, pulmonary, cardiac, peripheral or central nervous system manifestations, as well as less specific symptoms of fatigue (common) or fever and lymphadenopathy (less common) [7]. Clinical presentation usually guides the disease management, which is largely symptomatic for glandular manifestations and involves the use of immunosuppressive treatment approaches in patients with more severe organ involvement [8]. The evidence for the efficacy of various therapies currently recommended for the management of SS is modest overall [9,10], emphasising the need for better research.

As a direct consequence of the disease pathogenesis being centred around the process of autoimmune epithelitis, powered by the interplay between the cells of the innate and adaptive immune systems, and activated by interferons and other pro-inflammatory cytokines leading to chronic immune activation in a host with genetic susceptibility [11], various disease fingerprints can be identified from the peripheral blood, as well as serum, saliva, tears and salivary gland biopsies.

Significant progress has been achieved recently in clinical research in terms of better patient clinical and molecular characterisation [12], but despite this, the management of this condition remains challenging because of patient heterogeneity and various limitations of the way the disease activity and response to treatment are measured. These aspects very likely contribute to the lack of significant treatment advances in SS, despite preliminary signals of the efficacy of various biologic agents in clinical trials [13–15]. Better research into disease pathogenesis and distinct clinical and molecular phenotypes will hopefully enable better patient selection for available therapies as well as new target discoveries.

2. Materials and Methods

This review aimed as identifying the main papers published since 2000 investigating multi-omics (cytokine profiling in serum, tears, saliva, immunophenotyping, genomic, transcriptomic and metabolomic) studies in SS in an effort to identify distinct patient groups (endotypes) which can inform meaningful stratification for better disease characterisation and improved treatment strategies. Publications selected for this review followed these inclusion criteria: Sample size higher than 10, the inclusion of age and gender-matched healthy controls, and data on at least one omic analysis in any biologic sample relevant to SS (blood, serum, tears, saliva, salivary gland biopsy) and published in English. We presented the most informative papers found in the literature in tables, summarising the study design, sample size, control groups, main findings and their clinical relevance.

3. Results

3.1. Multi-Omic Biomarkers for Patient Stratification

3.1.1. Tear Biomarkers

One of the main symptoms of pSS is dry eye (xeropththalmia) as a result of lymphocytic infiltration of the lacrimal glands. Tears represent a valuable biological sample resource because of their proximity to the site of glandular inflammation and they might contain biomarkers that could help us to understand the pathogenesis of pSS, improve its diagnosis and have therapeutic implications.

Recently, several studies have concentrated their efforts on identifying those biomarkers through either cytokine, metabolomic or proteomic tear profiling (Table 1).

In this context, Chen et al. [16] determined the cytokine profile of tears, measured by a 27-plexcytokine assay in 29 pSS patients and 20 gender/age matched controls (non-SS sicca subjects and healthy controls—HCs). Elevated levels of pro-inflammatory cytokines, such as interleukin (IL)-1 receptor agonist (ra), IL-2, IL-17A, interferon (IFN)-γ, Macrophage inflammatory protein-1-β (MIP-1b), and Rantes (Regulated upon Activation, Normal T Cell Expressed and Presumably Secreted) and anti-inflammatory interleukin 4 IL-4, were found in pSS patients compared to controls. Interestingly, higher cytokine levels correlated positively with eye dryness severity and negatively with Schirmer's test which measures the volume of tears secreted over 5 min [16]. These findings are validated by another study Willems et al., which also found an increased concentration of IFN-γ, tumor necrosis factor alfa (TNF-α), IL-2, IL-4, IL-6, IL-10, IL-12p70 and IL-5 in tears of pSS patients compared to HCs [17]. Moreover, they also verified the negative correlation between the Schirmer test and the concentration of IL-2, IL-4, IL-10 and IL-12p70 in tears.

Tears are composed of water, electrolytes, mucins and hundreds of different proteins and metabolites. Urbanski et al. identified a metabolic signature of tears comprising nine metabolites specific to pSS, compared to patients with non-immune dry eye disease [18]. Metabolomic quantification by mass spectrometry and liquid chromatography showed that three metabolites, serine, aspartate and dopamine, had lower concentrations whereas six lipids (including pro-inflammatory lysophosphatidylcholine [19], sphingomyelin, and phoshatidylcholine diacyl) had increased concentrations in pSS patients compared to the non-pSS Sicca controls. Moreover, age, sex, use of anticholinergic drugs, or the presence of anti-Ro/SSA antibodies did not influence the association between the metabolomic signature and the pSS status, suggesting that it is a true disease signature.

Tear proteomic analysis by Das et al. [20] using high performance liquid chromatography (HPLC) and mass spectrometry revealed the upregulation of 83 proteins and downregulation of 112 proteins in pSS patients compared to HCs. Enrichment pathway analysis of upregulated proteins included leukocyte trans-endothelial migration, protein-lipid complex remodelling and collagen catabolic pathway. On the other hand, the analysis of downregulated proteins indicated that pathways, such as glycolysis and amino acid metabolism, were diminished in tears from pSS patients. The relationship between proteomic biomarkers and clinical outcomes was not explored.

3.1.2. Saliva Biomarkers

Dry mouth (xerostomia) is a key symptom of pSS occurring in more than 95% of patients as a consequence of autoimmune destruction of salivary glands [21,22]. Salivary gland pathology detected by salivary gland biopsy is included in the classification criteria for pSS, and in many patients, this is an essential diagnostic and prognostic tool. However, this is an invasive procedure that could lead to local complications in a minority of cases, [23] whereas the collection of saliva for research purposes is, in contrast, a non-invasive procedure. Biomarkers found in saliva could potentially reflect the pathogenesis of this disease. Therefore, many researchers have been aiming to identify those lately (Table 2).

One of the few studies examining the cytokine profile of unstimulated saliva of pSS patients using the Luminex platform found an increase in IFN-γ, IL-1, IL-4, IL-10, IL-12p40, IL-17, and TNF-α levels in pSS patients compared to non-SS and HCs [24]. Moreover, IL-6 levels were higher in pSS compared to HCs. Notably, unstimulated saliva flow rate correlated with INF-γ/IL-4 ratio and salivary gland biopsy focus score (the number of inflammatory infiltrates of at least 50 cells present in 4 mm^2 of salivary gland area) correlated with TNF-α/IL-4 ratio in pSS, suggesting a predominant Th1 saliva signature. Years later, Chen et al. [16] reported enhanced levels of IP-10 (Interferon gamma-induced protein 10 or C-X-C motif chemokine ligand 10, CXCL10) and MIP-1α in saliva samples from pSS compared to HCs and a negative correlation between MIP-1α levels and both unstimulated whole saliva as well as the stimulated whole saliva flow rates.

Table 1. Examples of studies investigating potential Sjögren's Syndrome biomarkers in tears.

Reference	Type of Study/Samples/Methods	Number (N) of pSS Patients and HCs Age (Mean ± SD)	Disease Signature Identified	Correlations with Clinical Outcomes
		BIOMARKERS IN TEARS		
		Cytokine profiling		
Chen et al., 2019 [16]	Cross-sectional Tear strips for Schirmer I test Unstimulated (UWS) and stimulated (SWS) saliva samples Method: Cytokine 27-plex Assay	N = 29 pSS 56.8 ± 13.0 years N = 20 sicca (non-SS) controls 51.7 ± 10.6 years N = 17 HCs 45.4 ± 10.9 years	Increased IL-1ra, IL-2, IL-4, IL-17A, IFN-γ, MIP-1b, and Rantes in pSS vs. non-SS/HCs ($p < 0.05$).	Increased dry eye severity level and ocular surface staining correlated with increased tear cytokine levels, except for IP-10. Negative correlations between Schirmer's test and tears IL-1ra, IL-2, IL-4, IL-8, IL-12p70, IL-17A, IFN-γ, MIP-1b, and Rantes (r = 0.26–0.61, $p < 0.05$).
Willems et al., 2021 [17]	Cross-sectional Tear samples Method: LUNARIS™ BioChip	N = 12 pSS 41.7 ± 13.3 years N = 13 HCs 43.0 ± 13.8 years	Tears: Increased I FN-γ, TNF-α, IL-2, IL-4, IL-6, IL-10 and IL-12p70 (left eye) and IL-5 (right eye) in pSS compared to non-SS and HCs ($p < 0.005$).	Schirmer test correlated to IL-2 (r = −0.702), IL-4 (r = −0.769), IL-10 (r = −0.839) and IL-12p70 (r = −0.753) left eye levels; IL-10 directly correlated with SPEED test score (r = 0.722; $p = 0.0001$) as well as NIKBUT score (r = −0.705; $p = 0.00002$).
		Metabolomic profiling		
Urbanski et al., 2021 [18]	Cross-sectional Tear strips for Schirmer I test Method: mass spectrometry/liquid chromatography	N = 40 female pSS 63 years N = 40 non pSS sicca controls 58 years	9 metabolites (serine, aspartate; dopamine and six lipids) defined a tear pSS metabolomic signature (ROC-AUC = 0.83) PCA analysis showed that 2 PC explained 74.5% variance defined by 8/9 metabolites; the six lipids were distributed in the PC 1 and the amino acids in the second one.	The association between the metabolomic signature and the pSS status was not altered by age, sex, use of anticholinergic drugs or presence of anti-SSA antibodies

Table 1. *Cont.*

Reference	Type of Study/Samples/Methods	Number (N) of pSS Patients and HCs Age (Mean ± SD)	Disease Signature Identified	Correlations with Clinical Outcomes
		BIOMARKERS IN TEARS		
		Proteomic profiling		
Das et al., 2021 [20]	Cross-sectional Tears, Tear washes Saliva Cryopreserved parotid gland biopsy samples Methods High performance liquid chromatography HPLC/mass spectrometry MS shotgun proteomics analysis Biopsy staining with anti-PRG4 mAb Bead-based immunoassay using the AlphaLISA	Tears N = 22 pSS (F:M = 10:1) 60.0 ± 16.5 years N = 20 HCs (F:M = 13:7) 31.2 ± 11.4 years Tear washes N = 14 pSS (F:M = 13:1) 59.5 ± 12.0 years N = 29 HCs (F:M = 17:12) 34.1 ± 14.2 years	Tears: 83 upregulated and 112 unique downregulated proteins in pSS vs. HCs. Enriched pathways in pSS: leukocyte trans-endothelial migration, protein-lipid complex remodelling and collagen catabolic. Enriched pathways in HCs: glycolysis/gluconeogenesis and glycolysis in senescence, amino acid metabolism and VEGFA/VEGFR2 signalling pathway. Overall, there was a loss of glycolysis and metabolism but an elevation of immune processes in pSS tears samples. PRG4 in tear washes was significantly decreased in pSS ($p < 0.01$).	Not explored

Legend: pSS—primary Sjögren's Syndrome, HC—healthy controls, Rantes—Regulated upon Activation, Normal T Cell Expressed and Presumably Secreted, MIP-1b—Macrophage Inflammatory Proteins, IFN—interferon, IL—interleukin, VEGFA—Vascular endothelial growth factor A, VEGFR2—VEGF receptor 2, PRG4—Proteoglycan 4.

Table 2. Examples of studies investigating potential Sjögren's Syndrome biomarkers in saliva/salivary glands.

Reference	Type of Study/Samples/Methods	Number (N) of pSS Patients and HCs Age (Mean ± SD)	Disease Signature Identified	Correlations with Clinical Outcomes
BIOMARKERS IN SALIVA/SALIVARY GLANDS				
Cytokine profiling				
Kang et al., 2011 [24]	Cross-sectional Unstimulated saliva samples Method: Luminex® bead-based assay	N = 30 pSS 49.9 ± 9.0 years N = 30 sicca (non-SS) 51.5 ± 10.0 years N = 25 HCs 49.4 ± 9.5 years	Saliva: Increased IFN-γ, IL-1, IL-4, IL-10, IL-12p40, IL-17, and TNF-α levels in pSS vs. non-SS and HCs ($p <$ 0.005). IL-6 levels higher in pSS vs. HCs ($p = 0.011$). INF-γ/IL-4; TNF-α/IL-4 higher in pSS vs. HCs ($p = 0.028$, $p = 0.038$, respectively).	No correlations were found between any salivary cytokine levels and clinical parameters. Unstimulated saliva flow rate correlated with INF-γ/IL-4 ratio (r = 0.411 $p = 0.024$) and focus score correlated with TNF-α/IL-4 ratio (r = 0.581, $p = 0.023$) in pSS, suggesting a predominant Th1 saliva signature.
Chen et al., 2019 [16]	Cross-sectional Tear strips for Schirmer I test Unstimulated (UWS) and stimulated (SWS) saliva samples Method: Cytokine 27-plex Assay	N = 29 pSS 56.8 ± 13.0 years N = 20 sicca (non-SS) controls 51.7 ± 10.6 years N = 17 HCs 45.4 ± 10.9 years	Saliva: increased IP-10 in pSS vs. non-SS/HCs. Both pSS and non-SS subjects had higher MIP-1α levels than HCs ($p < 0.05$).	UWS and SWS correlated negatively with MIP-1a saliva level (r = −0.276, $p = 0.046$ and r = −0.282, $p = 0.040$, respectively).
Metabolomic profiling				
Kageyama et al., 2015 [25]	Cross-sectional Unstimulated saliva samples Method: Gas chromatography mass spectrometry (GC-MS) analysis	N = 14 female pSS 44.2 ± 13.01 years N = 21 HCs 46.7± 10.24 years	41 of the metabolites were reduced in pSS patients compared to HCs ($p <$ 0.05). Decreased glycine, tyrosine, uric acid and fucose in pSS vs. HCs in PCA analysis.	Patient stratification based on saliva metabolome depicted two groups: one younger ($p = 0.082$) and with a lower SG biopsy score ($p = 0.014$). Glycine levels differentiate between the two groups.

Table 2. *Cont.*

Reference	Type of Study/Samples/Methods	Number (N) of pSS Patients and HCs Age (Mean ± SD)	Disease Signature Identified	Correlations with Clinical Outcomes
BIOMARKERS IN SALIVA/SALIVARY GLANDS				
Metabolomic profiling				
Herrala et al., 2020 [26]	Longitudinal study Stimulated saliva samples Method: proton nuclear magnetic resonance (1 H-NMR) spectroscopy	56 samples from N = 14 female pSS patients during four laboratory visits within 20 weeks. 48.6 years N = 15 HCs (mean age 49.8 years)	Increased choline in pSS patients at each time point ($p \leq 0.015$), taurine at the last three time points ($p \leq 0.013$), alanine at the last two time points ($p \leq 0.007$) and glycine at the last time point ($p = 0.005$). Inter-individual variation observed for glycine ($p \leq 0.007$, all time points), choline ($p \leq 0.033$, three last time points) and alanine ($p = 0.028$, baseline).	Not explored
Proteomic profiling				
Delaleu et al., 2015 [27]	Cross-sectional Unstimulated whole saliva Methods: 187-plex capture antibody-based assay	Saliva N = 48 pSS (females) 47 years N = 24 non-SS cohort (12 RA patients + 12 HCs) 51 years	Significant differences in 61 biomarkers in pSS vs. controls ($p < 0.001$). FGF-4 levels lower in pSS (the only decreased protein). A biomarker signature comprising clusterin, IL-5, FGF-4, and IL-4 yielded accurate group prediction for 93.8% of pSS and 100% of non-SS controls classification.	No biomarkers correlated with salivary flow rates
Das et al., 2021 [20]	Cross-sectional Tears, Tear washes Saliva Cryopreserved parotid gland biopsy samples Methods High performance liquid chromatography HPLC/mass spectrometry MS shotgun proteomics analysis Biopsy staining with anti-PRG4 mAb Bead-based immunoassay using the AlphaLISA	Saliva N = 30 pSS (F:M = 22:8) 45.2 ± 14.6 years N = 10 HCs (F:M =5:5) 46.8 ± 14.5 years	Saliva: 104 upregulated and 42 downregulated proteins in pSS vs. HCs. Enriched pathways in pSS: JAK-STAT signalling after IL-12 stimulation, superoxide metabolic process and phagocytosis. Enriched pathways in HCs: neutrophil degranulation, negative regulation of peptidase activity; 2.3-fold increase in PRG4 in pSS ($p < 0.05$). PRG4 expression was found in both the serous acini and the striated duct on parotid gland biopsies (N = 4).	Not explored

Table 2. *Cont.*

Reference	Type of Study/Samples/Methods	Number (N) of pSS Patients and HCs Age (Mean ± SD)	Disease Signature Identified	Correlations with Clinical Outcomes
BIOMARKERS IN SALIVA/SALIVARY GLANDS				
Salivary gland tissue transcriptomic profiling				
Vertstappen et al., 2021 [28]	Cross-sectional Parotid and labial gland biopsy Methods: RNAseq—HiSeq 2500 System (Illumina). Multiplexed bead-based immunoassays for cytokine profiling Assessment of CD45-positive infiltrates on SG biopsies	N = 34 pSS with 51 paired (parotid and labial) biopsies 21 biopsy positive 13 biopsy negative 52 years N = 20 non SS sicca controls 17 biopsy negative 50 years	Parotid glands: 1041 up-regulated and 194 down-regulated DEG and labial glands: 581 and 43, respectively, between biopsy positive pSS and controls. The top 20 up-regulated genes in both tissues were mostly B-cell or T cell related. No significant differences between biopsy negative pSS and controls. Transcript expression levels correlated between parotid and labial glands ($R^2 = 0.86$, $p < 0.0001$). Signatures enriched in biopsy-positive pSS compared with either biopsy-negative pSS or controls: IFN-α signalling, IL-12/IL-18 signalling, CD3/CD28 T cell activation, CD40 signalling in B-cells, double negative type-2 B-cells, and FcRL4$^+$ B-cells. Strong correlation between the IFN-α score in PBMCs and SGs.	No difference in ESSDAI, unstimulated salivary flow or ESSPRI in patient DEG clusters.

Legend: pSS—primary Sjögren's Syndrome, HC—healthy controls, IFN—interferon, IL—interleukin, TNF—Tumour necrosis factor, MIP—Macrophage Inflammatory Protein, FGF—Fibroblast growth factor, PRG—proteoglycan, FcRL-Fc Receptor Like, SGs—salivary gland, ESSDAI—EULAR Sjögren's syndrome (SS) disease activity index, ESSPRI—EULAR Sjogren's Syndrome Patient Reported Index.

Metabolomics analysis of saliva identified a total of 41 metabolites reduced in pSS patients compared to HCs [25]. Principal component analysis (PCA) revealed that saliva from pSS patients had less biological diversity compared to HCs. Two distinct groups of pSS patients were identified based on their metabolomic profile: the only clinical differences between the groups were older age and the presence of major salivary gland glanditis in one group compared to the other. Recently, a longitudinal study by Herrala et al. [26] investigated changes in the levels of salivary metabolites in pSS and HCs using proton nuclear magnetic resonance (NMR) spectroscopy over 20 weeks. Choline, taurine, alanine, and glycine were the most significantly different metabolites, all of them were found in higher concentrations in saliva samples from pSS patients than in HCs. Compared to the baseline of the HCs, choline was significantly elevated at each time point, taurine and glycine were significantly higher at weeks 1, 10 and 20, whereas alanine was higher at weeks 10 and 20, suggesting that the distinct saliva metabolic signature is relatively stable over time.

Saliva proteomic analysis by Delaleu et al. [27] using a multiplex capture antibody-based assay identified 61 differentially expressed proteins in pSS vs. non-SS controls including rheumatoid arthritis (RA) and HCs samples. Interestingly only one protein, fibroblast growth factor (FGF)-4, was found at a lower concentration in pSS while 60 different proteins were present at higher concentrations compared to controls. This comprehensive analysis recognised a proteomic signature based on the following proteins: clusterin, IL-5, FGF-4 and IL-4. The proteomic signature could correctly identify pSS patients with an accuracy of 93.8% and non-SS patients with an accuracy of 100%. However, none of the protein biomarkers correlated with saliva flow rates in pSS. A more recent study by Das et al. [20] identified the upregulation of 104 proteins and downregulation of 42 proteins in pSS compared to controls. Some enriched pathways in patients' saliva included JAK-STAT signalling after IL-12 stimulation, superoxide metabolic process and phagocytosis.

3.1.3. Potential pSS Biomarkers in Peripheral Blood

It is well known that pSS is characterised by an imbalance of immune cell types, including a loss of T cell tolerance and autoreactivity, increased infiltration of exocrine gland tissues, contributing to the inflammatory microenvironment, as well as B cell activation, which is crucial for ectopic lymphoid structure and germinal centre formation, which eventually leads to the irreversible glandular damage [29]. Table 3.

Mingueneau et al. [30] published a fascinating study whereby, using mass spectrometry and immunochemistry in paired blood and salivary gland biopsies, a SS disease signature was uncovered. These findings highlighted the presence of activated $CD8^+$ T cells, terminally differentiated plasma cells, and activated epithelial cells in biopsies, whereas in blood samples they observed a cell signature of low numbers of $CD4^+$ T cells, memory B cells, plasmacytoid dendritic cells and high numbers of activated $CD4^+$, $CD8^+$ T cells and plasmablasts. The blood signature observed correlated with clinical parameters and enabled patient stratification into different endotypes with distinct disease activity and degrees of glandular inflammation. In line with this result, Van der Kroef et al. [31] also observed reduced frequencies of memory B cells and plasmacytoid dendritic cells and increased frequencies of activated HLA-DR $CD4^+$ and $CD8^+$ T cells in pSS patients compared to HCs.

In 2021, Szabó et al. [32] published an article whose aim was to investigate whether the distribution of B cells in pSS could be affected by a change in the balance of circulating T follicular helper (Tfh) cell subsets and follicular regulatory T cells. Utilising multicolour flow cytometry, they discovered that pSS patients had a significant increase in activated Tfh cells compared to HCs. Interestingly, anti-La/SSB-positive patients had a higher frequency of T follicular regulatory cells compared to seronegative patients. In the B-cell compartment, they observed that memory B cells were decreased, and transitional and naïve B cells were significantly increased. Lastly, they identified a positive correlation between the proportion of activated Tfh cells and both the levels of anti-La/SSB autoantibody and the serum IgA titre. Moreover, they demonstrated the frequency of pro-inflammatory Tfh1 cells correlated positively with levels of serum IgG and anti-LA/SSB autoantibody, suggesting the potential implication of various immune cell subsets in the disease pathogenesis through correlations with serological markers.

Martin-Gutierrez et al. [33] identified an immune signature derived from the analysis of 29 different cell subsets including B and T cells, which was driven by five distinct cell subsets: transitional Bm2′ cells, late memory Bm5 cells, IgD-CD27-B cells, and $CD8^+$ naive and $CD8^+$ Tem which differentiated between pSS patients and matched HCs. Moreover, they identified a shared immunological profile across three disease phenotypes: systemic lupus erythematosus (SLE), pSS and SLE associated with SS. By applying machine learning approaches, they identified two patient endotypes based on immune cell alterations, irrespective of the underlying diagnosis, suggesting significant pathogenic commonalities between these three disease groups. Notably, correlations were found between clinical manifestations and the frequencies of the immune cell subsets driving the stratification. $CD8^+$ and $CD4^+$ T cell subsets and B cell populations correlated with the erythrocyte sedimentation rate (ESR) in pSS patients whereas haemoglobin levels correlated with the frequency of $CD8^+$central memory T cells. Disease damage scores also correlated with the frequency of $CD8^+$ TEMRA (effector memory T cells re-expresses CD45RA)cells, $CD8^+$ responder cells (CD25-$CD127^+$) and $CD8^+$CD25-CD127-T cells.

Single-cell RNA sequencing of peripheral blood mononuclear cells (PBMCs) identified the expansion of $CD4^+$ cytotoxic T lymphocytes and a population of $CD4^+$ T cells highly expressing the T cell receptor Alpha Variable 13-2gene, in pSS patients compared to HCs [34]. Pathway enrichment analysis revealed upregulation of genes involved with type I and II interferon signalling, TNF family signalling and antigen processing and presentation in pSS patients. Using flow cytometry, it was confirmed the percentages of $CD4^+$ Granzyme B^+ T cells in the $CD4^+$ T cell populations were significantly higher in pSS patients compared to the HCs. No correlations were found between the frequencies of $CD4^+$ T cells and clinical or serological parameters, including the disease activity index ESSDAI (EULAR Sjögren's syndrome (SS) disease activity index), ESR levels, or the presence of anti-Ro antibodies.

Disease-associated biomarkers can be detected in serum through proteomic or metabolomic technologies. Serum concentrations of proteins in pSS patients and HCs were measured by a high-throughput proteomic assay in a recent publication [35]. Using this complex assay 1110 proteins were quantified and, from those, 82 were found to be differentially expressed in pSS patients. Significant correlations between nine differently expressed serum proteins and the ESSDAI score were found. Using a second cohort of pSS patients, five proteins including CXCL13, TNF-receptor 2, CD48, B-cell activating factor (BAFF), and PD-L2 (Programmed cell death ligand 2) were validated as pSS-associated biomarkers. Another study investigated which serum protein biomarkers, measured by Bio-Plex, could distinguish pSS from other autoimmune diseases, such as SLE and RA [36]. Out of 63 proteins, they were able to identify eight and four proteins that could differentiate pSS from SLE and RA, respectively. A combination of four different proteins: BDNF (Brain Derived Neurotrophic Factor), I-TAC/CXCL11, soluble (s) CD163 and Fractalkine/CX3CL1 was identified as a pSS protein signature as it could discern pSS from other autoimmune diseases. A negative correlation between ESSDAI score and serum sCD163 concentrations was found.

Different reports [37–39] have also focused on analysing changes in serum metabolites by different techniques, such as mass spectrometry to find new molecules that could play a role in the pathogenesis of pSS and could become new drug targets [39]. Using a non-targeted gas chromatography-mass spectrometry (GC-MS) serum metabolic profile, the authors detected 21 metabolites that differentiated between pSS patients and controls, with 18 out of 21 metabolites further validated in another cohort. Two metabolites, stearic acid and linoleic acid had the adequate discriminatory capacity to separate pSS patients from HCs and correlated with clinical parameters, such as C-Reactive Protein (CRP), ESR, IgG, anti-Ro/SSA, anti-La/SSB, antinuclear antibodies, IgA and rheumatoid factor.

Table 3. Examples of studies investigating potential Sjögren's Syndrome biomarkers in saliva.

Reference	Type of Study/Samples/Methods	Number (N) of pSS Patients and HCs Age (Mean ± SD)	Disease Signature Identified	Correlations with Clinical Outcomes
BIOMARKERS IN PERIPHERAL BLOOD				
Immunophenotype profiling				
Mingueneau et al., 2016 [30]	Cross-sectional PBMC samples, a subset of paired SG biopsies Method: CyTOF, immunohistochemistry	N = 49 pSS 53 years N = 45 sicca (non-SS) and HCs 54 years	SG biopsies: increased activated $CD8^+$ T cells, terminally differentiated plasma cells, and activated epithelial cells Blood: 6-cell signature: decreased CD4, memory B-cells, plasmacytoid dendritic cell, and increased activated CD4 and CD8 T cells and plasmablasts.	The blood cellular components correlated with clinical parameters clustered patients into subsets with distinct disease activity and glandular inflammation.
Van der Kroef et al., 2020 [31]	Cross-sectional, PBMC samples Methods: Luminex, CyTOF	N = 88 SSc 54 years N = 31 SLE 43 years N = 23 pSS 56 years N = 44 HCs 50 years	pSS patients have increased HLA-DR $CD4^+$ and $CD8^+$ frequencies and reduced memory B cells and pDCs compared to HCs.	Not explored in pSS
Szabó et al., 2021 [32]	Cross-sectional PBMCs samples Methods: flow cytometry, functional analysis and ELISA.	N = 38 pSS 54 years N = 27 HCs 46 years	pSS patients showed a significant increase in activated T follicular helper cells. Frequencies of T follicular regulatory cells were increased in autoantibody La positive patients compared to seronegative pSS. Transitional and naïve B cells increased, memory B cells decreased,	The percentage of activated T follicular helper cells showed a positive correlation with the levels of anti-La/SSB autoantibody and with serum IgA titre. Frequency of Tfh1 positive correlation with levels of serum IgG and anti-LA/SSB autoantibody.
Martin-Gutierrez et al., 2021 [33]	Cross-sectional PBMC samples Method: Flow cytometry/ML	N = 45 pSS 59 (30–78) N = 29 SLE 48 (21–72) N = 14 SLE/SS 55 (26–56) N = 31 HCs	Patients with SS/SLE and SLE/SS shared immunological signatures. A signature comprising 5/29 immune cell subsets studied: transitional Bm2′ cells, late memory Bm5 cells, IgD-CD27-B cells, and $CD8^+$ naïve and $CD8^+$ Tem cells stratified patients	ESR correlated with 4 $CD8^+$ T cell, 3 $CD4^+$ T cell and 2 B cell subpopulations, which drove patient stratification. Hgb level correlated with % $CD8^+$ Tcm cells. Disease damage scores across correlated with %$CD8^+$ T cell, including $CD8^+CD25–CD127$, $CD8^+$ responder T cells, and $CD8^+$ Temra cells

Table 3. *Cont.*

Reference	Type of Study/Samples/Methods	Number (N) of pSS Patients and HCs Age (Mean ± SD)	Disease Signature Identified	Correlations with Clinical Outcomes
		BIOMARKERS IN PERIPHERAL BLOOD		
		Single-cell transcriptomic profile		
Hong et al., 2021 [34]	Cross-sectional PBMCs Methods: scRNAseq and Flow cytometry	N = 10 pSS patients 48.8 years N= 10 HCs 33 years	Two subpopulations expanded in pSS: one expressing cytotoxicity genes ($CD4^+$ CTLs cytotoxic T lymphocyte), and another highly expressing T cell receptor (TCR) variable gene ($CD4^+$ TRAV13-2^+ T cell). Total T cells significantly higher in pSS vs. HCs ($p = 0.008$). The IL-1β expression in macrophages, TCL1A in B cells, and IFN response genes in most cell subsets were upregulated in pSS. Susceptibility genes including HLA-DRB5, CTLA4, and AQP3 were highly expressed in pSS.	Correlation between the percentage of $CD4^+$ CTLs and clinical characteristics, such as ESR), anti-SSA positive, and ESSDAI but no significant correlation was found.
		Serum proteomics		
Nishikawa et al., 2016 [35]	Cross-sectional Methods: high-throughput proteomic analysis, ELISA.	Discovery cohort: N = 30 pSS 61 years N = 30 HCs 40 years Validation cohort: N = 50 pSS 60 years	A total of 82 (57 upregulated and 25 downregulated) serum proteins were differentially expressed in patients pSS vs. HCs. Enriched pathways: "extracellular region", "chemokine signalling pathway", "downstream of TNF-α", "platelet activation", and "platelet degranulation". Nine proteins correlated with disease activity in the discovery cohort. In the validation cohort five proteins: CXCL13, TNF-R2, CD48, BAFF, and PD-L2 showed a correlation with ESSDAI, and therefore, were proposed as disease activity-associated biomarkers.	Serum concentrations of CXCL13, TNF-R2, and CD48 were positively correlated with that of immunoglobulin (Ig) G. TNF-R2 was negatively correlated with unstimulated salivary. BAFF was negatively correlated with the excretion rate in the submandibular gland.

Table 3. *Cont.*

Reference	Type of Study/Samples/Methods	Number (N) of pSS Patients and HCs Age (Mean ± SD)	Disease Signature Identified	Correlations with Clinical Outcomes
BIOMARKERS IN PERIPHERAL BLOOD				
Serum proteomics				
Padern et al., 2021 [36]	Cross-sectional Methods: BioPlex, ELISA	N = 42 pSS 62.5 years N = 28 RA 60.5 years N = 25 SLE 40 years	Eight biomarkers could statistically discriminate samples from pSS versus SLE patients. Four could statistically discriminate pSS patients from RA patients. None of the studied biomarkers could simultaneously discriminate pSS from RA and SLE. We, therefore, determined the positive predictive value (PPV), sensitivity, and specificity of different combinations of BDNF, I-TAC/CXCL11, sCD163 and Fractalkine/CX3CL1 concentrations. These biomarkers were chosen because they were those most strongly associated with distinguishing pSS from the other AIDs.	Negative correlation between pSS activity according to the ESSDAI score and serum sCD163 concentrations.
Serum metabolomic profiling				
Xu et al., 2021 [39]	Cross-sectional Serum samples Method: non-targeted GC-MS	Discovery: 90 pSS patients, M:F = 1/10) 53 years 153 HCs (male/female: 1/10) 50.4 years Validation: 119 pSS, M:F = 1/10 52.9 years 143 HCs (M:F = 1/10 50.23 years	Increased alanine, tryptophan, glycolic acid, pelargonic acid, cis-1-2-dihydro-1-2-naphthalenediol, etc., and decrease in catechol, anabasine, 3-6-anhydro-D-galactose, beta-gentiobiose and ethanolamine in pSS patients vs. HCs. Stearic acid and linoleic acid distinguished pSS from HCs (ROC−AUC = 0.97−0.98)	Inflammatory markers, autoantibodies and Ig G levels correlated with various metabolite levels.

Legend: pSS—primary Sjögren's Syndrome, HC—healthy controls, SLE—systemic lupus erythematosus, pDCs—plasmacytoid dendritic cell, Tfh—T follicular helper cells, ESR—Erythrocyte Sedimentation Rate, Hgb—Haemoglobin, AQP- Aquaporin, CTLA—Cytotoxic T-Lymphocyte Associated, CXCL-C-X-C motif chemokine ligand, TNFR—Tumour Necrosis factor receptor, BAFF—B-cell activating factor, PDL2—Programmed cell death 1 ligand 2, BDNF—Brain-derived neurotrophic factor, sCD163—soluble CD163, ESSDAI—EULAR Sjögren's syndrome (SS) disease activity index.

3.1.4. Genetic and Epigenetic Studies

Although the aetiology of SS is unknown, it is considered that different factors, such as environmental, genetic and epigenetic, contribute to the disease pathogenesis. In this context, several studies [40–43], have focused on finding genetic and epigenetic factors that could be associated with SS. Transcriptomics, genome-wide association studies (GWAS) to identify genomic variants that are statistically associated with a risk of suffering the disease and epigenetic studies to determine whether gene expression is active or inactive based on DNA methylation are widely used nowadays.

Multi-omic pSS Signatures

It is well known that the diagnosis and treatment of SS, is challenging due to the existing molecular and clinical heterogenicity, which reflects different disease stages, variable types of organ involvement, disease severity and treatment, as well as patient-specific factors, such as age, environmental exposures and comorbidities. Thus, recent research (Table 4) has been focused on integrating genomic/epigenomic, transcriptomic, proteomic, metabolomic and immunophenotype characterisation and clinical data to gain more knowledge about the disease pathogenesis as well as being able to classify patients into groups defined by their molecular pattern.

Integrated transcriptomic and serum proteomic data with an immune signature comprising 24 different cell populations highlighted the presence of a pSS gene signature driven by interferon genes as well as ADAMs (a disintegrin and metalloprotease) substrates [44]. Interestingly, the genomic regions coding the genes identified as part of the disease signature were predominantly hypomethylated, therefore, transcriptionally activated. In addition, the proteomic analysis revealed some correlations between ADAMs substrates and ESSDAI scores. Relevantly, the authors confirmed that CD8+ T cells, especially TEMRA, produced the signature observed. Similarly, transcriptomic and cytokine profiling of pSS patients allowed the stratification of pSS patients into three distinct clusters, defined by IFN-responsive and inflammation-associated genes [45]. Interestingly, patients belonging to the cluster with the strongest IFN and inflammation gene signature also had high ESSDAI scores and elevated levels of anti-Ro/SSA and La/SSB autoantibodies. This cluster was also defined by a high serum concentration of cytokines, such as LIGHT and Blys and chemokine CXCL13 [45]. Soret et al. [46] and Barturen et al. [47] independently validated some of these findings, by showing a pSS patient transcriptomic stratification also driven by IFN-related pathways. Interestingly, in both studies, a cluster of patients with low disease activity was identified, characterised by a transcriptomic profile similar to that of HCs. Soret et al. [46] did not detect any differentially expressed genes, single nucleotide polymorphisms (SNPs) or differences in B and T cells, monocytes, basophils, eosinophils and neutrophils frequencies in pSS patients with low disease activity when compared to HCs.

Table 4. Examples of studies investigating potential Sjögren's Syndrome biomarkers using multi-omic approaches.

Reference	Type of Study/Samples/Methods	Number (N) of pSS Patients and HCs Age (Mean ± SD)	Disease Signature Identified	Correlations with Clinical Outcomes
		MULTIOMIC SIGNATURES		
Tasaki et al., 2017 [44]	Cross-sectional Methods: whole blood Transcriptomes microarrays, serum proteomes and peripheral immunophenotyping	N = 36 pSS patients 61 years N = 36 HCs 39 years	pSS gene signature predominantly involves the interferon signature including HERC5, EPSTI1 and CMPK2and ADAMs substrates. SGS was significantly overlapped with SS-causing genes indicated by a genome-wide association study as the regions that code genes in the SS gene signature were hypomethylated. Combining the molecular signatures with immunophenotypic profiles revealed that cytotoxic CD8 T cells were associated with SGS.	SGS positively correlated with the levels of autoantibodies, including anti-Ro/SSA and anti-La/SSB antigen–antibodies and serum IgG levels. Most ADAM substrates showed significant positive correlations with ESSDAI.
James et al., 2019 [45]	Cross-sectional Methods: RNAseq, Bioplex, ELISA, Luminex	N = 47 pSS patients 52 years	Three clusters of patients were identified based on transcriptomic analysis. No demographic differences between clusters. C1 weakest IFN signature and minimal activity of inflammation gene modules. C2 strongest IFN signature, strongest inflammation module. Higher ESSDAI scores. More patients presented anti-Ro and anti-La antibodies and higher levels. Higher levels of cytokines, such as LIGHT and Blys. CXCL13. C3 intermediate IFN signature, low activity of the inflammation modules.c3 higher levels of IL1, IL2 IL2RA.	C2 cluster presented higher ESSDAI scores

Table 4. *Cont.*

Reference	Type of Study/Samples/Methods	Number (N) of pSS Patients and HCs Age (Mean ± SD)	Disease Signature Identified	Correlations with Clinical Outcomes
MULTIOMIC SIGNATURES				
Soret et al., 2021 [46]	Cross-sectional Whole blood Methods: RNAseq, GWAS, Methylation and flow cytometry Serum sample Methods: Luminex, automated chemiluminescent immunoanalyzer (IDS-iSYS)	N = 304 pSS patients 58 years N = 330 HCs 53 years	Clustering of pSS samples based on transcriptomic data identified 4 different clusters (C1, C2, C3 and C4). C1 was enriched with IFN-related pathways, present an enriched up-regulated IFN signalling pathway; 35 SNPS were detected in genes associated with the immune system (HLA-DQB1, HLADQA1, HLA-DRA, HLA-C, HLA-G), signal transduction (NOTCH4), developmental biology (POU5F1), gene expression (DDX39B). Methylation in 87 genes. T cell lymphopenia. Increased in The IFNγ-induced protein (CXCL10/IP-10) as well as CCL8/MCP-2 and TNFα. IL-1 RII, was downregulated. No DEGs were noticed when comparing C2 to HCs. No SNPS were found in C2. Methylation of IFN genes MX1 and NLRC5. Frequency and the absolute number of T and B cells, monocytes, NK-like, DC, basophils, eosinophils, and neutrophils are similar to HCs. C3 was enriched with pathways related to B cell activation, and IFN signalling. SNPs detected in HLA-DQA, HLA-DRA (2 SNPs), BTNL2 and HCG23. Methylation in 56genes. Increased frequency of monocytes and lymphocytes. Increased in The IFNγ-induced protein (CXCL10/IP-10) as well as CCL8/MCP-2 and TNFα. IL-1 RII, was downregulated. C4 endotype with higher DEG including T and B activation, cytokine signalling and IL-15 production. The only SNPs identified in the intron LINC02571 gene and were previously associated with a risk for developing SLE. Methylation in 3000 genes. Decreased in B and T cells and monocytes. High percentage of neutrophils.	No statistically significant differences between the four clusters in ESSDAI or PGA mean scores. Statistically significant differences in the distribution of reported arthritis, rate of cancer history, coronary artery disease and chronic obstructive pulmonary disease were observed between the four clusters. Patients from C4 reported more severe clinical symptoms compared to the three other clusters. Some serological characteristics were significantly different across the four clusters, C1 and C3 have higher hypergammaglobulinemia, extractable nuclear antigen (ENA) antibodies, the presence of serum anti-SSA52/anti-SSA60 autoantibodies and higher circulating kappa and lambda free light chains (cFLC).

Table 4. *Cont.*

Reference	Type of Study/Samples/Methods	Number (N) of pSS Patients and HCs Age (Mean ± SD)	Disease Signature Identified	Correlations with Clinical Outcomes
MULTIOMIC SIGNATURES				
Barturen et al., 2021 [47]	Cross-sectional Follow-up Methods: Whole blood transcriptome and methylome	N = 955 cross-sectional patients with 7 autoimmune diseases 53.4 years N = 113 follow up patients. 47 years N = 267 HCs 46 years	Four clusters were identified and validated; 3 clusters represented inflammatory, lymphoid and interferon patterns; 1 cluster with low disease activity with no specific molecular pattern.	SLEDAI and ESSDAI scores were higher in all 3 clusters compared to the undefined cluster.

Legend: pSS—primary Sjögren's Syndrome, HC- healthy controls, HERC5—HECT And RLD Domain Containing E3 Ubiquitin Protein Ligase 5, EPSTI1—Epithelial Stromal Interaction 1, CMPK2—Cytidine/Uridine Monophosphate Kinase 2, ADAM—A Disintegrin And Metalloprotease, SGS-Sjögren´s gene signature, IFN—interferon, CXCL-C-X-C motif ligand, SNPs—Single nucleotide polymorphisms, Notch—Neurogenic locus notch homolog protein, MX1—myxovirus resistance protein 1,NLRC5—NLR Family CARD Domain Containing 5, CCL8/MCP2—monocyte chemotactic protein-2, SLEDAI—Systemic Lupus Erythematosus Disease Activity Index, ESSDAI—EULAR Sjögren's syndrome (SS) disease activity index.

4. Discussion

The significant progress made by high-throughput technologies, increased effort for large-scale academic and industry research collaborations to facilitate external validation, and advancement of computer algorithms for big data integration and cluster analysis provide unprecedented opportunities for better patient classification, improved pathogenic characterisation, prediction and therapeutic opportunities across all autoimmune diseases. It is increasingly recognised the need for better quality research, including both pre-clinical and clinical validation to enable meeting the ultimate goal of achieving clinical utility and patient benefit. Despite the impressive therapeutic advances leading to licensing of many new targeted therapies in autoimmune rheumatic disease (ARDs), such as inflammatory arthritis or SLE [48], patients with SS do not benefit from the same range of therapeutic options available for other conditions, despite shared pathogenesis [49]. Many of the signals of efficacy from early phase clinical trials of various biologics investigated in SS have not been replicated in larger studies and research is ongoing [50,51].

Efforts have been made in improving the way the response to treatment in clinical trials of patients with SS is assessed [52], while current treatment recommendations expanded to targeted biologic treatment options despite of lack of large phase 3 clinical trials [10]. In addition, novel approaches, such as advocating for a molecular classification of SS to drive precision medicine strategies have been proposed [46], which suggests that the future of clinical research in SS will likely involve multi-omic characterisation of patients (Figure 1). In this respect, good quality, reproducible research involving large cohort collaborations to capture the disease heterogeneity, as well as facilitate the validation of disease signatures, is required to improve knowledge about SS pathogenesis and facilitate the much-needed therapeutic advances.

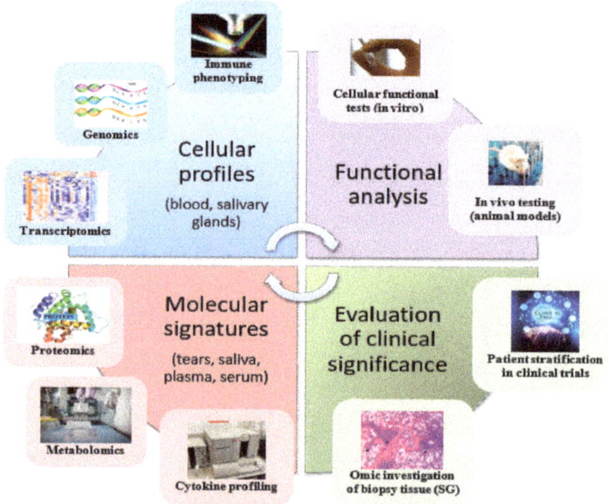

Figure 1. Potential multi-omic approaches taken in clinical research.

It is widely recognised that SS is associated with a genetic predisposition, similar to other ARDs, which has been confirmed in large GWAS studies which validated the associations with HLA, IRF5, STAT4 and BLK genetic loci, while also detecting novel susceptible loci [53]. The best characterised are the HLA genes, associated with an increased disease risk ranging from 1.85 to 3.41 as per a large meta-analysis [54]. Various non-HLA genes associated with the disease have also been described but very few have been validated across studies [11].

Research into the role of environmental factors and epigenetics currently supports the old hypothesis that a ubiquitous virus is a potential trigger for the mechanism of autoimmunity, with most data potentially implicating Epstein–Barr Virus (EBV), Human T cell Leukemia Virus-1 (HTLV-1), or Coxsackie virus in the development of SS [55], despite the lack of conclusive evidence for their causal role. The most well-defined epigenetic mechanisms likely to play a role in the pathogenesis of SS have been described as DNA methylation, histone modifications and non-coding RNAs [42].

SS is characterised by a pro-inflammatory environment and cytokine profiling of serum, tears and saliva identified a predominance of pro-inflammatory cytokines, such as MIP, IL-1, IFN-γ, TNF-α, IL-6, IL-12 or IL17 in various proportions, as well as increased anti-inflammatory molecules, such as IL-4 or IL-10. Some of these data have been validated across studies, while some of the biomarkers, including cytokine ratios suggesting a Th1 signature also correlated with clinically meaningful parameters, such as tear and saliva secretion [16,17,24].

Dysregulation of various immune cell populations has been hypothesised as one of the key factors implicated in disease pathogenesis. Immune profiling of patients with SS found distinct immune signatures in the salivary gland tissue compared to peripheral blood, as expected. The main players seem to be activated $CD8^+$ T cells, terminally differentiated plasma cells, and activated epithelial cells in biopsies, whereas the peripheral blood signatures comprised high numbers of activated $CD4^+$, $CD8^+$ T cells. Although these signatures were not perfectly validated across various studies [30,33], some correlated with serological and clinical parameters, suggesting a potential clinical utility.

Proteomic analysis revealed distinct signatures in tears and saliva compared to the serum of patients SS, with the majority of signatures being able to differentiate, with high accuracy, SS patients from controls, and a few correlating with clinical meaningful parameters. Enriched pathway analysis also overlapped with some cytokine signature findings, such as the upregulation of the JAK-STAT signalling after IL-12 stimulation in saliva [27]. The protein patterns identified in saliva were associated with B cell immune responses, macrophage differentiation and T cell chemotaxis, which showed similarities with salivary gland histopathological features [27], suggesting a potential role for saliva analysis as a proxy measure of glandular inflammation.

Transcriptomic profiling of salivary gland tissue was characterised by the upregulation of IFN-α and IL-12/IL-18 signalling, as well as CD3/CD28 T cell activation, CD40 signalling in B-cells, as well as significant correlation with the IFN-α score in PBMCs [28], which shows similarities with proteomic profiles of saliva. IFN response genes were also upregulated in most cell subsets when assessed by single-cell blood transcriptomic analysis, highlighting the role of the IFN activation pathway in the pathogenesis of the disease. In terms of potential clinical implications, the IFNγ/IFNα mRNA ratio in salivary gland tissue was shown to have the best discriminative capacity for lymphoma development in patients with pSS [56]. Patient stratification based on transcriptomic signatures identified distinct clusters driven by IFN and B cell activation, as well as SNPs in HLA genes and epigenetic modifications including gene hypomethylation [46], all processes recognised as involved in the disease pathogenesis, although the clinical significance of patient stratification was less clear.

Metabolomic characterisation of serum, tears and saliva of patients with pSS identified distinct signatures with almost no overlap between various biologic fluids [18,26,39]. Further research exploring the inter-individual variability and its stability over time is required [26].

The power of integrating several omic technologies in the investigation of the disease fingerprints harnessed evidence for the role of cytotoxic CD8 T cells in the disease pathogenesis [44] as well as enabled the identification of inflammatory, lymphoid and IFN-driven patient clusters generated by a combination of the transcriptome, methylome and cytokine profilin [45,47]. Patient clusters driven by high IFN and pro-inflammatory signatures were also associated with high disease activity suggesting that these pathways are relevant to the disease pathogenesis.

5. Conclusions

Omic investigation of SS provides a valuable insight into the disease pathogenesis and patient molecular heterogeneity which has implications for SS prognosis and better management strategies to address the unmet patient needs. Further research into standardising technologies and validating findings across large patient populations, as well as further exploration of potential correlations with clinical significance, are required to establish which are the strongest molecular signals that could be potentially translated into research with patient benefit. Ultimately, integrating data provided by multiple omics analysis can provide the much-required complementary knowledge related to the interplay between genes, environment, immune cell activation and pro-inflammatory milieu which all sustain the pathogenic processes associated with SS.

Understanding how the disease's natural course or treatment impacts these molecular signatures, as well as which pathways can be targeted by available and novel treatments will open a new era for research in SS.

Author Contributions: L.M.-G. and C.C. conceptualised the paper search strategy. All authors reviewed the literature and contributed to data collection. L.M.-G. and C.C. wrote the first draft of the manuscript. All authors reviewed the manuscript, provided intellectual input in the presentation of findings, conclusions, discussions, and limitations. All authors have read and agreed to the published version of the manuscript.

Funding: This work was supported by grants from the MRC EMINENT project grant (MR/X004694/1), NIHR UCLH Biomedical Research Centre grant BRC772/III/EJ/101350, BRC773/III/CC/101350. This work was performed within the Centre for Adolescent Rheumatology Versus Arthritis at UCL UCLH and GOSH supported by grants from Versus Arthritis (21593, 22908 and 20164). The views expressed are those of the authors and not necessarily those of the NHS, the NIHR or the Department of Health.

Institutional Review Board Statement: No institutional review board was required as this is a literature review.

Informed Consent Statement: Not required as this study is a literature review.

Conflicts of Interest: The authors declare that the perspective was conducted in the absence of any commercial or financial relationships that could be construed as a potential conflict of interest.

References

1. Acar-Denizli, N.; Horváth, I.F.; Mandl, T.; Priori, R.; Vissink, A.; Hernandez-Molina, G.; Armagan, B.; Praprotnik, S.; Sebastian, A.; Bartoloni, E.; et al. Systemic phenotype related to primary Sjögren's syndrome in 279 patients carrying isolated anti-La/SSB antibodies. *Clin. Exp. Rheumatol.* **2020**, *38*, 85–94. [PubMed]
2. Alani, H.; Henty, J.R.; Thompson, N.L.; Jury, E.; Ciurtin, C. Systematic review and meta-analysis of the epidemiology of polyautoimmunity in Sjögren's syndrome (secondary Sjögren's syndrome) focusing on autoimmune rheumatic diseases. *Scand. J. Rheumatol.* **2018**, *47*, 141–154. [CrossRef] [PubMed]
3. Shiboski, C.H.; Shiboski, S.C.; Seror, R.; Criswell, L.A.; Labetoulle, M.; Lietman, T.M.; Rasmussen, A.; Scofield, H.; Vitali, C.; Bowman, S.J. American College of Rheumatology/European League Against Rheumatism classification criteria for primary Sjögren's syndrome: A consensus and data-driven methodology involving three international patient cohorts. *Arthritis Rheumatol.* **2017**, *69*, 35–45. [CrossRef] [PubMed]
4. Mavragani, C.P.; Moutsopoulos, H.M. Primary versus secondary Sjögren syndrome: Is it time to reconsider these terms? *J. Rheumatol.* **2019**, *46*, 665–666. [CrossRef]
5. Basiaga, M.L.; Stern, S.M.; Mehta, J.J.; Edens, C.; Randell, R.L.; Pomorska, A.; Irga-Jaworska, N.; Ibarra, M.F.; Bracaglia, C.; Nicolai, R.; et al. Childhood Sjogren syndrome: Features of an international cohort and application of the 2016 ACR/EULAR classification criteria. *Rheumatology* **2021**, *60*, 3144–3155. [CrossRef]
6. Ciurtin, C.; Cho, Y.; Al-Obaidi, M.; Jury, E.C.; Price, E.J. Barriers to translational research in Sjögren's syndrome with childhood onset: Challenges of recognising and diagnosing an orphan rheumatic disease. *Lancet Rheumatol.* **2021**, *3*, e138–e148. [CrossRef]

7. Seror, R.; Bowman, S.J.; Brito-Zeron, P.; Theander, E.; Bootsma, H.; Tzioufas, A.; Gottenberg, J.E.; Ramos-Casals, M.; Dörner, T.; Ravaud, P.; et al. EULAR Sjogren's syndrome disease activity index (ESSDAI): A user guide. *RMD Open* **2015**, *1*, e000022. [CrossRef]
8. Price, E.; Allen, A.; Rauz, S.; Tappuni, A.; Sutcliffe, N.; Bombardieri, M.; Carty, S.; Ciurtin, C.; Crampton, B.; Duncalfe, L.; et al. The management of Sjogren's syndrome: British Society for Rheumatology guideline scope. *Rheumatology* **2021**, *60*, 2122–2127. [CrossRef]
9. Doolan, G.; Faizal, N.M.; Foley, C.; Al-Obaidi, M.; Jury, E.C.; Price, E.; Ramanan, A.V.; Lieberman, S.M.; Ciurtin, C. Treatment strategies for Sjogren's syndrome with childhood onset: A systematic review of the literature. *Rheumatology* **2022**, *61*, 892–912. [CrossRef]
10. Ramos-Casals, M.; Brito-Zerón, P.; Bombardieri, S.; Bootsma, H.; De Vita, S.; Dörner, T.; Fisher, B.A.; Gottenberg, J.E.; Hernandez-Molina, G.; Kocher, A.; et al. EULAR recommendations for the management of Sjögren's syndrome with topical and systemic therapies. *Ann. Rheum. Dis.* **2020**, *79*, 3–18. [CrossRef]
11. Sandhya, P.; Theyilamannil Kurien, B.; Danda, D.; Hal Scofield, R. Update on pathogenesis of Sjogren's syndrome. *Curr. Rheumatol. Rev.* **2017**, *13*, 5–22. [CrossRef] [PubMed]
12. Ng, W.F. Primary Sjögren's syndrome research and therapy: Has a new dawn arrived? *Curr. Pharm. Biotechnol.* **2012**, *13*, 1987–1988. [CrossRef] [PubMed]
13. Mariette, X.; Seror, R.; Quartuccio, L.; Baron, G.; Salvin, S.; Fabris, M.; Desmoulins, F.; Nocturne, G.; Ravaud, P.; De Vita, S. Efficacy and safety of belimumab in primary Sjogren's syndrome: Results of the BELISS open-label phase II study. *Ann. Rheum. Dis.* **2015**, *74*, 526–531. [CrossRef]
14. Meiners, P.M.; Vissink, A.; Kroese, F.G.M.; Spijkervet, F.K.L.; Smitt-Kamminga, N.S.; Abdulahad, W.H.; Bulthuis-Kuiper, J.; Brouwer, E.; Arends, S.; Bootsma, H. Abatacept treatment reduces disease activity in early primary Sjogren's syndrome (open-label proof of concept ASAP study). *Ann. Rheum. Dis.* **2014**, *73*, 1393–1396. [CrossRef] [PubMed]
15. Fisher, B.A.; Everett, C.C.; Rout, J.; O'Dwyer, J.L.; Emery, P.; Pitzalis, C.; Ng, W.F.; Carr, A.; Pease, C.T.; Price, E.J.; et al. Effect of rituximab on a salivary gland ultrasound score in primary Sjogren's syndrome: Results of the TRACTISS randomised double-blind multicentre substudy. *Ann. Rheum. Dis.* **2018**, *77*, 412–416. [CrossRef] [PubMed]
16. Chen, X.; Aqrawi, L.A.; Utheim, T.P.; Tashbayev, B.; Utheim, Ø.A.; Reppe, S.; Hove, L.H.; Herlofson, B.B.; Singh, P.B.; Palm, Ø.; et al. Elevated cytokine levels in tears and saliva of patients with primary Sjogren's syndrome correlate with clinical ocular and oral manifestations. *Sci. Rep.* **2019**, *9*, 7319. [CrossRef]
17. Willems, B.; Tong, L.; Minh, T.D.T.; Pham, N.D.; Nguyen, X.H.; Zumbansen, M. Novel cytokine multiplex assay for tear fluid analysis in Sjogren's syndrome. *Ocul. Immunol. Inflamm.* **2021**, *29*, 1639–1644. [CrossRef]
18. Urbanski, G.; Assad, S.; Chabrun, F.; de la Barca, J.M.C.; Blanchet, O.; Simard, G.; Lenaers, G.; Prunier-Mirebeau, D.; Gohier, P.; Lavigne, C.; et al. Tear metabolomics highlights new potential biomarkers for differentiating between Sjogren's syndrome and other causes of dry eye. *Ocul. Surf.* **2021**, *22*, 110–116. [CrossRef]
19. Knuplez, E.; Marsche, G. An updated review of pro-and anti-inflammatory properties of plasma lysophosphatidylcholines in the vascular system. *Int. J. Mol. Sci.* **2020**, *21*, 4501. [CrossRef]
20. Das, N.; Menon, N.G.; De Almeida, L.G.; Woods, P.S.; Heynen, M.L.; Jay, G.D.; Caffery, B.; Jones, L.; Krawetz, R.; Schmidt, T.A.; et al. Proteomics analysis of tears and saliva from Sjogren's syndrome patients. *Front. Pharmacol.* **2021**, *12*, 787193. [CrossRef]
21. Ramos-Casals, M.; Brito-Zerón, P.; Sisó-Almirall, A.; Bosch, X. Primary Sjogren syndrome. *BMJ* **2012**, *344*, e3821. [CrossRef]
22. Brito-Zeron, P.; Baldini, C.; Bootsma, H.; Bowman, S.J.; Jonsson, R.; Mariette, X.; Sivils, K.; Theander, E.; Tzioufas, A.; Ramos-Casals, M. Sjögren syndrome. *Nat. Rev. Dis. Primers* **2016**, *2*, 16047. [CrossRef]
23. Olsson, P.; Ekblad, F.; Hassler, A.; Bengtsson, M.; Warfvinge, G.; Mandl, T.; Kvarnström, M. Complications after minor salivary gland biopsy: A retrospective study of 630 patients from two Swedish centres. *Scand. J. Rheumatol.* **2022**, 1–9. [CrossRef] [PubMed]
24. Kang, E.H.; Lee, Y.J.; Hyon, J.Y.; Yun, P.Y.; Song, Y.W. Salivary cytokine profiles in primary Sjogren's syndrome differ from those in non-Sjogren sicca in terms of TNF-alpha levels and Th-1/Th-2 ratios. *Clin. Exp. Rheumatol.* **2011**, *29*, 970–976.
25. Kageyama, G.; Saegusa, J.; Irino, Y.; Tanaka, S.; Tsuda, K.; Takahashi, S.; Sendo, S.; Morinobu, A. Metabolomics analysis of saliva from patients with primary Sjogren's syndrome. *Clin. Exp. Immunol.* **2015**, *182*, 149–153. [CrossRef]
26. Herrala, M.; Mikkonen, J.J.; Pesonen, P.; Lappalainen, R.; Tjäderhane, L.; Niemelä, R.K.; Seitsalo, H.; Salo, T.; Myllymaa, S.; Kullaa, A.M. Variability of salivary metabolite levels in patients with Sjogren's syndrome. *J. Oral. Sci.* **2020**, *63*, 22–26. [CrossRef] [PubMed]
27. Delaleu, N.; Mydel, P.; Kwee, I.; Brun, J.G.; Jonsson, M.V.; Jonsson, R. High fidelity between saliva proteomics and the biologic state of salivary glands defines biomarker signatures for primary Sjogren's syndrome. *Arthritis Rheumatol.* **2015**, *67*, 1084–1095. [CrossRef]
28. Verstappen, G.M.; Gao, L.; Pringle, S.; Haacke, E.A.; van der Vegt, B.; Liefers, S.C.; Patel, V.; Hu, Y.; Mukherjee, S.; Carman, J.; et al. The transcriptome of paired major and minor salivary gland tissue in patients with primary Sjögren's Syndrome. *Front. Immunol.* **2021**, *12*, 681941. [CrossRef]
29. Yamamoto, K. Pathogenesis of Sjogren's syndrome. *Autoimmun. Rev.* **2003**, *2*, 13–18. [CrossRef]
30. Mingueneau, M.; Boudaoud, S.; Haskett, S.; Reynolds, T.L.; Nocturne, G.; Norton, E.; Zhang, X.; Constant, M.; Park, D.; Wang, W.; et al. Cytometry by time-of-flight immunophenotyping identifies a blood Sjogren's signature correlating with disease activity and glandular inflammation. *J. Allergy Clin. Immunol.* **2016**, *137*, 1809–1821. [CrossRef]

31. van der Kroef, M.; van den Hoogen, L.L.; Mertens, J.S.; Blokland, S.L.; Haskett, S.; Devaprasad, A.; Carvalheiro, T.; Chouri, E.; Vazirpanah, N.; Cossu, M.; et al. Cytometry by time of flight identifies distinct signatures in patients with systemic sclerosis, systemic lupus erythematosus and Sjögrens syndrome. *Eur. J. Immunol.* **2020**, *50*, 119–129. [CrossRef] [PubMed]
32. Szabo, K.; Jambor, I.; Szanto, A.; Horvath, I.F.; Tarr, T.; Nakken, B.; Szodoray, P.; Papp, G. The imbalance of circulating follicular T helper cell subsets in primary Sjögren's syndrome associates with serological alterations and abnormal B-cell distribution. *Front. Immunol.* **2021**, *12*, 639975. [CrossRef] [PubMed]
33. Martin-Gutierrez, L.; Peng, J.; Thompson, N.L.; Robinson, G.A.; Naja, M.; Peckham, H.; Wu, W.; J'bari, H.; Ahwireng, N.; Waddington, K.E.; et al. Stratification of patients with Sjögren's syndrome and patients with systemic lupus erythematosus according to two shared immune cell signatures, with potential therapeutic implications. *Arthritis Rheumatol.* **2021**, *73*, 1626–1637. [CrossRef] [PubMed]
34. Hong, X.; Meng, S.; Tang, D.; Wang, T.; Ding, L.; Yu, H.; Li, H.; Liu, D.; Dai, Y.; Yang, M. Single-cell RNA sequencing reveals the expansion of cytotoxic CD4+ T lymphocytes and a landscape of immune cells in primary Sjögren's syndrome. *Front. Immunol.* **2020**, *11*, 594658. [CrossRef]
35. Nishikawa, A.; Suzuki, K.; Kassai, Y.; Gotou, Y.; Takiguchi, M.; Miyazaki, T.; Yoshimoto, K.; Yasuoka, H.; Yamaoka, K.; Morita, R.; et al. Identification of definitive serum biomarkers associated with disease activity in primary Sjogren's syndrome. *Arthritis Res. Ther.* **2016**, *18*, 106. [CrossRef] [PubMed]
36. Padern, G.; Duflos, C.; Ferreira, R.; Assou, S.; Guilpain, P.; Maria, A.T.J.; Goulabchand, R.; Galea, P.; Jurtela, M.; Jorgensen, C.; et al. Identification of a novel serum proteomic signature for primary Sjögren's syndrome. *Front. Immunol.* **2021**, *12*, 631539. [CrossRef]
37. Fernández-Ochoa, Á.; Borrás-Linares, I.; Quirantes-Piné, R.; Alarcón-Riquelme, M.E.; Beretta, L.; Segura-Carretero, A.; Consortium, P.C. Discovering new metabolite alterations in primary sjögren's syndrome in urinary and plasma samples using an HPLC-ESI-QTOF-MS methodology. *J. Pharm. Biomed. Anal.* **2020**, *179*, 112999. [CrossRef]
38. Bengtsson, A.A.; Trygg, J.; Wuttge, D.M.; Sturfelt, G.; Theander, E.; Donten, M.; Moritz, T.; Sennbro, C.J.; Torell, F.; Lood, C.; et al. Metabolic profiling of systemic lupus erythematosus and comparison with primary Sjögren's syndrome and systemic sclerosis. *PLoS ONE* **2016**, *11*, e0159384. [CrossRef]
39. Xu, T.; Guo, Y.; Lu, J.; Shan, J.; Lin, L.; Qian, W.; Chen, W.; Wang, J.; Lv, X.; Ke, M.; et al. Untargeted serum metabolomics and potential biomarkers for Sjogren's syndrome. *Clin. Exp. Rheumatol.* **2021**, *39*, 23–29. [CrossRef]
40. Bordron, A.; Devauchelle-Pensec, V.; Le Dantec, C.; Capdeville, A.; Brooks, W.H.; Renaudineau, Y. Epigenetics in primary Sjögren's syndrome. *Adv. Exp. Med. Biol.* **2020**, *1253*, 285–308.
41. Nezos, A.; Mavragani, C.P. Contribution of genetic factors to Sjögren's syndrome and Sjögren's syndrome related lymphomagenesis. *J. Immunol. Res.* **2015**, *2015*, 754825. [CrossRef] [PubMed]
42. Imgenberg-Kreuz, J.; Rasmussen, A.; Sivils, K.; Nordmark, G. Genetics and epigenetics in primary Sjogren's syndrome. *Rheumatology* **2021**, *60*, 2085–2098. [CrossRef] [PubMed]
43. Teruel, M.; Barturen, G.; Martínez-Bueno, M.; Castellini-Pérez, O.; Barroso-Gil, M.; Povedano, E.; Kerick, M.; Català-Moll, F.; Makowska, Z.; Buttgereit, A.; et al. Integrative epigenomics in Sjögren´s syndrome reveals novel pathways and a strong interaction between the HLA, autoantibodies and the interferon signature. *Sci. Rep.* **2021**, *11*, 23292. [CrossRef] [PubMed]
44. Tasaki, S.; Suzuki, K.; Nishikawa, A.; Kassai, Y.; Takiguchi, M.; Kurisu, R.; Okuzono, Y.; Miyazaki, T.; Takeshita, M.; Yoshimoto, K.; et al. Multiomic disease signatures converge to cytotoxic CD8 T cells in primary Sjögren's syndrome. *Ann. Rheum. Dis.* **2017**, *76*, 1458–1466. [CrossRef] [PubMed]
45. James, J.A.; Guthridge, J.M.; Chen, H.; Lu, R.; Bourn, R.L.; Bean, K.; Munroe, M.E.; Smith, M.; Chakravarty, E.; Baer, A.N.; et al. Unique Sjögren's syndrome patient subsets defined by molecular features. *Rheumatology* **2020**, *59*, 860–868. [CrossRef] [PubMed]
46. Soret, P.; Le Dantec, C.; Desvaux, E.; Foulquier, N.; Chassagnol, B.; Hubert, S.; Jamin, C.; Barturen, G.; Desachy, G.; Devauchelle-Pensec, V.; et al. A new molecular classification to drive precision treatment strategies in primary Sjogren's syndrome. *Nat. Commun.* **2021**, *12*, 3523. [CrossRef]
47. Barturen, G.; Babaei, S.; Català-Moll, F.; Martínez-Bueno, M.; Makowska, Z.; Martorell-Marugán, J.; Carmona-Sáez, P.; Toro-Domínguez, D.; Carnero-Montoro, E.; Teruel, M.; et al. Integrative analysis reveals a molecular stratification of systemic autoimmune diseases. *Arthritis Rheumatol.* **2021**, *73*, 1073–1085. [CrossRef]
48. Tsang-A-Sjoe, M.W.; Bultink, I.E. New developments in systemic lupus erythematosus. *Rheumatology* **2021**, *60*, vi21–vi28. [CrossRef]
49. Shah, N.R.; Noll, B.D.; Stevens, C.B.; Brennan, M.T.; Mougeot, F.B.; Mougeot, J.L.C. Biosemantics guided gene expression profiling of Sjogren's syndrome: A comparative analysis with systemic lupus erythematosus and rheumatoid arthritis. *Arthritis Res. Ther.* **2017**, *19*, 192. [CrossRef]
50. Sada, P.R.; Isenberg, D.; Ciurtin, C. Biologic treatment in Sjogren's syndrome. *Rheumatology* **2015**, *54*, 219–230. [CrossRef]
51. Fox, R.I.; Fox, C.M.; Gottenberg, J.E.; Dörner, T. Treatment of Sjogren's syndrome: Current therapy and future directions. *Rheumatology* **2021**, *60*, 2066–2074. [CrossRef]
52. Seror, R.; Baron, G.; Camus, M.; Cornec, D.; Perrodeau, E.; Bowman, S.J.; Bombardieri, M.; Bootsma, H.; Gottenberg, J.E.; Fisher, B.; et al. Development and preliminary validation of the Sjogren's Tool for Assessing Response (STAR): A consensual composite score for assessing treatment effect in primary Sjogren's syndrome. *Ann. Rheum. Dis.* **2022**, *81*, 979–989. [CrossRef] [PubMed]

53. Lessard, C.J.; Li, H.; Adrianto, I.; Ice, J.A.; Rasmussen, A.; Grundahl, K.M.; Kelly, J.A.; Dozmorov, M.G.; Miceli-Richard, C.; Bowman, S.; et al. Variants at multiple loci implicated in both innate and adaptive immune responses are associated with Sjogren's syndrome. *Nat. Genet.* **2013**, *45*, 1284–1292. [CrossRef] [PubMed]
54. Cruz-Tapias, P.; Rojas-Villarraga, A.; Maier-Moore, S.; Anaya, J.M. HLA and Sjogren's syndrome susceptibility. A meta-analysis of worldwide studies. *Autoimmun. Rev.* **2012**, *11*, 281–287. [CrossRef] [PubMed]
55. Lucchesi, D.; Pitzalis, C.; Bombardieri, M. EBV and other viruses as triggers of tertiary lymphoid structures in primary Sjogren's syndrome. *Expert Rev. Clin. Immunol.* **2014**, *10*, 445–455. [CrossRef] [PubMed]
56. Nezos, A.; Gravani, F.; Tassidou, A.; Kapsogeorgou, E.K.; Voulgarelis, M.; Koutsilieris, M.; Crow, M.K.; Mavragani, C.P. Type I and II interferon signatures in Sjogren's syndrome pathogenesis: Contributions in distinct clinical phenotypes and Sjogren's related lymphomagenesis. *J. Autoimmun.* **2015**, *63*, 47–58. [CrossRef] [PubMed]

biomedicines

Review

The Current Status of Molecular Biomarkers for Inflammatory Bowel Disease

Zahra Alghoul [1,2], Chunhua Yang [1,3,*] and Didier Merlin [1,3]

1 Digestive Diseases Research Group, Institute for Biomedical Sciences, Georgia State University, Atlanta, GA 30303, USA; zalghoul@gsu.edu (Z.A.); dmerlin@gsu.edu (D.M.)
2 Department of Chemistry, College of Arts & Sciences, Georgia State University, Atlanta, GA 30303, USA
3 Atlanta Veterans Affairs Medical Center, Decatur, GA 30033, USA
* Correspondence: cyang16@gsu.edu; Tel.: +1-404-413-3598

Abstract: Diagnosis and prognosis of inflammatory bowel disease (IBD)—a chronic inflammation that affects the gastrointestinal tract of patients—are challenging, as most clinical symptoms are not specific to IBD, and are often seen in other inflammatory diseases, such as intestinal infections, drug-induced colitis, and monogenic diseases. To date, there is no gold-standard test for monitoring IBD. Endoscopy and imaging are essential diagnostic tools that provide information about the disease's state, location, and severity. However, the invasive nature and high cost of endoscopy make it unsuitable for frequent monitoring of disease activity in IBD patients, and even when it is possible to replace endoscopy with imaging, high cost remains a concern. Laboratory testing of blood or feces has the advantage of being non-invasive, rapid, cost-effective, and standardizable. Although the specificity and accuracy of laboratory testing alone need to be improved, it is increasingly used to monitor disease activity or to diagnose suspected IBD cases in combination with endoscopy and/or imaging. The literature survey indicates a dearth of summarization of biomarkers for IBD testing. This review introduces currently available non-invasive biomarkers of clinical importance in laboratory testing for IBD, and discusses the trends and challenges in the IBD biomarker studies.

Keywords: proteomics; epigenetics; endoscopy; imaging; laboratory testing

Citation: Alghoul, Z.; Yang, C.; Merlin, D. The Current Status of Molecular Biomarkers for Inflammatory Bowel Disease. *Biomedicines* **2022**, *10*, 1492. https://doi.org/10.3390/biomedicines10071492

Academic Editor: Marianna Christodoulou

Received: 24 May 2022
Accepted: 21 June 2022
Published: 24 June 2022

1. Introduction

Inflammatory bowel disease (IBD) is a set of chronic and idiopathic inflammatory conditions that affect more than 3.5 million patients worldwide. The two major forms of IBD are Crohn's disease (CD), in which inflammation affects any segment of the gastrointestinal (GI) tract [1], and ulcerative colitis (UC), in which inflammation affects the inner lining of the colon or rectum [2]. Patients with IBD are up to six times more likely to develop colorectal cancer than the general population [3,4]. In addition to the molecular alterations (such as chromosomal instability, microsatellite instability, and hypermethylation) that contribute to sporadic colorectal cancer, IBD-related colorectal cancer is linked to inflammation that induces the transcription of mutated cancer genes [5]. Loss-of-function mutations in tumor-suppressor protein p53 occur in both sporadic and IBD-related colorectal cancer, but they occur earlier in the non-dysplastic mucosa of IBD-related colorectal cancer than in sporadic colorectal cancer [4,5]. Another mutation observed in both types of cancer is the nonfunctional adenomatous polyposis coli (APC) gatekeeper gene. Unlike the p53 mutation, APC mutation occurs just prior to carcinoma in IBD-related colorectal cancer, but at a much earlier stage in sporadic colorectal cancer [4]. Other gene mutations linked to IBD-related colorectal cancer include p27, k-Ras (12p12) oncogene, human mismatch repair genes (e.g., hMLH1, hMSH2), and p16 [4].

CD and UC are both characterized by mucosal inflammation, with occasional flares and remittance. Inflammation in CD can affect any segment of the GI tract, and spreads in a non-continuous pattern [1,6]. CD commonly involves the formation of strictures, abscesses, and fistulas [6]. Its histological features include thickened submucosa, fissuring ulceration, transmural inflammation, and non-caseating granulomas [6]. Inflammation in UC affects the inner lining of the colon or rectum, and spreads in a continuous pattern [2,6]. It shows superficial inflammatory changes in the mucosa and submucosa, and involves the formation of cryptitis and crypt abscesses [6]. The clinical symptoms of IBD include abdominal pain, diarrhea, rectal bleeding, weight loss, nausea, intestinal pain and, in some cases, fever [7,8]. As these symptoms are not specific to IBD, the clinical diagnostic process must consist of using a combination of endoscopic, radiological, clinical, histological, and laboratory tests [9]; a single technique is often insufficient for the diagnosis.

Endoscopy and imaging are essential techniques for the diagnosis, management, and treatment of IBD. They are used in the initial evaluation of patients with suspected IBD, as well as in making a differential diagnosis of UC versus CD in confirmed IBD cases [10]. The strength of endoscopy as a diagnostic tool lies primarily in its ability to visually observe different bowel segments, allowing clinicians to assess disease severity and monitor disease activity over time. Ileocolonoscopy has traditionally been the most used form of endoscopy in IBD. The initial evaluation of patients presenting with clinical symptoms suggestive of IBD should be carried out with ileocolonoscopy, as recommended by the American Society for Gastrointestinal Endoscopy (ASGE) Standards of Practice Committee [11]. In addition to providing a visual of the colon and the terminal ileum, ileocolonoscopy can be used to obtain biopsy specimens for further analysis. The ASGE suggests obtaining at least two biopsy specimens from five sites throughout the bowel during the initial evaluation [12]. However, the invasiveness and high cost of ileocolonoscopy are major drawbacks that have limited its frequent use for monitoring disease activity.

New, less-invasive endoscopic techniques that can more accurately diagnose IBD, while also providing a differential diagnosis of CD and UC, have emerged in the past few years. These include video capsule endoscopy (VCE), confocal laser endomicroscopy (CLE), and single- or double-balloon-assisted enteroscopy (SBE and DBE, respectively). VCE provides imaging of the whole bowel via ingestion of a wireless capsule endoscope [13]. This technique is particularly useful for inspecting areas in the GI tract that cannot be visualized by colonoscopy [14]. Although the risk of capsule retention is low, it remains the primary concern in patients with suspected or known IBD [15]. VCE is less invasive and more cost-effective than ileocolonoscopy, but it cannot be used in performing biopsies. In CLE, a confocal laser microscope is used in vivo to obtain living tissue images during colonoscopy [16]. CLE has the advantage of offering a faster diagnosis than a traditional colonoscopy. Enteroscopy in both of its forms (SBE and DBE) allows access to small bowel areas that standard endoscopy cannot reach. Additionally, enteroscopy can be used in performing histological analysis. However, due to its technical complexity and time-consuming preparation, enteroscopy is not recommended for the initial evaluation of suspected IBD cases [17].

In confirmed IBD cases, clinical symptoms alone are insufficient for clinicians to determine the extent of mucosal inflammation, or to make a differential diagnosis between UC and CD. There has been a growing interest in the use of cross-sectional imaging modalities such as magnetic resonance enterography (MRE), ultrasonography (US), and computed tomography (CT) as tools to supplement endoscopy in the diagnosis and monitoring of IBD [18]. These techniques are instrumental in detecting mural and extramural complications and assessing laminal inflammation in areas affected by CD in the small bowel that are beyond the reach of colonoscopy [19]. Due to their ability to diagnose CD with high accuracy, cross-sectional imaging modalities are used to make differential diagnoses in suspected cases of UC [20]. This aspect is critical because these diseases differ in their prognosis and required treatments.

Although imaging techniques offer highly accurate IBD diagnosis, they require experienced personnel, sophisticated instruments, and high costs, hampering their routine application. Laboratory testing's advantage lies in the fact that these tests can be standardized, rapid, and cost-effective, but they can also be applied to the already established patient sample libraries to process independent investigations. An increasing number of laboratory tests, combined with endoscopy or imaging, are used to monitor disease activity or diagnose suspected IBD cases. As good laboratory test results rely on the proper use of molecular biomarkers from the patients' tissue, blood (serum), or fecal samples, this review summarizes currently available biomarkers of clinical importance in laboratory testing of IBD, discusses the possible involved genetic and epigenetic factors, and envisions the trends and challenges of biomarker discovery in IBD.

2. Non-Invasive Molecular Biomarkers of IBD: Serum Proteins, Serological Antibodies, and Fecal Proteins

Biomarkers play critical roles in the early detection and monitoring of disease progression and therapeutic responses (Figure 1). Disease activity can be monitored with laboratory tests that measure circulating biomarkers in the blood (serum or plasma), tissue, or feces. A biomarker is defined as "a characteristic that is objectively measured and evaluated as an indicator of normal biological processes, pathogenic processes, or pharmacological responses to a therapeutic intervention" [21]. Identifying a biomarker or several biomarkers of a given condition's pathologies might help to diagnose, prognose, and assess therapeutic responses. For a biomarker to be effective, it should possess several attributes, such as being non-invasive, inexpensive, convenient for sampling, reproducible, and disease-specific (i.e., accurate and precise). An ideal biomarker also needs to have a rapid test-to-result turnaround time, be standardizable to provide comparable test results across different assays, be widely available and stable for storage, have a wide dynamic range, use defined thresholds to determine the absence/presence or extent of inflammation, and be responsive to changes in the state of inflammation [22].

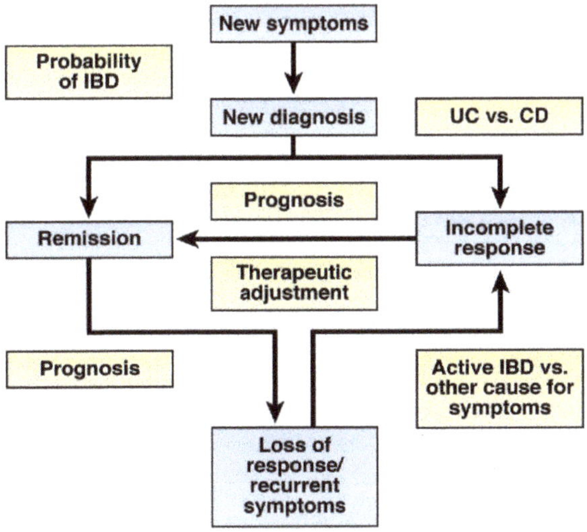

Figure 1. The potential role of biomarker assays in the care of patients with suspected or established IBD: Biomarkers may be used in all phases of the care. For patients with suspected IBD, biomarkers can be used to select which patients are unlikely to have IBD and could forgo further testing. Once patients are diagnosed, biomarkers can determine which patients have CD or UC and predict the disease course. Biomarkers can be used to determine which patients are most likely to respond to therapies, determine prognosis, and identify those who require more aggressive therapies. In patients with recurrent symptoms, biomarkers can differentiate patients with active inflammation from those likely to have symptoms from other causes. Adapted from James D. Lewis's review [23]; *Gastroenterology*, Volume 140 Issue 6, Pages 1817–1826.e2; https://doi.org/10.1053/j.gastro.2010.11.058.

Several molecular biomarkers have been established as reliable measures for disease activity in IBD [22,24]. They are minimally invasive and relatively inexpensive compared to colonoscopy and imaging techniques. They can also assist in identifying patients who require diagnosis with endoscopy and biopsies. However, many of these biomarkers have limitations in terms of their specificity, sensitivity, responsiveness, and/or other desirable attributes of IBD biomarkers [22]. There are currently three major types of molecular biomarkers available for IBD: serum biomarkers, serological antibodies, and fecal biomarkers.

2.1. Serum Biomarkers

Several inflammatory serum biomarkers have become part of routine laboratory testing for the diagnosis of IBD. Although they are not specific to IBD, these serum biomarkers are commonly used for initial diagnosis due to their ease of use, low cost, and well-established protocols. The most common of these tests are those for C-reactive protein (CRP) and the erythrocyte sedimentation rate (ESR).

CRP is a pentameric protein that is produced in the liver by hepatocytes. It is found in serum at <1 mg/L under physiological conditions. Its concentration increases during an acute-phase response, as pro-inflammatory cytokines such as IL-6, tumor necrosis factor α (TNF-α), and IL-1β stimulate its production in the hepatocytes [25–27]. CRP has a relatively short half-life (about 19 h) [28], making it a better indicator of inflammation than most acute-phase proteins. Elevated CRP levels are observed in most active CD cases, whereas the CRP levels of UC patients show little-to-no increase in the case of active disease [27,29]. This may reflect the production of CRP by mesenteric adipocytes in patients with CD [30]. Although CRP is widely used as a biomarker for IBD, it lacks specificity; elevated CRP levels are also observed in autoimmune disorders, infections, and malignancies [25].

ESR is a measure of how quickly erythrocytes sediment through plasma in a column, with a higher rate taken as indicating more inflammation. ESR values are affected by physiological factors such as pregnancy, age, and gender, as well as changes in hematocrit levels in patients with anemia and polycythemia [31]. Medications that cause changes in the size of erythrocytes can also affect ESR values [32]. Changes in ESR values are not specific to IBD, and can be due to any inflammatory stimulus. Unlike CRP, ESR values are altered in both UC and CD, and we cannot distinguish them. ESR values peak more slowly than CRP, and take longer to return to normal after the end of an inflammatory flare [28].

CRP and ESR have been studied long enough to become established in IBD diagnosis. While both tests lack the specificity and accuracy to be considered a gold-standard diagnosis, CRP has some advantages over ESR. For example, the CRP concentration changes faster than the ESR value upon a change in disease activity, CRP has a broader range of abnormal values than ESR, and (unlike ESR) CRP does not show age-related variation [33].

Leucine-rich alpha-2 glycoprotein (LRG) is a 50 kD protein that is secreted by hepatocytes, neutrophils, macrophages, and intestinal epithelial cells [34–36]. It has recently emerged as a novel serological biomarker for IBD and rheumatoid arthritis. Studies have found that levels of LRG are elevated in patients with active UC, and decrease with a decline in disease activity [37,38]. Notably, elevated levels of LRG correlate better than CRP with clinical and endoscopic scores in patients with active UC and CD [38–40]. LRG has been also found to predict mucosal healing in both UC and CD patients with normal CRP levels [41].

2.2. Serological Antibodies

Serological testing is a well-established diagnostic tool for a variety of immune diseases. Its use in IBD has been mainly focused on patients with a confirmed diagnosis; little work has been done on its potential as a primary diagnostic tool in patients with suspected IBD. Perinuclear anti-neutrophil cytoplasmic antibodies (p-ANCAs) and anti-*Saccharomyces cerevisiae* antibodies (ASCAs) are the two primary antibodies currently examined in IBD studies. ANCAs are a group of antibodies produced against antigens in the cytoplasm of neutrophils. ASCAs are produced against mannan and other yeast cell wall components. Both have been reported to provide clinically useful positive or negative predictive values: p-ANCA+/ASCA− is reported in patients with UC, while p-ANCA−/ASCA+ is seen in patients with CD. Although each of these biomarker antibodies can be used to discriminate UC from CD, they both have low accuracy and sensitivity [42]. Positive results for either antibody are not unique to IBD, and may be related to several other GI and inflammatory conditions, such as celiac disease, Behcet's disease, cystic fibrosis, and rheumatoid arthritis [42,43].

2.3. Fecal Biomarkers

Fecal biomarkers are the proteins that are explicitly found in stool samples of patients with IBD. The fecal biomarkers for IBD reported to date are mainly fecal leukocyte proteins. These include calprotectin, calgranulin C, lactoferrin, and lipocalin-2. They have several advantages over blood biomarkers, including the ease of sample accessibility, high biomarker concentration due to the direct contact of the fecal sample with the site of inflammation, and higher specificity for IBD because they reflect GI inflammation (unlike serum biomarkers, which are increased by various types of inflammation) [44].

Calprotectin is the most widely used fecal biomarker for IBD. It is a calcium- and zinc-binding protein that is abundant in neutrophils, eosinophils, and macrophages. Changes in its concentration are observed in various secretory and excretory products in the body upon activation of granulocytes and mononuclear phagocytes [45]. Elevated fecal calprotectin levels are expected in patients with active IBD, due to the presence of a high number of neutrophils in the GI tract, which is characteristic of the disease [28]. Calprotectin is resistant to degradation, and is stable for 7 days in fecal samples stored at room temperature [46]. Changes in fecal calprotectin levels are not exclusive to IBD; alterations are also observed in various colon and intestine diseases [47].

Calgranulin C (S100A12) belongs to the S100 family of low-molecular-weight calcium-binding proteins, which activate the NF-κB pathway and increase cytokine release during pro-inflammatory processes [31]. The serum concentration of calgranulin C is high in IBD [48], but the fecal concentration is higher, making the fecal assay more sensitive to IBD. Elevated levels of calgranulin C have been reported in other inflammatory conditions, such as arthritis [49].

Lactoferrin is another biomarker whose levels are significantly elevated in active IBD. It is an iron-binding glycoprotein that is found specifically in neutrophils; in this respect, it contrasts with calprotectin, which is found in several types of cells. Lactoferrin has high specificity and sensitivity for diagnosing active IBD [50].

Lipocalin-2 (LCN-2), also known as neutrophil gelatinase-associated lipocalin (NGAL) or siderocalin (Scn), is a bacteriostatic protein stored in neutrophil granules [51,52]. LCN-2 is involved in innate immunity by secluding iron from pathogenic bacteria, limiting their invasion. It is a highly stable protein whose elevated expression by gut epithelial cells has been demonstrated in colonic biopsies from inflamed areas of patients with IBD. Serum LCN-2 has been proven to be an active biomarker in UC patients, and it is widely used as a fecal biomarker of acute inflammation in the animal model of UC, indicating that it can potentially be used as a fecal biomarker of human UC. Upregulation of LCN-2 is believed to be induced by IL-22 and IL-17A [53].

2.4. Diagnostic/Prognostic Accuracy

The major concern about diagnosis and prognosis of IBD that solely rely on singular molecular biomarkers is their detection accuracy. A study showed that the biomarkers' correlation coefficients with endoscopy could vary from 0.48 to 0.83 (for calprotectin) and from 0.19 to 0.87 (for lactoferrin) in IBD patients [23] (Table 1). IBD detection methods that combine endoscopy with histopathology biomarkers can be highly accurate, such as in the context of oncostatin M (OSM) or oncostatin M receptor (OSMR), which are found to be highly overexpressed in the inflamed intestinal tissue of active IBD patients, with a *p*-value < 0.001 for OSM (*n* = 42) and a *p*-value < 0.05 for OSMR (*n* = 86) at a false discovery rate (FDR) of 1% [54].

Table 1. Correlation of biomarkers with disease activity, determined by endoscopy.

Patient Population	Assessment of Endoscopic Disease Activity	Lactoferrin (Correlation Coefficient)	Calprotectin (Correlation Coefficient)	CRP (Correlation Coefficient)
CD	CDEIS *	0.77	0.73	0.55
CD	SES-CD **	0.19	0.48	
UC	Mayo score	0.35	0.51	
UC	Matt's index		0.81	
CD	SES-CD	0.63	0.64	0.52
IBD			0.52	
UC	Mayo score		0.57	
CD	SES-CD	0.76	0.72	0.46
CD	CDEIS	0.87	0.83	0.61
UC	Rachmilewitz index		0.83	0.50
CD	CDEIS		0.75	0.53

* CDEIS, Crohn's Disease Endoscopic Index of Severity; ** SES-CD, Simple Endoscopic Score for Crohn's Disease. Adapted from James D. Lewis's review [23]; *Gastroenterology*, Volume 140 Issue 6 Pages 1817–1826.e2; https://doi.org/10.1053/j.gastro.2010.11.058.

To date, C-reactive protein and fecal calprotectin are considered reliable markers of disease activity, with demonstrated utility in IBD management [55]. However, single-biomarker-based detections often present a larger ambiguous "grey zone" than detections made using composite biomarkers (Figure 2). Composite biomarkers are defined as "a combination of ≥2 biomarkers", and are selected using an optimized algorithm to render a single interpretive output. The combination of different biomarkers has shown higher accuracy, and is expected to reduce the "grey zone" of each biomarker and replace single-marker approaches in the future of research and clinical practice [55] (Figure 2).

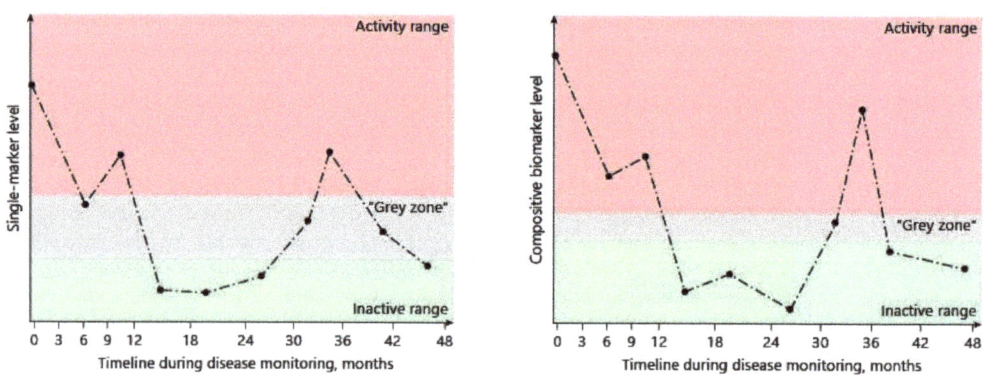

Figure 2. Improvements are provided by composite biomarkers. Careful selection of markers and their integration can optimize the diagnostic accuracy of single biomarkers of disease activity and drastically reduce the blind spot resulting from the "grey zone". Adapted from Dragoni G. et al.'s review [55]; *Digestive Diseases*, https://doi.org/10.1159/000511641.

3. Trends in IBD Biomarker Discovery

3.1. Proteomics

Proteomics, the study of the set of gene-encoded proteins known as the proteome, uses a range of techniques for separating, identifying, and structurally characterizing proteins. Proteomics goes beyond the study of proteins in a given cell, including their isoforms, post-translational modifications, and protein–protein interactions [56]. Depending on the analysis method, proteomic approaches can be bottom-up or top-down. In bottom-up proteomics, proteolytic digestion breaks the extracted proteins into peptides, which are then analyzed by mass spectrometry (MS). In top-down proteomics, intact proteins are analyzed. The samples used in IBD-related studies are usually obtained from blood (serum or plasma) or colonic biopsies. Liquid chromatography coupled with electrospray tandem mass spectrometry (LC–ESI-MS/MS) is the most widely used proteomic technique in IBD research. Other commonly used techniques include two-dimensional gel electrophoresis coupled with matrix-assisted laser desorption/ionization (MALDI)-MS screening and immunofluorescence microscopy.

Due to the strong connections between protein expression and disease activity, the application of proteomics in biomarker discovery is a promising emerging field. Advances in LC–MS instrumentation, such as the combination of ultrahigh-performance liquid chromatography (UPLC) with nano-electrospray ionization and high-resolution mass spectrometry (HRMS), have revealed the potential of MS-based proteomics to compete with or even replace traditional immunoassay techniques. It is hoped that proteomics may help to develop personalized and precision medicine [57]. Instead of focusing on finding a single biomarker, current proteomic biomarker research aims to identify protein biomarker panels representing an individual's disease state. In this context, three approaches have emerged over the past few years: (1) Proteotyping—a multiprotein approach used to determine an individual's unique proteome [58]. (2) Proteogenomics—a multi-omics approach in which genomic and proteomic analyses are performed on the same sample; data obtained from this pairing contain information that would not be obtained using either technique alone [59,60]. (3) Proteoforms—protein variants that result from post-translational modifications of proteins, genetic mutations, or truncations. MS immunoassays are often used to map a specific protein's proteoforms to distinguish between normal and clinical fluctuations [61,62].

To date, the use of proteomics in IBD has focused on three areas: identifying novel protein biomarkers for diagnosis, understanding the pathological mechanisms underlying disease activity, and monitoring the response to treatment. Berndt et al. pioneered the use of proteomics in IBD by performing proteomic analysis of normal and inflamed intestinal mucosa using multi-epitope ligand cartography immunofluorescence microscopy. The authors found that different T-cell populations in the mucosa expressed distinct proteins in each form of IBD [63]. An experimental approach based on combining discovery proteomics with targeted verification experiments successfully assessed transmural intestinal complications in CD, with 70% sensitivity and 72.5% specificity. This approach, which used label-free LC–MS/MS, identified a serological biomarker panel that could discriminate complicated CD from uncomplicated CD, rheumatoid arthritis, UC, and healthy controls [64]. Another study that used LC–MS identified a panel of four proteins that could distinguish active pediatric IBD from non-IBD with high sensitivity and specificity.

Additionally, the study found that two of the identified proteins were elevated in IBD stool samples, demonstrating that fecal samples can be used for measuring these biomarkers [65]. Several studies attempted to identify differentially expressed proteins in patients with UC and CD through proteomic profiling of serum or colonic biopsies. Proteomic profiling of colon biopsies using MALDI-MS identified distinct protein peaks for UC and CD specimens, indicating that it could be possible to differentially diagnose these IBD forms using protein profiles [66–68]. In a study that compared the proteomic spectra of submucosal samples from inflamed UC versus CD and uninflamed UC versus CD, two distinct peaks were identified in the first case, and three in the second [66]. Another study identified a set of 25 proteins as differentiators for UC and CD in colonic mucosal tissue samples obtained from 62 patients with confirmed UC/CD [67]. Screening of mucosal biopsies obtained from children with suspected IBD identified two distinct biomarker panels: one consisted of 5 proteins that were reported to discriminate IBD from control patients, while the other consisted of 12 proteins reported to allow the differential diagnosis of CD and UC patients [68]. Protein profiling of 120 serum samples from patients with

CD or UC and inflammatory and healthy controls was performed using surface-enhanced laser desorption/ionization–time-of-flight mass spectrometry (SELDI-TOF-MS). This work identified four diagnostic protein biomarkers for IBD, one of which could reportedly discriminate UC from CD with accuracies similar to or higher than those of the ANCA and ASCA serological tests [69]. Proteomic profiling of stricturing CD, non-stricturing CD, and UC patients identified a smaller set of peptides for differentiating stricture versus non-stricture CD in IBD [70].

In addition to diagnostic biomarkers, several studies have used proteomics to identify biomarkers that could be used to assess treatment responses in IBD. One study monitored the treatment response to infliximab in IBD patients by measuring the levels of circulating chemokines and monocyte activation using LC–nano-ESI-MS/MS. The study found that 2 weeks from the start of treatment, decreases were evident in the levels of macrophage-derived CD14 and CD86, as well as the chemokine, CCL2 potentially providing a mechanistic explanation for why not all patients respond to this treatment [71]. Another study investigated the treatment response to infliximab and prednisone in children with IBD. The study identified 18 proteins and 3 miRNAs that were responsive to both drugs; some were downregulated with inflammation, while others were upregulated as the inflammation was resolved [72].

3.2. Genetics

Pathological studies of IBD and its two subtypes suggest a genetic risk factor behind the immune response to the intestinal microbiota. Genome-wide association studies (GWASs) have identified approximately 240 gene loci associated with susceptibility to IBD [73]. Several studies have used genetic profiling of blood samples to identify gene panels that may help to differentiate IBD from healthy controls [74], active from inactive CD [75], and CD from UC [76–78]. Distinct gene panels were also identified in peripheral blood samples from pediatric IBD patients in clinical remission compared to healthy controls [79]. Other studies performed gene expression analysis on mucosal biopsies from IBD patients, and identified distinct gene panels for IBD versus healthy controls [80] and UC versus healthy controls [81]. The use of genetics to identify loci associated with IBD can potentially define causal disease mechanisms, which could, in turn, advance the biomarker discovery process [82].

3.3. Epigenetics

Epigenetics, which describes changes in gene function caused by gene–environment interactions rather than changes in the DNA sequence, is gaining research interest among scientists seeking to study the pathogenesis and diagnosis of IBD [83,84]. DNA methylation and RNA interference are the two most heavily researched areas in IBD epigenetic studies.

DNA methylation refers to adding a methyl group to cytosine residues in the CpG dinucleotide sequence [85]. Early studies of DNA methylation changes in the mucosa of IBD patients focused primarily on their use as predictors of malignancy [86]. Recent studies have shown that the DNA methylation of specific genes plays a role in the pathogenesis of IBD, suggesting that they could be useful as biomarkers [87,88]. A genome-wide methylation profiling conducted on rectal biopsies identified panels of genes (e.g., *THRAP2*, *FANCC*, *GBGT1*, *DOK2*, *TNFSF4*, *TNFSF12*, and *FUT7*) that showed evidence of differential methylation in CD and UC specimens in comparison to those from healthy controls [88]. Another study identified seven differentially methylated CpG sites in the diseased intestinal tissue of IBD patients compared to normal intestinal tissue from the same patients [89]. Genome-wide changes in DNA methylation have also been analyzed using the peripheral blood of patients with IBD. Analysis of the DNA methylation changes using peripheral blood from CD patients identified 50 genes that showed significant differential methylation compared to that in healthy controls [87]. Site-specific DNA methylation changes in genes associated with IBD pathways have also been identified, with the results showing a 45% overlap of the differentially methylated positions in UC and CD [90].

MicroRNAs (miRNAs) are non-coding, single-stranded RNA species that consist of 18–25 nucleotides. Disruptions in their expression profiles and function are observed in human diseases such as cancer and neurological, cardiovascular, and autoimmune diseases [91]. The potential of miRNAs as diagnostic biomarkers and treatment options in IBD has garnered growing interest in the past few years. Colonic tissue and circulating miRNAs (e.g., serum, feces) are the two types of samples used in most of these studies.

Several studies have successfully identified distinct miRNA profiles reflecting the up- or downregulation of one or more miRNAs in colonic biopsy specimens of IBD patients [92] (Table 2). One of the pioneering studies in this area identified the differential expression of 11 miRNAs in the mucosal tissue samples of patients with active UC [93]. Other studies that examined the colonic mucosa of patients with active UC reported upregulation of one or more miRNAs (such as miR-21 [94], miR-150 [95], and miR-155 [94]) and downregulation of others (such as miR-143 and miR-145 [96]), in comparison to healthy controls. Similarly, some studies compared the colonic mucosa of patients with active CD to healthy controls, and reported upregulation of miR-196 [97] and downregulation of miR-7 [98]. Other studies assayed the expression of hundreds of miRNAs, and identified panels differentially expressed in the colonic tissues of patients with UC and CD versus controls [99–101].

Table 2. A summary of microRNAs that are correlated with ulcerative colitis (UC#1–12) or Crohn's disease (CD, #13–22).

#	MiRNAs	Disease Subtype	Sample Type	Techniques Used	Outcome
1	miR-19a	UC, HC	Biopsy, murine tissue	RT-qPCR	Reduced expression of miR-19a in human colon tissue with UC and DSS-treated murine colitis.
2	miR-21	UC, HC	Biopsy	RT-qPCR, ISH	Overexpression of miR-21 in UC.
3	miR-21-5p	UC, HC	Sera, rat tissue	RT-qPCR, Transfection	MiR-21-5p was downregulated in the sera and colon tissue of UC compared with healthy people and the control group.
4	miR-124	UC, HC	Biopsy	RT-qPCR	MiR-124 regulated the expression of STAT3. Reduced levels of miR-124 in colon tissues of children with active UC appeared to increase the expression and activity of STAT3.
5	miR-141	UC, HC	Biopsy	Microarray, RT-qPCR	MiR-141 played a role in the bowel inflammation of individuals with active UC via downregulation of CXCL5 expression.
6	miR-150	UC, HC	Murine model	RT-qPCR	MiR-150 was elevated and c-Myb was downregulated in the human colon with active UC compared to HC.
7	miR-155	Colitis	Murine tissue, cell culture	RT-qPCR, transfection	MiR-155 promoted the pathogenesis of experimental colitis by repressing SHIP-1 expression.
8	miR-193a-3p	UC, HC	Cell culture, biopsy	RT-qPCR, ISH	MiR-193a-3p reduced intestinal inflammation in response to microbiota.
9	miR-206	UC, HC	Cell culture, biopsy	RT-qPCR,	MiR-206 as a biomarker for response to mesalamine treatment in UC.
10	miR-21, miR-155	UC, HC	Biopsy	RT-qPCR	MiR-21 and miR-155 were highly expressed in UC.
11	miR-143, miR-145	UC, HC	Biopsy	RT-qPCR, ISH	MiR-143 and miR-145 were downregulated in UC.
12	miR-125b, miR-155, miR-223 and miR-138	UC	Biopsy	RT-qPCR, microarray	Differential expression of miR-223, miR-125b, miR-138, and miR-155 in the inflamed mucosa compared to non-inflamed mucosa and controls.
13	miR-7	CD, HC	Cell culture, biopsy	Transfection, RT-qPCR	MiR-7 modulated CD98 expression during intestinal epithelial cell differentiation.
14	miR-19b	CD, HC	Biopsy, cell culture	RT-qPCR, ISH	MiR-19b suppressed the inflammation and prevented the pathogenesis of CD.
15	miR-29b	CD	Fibroblasts	RT-qPCR	MCL-1 was modulated in CD fibrosis by miR-29b via IL-6 and IL-8

Table 2. *Cont.*

#	MiRNAs	Disease Subtype	Sample Type	Techniques Used	Outcome
16	miR-122	CD, HC	Cell culture, biopsy	RT-qPCR, Transfection	MiR-122 reduced the expression of pro-inflammatory cytokines (TNF and IFN-γ) and promoted the release of anti-inflammatory cytokines (e.g., IL-4 and IL-10). Significant increase in miR-122 expression in cells treated with 5′-AZA.
17	miR-141	CD	Murine models, biopsy	Microarray, RT-qPCR	MiR-141 regulated colonic leukocytic trafficking by targeting CXCL12β during murine colitis and human CD.
18	miR-155	CD, HC	PBMC	RT-qPCR, transfection	MiR-155 regulated IL-10-producing CD24 CD27+ B Cells.
19	miR-200b	CD, HC	Biopsy, serum. cell culture	RT-qPCR	MiR-200b was involved in intestinal fibrosis of CD.
20	miR-590-5p	CD, HC	Human and murine tissues	RT-qPCR	Decreased miR-590-5p levels in CD.
21	miR-146a, miR-155	CD	Biopsy	RT-qPCR	MiR-146a and -155 showed increased duodenal expression in pediatric CD.
22	miR-223-3p, miR-31-5p	CD, HC	Biopsy	Nanostring	Mir-223-3p expression showed age- and sex-related effects and miR-31-5p expression was driven by location

HC: healthy controls, RT-qPCR: quantitative real-time polymerase chain reaction, Biopsy: colon tissue biopsy, ISH: in situ hybridization, PBMCs: peripheral blood mononuclear cells, DSS: dextran sodium sulfate, TNF: Tumor necrosis factor alpha. Adapted and modified from Jaslin P. James et al.'s review [92]; *Int. J. Mol. Sci.* 2020, 21, 7893; doi:10.3390/ijms21217893.

Distinct profiles of circulating miRNAs have also been identified in blood samples of IBD patients. Several studies identified many upregulated or downregulated miRNAs in peripheral blood samples from patients with IBD. Samples were obtained from patients with UC or CD versus healthy controls [101–104] and pediatric CD versus healthy controls [105]. Distinct panels of miRNAs have also been identified in fecal samples of IBD patients [106–108]. More investigation into the specificity of miRNAs for IBD is required before they can be used as diagnostic tools, as some miRNAs are known to be associated with other conditions. For example, miR-21 is significantly high in the blood of UC patients [103], but is also upregulated in patients with colorectal cancer [109]. One study examined the differential expression of miRNAs between UC and CD in saliva, in addition to blood and colon tissue samples [110]. The study identified several miRNAs (i.e., miR-21, miR-31, miR-142-3p, miR-142-5p) whose expression levels in all three types of samples were significantly altered between IBD and non-IBD patients.

4. Challenges and Future Directions

4.1. Proteomic Biomarker Discovery

The typical protein biomarker discovery and validation process consists of six phases: discovery, qualification, verification, assay optimization, clinical evaluation/validation, and commercialization [111]. During the discovery phase, researchers identify a list of 20 to several hundred proteins that are differentially expressed between healthy and disease-confirmed samples. This identification process is based on an unbiased, semi-quantitative assessment of peptide abundances in both samples. In the next phase, qualification, this unbiased approach is replaced with a targeted analysis to confirm the differential expression of the candidate proteins identified in the discovery phase. In the verification phase, a more significant number of samples are used to account for the variations in the proteomes of the different studied sets. At this stage, specificity and sensitivity acquire particular importance when the researchers select the few protein biomarkers used in the assay optimization and clinical evaluation phases. In the assay optimization phase, an antibody is selected for each biomarker candidate and used to develop an immunoassay to replace the MS step in protein quantification. During the evaluation/validation phase, the assay is evaluated for analytical parameters, such as accuracy and precision. If clinical validation is successful, the protein biomarker moves to the commercialization state [111].

The path to successful protein biomarker discovery through this multistage process faces several challenges. As a result, the introduction of new protein biomarkers has been slow, and has not met the clinical need for proteomic tests [112]. Some relevant challenges include the low number of samples under study and the lack of well-designed study methods and standard protocols [113]. These variables can be optimized through more careful choices of sample types and sizes. Sample selection and processing require special consideration when performing a proteomic analysis. For example, human plasma contains tens of thousands of proteins that differ in their structures and abundances [114]. It is not always possible to identify a single or multiple disease-specific proteins that could be used as markers for a particular disease. The proteins selected in the discovery phase are often classified as false positives. This is primarily due to the low frequency of selecting low-abundance proteins and limitations in their detection [111]. Even using other biofluids—such as urine, cerebrospinal fluid, cell line homogenates, or tissue lysates—has not eliminated this complexity [111]. There are also considerations more specific to the study of IBD. Intestinal mucosal biopsies are widely used in IBD studies. Protein degradation during and after extraction might lead to the under- or over-representation of specific proteins [115]. The use of protease inhibitors that minimize protein degradation can keep this variable under control. Cell heterogeneity of the mucosal specimens is another variable that could lead to an inaccurate proteome analysis [115]. Enriching samples for specific cell types and/or organelles can lower the sample's complexity and improve the protein identification efficiency [115,116]. The statistical power of a proteomic study is another factor that requires special attention in the biomarker discovery pipeline, especially in the discovery and verification stages. Skates et al. proposed a statistical framework for increasing the probability of identifying a biomarker that can reach the clinical validation stage [117]. According to their framework, the success of a biomarker in reaching clinical validation depends on the number of candidate proteins examined at each stage, the separation in biomarker signal between cases and controls (as measured by standard deviation), and the percentage of cases in which the biomarker is expressed. The authors provided probability tables that can be used in determining the proper sample size for a given study.

Although significant progress has been achieved in the instrumentation and sample preparation of proteomic techniques, proteomics in biomarker discovery is still in its early stages. Compared to molecular biomarkers, significant work is required to prove the utility of any protein panel as a new biomarker for IBD.

4.2. Epigenetics in Diagnostic Biomarkers

Epigenetic signatures are tissue- and cell-type-specific. A major challenge in IBD epigenetic studies using peripheral blood or mucosal biopsies is the cell-type heterogeneity of these specimens. Additional non-disease-specific cell types can lead to complications in interpreting the data due to interference from the different individual epigenetic features. Thus, disease-specific cell types should be purified from the mixed cell or tissue samples before analysis. However, several cell types have been linked to the pathogenesis of IBD, making the selection of disease-specific cell types in IBD a challenge. Although the techniques used in epigenetic studies are well established, they also have their limitations. Most microRNA studies use real-time quantitative PCR followed by microarrays. Although these techniques can identify a wide number of miRNAs, they are not sensitive to functionally distinct microRNA variants and slight nucleotide variations between microRNAs in the same families. They also have a low dynamic range, and cannot detect miRNAs with low expression levels [118]. Next-generation sequencing (NGS) is a high-throughput and fast method that has emerged lately as a more effective technique for identifying novel microRNAs [119].

Other challenges emerge from environmental factors, such as age, diet, and smoking, which can affect the epigenome. Hence, a well-designed study seeking to identify disease-specific variations selectively would require a careful selection of patients and controls.

5. Conclusions

The role of endoscopy and inflammatory biomarkers in the diagnosis of IBD has been extensively studied over the years, improving our understanding of the utility and limitations of each diagnostic tool in clinical settings. Although the combination of endoscopy and molecular tests has become a well-established diagnostic tool for IBD, there is continuing effort to find an ideal diagnostic tool that can overcome the challenges limiting the current tools. Lately, there has been growing interest in switching from using a single biomarker to the biomarker panel approach, in an effort to identify biomarkers that, together, are specific to IBD and can enable differential diagnosis of UC versus CD. This shift in research focus is evident from the increasing number of studies looking into the use of proteomics and genomics for identifying biomarker signatures. As the causes of IBD are still undetermined, with immunological, genetic, and environmental triggers having been found to contribute to disease progression [120–123], researchers also continue to search for new molecular biomarkers that are associated with these factors—especially in the context of new fecal biomarkers and serological antibodies.

Author Contributions: C.Y. and D.M. developed the concept, and Z.A. wrote the draft. The manuscript was then critically revised by D.M. and C.Y. All authors have read and agreed to the published version of the manuscript.

Funding: This work was funded by the Department of Veterans Affairs (Merit Award BX002526 to D. Merlin) and by the National Institute of Diabetes and Digestive and Kidney Diseases (RO1-DK-116306 and RO1-DK-107739 to D. Merlin). D. Merlin is a recipient of a Senior Research Career Scientist Award (BX004476) from the Department of Veterans Affairs.

Data Availability Statement: No new data were generated or analyzed in support of this research.

Acknowledgments: The authors appreciate the support from the Elsevier for reusing the Figure and table from Gastroenterology (Volume 140, Issue 6, pages 1817–1826.e2) for our review's Figure 1 and Table 1. We also appreciate the S. Karger AG, Basel publishers for reusing the Figure for our review's Figure 2 from the journal Digestive Diseases (2021; 39: 190–203, https://doi.org/10.1159/000511641), and thank the MDPI for reusing table unit from Int. J. Mol. Sci. (2020, 21, 7893; doi:10.3390/ijms21217893) for our review's Table 2.

Conflicts of Interest: The authors declare no conflict of interest.

References

1. Baumgart, D.C.; Sandborn, W.J. Crohn's disease. *Lancet* **2012**, *380*, 1590–1605. [CrossRef]
2. Ordas, I.; Eckmann, L.; Talamini, M.; Baumgart, D.C.; Sandborn, W.J. Ulcerative colitis. *Lancet* **2012**, *380*, 1606–1619. [CrossRef]
3. Ekbom, A.; Helmick, C.; Zack, M.; Adami, H.O. Ulcerative colitis and colorectal cancer. A population-based study. *N. Engl. J. Med.* **1990**, *323*, 1228–1233. [CrossRef] [PubMed]
4. Mattar, M.C.; Lough, D.; Pishvaian, M.J.; Charabaty, A. Current management of inflammatory bowel disease and colorectal cancer. *Gastrointest Cancer Res.* **2011**, *4*, 53–61.
5. Keller, D.S.; Windsor, A.; Cohen, R.; Chand, M. Colorectal cancer in inflammatory bowel disease: Review of the evidence. *Tech. Coloproctol.* **2019**, *23*, 3–13. [CrossRef]
6. Khor, B.; Gardet, A.; Xavier, R.J. Genetics and pathogenesis of inflammatory bowel disease. *Nature* **2011**, *474*, 307–317. [CrossRef]
7. Gomollon, F.; Dignass, A.; Annese, V.; Tilg, H.; Van Assche, G.; Lindsay, J.O.; Peyrin-Biroulet, L.; Cullen, G.J.; Daperno, M.; Kucharzik, T.; et al. 3rd European Evidence-based Consensus on the Diagnosis and Management of Crohn's Disease 2016: Part 1: Diagnosis and Medical Management. *J. Crohn's Colitis* **2017**, *11*, 3–25. [CrossRef]
8. Rubin, D.T.; Ananthakrishnan, A.N.; Siegel, C.A.; Sauer, B.G.; Long, M.D. ACG Clinical Guideline: Ulcerative Colitis in Adults. *Am. J. Gastroenterol.* **2019**, *114*, 384–413. [CrossRef]
9. Gecse, K.B.; Vermeire, S. Differential diagnosis of inflammatory bowel disease: Imitations and complications. *Lancet Gastroenterol. Hepatol.* **2018**, *3*, 644–653. [CrossRef]
10. Simpson, P.; Papadakis, K.A. Endoscopic evaluation of patients with inflammatory bowel disease. *Inflamm. Bowel Dis.* **2008**, *14*, 1287–1297. [CrossRef]

11. Shergill, A.K.; Lightdale, J.R.; Bruining, D.H.; Acosta, R.D.; Chandrasekhara, V.; Chathadi, K.V.; Decker, G.A.; Early, D.S.; Evans, J.A.; Fanelli, R.D.; et al. The role of endoscopy in inflammatory bowel disease. *Gastrointest. Endosc.* **2015**, *81*, 1101–1121.e1113. [CrossRef] [PubMed]
12. Lamb, C.A.; Kennedy, N.A.; Raine, T.; Hendy, P.A.; Smith, P.J.; Limdi, J.K.; Hayee, B.H.; Lomer, M.C.E.; Parkes, G.C.; Selinger, C.; et al. British Society of Gastroenterology consensus guidelines on the management of inflammatory bowel disease in adults. *Gut* **2019**, *68* (Suppl. 3), s1–s106. [CrossRef] [PubMed]
13. Iddan, G.; Meron, G.; Glukhovsky, A.; Swain, P. Wireless capsule endoscopy. *Nature* **2000**, *405*, 417. [CrossRef] [PubMed]
14. Mustafa, B.F.; Samaan, M.; Langmead, L.; Khasraw, M. Small bowel video capsule endoscopy: An overview. *Expert Rev. Gastroenterol. Hepatol.* **2013**, *7*, 323–329. [CrossRef] [PubMed]
15. Rezapour, M.; Amadi, C.; Gerson, L.B. Retention associated with video capsule endoscopy: Systematic review and meta-analysis. *Gastrointest. Endosc.* **2017**, *85*, 1157–1168.e1152. [CrossRef]
16. Kiesslich, R.; Burg, J.; Vieth, M.; Gnaendiger, J.; Enders, M.; Delaney, P.; Polglase, A.; McLaren, W.; Janell, D.; Thomas, S.; et al. Confocal laser endoscopy for diagnosing intraepithelial neoplasias and colorectal cancer in vivo. *Gastroenterology* **2004**, *127*, 706–713. [CrossRef]
17. Bharadwaj, S.; Narula, N.; Tandon, P.; Yaghoobi, M. Role of endoscopy in inflammatory bowel disease. *Gastroenterol. Rep.* **2018**, *6*, 75–82. [CrossRef]
18. Maaser, C.; Sturm, A.; Vavricka, S.R.; Kucharzik, T.; Fiorino, G.; Annese, V.; Calabrese, E.; Baumgart, D.C.; Bettenworth, D.; Borralho Nunes, P.; et al. ECCO-ESGAR Guideline for Diagnostic Assessment in IBD Part 1: Initial diagnosis, monitoring of known IBD, detection of complications. *J. Crohn's Colitis* **2018**, *13*, 144–164. [CrossRef]
19. Panes, J.; Jairath, V.; Levesque, B.G. Advances in Use of Endoscopy, Radiology, and Biomarkers to Monitor Inflammatory Bowel Diseases. *Gastroenterology* **2017**, *152*, 362–373.e363. [CrossRef]
20. Walsh, A.J.; Bryant, R.V.; Travis, S.P.L. Current best practice for disease activity assessment in IBD. *Nat. Rev. Gastroenterol. Hepatol.* **2016**, *13*, 567–579. [CrossRef]
21. Biomarkers Definitions Working Group; Atkinson, A.J., Jr.; Colburn, W.A.; DeGruttola, V.G.; DeMets, D.L.; Downing, G.J.; Hoth, D.F.; Oates, J.A.; Peck, C.C.; Spilker, B.A.; et al. Biomarkers and surrogate endpoints: Preferred definitions and conceptual framework. *Clin. Pharmacol. Ther.* **2001**, *69*, 89–95. [CrossRef]
22. Sands, B.E. Biomarkers of Inflammation in Inflammatory Bowel Disease. *Gastroenterology* **2015**, *149*, 1275–1285.e1272. [CrossRef] [PubMed]
23. Lewis, J.D. The utility of biomarkers in the diagnosis and therapy of inflammatory bowel disease. *Gastroenterology* **2011**, *140*, 1817–1826.e1812. [CrossRef]
24. Viennois, E.; Zhao, Y.; Merlin, D. Biomarkers of Inflammatory Bowel Disease: From Classical Laboratory Tools to Personalized Medicine. *Inflamm. Bowel Dis.* **2015**, *21*, 2467–2474. [CrossRef]
25. Pepys, M.B.; Hirschfield, G.M. C-reactive protein: A critical update. *J. Clin. Investig.* **2003**, *111*, 1805–1812. [CrossRef]
26. Darlington, G.J.; Wilson, D.R.; Lachman, L.B. Monocyte-conditioned medium, interleukin-1, and tumor necrosis factor stimulate the acute phase response in human hepatoma cells in vitro. *J. Cell Biol.* **1986**, *103*, 787–793. [CrossRef]
27. Vermeire, S.; Van Assche, G.; Rutgeerts, P. C-Reactive Protein as a Marker for Inflammatory Bowel Disease. *Inflamm. Bowel Dis.* **2004**, *10*, 661–665. [CrossRef]
28. Vermeire, S.; Van Assche, G.; Rutgeerts, P. Laboratory markers in IBD: Useful, magic, or unnecessary toys? *Gut* **2006**, *55*, 426. [CrossRef]
29. Saverymuttu, S.H.; Hodgson, H.J.; Chadwick, V.S.; Pepys, M.B. Differing acute phase responses in Crohn's disease and ulcerative colitis. *Gut* **1986**, *27*, 809–813. [CrossRef]
30. Peyrin-Biroulet, L.; Gonzalez, F.; Dubuquoy, L.; Rousseaux, C.; Dubuquoy, C.; Decourcelle, C.; Saudemont, A.; Tachon, M.; Béclin, E.; Odou, M.-F.; et al. Mesenteric fat as a source of C reactive protein and as a target for bacterial translocation in Crohn's disease. *Gut* **2012**, *61*, 78–85. [CrossRef]
31. Mendoza, J.L.; Abreu, M.T. Biological markers in inflammatory bowel disease: Practical consideration for clinicians. *Gastroentérologie Clin. Biol.* **2009**, *33*, S158–S173. [CrossRef]
32. Barnes, B.H.; Borowitz, S.M.; Saulsbury, F.T.; Hellems, M.; Sutphen, J.L. Discordant Erythrocyte Sedimentation Rate and C-reactive Protein in Children with Inflammatory Bowel Disease Taking Azathioprine or 6-Mercaptopurine. *J. Pediatric Gastroenterol. Nutr.* **2004**, *38*, 509–512. [CrossRef]
33. Gabay, C.; Kushner, I. Acute-phase proteins and other systemic responses to inflammation. *N. Engl. J. Med.* **1999**, *340*, 448–454. [CrossRef]
34. O'Donnell, L.C.; Druhan, L.J.; Avalos, B.R. Molecular characterization and expression analysis of leucine-rich alpha2-glycoprotein, a novel marker of granulocytic differentiation. *J. Leukoc. Biol.* **2002**, *72*, 478–485.
35. Shirai, R.; Hirano, F.; Ohkura, N.; Ikeda, K.; Inoue, S. Up-regulation of the expression of leucine-rich alpha(2)-glycoprotein in hepatocytes by the mediators of acute-phase response. *Biochem. Biophys. Res. Commun.* **2009**, *382*, 776–779. [CrossRef]
36. Naka, T.; Fujimoto, M. LRG is a novel inflammatory marker clinically useful for the evaluation of disease activity in rheumatoid arthritis and inflammatory bowel disease. *Immunol. Med.* **2018**, *41*, 62–67. [CrossRef]

37. Serada, S.; Fujimoto, M.; Terabe, F.; Iijima, H.; Shinzaki, S.; Matsuzaki, S.; Ohkawara, T.; Nezu, R.; Nakajima, S.; Kobayashi, T.; et al. Serum leucine-rich alpha-2 glycoprotein is a disease activity biomarker in ulcerative colitis. *Inflamm. Bowel Dis.* **2012**, *18*, 2169–2179. [CrossRef]
38. Shinzaki, S.; Matsuoka, K.; Iijima, H.; Mizuno, S.; Serada, S.; Fujimoto, M.; Arai, N.; Koyama, N.; Morii, E.; Watanabe, M.; et al. Leucine-rich Alpha-2 Glycoprotein is a Serum Biomarker of Mucosal Healing in Ulcerative Colitis. *J. Crohn's Colitis* **2017**, *11*, 84–91. [CrossRef]
39. Kawamoto, A.; Takenaka, K.; Hibiya, S.; Ohtsuka, K.; Okamoto, R.; Watanabe, M. Serum Leucine-Rich alpha2 Glycoprotein: A Novel Biomarker For Small Bowel Mucosal Activity in Crohn's Disease. *Clin. Gastroenterol. Hepatol.* **2022**, *20*, e1196–e1200. [CrossRef]
40. Kawamura, T.; Yamamura, T.; Nakamura, M.; Maeda, K.; Sawada, T.; Ishikawa, E.; Iida, T.; Mizutani, Y.; Ishikawa, T.; Kakushima, N.; et al. Accuracy of Serum Leucine-Rich Alpha-2 Glycoprotein in Evaluating Endoscopic Disease Activity in Crohn's Disease. *Inflamm. Bowel Dis.* 2022, *Epub Ahead of Print*. [CrossRef]
41. Yasutomi, E.; Inokuchi, T.; Hiraoka, S.; Takei, K.; Igawa, S.; Yamamoto, S.; Ohmori, M.; Oka, S.; Yamasaki, Y.; Kinugasa, H.; et al. Leucine-rich alpha-2 glycoprotein as a marker of mucosal healing in inflammatory bowel disease. *Sci. Rep.* **2021**, *11*, 11086. [CrossRef]
42. Sellin, J.H.; Shah, R.R. The Promise and Pitfalls of Serologic Testing in Inflammatory Bowel Disease. *Gastroenterol. Clin. N. Am.* **2012**, *41*, 463–482. [CrossRef] [PubMed]
43. Iskandar, H.N.; Ciorba, M.A. Biomarkers in inflammatory bowel disease: Current practices and recent advances. *Transl. Res.* **2012**, *159*, 313–325. [CrossRef] [PubMed]
44. Lopez, R.N.; Leach, S.T.; Lemberg, D.A.; Duvoisin, G.; Gearry, R.B.; Day, A.S. Fecal biomarkers in inflammatory bowel disease. *J. Gastroenterol. Hepatol.* **2017**, *32*, 577–582. [CrossRef] [PubMed]
45. Johne, B.; Fagerhol, M.K.; Lyberg, T.; Prydz, H.; Brandtzaeg, P.; Naess-Andresen, C.F.; Dale, I. Functional and clinical aspects of the myelomonocyte protein calprotectin. *Mol. Pathol.* **1997**, *50*, 113–123. [CrossRef]
46. RØseth, A.G.; Fagerhol, M.K.; Aadland, E.; Schjønsby, H. Assessment of the Neutrophil Dominating Protein Calprotectin in Feces: A Methodologic Study. *Scand. J. Gastroenterol.* **1992**, *27*, 793–798. [CrossRef]
47. Poullis, A.; Foster, R.; Mendall, M.A.; Fagerhol, M.K. Emerging role of calprotectin in gastroenterology. *J. Gastroenterol. Hepatol.* **2003**, *18*, 756–762. [CrossRef]
48. Foell, D.; Kucharzik, T.; Kraft, M.; Vogl, T.; Sorg, C.; Domschke, W.; Roth, J. Neutrophil derived human S100A12 (EN-RAGE) is strongly expressed during chronic active inflammatory bowel disease. *Gut* **2003**, *52*, 847–853. [CrossRef]
49. Perera, C.; McNeil, H.P.; Geczy, C.L. S100 Calgranulins in inflammatory arthritis. *Immunol. Cell Biol.* **2010**, *88*, 41–49. [CrossRef]
50. Kane, S.V.; Sandborn, W.J.; Rufo, P.A.; Zholudev, A.; Boone, J.; Lyerly, D.; Camilleri, M.; Hanauer, S.B. Fecal lactoferrin is a sensitive and specific marker in identifying intestinal inflammation. *Am. J. Gastroenterol.* **2003**, *98*, 1309–1314. [CrossRef]
51. Krzystek-Korpacka, M.; Kempiński, R.; Bromke, M.; Neubauer, K. Biochemical Biomarkers of Mucosal Healing for Inflammatory Bowel Disease in Adults. *Diagnostics* **2020**, *10*, 367. [CrossRef]
52. Buisson, A.; Vazeille, E.; Minet-Quinard, R.; Goutte, M.; Bouvier, D.; Goutorbe, F.; Pereira, B.; Barnich, N.; Bommelaer, G. Fecal Matrix Metalloprotease-9 and Lipocalin-2 as Biomarkers in Detecting Endoscopic Activity in Patients With Inflammatory Bowel Diseases. *J. Clin. Gastroenterol.* **2018**, *52*, e53–e62. [CrossRef]
53. Stallhofer, J.; Friedrich, M.; Konrad-Zerna, A.; Wetzke, M.; Lohse, P.; Glas, J.; Tillack-Schreiber, C.; Schnitzler, F.; Beigel, F.; Brand, S. Lipocalin-2 Is a Disease Activity Marker in Inflammatory Bowel Disease Regulated by IL-17A, IL-22, and TNF-alpha and Modulated by IL23R Genotype Status. *Inflamm. Bowel Dis.* **2015**, *21*, 2327–2340. [CrossRef]
54. West, N.R.; Hegazy, A.N.; Owens, B.M.J.; Bullers, S.J.; Linggi, B.; Buonocore, S.; Coccia, M.; Gortz, D.; This, S.; Stockenhuber, K.; et al. Oncostatin M drives intestinal inflammation and predicts response to tumor necrosis factor-neutralizing therapy in patients with inflammatory bowel disease. *Nat. Med.* **2017**, *23*, 579–589. [CrossRef]
55. Dragoni, G.; Innocenti, T.; Galli, A. Biomarkers of Inflammation in Inflammatory Bowel Disease: How Long before Abandoning Single-Marker Approaches? *Dig. Dis.* **2020**, *39*, 190–203. [CrossRef]
56. Tyers, M.; Mann, M. From genomics to proteomics. *Nature* **2003**, *422*, 193–197. [CrossRef]
57. Wright, I.; Van Eyk, J.E. A Roadmap to Successful Clinical Proteomics. *Clin. Chem.* **2017**, *63*, 245–247. [CrossRef]
58. Shillingford, J.M.; Miyoshi, K.; Robinson, G.W.; Bierie, B.; Cao, Y.; Karin, M.; Hennighausen, L. Proteotyping of Mammary Tissue from Transgenic and Gene Knockout Mice with Immunohistochemical Markers: A Tool To Define Developmental Lesions. *J. Histochem. Cytochem.* **2003**, *51*, 555–565. [CrossRef]
59. Ellis, M.J.; Gillette, M.; Carr, S.A.; Paulovich, A.G.; Smith, R.D.; Rodland, K.K.; Townsend, R.R.; Kinsinger, C.; Mesri, M.; Rodriguez, H.; et al. Connecting Genomic Alterations to Cancer Biology with Proteomics: The NCI Clinical Proteomic Tumor Analysis Consortium. *Cancer Discov.* **2013**, *3*, 1108–1112. [CrossRef]
60. Zhang, B.; Whiteaker, J.R.; Hoofnagle, A.N.; Baird, G.S.; Rodland, K.D.; Paulovich, A.G. Clinical potential of mass spectrometry-based proteogenomics. *Nat. Rev. Clin. Oncol.* **2019**, *16*, 256–268. [CrossRef]
61. Trenchevska, O.; Nelson, R.W.; Nedelkov, D. Mass Spectrometric Immunoassays in Characterization of Clinically Significant Proteoforms. *Proteomes* **2016**, *4*, 13. [CrossRef]
62. Smith, L.M.; Kelleher, N.L.; Linial, M.; Goodlett, D.; Langridge-Smith, P.; Ah Goo, Y.; Safford, G.; Bonilla, L.; Kruppa, G.; Zubarev, R.; et al. Proteoform: A single term describing protein complexity. *Nat. Methods* **2013**, *10*, 186–187. [CrossRef]

63. Berndt, U.; Bartsch, S.; Philipsen, L.; Danese, S.; Wiedenmann, B.; Dignass, A.U.; Hämmerle, M.; Sturm, A. Proteomic Analysis of the Inflamed Intestinal Mucosa Reveals Distinctive Immune Response Profiles in Crohn's Disease and Ulcerative Colitis. *J. Immunol.* **2007**, *179*, 295–304. [CrossRef]
64. Yau, Y.Y.; Leong, R.W.L.; Pudipeddi, A.; Redmond, D.; Wasinger, V.C. Serological Epithelial Component Proteins Identify Intestinal Complications in Crohn's Disease. *Mol. Cell. Proteom.* **2017**, *16*, 1244–1257. [CrossRef]
65. Deeke, S.A.; Starr, A.E.; Ning, Z.; Ahmadi, S.; Zhang, X.; Mayne, J.; Chiang, C.-K.; Singleton, R.; Benchimol, E.I.; Mack, D.R.; et al. Open:Mucosal-luminal interface proteomics reveals biomarkers of pediatric inflammatory bowel disease-associated colitis. *Am. J. Gastroenterol.* **2018**, *113*, 713–724. [CrossRef]
66. M'Koma, A.E.; Seeley, E.H.; Washington, M.K.; Schwartz, D.A.; Muldoon, R.L.; Herline, A.J.; Wise, P.E.; Caprioli, R.M. Proteomic profiling of mucosal and submucosal colonic tissues yields protein signatures that differentiate the inflammatory colitides. *Inflamm. Bowel Dis.* **2010**, *17*, 875–883. [CrossRef]
67. Seeley, E.H.; Washington, M.K.; Caprioli, R.M.; M'Koma, A.E. Proteomic patterns of colonic mucosal tissues delineate Crohn's colitis and ulcerative colitis. *Proteom. Clin. Appl.* **2013**, *7*, 541–549. [CrossRef]
68. Starr, A.E.; Deeke, S.A.; Ning, Z.; Chiang, C.-K.; Zhang, X.; Mottawea, W.; Singleton, R.; Benchimol, E.I.; Wen, M.; Mack, D.R.; et al. Proteomic analysis of ascending colon biopsies from a paediatric inflammatory bowel disease inception cohort identifies protein biomarkers that differentiate Crohn's disease from UC. *Gut* **2017**, *66*, 1573–1583. [CrossRef]
69. Meuwis, M.-A.; Fillet, M.; Geurts, P.; de Seny, D.; Lutteri, L.; Chapelle, J.-P.; Bours, V.; Wehenkel, L.; Belaiche, J.; Malaise, M.; et al. Biomarker discovery for inflammatory bowel disease, using proteomic serum profiling. *Biochem. Pharmacol.* **2007**, *73*, 1422–1433. [CrossRef]
70. Townsend, P.; Zhang, Q.; Shapiro, J.; Webb-Robertson, B.-J.; Bramer, L.; Schepmoes, A.A.; Weitz, K.K.; Mallette, M.; Moniz, H.; Bright, R.; et al. Serum Proteome Profiles in Stricturing Crohn's Disease: A Pilot Study. *Inflamm. Bowel Dis.* **2015**, *21*, 1935–1941. [CrossRef]
71. Magnusson, M.K.; Strid, H.; Isaksson, S.; Bajor, A.; Lasson, A.; Ung, K.-A.; Öhman, L. Response to Infliximab Therapy in Ulcerative Colitis is Associated with Decreased Monocyte Activation, Reduced CCL2 Expression and Downregulation of Tenascin C. *J. Crohn's Colitis* **2014**, *9*, 56–65. [CrossRef]
72. Heier, C.R.; Fiorillo, A.A.; Chaisson, E.; Gordish-Dressman, H.; Hathout, Y.; Damsker, J.M.; Hoffman, E.P.; Conklin, L.S. Identification of Pathway-Specific Serum Biomarkers of Response to Glucocorticoid and Infliximab Treatment in Children with Inflammatory Bowel Disease. *Clin. Transl. Gastroenterol.* **2016**, *7*, e192. [CrossRef]
73. de Lange, K.M.; Moutsianas, L.; Lee, J.C.; Lamb, C.A.; Luo, Y.; Kennedy, N.A.; Jostins, L.; Rice, D.L.; Gutierrez-Achury, J.; Ji, S.-G.; et al. Genome-wide association study implicates immune activation of multiple integrin genes in inflammatory bowel disease. *Nat. Genet.* **2017**, *49*, 256–261. [CrossRef]
74. Mesko, B.; Poliskal, S.; Szegedi, A.; Szekanecz, Z.; Palatka, K.; Papp, M.; Nagy, L. Peripheral blood gene expression patterns discriminate among chronic inflammatory diseases and healthy controls and identify novel targets. *BMC Med. Genom.* **2010**, *3*, 15. [CrossRef]
75. Burakoff, R.; Hande, S.; Ma, J.; Banks, P.A.; Friedman, S.; Makrauer, F.; Liew, C.-C. Differential Regulation of Peripheral Leukocyte Genes in Patients with Active Crohn's Disease and Crohn's Disease in Remission. *J. Clin. Gastroenterol.* **2010**, *44*, 120–126. [CrossRef]
76. Burakoff, R.; Chao, S.; Perencevich, M.; Ying, J.; Friedman, S.; Makrauer, F.; Odze, R.; Khurana, H.; Liew, C.-C. Blood-based biomarkers can differentiate ulcerative colitis from crohn's disease and noninflammatory diarrhea. *Inflamm. Bowel Dis.* **2011**, *17*, 1719–1725. [CrossRef]
77. Burakoff, R.; Pabby, V.; Onyewadume, L.; Odze, R.; Adackapara, C.; Wang, W.; Friedman, S.; Hamilton, M.; Korzenik, J.; Levine, J.; et al. Blood-based Biomarkers Used to Predict Disease Activity in Crohn's Disease and Ulcerative Colitis. *Inflamm. Bowel Dis.* **2015**, *21*, 1132–1140. [CrossRef]
78. Burczynski, M.E.; Peterson, R.L.; Twine, N.C.; Zuberek, K.A.; Brodeur, B.J.; Casciotti, L.; Maganti, V.; Reddy, P.S.; Strahs, A.; Immermann, F.; et al. Molecular Classification of Crohn's Disease and Ulcerative Colitis Patients Using Transcriptional Profiles in Peripheral Blood Mononuclear Cells. *J. Mol. Diagn.* **2006**, *8*, 51–61. [CrossRef]
79. van Lierop, P.P.E.; Swagemakers, S.M.; de Bie, C.I.; Middendorp, S.; van Baarlen, P.; Samsom, J.N.; van Ijcken, W.F.J.; Escher, J.C.; van der Spek, P.J.; Nieuwenhuis, E.E.S. Gene Expression Analysis of Peripheral Cells for Subclassification of Pediatric Inflammatory Bowel Disease in Remission. *PLoS ONE* **2013**, *8*, e79549. [CrossRef]
80. Wu, F.; Dassopoulos, T.; Cope, L.; Maitra, A.; Brant, S.R.; Harris, M.L.; Bayless, T.M.; Parmigiani, G.; Chakravarti, S. Genome-wide gene expression differences in Crohn's disease and ulcerative colitis from endoscopic pinch biopsies: Insights into distinctive pathogenesis. *Inflamm. Bowel Dis.* **2007**, *13*, 807–821. [CrossRef]
81. Dieckgraefe, B.K.; Stenson, W.F.; Korzenik, J.R.; Swanson, P.E.; Harrington, C.A. Analysis of mucosal gene expression in inflammatory bowel disease by parallel oligonucleotide arrays. *Physiol. Genom.* **2000**, *4*, 1–11. [CrossRef]
82. Graham, D.B.; Xavier, R.J. Pathway paradigms revealed from the genetics of inflammatory bowel disease. *Nature* **2020**, *578*, 527–539. [CrossRef] [PubMed]
83. Ventham, N.T.; Kennedy, N.A.; Nimmo, E.R.; Satsangi, J. Beyond Gene Discovery in Inflammatory Bowel Disease: The Emerging Role of Epigenetics. *Gastroenterology* **2013**, *145*, 293–308. [CrossRef] [PubMed]

84. Jenke, A.C.; Zilbauer, M. Epigenetics in inflammatory bowel disease. *Curr. Opin. Gastroenterol.* **2012**, *28*, 577–584. [CrossRef] [PubMed]
85. Jones, P.A. Functions of DNA methylation: Islands, start sites, gene bodies and beyond. *Nat. Rev. Genet.* **2012**, *13*, 484–492. [CrossRef] [PubMed]
86. Issa, J.-P.J.; Ahuja, N.; Toyota, M.; Bronner, M.P.; Brentnall, T.A. Accelerated Age-related CpG Island Methylation in Ulcerative Colitis. *Cancer Res.* **2001**, *61*, 3573–3577.
87. Nimmo, E.R.; Prendergast, J.G.; Aldhous, M.C.; Kennedy, N.A.; Henderson, P.; Drummond, H.E.; Ramsahoye, B.H.; Wilson, D.C.; Semple, C.A.; Satsangi, J. Genome-wide Methylation Profiling in Crohn's Disease Identifies Altered Epigenetic Regulation of Key Host Defense Mechanisms Including the Th17 Pathway. *Inflamm. Bowel Dis.* **2011**, *18*, 889–899. [CrossRef]
88. Cooke, J.; Zhang, H.; Greger, L.; Silva, A.-L.; Massey, D.; Dawson, C.; Metz, A.; Ibrahim, A.; Parkes, M. Mucosal Genome-wide Methylation Changes in Inflammatory Bowel Disease. *Inflamm. Bowel Dis.* **2012**, *18*, 2128–2137. [CrossRef]
89. Lin, Z.; Hegarty, J.; Cappel, J.; Yu, W.; Chen, X.; Faber, P.; Wang, Y.; Kelly, A.; Poritz, L.; Peterson, B.; et al. Identification of disease-associated DNA methylation in intestinal tissues from patients with inflammatory bowel disease. *Clin. Genet.* **2011**, *80*, 59–67. [CrossRef]
90. McDermott, E.; Ryan, E.J.; Tosetto, M.; Gibson, D.; Burrage, J.; Keegan, D.; Byrne, K.; Crowe, E.; Sexton, G.; Malone, K.; et al. DNA Methylation Profiling in Inflammatory Bowel Disease Provides New Insights into Disease Pathogenesis. *J. Crohn's Colitis* **2015**, *10*, 77–86. [CrossRef]
91. Esteller, M. Non-coding RNAs in human disease. *Nat. Rev. Genet.* **2011**, *12*, 861–874. [CrossRef]
92. James, J.P.; Riis, L.B.; Malham, M.; Hogdall, E.; Langholz, E.; Nielsen, B.S. MicroRNA Biomarkers in IBD-Differential Diagnosis and Prediction of Colitis-Associated Cancer. *Int. J. Mol. Sci.* **2020**, *21*, 7893. [CrossRef] [PubMed]
93. Wu, F.; Zikusoka, M.; Trindade, A.; Dassopoulos, T.; Harris, M.L.; Bayless, T.M.; Brant, S.R.; Chakravarti, S.; Kwon, J.H. MicroRNAs Are Differentially Expressed in Ulcerative Colitis and Alter Expression of Macrophage Inflammatory Peptide-2α. *Gastroenterology* **2008**, *135*, 1624–1635.e1624. [CrossRef] [PubMed]
94. Takagi, T.; Naito, Y.; Mizushima, K.; Hirata, I.; Yagi, N.; Tomatsuri, N.; Ando, T.; Oyamada, Y.; Isozaki, Y.; Hongo, H.; et al. Increased expression of microRNA in the inflamed colonic mucosa of patients with active ulcerative colitis. *J. Gastroenterol. Hepatol.* **2010**, *25*, S129–S133. [CrossRef] [PubMed]
95. Bian, Z.; Li, L.; Cui, J.; Zhang, H.; Liu, Y.; Zhang, C.-Y.; Zen, K. Role of miR-150-targeting c-Myb in colonic epithelial disruption during dextran sulphate sodium-induced murine experimental colitis and human ulcerative colitis. *J. Pathol.* **2011**, *225*, 544–553. [CrossRef]
96. Pekow, J.R.; Dougherty, U.; Mustafi, R.; Zhu, H.; Kocherginsky, M.; Rubin, D.T.; Hanauer, S.B.; Hart, J.; Chang, E.B.; Fichera, A.; et al. miR-143 and miR-145 are Downregulated in Ulcerative Colitis: Putative Regulators of Inflammation and Protooncogenes. *Inflamm. Bowel Dis.* **2011**, *18*, 94–100. [CrossRef]
97. Brest, P.; Lapaquette, P.; Souidi, M.; Lebrigand, K.; Cesaro, A.; Vouret-Craviari, V.; Mari, B.; Barbry, P.; Mosnier, J.-F.; Hébuterne, X.; et al. A synonymous variant in IRGM alters a binding site for miR-196 and causes deregulation of IRGM-dependent xenophagy in Crohn's disease. *Nat. Genet.* **2011**, *43*, 242–245. [CrossRef]
98. Nguyen, H.T.T.; Dalmasso, G.; Yan, Y.; Laroui, H.; Dahan, S.; Mayer, L.; Sitaraman, S.V.; Merlin, D. MicroRNA-7 Modulates CD98 Expression during Intestinal Epithelial Cell Differentiation. *J. Biol. Chem.* **2010**, *285*, 1479–1489. [CrossRef]
99. Wu, F.; Zhang, S.; Dassopoulos, T.; Harris, M.L.; Bayless, T.M.; Meltzer, S.J.; Brant, S.R.; Kwon, J.H. Identification of microRNAs associated with ileal and colonic Crohn's disease. *Inflamm. Bowel Dis.* **2010**, *16*, 1729–1738. [CrossRef]
100. Fasseu, M.; Tréton, X.; Guichard, C.; Pedruzzi, E.; Cazals-Hatem, D.; Richard, C.; Aparicio, T.; Daniel, F.; Soulé, J.-C.; Moreau, R.; et al. Identification of Restricted Subsets of Mature microRNA Abnormally Expressed in Inactive Colonic Mucosa of Patients with Inflammatory Bowel Disease. *PLoS ONE* **2010**, *5*, e13160. [CrossRef]
101. Iborra, M.; Bernuzzi, F.; Correale, C.; Vetrano, S.; Fiorino, G.; Beltrán, B.; Marabita, F.; Locati, M.; Spinelli, A.; Nos, P.; et al. Identification of serum and tissue micro-RNA expression profiles in different stages of inflammatory bowel disease. *Clin. Exp. Immunol.* **2013**, *173*, 250–258. [CrossRef]
102. Duttagupta, R.; DiRienzo, S.; Jiang, R.; Bowers, J.; Gollub, J.; Kao, J.; Kearney, K.; Rudolph, D.; Dawany, N.B.; Showe, M.K.; et al. Genome-Wide Maps of Circulating miRNA Biomarkers for Ulcerative Colitis. *PLoS ONE* **2012**, *7*, e31241. [CrossRef] [PubMed]
103. Paraskevi, A.; Theodoropoulos, G.; Papaconstantinou, I.; Mantzaris, G.; Nikiteas, N.; Gazouli, M. Circulating MicroRNA in inflammatory bowel disease. *J. Crohn's Colitis* **2012**, *6*, 900–904. [CrossRef] [PubMed]
104. Wu, F.; Guo, N.J.; Tian, H.; Marohn, M.; Gearhart, S.; Bayless, T.M.; Brant, S.R.; Kwon, J.H. Peripheral blood MicroRNAs distinguish active ulcerative colitis and Crohn's disease. *Inflamm. Bowel Dis.* **2010**, *17*, 241–250. [CrossRef] [PubMed]
105. Zahm, A.M.; Thayu, M.; Hand, N.J.; Horner, A.; Leonard, M.B.; Friedman, J.R. Circulating microRNA is a biomarker of pediatric Crohn disease. *J. Pediatric Gastroenterol. Nutr.* **2011**, *53*, 26–33. [CrossRef] [PubMed]
106. Verdier, J.; Breunig, I.R.; Ohse, M.C.; Roubrocks, S.; Kleinfeld, S.; Roy, S.; Streetz, K.; Trautwein, C.; Roderburg, C.; Sellge, G. Faecal Micro-RNAs in Inflammatory Bowel Diseases. *J. Crohn's Colitis* **2019**, *14*, 110–117. [CrossRef]
107. Schönauen, K.; Le, N.; von Arnim, U.; Schulz, C.; Malfertheiner, P.; Link, A. Circulating and Fecal microRNAs as Biomarkers for Inflammatory Bowel Diseases. *Inflamm. Bowel Dis.* **2018**, *24*, 1547–1557. [CrossRef] [PubMed]
108. Wohnhaas, C.T.; Schmid, R.; Rolser, M.; Kaaru, E.; Langgartner, D.; Rieber, K.; Strobel, B.; Eisele, C.; Wiech, F.; Jakob, I.; et al. Fecal MicroRNAs Show Promise as Noninvasive Crohn's Disease Biomarkers. *Crohn's Colitis 360* **2020**, *2*, otaa003. [CrossRef]

109. Kanaan, Z.; Rai, S.N.; Eichenberger, M.R.; Roberts, H.; Keskey, B.; Pan, J.; Galandiuk, S. Plasma MiR-21: A Potential Diagnostic Marker of Colorectal Cancer. *Ann. Surg.* **2012**, *256*, 544–551. [CrossRef]
110. Schaefer, J.S.; Attumi, T.; Opekun, A.R.; Abraham, B.; Hou, J.; Shelby, H.; Graham, D.Y.; Streckfus, C.; Klein, J.R. MicroRNA signatures differentiate Crohn's disease from ulcerative colitis. *BMC Immunol.* **2015**, *16*, 5. [CrossRef]
111. Rifai, N.; Gillette, M.A.; Carr, S.A. Protein biomarker discovery and validation: The long and uncertain path to clinical utility. *Nat. Biotechnol.* **2006**, *24*, 971–983. [CrossRef]
112. Anderson, N.L. The Clinical Plasma Proteome: A Survey of Clinical Assays for Proteins in Plasma and Serum. *Clin. Chem.* **2010**, *56*, 177–185. [CrossRef] [PubMed]
113. van der Burgt, Y.E.M. Protein biomarker discovery is still relevant and has entered a new phase. *EBioMedicine* **2019**, *43*, 15. [CrossRef] [PubMed]
114. Anderson, N.L.; Anderson, N.G. The Human Plasma Proteome: History, Character, and Diagnostic Prospects. *Mol. Cell. Proteom.* **2002**, *1*, 845–867. [CrossRef]
115. Alex, P.; Gucek, M.; Li, X. Applications of proteomics in the study of inflammatory bowel diseases: Current status and future directions with available technologies. *Inflamm. Bowel Dis.* **2009**, *15*, 616–629. [CrossRef] [PubMed]
116. Pisanu, S.; Biosa, G.; Carcangiu, L.; Uzzau, S.; Pagnozzi, D. Comparative evaluation of seven commercial products for human serum enrichment/depletion by shotgun proteomics. *Talanta* **2018**, *185*, 213–220. [CrossRef] [PubMed]
117. Skates, S.J.; Gillette, M.A.; LaBaer, J.; Carr, S.A.; Anderson, L.; Liebler, D.C.; Ransohoff, D.; Rifai, N.; Kondratovich, M.; Težak, Ž.; et al. Statistical Design for Biospecimen Cohort Size in Proteomics-based Biomarker Discovery and Verification Studies. *J. Proteome Res.* **2013**, *12*, 5383–5394. [CrossRef]
118. Peck, B.C.; Weiser, M.; Lee, S.E.; Gipson, G.R.; Iyer, V.B.; Sartor, R.B.; Herfarth, H.H.; Long, M.D.; Hansen, J.J.; Isaacs, K.L.; et al. MicroRNAs Classify Different Disease Behavior Phenotypes of Crohn's Disease and May Have Prognostic Utility. *Inflamm. Bowel Dis.* **2015**, *21*, 2178–2187. [CrossRef]
119. Kalla, R.; Ventham, N.T.; Kennedy, N.A.; Quintana, J.F.; Nimmo, E.R.; Buck, A.H.; Satsangi, J. MicroRNAs: New players in IBD. *Gut* **2015**, *64*, 504–517. [CrossRef]
120. Ungaro, R.; Mehandru, S.; Allen, P.B.; Peyrin-Biroulet, L.; Colombel, J.-F. Ulcerative colitis. *Lancet* **2017**, *389*, 1756–1770. [CrossRef]
121. Torres, J.; Mehandru, S.; Colombel, J.F.; Peyrin-Biroulet, L. Crohn's disease. *Lancet* **2017**, *389*, 1741–1755. [CrossRef]
122. van der Sloot, K.W.J.; Amini, M.; Peters, V.; Dijkstra, G.; Alizadeh, B.Z. Inflammatory Bowel Diseases: Review of Known Environmental Protective and Risk Factors Involved. *Inflamm. Bowel Dis.* **2017**, *23*, 1499–1509. [CrossRef] [PubMed]
123. Xavier, R.J.; Podolsky, D.K. Unravelling the pathogenesis of inflammatory bowel disease. *Nature* **2007**, *448*, 427–434. [CrossRef] [PubMed]

biomedicines MDPI

Review

Cell Death-Related Ubiquitin Modifications in Inflammatory Syndromes: From Mice to Men

Nieves Peltzer [1,2,3,*,†] and Alessandro Annibaldi [1,*,†]

1 Center for Molecular Medicine Cologne (CMMC), University of Cologne, Robert-Koch-Strasse 21, 50931 Köln, Germany
2 Department of Translational Genomics, University of Cologne, Weyertal 115b, 50931 Köln, Germany
3 Cologne Excellence Cluster on Cellular Stress Responses in Aging-Associated Diseases (CECAD) Research Center, University of Cologne, Joseph-Steltzmann-Strasse 26, 50931 Köln, Germany
* Correspondence: m.peltzer@uni-koeln.de (N.P.); a.annibaldi@uni-koeln.de (A.A.)
† These authors contributed equally to this work.

Abstract: Aberrant cell death can cause inflammation and inflammation-related diseases. While the link between cell death and inflammation has been widely established in mouse models, evidence supporting a role for cell death in the onset of inflammatory and autoimmune diseases in patients is still missing. In this review, we discuss how the lessons learnt from mouse models can help shed new light on the initiating or contributing events leading to immune-mediated disorders. In addition, we discuss how multiomic approaches can provide new insight on the soluble factors released by dying cells that might contribute to the development of such diseases.

Keywords: cell death; apoptosis; necroptosis; pyroptosis; ubiquitin; LUBAC; OTULIN; A20; inflammation; autoimmunity; human genetics

Citation: Peltzer, N.; Annibaldi, A. Cell Death-Related Ubiquitin Modifications in Inflammatory Syndromes: From Mice to Men. *Biomedicines* **2022**, *10*, 1436. https://doi.org/10.3390/biomedicines10061436

Academic Editor: Marianna Christodoulou

Received: 20 May 2022
Accepted: 15 June 2022
Published: 17 June 2022

Publisher's Note: MDPI stays neutral with regard to jurisdictional claims in published maps and institutional affiliations.

1. Introduction

The field of cell death has undergone quite a substantial evolution over the past three decades. During the 1990s and the first years of the new century, scientists were mainly focused on how the knowledge of cell death pathways could have helped improve cancer therapy, to eliminate as many cancer cells as possible, with little interest for the pathophysiological roles of cell death [1,2]. In the last decade or so, the field has moved towards a new direction, with the purpose to understand how the decision between life and death regulates tissue homeostasis and inflammatory responses during tissue damage or pathogen infection [3,4]. It became increasingly clear that cell death-regulating molecules are not always programmed to kill and can fulfil cell death-independent functions [5,6]. In addition, cell death is not a biological end point and, in the process of dying or after their death, cells can still emit signals in a programmed manner [7]. These signals evoke inflammatory programs that are essential for the ability of tissues to recover from different types of insults and restore homeostasis [8]. However, aberrantly regulated cell death can exacerbate inflammatory processes that can in turn cause tissue failure and inflammatory disorders and autoimmunity [9]. Therefore, the magnitude of cell death processes is always fine-tuned by multiple control mechanisms that are in place to prevent the detrimental effects of uncontrolled cell death [10].

Cytokines of the tumour necrosis factor (TNF) family are crucial regulators of cell death, inflammation and autoimmunity [11]. TNF receptor 1 (TNFR1) is a member of the death receptor (DR) family. These receptors are characterised by the presence, in their intracellular portion, of a death domain (DD) that is able to initiate cell death cascades [11]. Other well-studied members of this family are CD95 (Fas/APO-1), TNF-related apoptosis-inducing ligand (TRAIL)-R1 (DR4) and TRAIL-R2 (DR5) [12]. It is the research conducted on the TNF-TNFR1 system during the past two decades that allowed scientists to discover the previously unappreciated link between cell death and inflammation. Indeed, the numerous mouse models developed in the past 10 years, bearing inactivating mutation in players on the TNFR1-signalling pathway, revealed that cell death could be the triggering event in inflammatory and autoimmune diseases [9]. In this review, we focus on different mouse models that led to the discovery of the relationship between cell death and inflammation, how they contributed to establish the link between cell death and inflammation-related disorders and how these disorders resemble human autoinflammatory and autoimmune diseases. In addition, we review the latest omic-based approaches adopted to elucidate the inflammatory potential of dying cells.

2. TNFR1-Signalling Pathway

Upon binding to its cognate ligand, TNF, TNFR1 trimerizes and initiates the formation of a receptor-bound complex called complex-I or TNFR1-associated signalling complex (TNFR1-RSC) [13]. The first event for complex-I formation is the recruitment of the adaptor protein TRADD, and the kinase RIPK1, via DD-interaction [14]. Subsequently, TRADD recruits TRAF2 that in turn binds to the E3 ligases cIAP1 and cIAP2 [15]. These cIAPs synthesise ubiquitin chains of different topologies (i.e., K63, K48 and K11) on different components of complex-I, including themselves and RIPK1 [16,17]. These ubiquitin chains serve as a scaffold to recruit another E3 ligase, the linear ubiquitin chain assembly complex, LUBAC [18]. LUBAC is a tripartite E3 ligase complex formed of HOIP, HOIL-1 and SHARPIN [19,20]. LUBAC conjugates linear ubiquitin chains, also called methionine (M) 1 chains, to several complex-I components, including RIPK1 and TNFR1 [21]. The ubiquitin chains formed by cIAP1/2 and LUBAC stabilise complex-I and favour the recruitment of different kinase complexes: TAB2/TAB3/TAK1 [22–24], NEMO/IKKα/IKKβ and NEMO/TANK1/NAP1/TBK1/IKKε [25]. The TAK1 and IKKα/IKKβ kinases are required for the activation of NF-κB and MAPKs for the expression of pro-survival and pro-inflammatory genes that are required to mount an innate immune response [26]. TAK1 also controls MK2 activation. The M1 chains synthesised by LUBAC also serve to recruit deubiquitinating enzymes (DUBs), namely A20 and CYLD, this last via the adapter SPATA2, which have opposing effects in complex-I [21,27]. While CYLD hydrolyses ubiquitin chains, prevalently K63 chains [28], to control the extent of NF-κB activation, A20 shields them and prevents their removal, ensuring complex-I stability [21] (Figure 1).

The TNF-signalling pathway is tightly controlled by a number of checkpoints that rely on ubiquitination-, phosphorylation-, gene-expression-dependent and proteolytic events [10,29]. In conditions where any of these checkpoints is disabled, there is the formation of a secondary cytoplasmic complex, referred to as complex-II, which is composed of FADD, Caspase-8, cFLIP, RIPK1 and RIPK3, and that has cytotoxic activity [10,13,29]. Complex II can induce (i) apoptosis, via the activation of the initiator Caspase-8, which, in turn, cleaves and activates the executioner Caspase-3 and Caspase-7 [30]; (ii) necroptosis, mediated by the kinase activity of RIPK1 and RIPK3 and the pseudokinase MLKL, which, following RIPK3-mediated activation, forms pores in the plasma membrane [31–33]; and (iii) pyroptosis, following Caspase-8-mediated cleavage of Gasdermin D that, similarly to MLKL, has the ability to form pores on the plasma membrane, leading to a lytic type of death [34,35]. While in non-immune cells Caspase-8 activation mainly leads to apoptosis, in innate immune cells, such as macrophages, Caspase-8 activation can induce both caspase-dependent apoptosis and Gasdermin D-mediated pyroptosis-like death [35] (Figure 1).

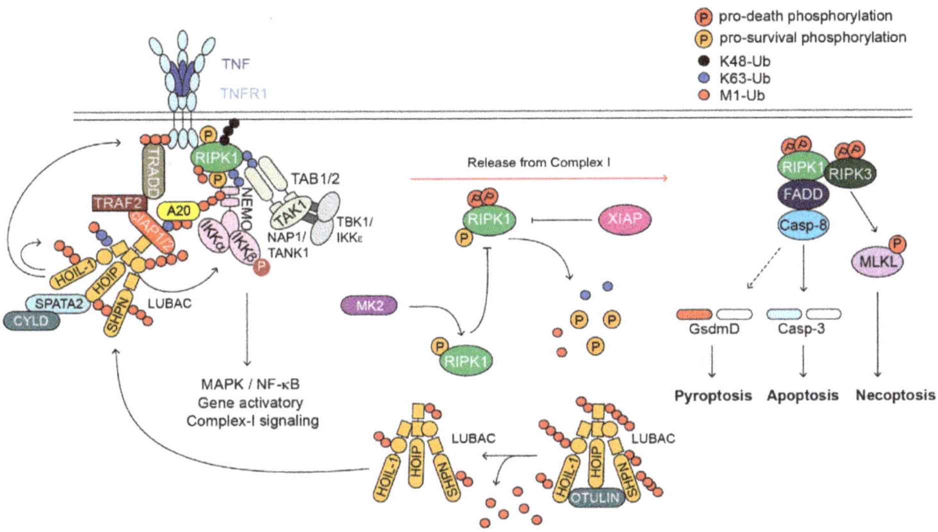

Figure 1. TNFR1-induced-signalling pathway. Cartoon depicting the TNFR1-induced-signalling pathway. Upon binding of TNF to TNFR1, a membrane-bound complex referred to as complex-I forms. This complex is characterised by the presence of adaptor proteins (e.g., TRADD, TRAF2, SPATA2, TAB1/2 and NAP1/TANK1), E3 ligases (e.g., cIAP1/2 and LUBAC), which conjugate poly-ubiquitin chains of different topology (i.e., K63, K48, K11 and M1) to different proteins of the complex, the deubiquitinases (DUBs) A20 and CYLD, and protein kinases such as RIPK1, IKK1/2, TAK1 and TBK1/IKKε. Complex-I promotes the activation of NF-κB and MAPKs that in turn mediate the expression of pro-survival as well as pro-inflammatory genes. Under certain circumstances, a secondary cytoplasmic complex originates in the cytosol from complex-I, referred to as complex-II. This complex is composed of FADD, cFLIP, Caspase-8, RIPK1 and RIPK3. Caspase-8 can trigger apoptosis via activation of Caspase-3 or, in some cell type, pyroptosis, via cleavage of Gasdermin D. Upon Caspase-8 inhibition by the means of synthetic- or viral-encoded caspase inhibitor, RIPK1 activates RIPK3 that in turn phosphorylates MLKL that undergoes activation and executes necroptosis. Of note, OTULIN, XIAP and MK2, have important regulatory functions in the TNFR1-signalling pathway, despite they are not directly recruited to complex-I or complex-II. OTULIN regulates the availability of the LUBAC components for their recruitment to complex-I. XIAP controls RIPK1 ubiquitination status outside complex-II and potentially its cytotoxic activity. MK2, by phosphorylating RIPK1, modulates its killing activity.

Both the conjugation and hydrolysis of ubiquitin moieties represent one of the most studied checkpoint mechanisms that control complex-II formation and restrain TNF cytotoxicity [36]. Amongst the key proteins and protein complexes responsible for these ubiquitin-system-mediated control mechanisms, are LUBAC (E3 ligase complex) [37], XIAP (X-linked IAP) [38], A20 and OTULIN (DUBs) [21] (Figure 2). In the next sections we focus on the pathological consequences of mutations disrupting the activity of the above indicated E3s and DUBs in genetically modified mice and human patients, and highlight the similarities and differences between the two systems.

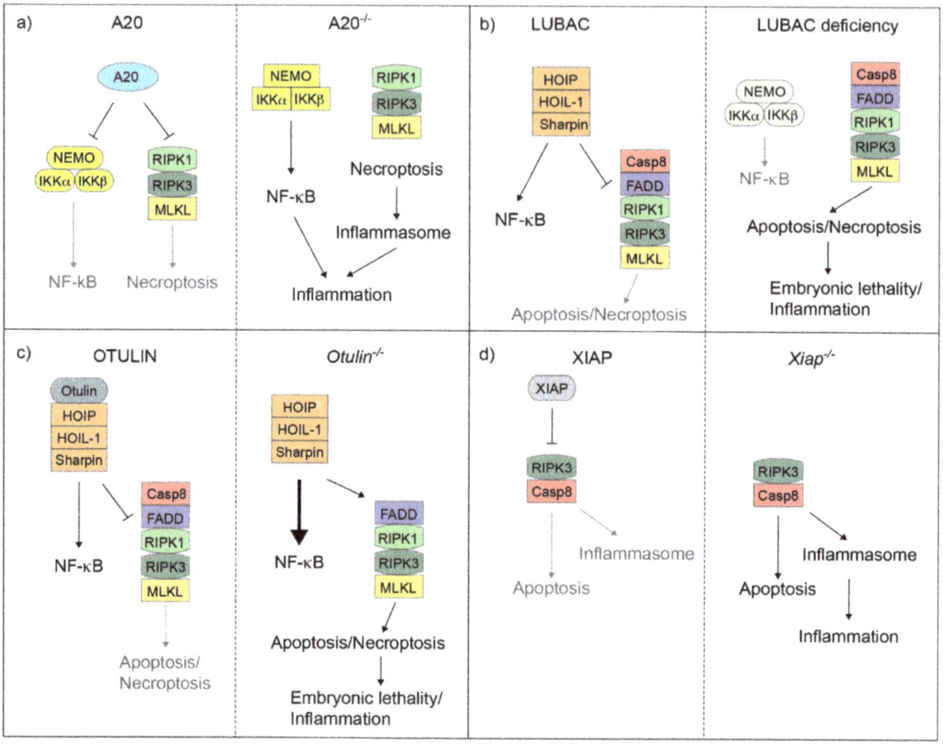

Figure 2. A20, LUBAC, OTULIN and XIAP regulate the balance between cell death and inflammation in mice. Cartoons illustrating how A20 (**a**), LUBAC (**b**), OTULIN (**c**) and XIAP (**d**) control the balance between NF-κB-mediated gene activation and complex-II-mediated cell death in mouse cells. A20 deletion results in the deregulation of both NF-κB response and RIPK1/RIPK3/MLKL-induced necroptosis, which, in turn, triggers inflammasome activation (**a**). Individual deletion of the LUBAC components causes an attenuation of NF-κB response, but an exacerbation of complex-II-mediated cell death, which can result in embryonic lethality or cell death-dependent inflammation in adult mice (**b**). OTULIN deletion leads to hyperactivation of NF-κB and, at the same time, can unleash complex-II-mediated cell death. This can in turn cause embryonic lethality or cell death-dependent inflammation in adult mice (**c**). XIAP deletion causes both Caspase-8-dependent apoptosis and RIPK3-dependent inflammasome activation, which eventually triggers inflammation (**d**).

3. LUBAC

Genetic deletion of any of the three LUBAC components causes either absence of linear chains (HOIP and HOIL-1) or reduction in linear chains (SHARPIN) in TNFR1-induced complex-I [39]. The phenotype of mice bearing a naturally occurring mutation in the SHARPIN gene, referred to as *cpdm* (chronic proliferative dermatitis mice) (Table 1), which causes its deletion, initially puzzled scientists of the cell death field [40]. Although SHARPIN deletion caused reduced NF-κB-mediated gene activation in vitro, *cpdm* mice developed chronic dermatitis and multiorgan inflammation [20]. This conundrum was solved by findings showing that (i) attenuation of linear ubiquitination in complex-I on the one side impairs NF-κB activation and reduces gene expression, but on the other side causes cell death by favouring complex-II formation in response to TNF stimulation [20]; and (ii) TNF-mediated cell death can be a potent trigger of inflammation. The latter was supported by the evidence that genetic deletion of TNF rescues *cpdm* mice from developing dermatitis and represented a watershed in the cell death field [20]. Indisputable evidence confirming that cell death was the cause of inflammation in *cpdm* mice came from the fact that combined deletion of FADD in keratinocytes or Caspase-8, to suppress apoptosis, and RIPK3 or MLKL to suppress necroptosis, prevented the inflammatory phenotype of SHARPIN mutant-mice [41,42]. Around the same years, animal models of cell death-induced inflammation, some of which we discuss here, boomed, further corroborating the notion that cell death can be the etiological agent of inflammatory syndromes. Differently from mice bearing the SHARPIN mutation, mice lacking HOIP or HOIL-1 are embryonically lethal due to exacerbated endothelial cell death and heart defects [39,43] (Table 1). Yet, the selective deletion of HOIP or HOIL-1 in keratinocyte causes severe skin inflammation, which is cell death dependent [44,45]. However, differently from *cpdm*, lethal dermatitis was found to be only partially TNF-driven [44]. Indeed, while concomitant deletion of RIPK3 and Caspase-8 completely prevents the inflammatory lesions, loss of TNFR1 delays the onset of dermatitis, but mutant mice still succumb later in life due to severe skin inflammation [44]. What triggers cell death in the absence of linear ubiquitination beyond TNF? It was reported that the inflammatory phenotype occurring as a consequence of HOIL-1 deficiency could be significantly delayed by the simultaneous deletion of TNFR1 and the DD of CD95 and TRAIL-R. This indicates that these three death receptors act in concert to induce cell death in the skin when LUBAC activity is completely abrogated [44]. In the liver, HOIP deletion causes hepatocellular carcinoma that arises from inflammation caused by hepatocyte death [46]. The role of LUBAC in immune cells has also been described. T cell-specific deletion of HOIP or HOIL-1 leads to an almost complete depletion of CD4/CD8+ T cells in mice [47]. HOIP- and HOIL-1-deficient T cells exhibited delayed NF-κB activation upon TCR and TNFR1 stimulation, consistently with a role for LUBAC in the activation of NF-κB. However, enforced NF-κB activation via the overexpression of a constitutively active version of IKKβ (IKKβca), does not restore normal T cell numbers [47]. Equally intriguing was the fact that T cell developmental defects also seem to be independent of cell death activation. Similarly, loss of LUBAC in B cells impairs signalling via the TNFR superfamily member CD40, highlighting an important role of linear ubiquitination in B cell activation [48]. Notably, mice with full body deficiency in HOIP or HOIL-1 displayed severe defects in haematopoietic progenitors, which affected erythropoiesis, and this was independent of cell death [39,49] (Table 1). This evidence would suggest that, differently from keratinocytes or other cell types, LUBAC activity does not primarily inhibit cell death in immune cells, but it rather plays an important role in coordinating different signals required for cell development or differentiation.

Last, although HOIL-1 also bears E3 ligase activity, it is not essential for LUBAC activity as E3 catalytic inactive mice are viable [18,50,51]. Instead, HOIL-1 catalytic activity limits linear ubiquitination. Indeed, mouse embryonic fibroblasts (MEFs) expressing catalytically inactive HOIL-1 are protected from cell death and have enhanced NF-κB activation in response to TNF due to increased levels of linear ubiquitin chains [39,51]. As a consequence, HOIL-1 catalytic inactive mice are protected from hepatocyte death in a model of liver damage and are also protected from dermatitis in a *cpdm* background [51]. Intriguingly, mice harbouring HOIL-1 catalytic inactivity displayed increased glycogen deposition in muscle [52].

To date, only a limited number of patients bearing LUBAC mutations have ever been reported. The amino acid sequence similarity between human and mouse is 91.6% for HOIL-1, 86.5% for HOIP and 73.6% for SHARPIN. The first reported LUBAC-mutant patients, in 2012, were two sisters from a non-consanguineous marriage, with compound heterozygous mutation of HOIL-1, consisting of deletion and nonsense mutation (p.Q185X) and one boy from a consanguineous marriage, with homozygous deletion of two nucleotides (c.121_122delCT) in HOIL-1 [53]. Few weeks after their birth, they started developing a series of disorders including autoinflammation (e.g., abdominal pain), immunodeficiency, which rendered the three patients susceptible to bacterial infections, amylopectin-like deposits in muscle and cardiomyopathy [53]. They all died by the age of eight years [53]. Immune cells, in particular the monocytes, lacking HOIL-1 are hyperresponsive to IL-1β stimulation, with exaggerated cytokine production, including IL-6 and IL-8. On the other hand, non-immune cells, such as fibroblasts, exhibited a delayed NF-κB response following IL-1β and TNF stimulation and severely impaired cytokine and chemokine production [53] (Table 1). This dichotomy most likely explains the paradoxical clinical phenotype observed in the three HOIL-1 mutant patients. Indeed, while the hyperresponsiveness of monocytes is the underlying cause of autoinflammation, the refractoriness of non-immune cells to mount an innate immune response could explain the immunodeficiency and susceptibility to bacterial infections. One year later, two independent studies reported a few more cases of patients with homozygous or heterozygous compound truncating or missense mutations of HOIL-1 [54]. These patients suffered from progressive muscular weakness, abnormal accumulation of glycogen in muscles and cardiomyopathy. Intriguingly, they presented no sign of autoinflammation and immunodeficiency [54]. Even more intriguingly was the discovery of two more patients, in 2018, carrying HOIL-1 mutations with both autoinflammatory/immunodeficiency and myopathic features [55]. At present, it is not known what determines whether HOIL-1 mutant patients preponderantly have one or the other phenotype, or both. It is also difficult to understand whether the mutations present in these patients would result in reduced HOIL-1 E3 ligase activity or whether they would rather behave as a linear ubiquitin-null mutant.

The first HOIP homozygous missense mutation (L72P) patient was reported in 2015 [56]. This patient was born from consanguineous parents. The second one, born from non-consanguineous parents, carrying biallelic variants, was identified in 2019 [57]. In both cases, the authors reported that, similarly to some HOIL-1 mutant patients, fibroblasts have impaired NF-κB activation upon TNF and IL-1β stimulation while monocytes are hyperresponsive to IL-1β stimulation [56]. In both cases, patients presented clinical features characteristic of multiorgan autoinflammation and immunodeficiency (recurrent bacterial infection). Lymphopenia (T cell depletion) was only present in the first observed patient [56]. The similar clinical manifestations between HOIL-1 and HOIP mutant patients can probably be attributed to the loss of linear ubiquitination, which is a common feature of loss of these two proteins, at least in mice [39] (Table 1).

Table 1. Overview of the pathological consequences of the deletion or mutation of the indicated genes in mice and human patients.

	Mouse		Human	
Gene	**Tissue**	**Phenotype**	**Mutation**	**Phenotype**
Hoip/ Hoil-1	Full body deletion	-Embryonic lethality -Excessive endothelial cell death [39,43]	Deletion/ nonsense/ missense	-Autoinflammation (e.g., abdominal pain) -Immunodeficiency (recurrent bacterial infections) -Amylopectin-like deposit in the muscle -Cardiomyopathy -Progressive muscular weakness -Immune cells hyperresponsive to IL-1β stimulation -Delayed NF-kB activation in non-immune cells -Lymphopenia [53–56]
	Skin-specific deletion	-Excessive keratinocyte death -Severe skin inflammation [44,45]		
	T cell-specific deletion	-Severe T cell depletion -Delayed NF-κB activation in TNF- and TCR-induced-signalling pathways [47]		
Hoip	Liver-specific deletion	-Hepatocyte death-driven inflammation -Hepatocellular carcinoma [46]		
	B cell-specific deletion	-Impaired CD40 signalling and antibody production [48]		
Hoil-1	Catalytic inactive (Full body)	-Increased NF-κB activation -Protection from cell death -Glycogen deposition [51]		

Table 1. *Cont.*

	Mouse		Human	
Gene	**Tissue**	**Phenotype**	**Mutation**	**Phenotype**
Sharpin	Full body mutation	-Chronic proliferative dermatitis -Liver inflammation -Peyer's patches loss -Splenomegaly [40]	Missense	Risk factor for LOAD (Late-onset Alzheimer disease) [58,59]
Otulin	Full body deletion	-Embryonic lethality -Loss of vascularization [60]	Loss of function	-ORAS (OTULIN-related autoinflammatory syndrome) or Otulipenia -Systemic sterile inflammation (e.g., joint swelling, prolonged fever, diarrhoea, panniculitis) -Developmental delay -Increased linear ubiquitination and NF-κB activation -Infliximab (monoclonal anti-TNF antibody) ameliorates the symptoms [61–63]
	Full body conditional deletion	Decreased survival [61]		
	Myeloid-cell specific deletion	-Severe acute systemic inflammation -NF-κB hyperactivation -Excessive cytokine production [61]		
	Catalytic inactivation (full body)	-Embryonic lethality -Deregulated endothelial cell death [64]		
	Skin-specific deletion	-Deregulated keratinocyte death -Severe skin inflammation [65,66]		
	Liver-specific deletion	-TNFR1-driven hepatocyte death -Compensatory proliferation -Hepatocellular carcinoma [67]		
A20	Full body deletion	-Perinatal lethality [68] -Cell death-dependent multiorgan inflammation (e.g., liver, kidneys, joints)	Nonsense	-Early onset systemic inflammation (e.g., arthritis, oral and genital ulcers, SLE-like disease, central nervous system vasculitis) -Patients' derived immune cells have elevated cytokine levels, are hyperresponsive to inflammasome activation. Fibroblasts have increased NF-kB activation following TNF stimulation [69–75]
	Myeloid cell-specific deletion	-Cell death-dependent joint inflammation [76]		
	ZnF7 mutation (full body)	-Cell death-dependent joint inflammation [76]		
Xiap	Full body deletion	-Low grade ileal inflammation, TNFR1- and TNFR2-dependent -Increased sensitivity to pathogenic bacterial strain -Cell death-dependent reduction in Paneth cells and dendritic cells -Cell death is the trigger of ileal inflammation [77,78]	Deletion/ insertion/ nonsense/ missense/ frameshift/ intronic	-Familia hemophagocytic lymphohystiocystiosis (FHLH) or XLP2 -High risk IBD -Excessive activation of macrophages and dendritic cells following EBV infection [79,80]

Very recently, a non-synonymous variant of SHARPIN was identified as genetic risk factor for LOAD (late-onset Alzheimer's disease) in a cohort of 202 Japanese individuals [58]. This variant has an amino acid substitution (G186R) that seemingly affects its subcellular localisation and NF-κB activation. In a follow-up study, six more SHARPIN variants were identified from a cohort of 180 patients with LOAD and 184 patients with mild cognitive impairment (Table 1). This, at present, is the only reported association between SHARPIN mutation and human diseases [59].

The different LUBAC mutant mouse models described above have been instrumental in unveiling the physiological role of linear ubiquitination and how linear ubiquitin chains orchestrate inflammatory programs. Equally important was the fact that they allowed to unravel the link between cell death and inflammation and the potent inflammatory potential of cell death activation. Mouse work revealed that the individual LUBAC components contribute to optimal gene activation following stimulation of immune receptors, including TNFR1, but, most importantly, they limit the killing activity of TNF. Therefore, the predominant phenotypic effect triggered in mice by their absence is cell death and cell death-induced inflammation. However, the phenotypic similarities between LUBAC mutant mice and patients are limited to some features of glycogen deposition and heart defects. In the human setting, loss of function mutations of HOIP, HOIL-1 or SHARPIN do not only result in autoinflammation, but also in immunodeficiency, glycogen storage disorders (HOIP and HOIL-1) and neurodegeneration (SHARPIN). This might indicate that (i) in humans, the gene activatory functions of LUBAC are as or more predominant than its cell death inhibitory functions, although the occurrence of cell death has not been fully analysed in patients; and (ii) those individuals carrying LUBAC mutations that escape lethality in utero, might have backup systems in place to regulate cell death and inflammation that are not completely dependent on LUBAC. An extensive cell death analysis in LUBAC mutant patients, using classical cell death markers might help elucidate the role of LUBAC in controlling organism homeostasis, and find therapeutic strategies to improve the care of LUBAC mutant patients.

4. OTULIN

OTULIN is the only known linear chain-specific deubiquitinase [81]. The amino acid sequence similarity between human and mouse OTULIN is 90.1%. In 2016, homozygous missense mutation of OTULIN was identified in three siblings, from a consanguineous family, affected by a severe sterile form of autoinflammation, which was named OTULIN-related autoinflammatoy syndrome (ORAS) [61]. In the same year, another group identified three more patients, from three different consanguineous families, carrying OTULIN biallelic mutations and symptoms of systemic sterile inflammation (e.g., prolonged fevers and diarrhoea) that they called Otulipenia [63] (Table 1). Few more patients carrying compound heterozygous mutations on OTULIN were identified, with similar clinical manifestations [62]. This prompted different groups of scientists to investigate the molecular basis of this mutant OTULIN-driven inflammatory disorder using mouse models. The most intuitive explanation as to why OTULIN-mutant patients suffered from a severe inflammatory disease was that, without functional OTULIN, there would be an excess of linear chains that would in turn exacerbate NF-κB responses and the consequent production of pro-inflammatory factors. Indeed, both OTULIN-mutant patients and OTULIN-deficient mice (conditional full body and myeloid cell-specific) exhibited linear chains accumulation, increased NF-κB activation and excessive cytokine production [61,63]. The fact that Infliximab (TNF-neutralising antibody) drastically reduced the inflammatory syndrome in patients and mice [61] indicated that the main instigator of the excessive NF-κB activation and cytokine production observed in absence of functional OTULIN (patients) and full-length protein (mice) is TNF (Table 1). Intriguingly, it was concluded that, differently from LUBAC deficient mice, it was the gene activation ability of TNF rather than its cell death-promoting potential that caused the inflammatory phenotype in mice lacking OTULIN.

This view was subsequently challenged by another study that, using a catalytically inactive mutant of OTULIN in mice, showed that the primary effect of linear chains accumulation is not NF-κB hyperactivation but rather complex-II-mediated cell death, in the form of apoptosis and necroptosis [64] (Table 1), the reason for this being that absence of OTULIN causes hyperubiquitination of the LUBAC components that impairs their recruitment to complex-I (Figure 1) [21,64]. This would in turn favour complex-II formation and cell death. This scenario was later confirmed by reports showing that OTULIN deletion in the liver causes TNFR1-driven, apoptosis- and compensatory proliferation-mediated liver pathology, while OTULIN deletion in keratinocytes causes TNFR1-driven, RIPK1 kinase activity-mediated, cell death-dependent skin inflammation [65,66] (Table 1). Of note, some OTULIN-mutant patients display signs of liver dysfunction and skin inflammation in the form of panniculitis and neutrophilic dermatosis [67]. Despite the contrasting results concerning the etiological agent of the systemic inflammatory syndrome that characterises OTULIN mutant mice (NF-κB hyperactivation vs. cell death), the common denominator of these mouse models is that, similarly to human settings, the blockade of the TNF/TNFR1 system significantly ameliorates the disease.

Importantly, OTULIN has also been implicated in signalling events that are different from cell death and inflammation per se. For example, the *Gumby* mutation, which is a spontaneous mutation in OTULIN (W96R), results in embryonic lethality resembling the report on OTULIN catalytically inactive mice [60]. Gumby mice display increased Wnt signalling [60]. Whether this is the cause for lethality remains unresolved. In addition, the pathology of OTULIN deficiency in the liver seems to be independent of TNFR1 signalling but dependent on aberrant mTOR activation [67].

The OTULIN-mutant mouse models were extremely useful to unveil the link between OTULIN mutations and excessive linear chains in ORAS/Otulipenia. In addition, they allowed to understand why infliximab has such therapeutic benefits in patients. At present, and similar to LUBAC-mutant patients, evidence that OTULIN absence can unleash inflammatory cell death is still missing. One could speculate that absence of OTULIN activity can induce cell death only in some cell types, while in others sustained NF-κB activation would be the main outcome. Future analysis employing cell death-specific stainings will be required to understand the mechanisms of hyperinflammation and the respective contributions of cell death and NF-κB in OTULIN mutant patients.

5. A20

A20 is a deubiquitinase enzyme that exhibits 88.1% amino acid sequence conservation between human and mouse. It was identified in 1990 as a NF-κB target gene [82], which had the ability of preventing TNF-induced cytotoxicity [83]. It was subsequently discovered that A20 is not only a NF-κB target gene but also an inhibitor of the NF-κB-signalling pathway [84]. Consistent with this idea, A20 null mice die perinatally due to multiorgan inflammation [84]. Intriguingly, the ability of A20 to suppress inflammation does not reside in its deubiquitinase activity, since mice carrying a point mutation in A20 catalytic domain do not display inflammation [85]. After the realisation that A20 deubiquitinase activity is dispensable to control inflammation, different groups have tried to uncover which other domains of A20 are responsible for this. A20 is a ubiquitin-editing enzyme, which not only possesses DUB activity but also E3 ligase activity, mediated by the fourth zinc finger domain (ZnF4) [68]. Surprisingly, the E3 ligase activity is not required to keep inflammation in check, since ZnF4 mutant mice are viable and healthy [86,87]. A major advance in the understanding of the role of A20 in repressing inflammatory processes came from studies whereby the ZnF7 was mutated. It was shown that the ZnF7 is required for the ability of A20 to suppress NF-κB activation, in a non-catalytic fashion [88]. A subsequent work proved that A20 stabilises linear chains by direct ZnF7-mediated binding; indeed, in the absence of A20, complete absence of linear chains was observed in complex-I [21,89]. Therefore, a model was proposed whereby A20, via the ZnF7 domain, binds to and shields linear chains, preventing their hydrolysis by other DUBs (e.g., CYLD), ensuring complex-I stability. At the same time, this shielding prevents the excessive recruitment of NF-κB-activating molecules to complex-I, such as the NEMO/IKK complex, thereby controlling NF-κB activation [21,90]. Consistently with this model, ZnF7 mutant mice display spontaneous inflammation [76]. Similar to the various LUBAC-mutant mice, the *Tnfaip3*$^{-/-}$, the *Tnfaip3*$^{myel\text{-}KO}$ (A20 full knock-out or selectively in myeloid cells, respectively) and *Tnfaip3*^{ZnF7mut} mutant mice were extremely useful to strengthen the link between cell death and inflammation, and to gain a better understanding of the cause of the inflammatory diseases displayed by A20 mutant patients. A20-deficient mice suffered from severe multiorgan inflammation, including, but not limited to, liver, kidneys and joints. Similarly, although milder, mice lacking A20 in myeloid cells and those bearing a ZnF7 mutant version of A20 developed arthritis [76]

(Table 1). Importantly, it has been proven at the genetic level that the cause of arthritis in these mice is not hyperactivation of NF-κB, but rather RIPK1/RIPK3/MLKL-dependent necroptosis of macrophages. This leads in turn to NLRP3 inflammasome activation within the same dying macrophages with the consequent release of IL-1β. Excessive production of IL-1β then causes cartilage erosion and joint inflammation [76,91].

In humans, it has been known for two decades that A20 is a susceptibility gene for autoinflammatory diseases such as systemic lupus erythematosus (SLE), rheumatoid arthritis, psoriasis and diabetes [70–75]. However, it was only in 2016 that it was proven that loss-of-function germline mutations in A20 cause systemic autoinflammatory disease [69]. The authors of this study identified five heterozygous truncating mutations in five families. Patients carrying the mutations displayed a range of clinical manifestations including early-onset systemic inflammation, arthritis, oral and genital ulcers, SLE-like disease and central nervous system vasculitis. This systemic inflammatory disease caused by A20 haploinsufficiency was named HA20. Patient-derived immune cells had a strong inflammatory signature (e.g., elevated levels of TNF, IL-6 and IL-17) and were hyperresponsive to inflammasome activation following LPS stimulation. Along the same line, patient-derived fibroblasts exhibited increased NF-κB activation upon TNFR1 stimulation [69] (Table 1).

A20 is another example of how mouse models can be extremely valuable to accelerate the understanding of (i) the genetic cause of a human pathology, or group of pathologies, and (ii) the molecular mechanisms driving the pathology. At the same time, the genetic studies in humans can indicate how to refine the existing mouse models to develop better preclinical disease models. For example, although the A20 mutant mice develop a set of diseases that quite closely recapitulate the patient's clinical features, it is surprising that almost all the human mutations are found in the OTU catalytic domain (catalytic inactive A20 mutant mice are normal), and no mutation has ever been found in the ZnF7 domain [92]. One explanation could be that in patients there is very little to no A20 detected in lysates from fibroblasts or PBMCs, suggesting that the mutations in the OTU destabilise A20 rather than solely killing its catalytic activity [69]. This possibility could be addressed by generating new genetically modified mice bearing the human corresponding A20 mutations. In addition, the fact that in mice the individual mutations in the OTU and ZnF4 do not trigger the inflammatory phenotype [86,87] could suggest that the combination of the two mutations might induce the phenotype observed in $Tnfaip3^{-/-}$ and $Tnfaip3^{ZnF7mut}$ mice. These new potential mouse models would broaden the possibility to study HA20 and find novel therapeutic approaches.

Finally, while the mouse models clearly indicated that the aetiology of the systemic inflammation observed in A20 mutant mice is cell death, in patients there is still lack of evidence supporting this possibility. Similar to what was highlighted for LUBAC and OTULIN mutant patients, cell death marker stainings, (phospho)proteomic and ubiquitinome analysis on patients' samples might help determine the contribution of cell death vs. NF-κB hyperactivation to the disease.

6. XIAP

XIAP is an E3 ligase enzyme that belongs to the IAP (inhibitor of apoptosis) family [93]. Human and mouse XIAP share 89.3% of their amino acid sequence. It was initially characterised as able to inhibit Caspase-9, by preventing its dimerization, and Caspase-3 and Caspase-7, by blocking their active site [94–97]. However, it subsequently became clear that XIAP is an important immune regulator, both in a cell death-independent and -dependent manner. Indeed, its E3 ligase activity plays a crucial role in pathogen responses mediated by NOD2, a member of the NLR family [98]. In the NOD2-signalling pathway, XIAP-mediated ubiquitination of RIPK2 is crucial for the correct activation of the pathway and secretion of the cytokines needed for the pathogen response [98]. Therefore, the importance of XIAP in the NOD2 pathway is independent of its ability to regulate cell death. By contrast, following TLR activation (e.g., TLR2 or TLR4), XIAP is required to prevent, in myeloid cells, RIPK3-mediated necroptosis and the concomitant NLRP3 activation and

IL-1β release [99] (Figure 2 and Table 1). Additionally, although XIAP is not recruited to TNF-induced complex-I, it regulates RIPK1 ubiquitination outside complex-I, therefore contributing to the regulation of complex-II formation [38] (Figure 1). In 2006, mutations in XIAP were found in 12 individuals belonging to three different families affected by X-linked lymphoproliferative syndrome (XLP) [79]. This was the first time that XIAP mutations were associated to a human disease. To date, many XIAP mutations have been identified, recently summarised in [100], which include nonsense and missense mutations, exon deletion, and small insertions and deletions, often leading to premature stop codon and protein deficiency. XLP is a rare immunodeficiency that is characterised by hemophagocytic lymphohystiocystosis (HLH), hypogammaglobulinaemia and lymphoma [101]. This syndrome normally develops following Epstein–Barr virus infection. The identification of more XIAP-mutant patients in the following years prompted clinicians to classify XIAP deficiency-caused disease as familial HLH (FHHL) or XLP2 [102] (Table 1). XLP2 differs from XLP1 in some immunological features, including absence of lymphoma development and high risk of IBD (inflammatory bowel disease). In particular, IBD is observed in about 25% of XIAP mutant patients and it is often refractory to treatment and lethal in 10% of the cases. Moreover, XIAP mutations are detected in up to 4% of male paediatric patients with very early onset IBD [80,103]. Another immunologic feature of XLP2 patients is the excessive activation of macrophages and dendritic cells to EBV and other viruses [101]. However, the mechanisms linking XIAP mutations to XLP2 disease and intestinal epithelial barrier damage are not yet entirely understood. Given the fact that XIAP plays a crucial role in the NOD2 pathway and NOD2 was the first ever identified risk gene for IBD, one could argue that XIAP deletion would predispose patients to IBD by impairing the NOD2 pathway. This speculation would be supported by the evidence that the majority of the missense mutations on XIAP map either in the BIR2 or the RING domain, both crucial for correct activation of the NOD2-signalling pathway [100]. However, while the penetrance of IBD in XIAP-mutant patients is 23%, only 1.5% of individuals carrying homozygous NOD2 risk variants develop IBD [78]. This suggests that the role of XIAP in ensuring intestinal homeostasis goes beyond its function in the NOD2 pathway, perhaps to its cell death regulatory functions.

Mouse models were again of great help to gain insights into the role of XIAP in the regulation of immune system responses (Table 1). Indeed, two recent works have shed new light on the role of XIAP deficiency in IBD pathogenesis, using *Xiap*$^{-/-}$ mice as a model [77,78]. In one case, the authors showed that XIAP-deficient mice have a reduced number of Paneth cells, as consequence of their death, which is TNF- and microbiota-dependent and RIPK1/RIPK3-mediated. Decrease in Paneth cells correlates with a decrease in production and secretion of antimicrobial peptides and change in the structure of the microbiota, termed dysbiosis. These changes, per se, are not enough to elicit intestinal inflammation in mice. However, *Xiap*$^{-/-}$ mice are very sensitive to intestinal inflammation triggers, such as DSS or the pathogenic bacterial strain *Helicobacter hepaticus*. Importantly, delivery of antimicrobial peptide to the intestine, by means of adenoviruses, allowed XIAP-deficient mice to clear the *H. hepaticus* [78]. These findings are in line with the fact that in XIAP-mutant patients, similar to the mouse, there is a reduction in Paneth cells in the ileum and often the intestinal inflammation is triggered by bacterial or viral infection. In the other study, the authors have demonstrated that, unlike the abovementioned work, XIAP-deficient mice develop spontaneous ileal inflammation, which is microbiome and TNF dependent. In addition, the authors showed that both TNFR1 and TNFR2 contributed to the inflammation since their individual genetic ablation abrogates the inflammation. Furthermore, they proved that the death of dendritic cells mediated by the TLR/TNFR2/RIPK3 axis ignites the intestinal inflammation in XIAP-null mice [78]. Very interestingly, elevated levels of TNFR2 correlates with disease severity in paediatric patients affected by IBD. These studies are attractive representative examples of how mouse models can help take human genetic studies forward to understand the molecular mechanisms underlying pathological clinical cases. For example, they corroborate the central role of Paneth cell damage in IBD and the importance of TNF as the instigator of the disease. They also hint at the importance of cell death, of both Paneth cells and dendritic cells, in the initiation of the intestinal inflammation in XIAP-mutant patients. Equally important, these mouse-based studies can help devise new therapeutic intervention, such as cell death inhibition, or delivery to the intestine of specific antimicrobial compounds, which will ultimately help improve the patient's care.

7. Omics

Genomic sequencing approaches have decisively contributed to identify the cause of many inflammation-related genetic diseases. The combination of genetic studies with biochemical studies has then helped dissect the molecular mechanisms underlying these genetic diseases and provided sound scientific basis for therapeutic intervention. The question that can be raised now is: How can different omic approaches further help understand the aetiology of human inflammatory diseases? It is established that TNF often plays an eminent role in inflammatory syndromes and that TNF-induced cell death, rather than TNF-induced gene activation, might be the decisive factor for the development of these syndromes. Therefore, the next question to be addressed now relates to the nature of factors released by dying cells that can instigate inflammatory processes.

Different cell death forms have different inflammatory potential. It is widely accepted that, while necroptosis and pyroptosis are inflammatory cell death forms, because of membrane rupture and intracellular content spillage, apoptosis is immunologically silent [3]. Interestingly, it has been demonstrated that necroptotic cells do not only passively release soluble factors as a consequence of plasma membrane rupture, but they are transcriptionally and translationally active for the production of pro-inflammatory cytokines [7]. Conversely, it was reported that apoptotic cells shut down translation via caspases, hence their scarce inflammatory potential [104]. However, some mouse models seem to challenge, at least partially, the current dogma. For example, the skin inflammatory phenotype observed in SHARPIN mutant mice has, at present, to be solely accounted to apoptosis [41,42]. In order to ascertain whether the different inflammatory potential of apoptosis and necroptosis is to be accounted to differences in the soluble factors that are released by the dying cells, Tanzer and colleagues took an unbiased, mass-spectrometry-based approach [105]. With this approach they analysed supernatant of human lymphoma cells and primary human macrophages undergoing TNF-induced apoptosis or necroptosis. As expected, a large number of proteins were significantly associated with either cell death type. Surprisingly, there was no significant qualitative or quantitative difference in terms of conventional cytokines between apoptosis and necroptosis. However, intriguingly, the authors observed that the supernatant of apoptotic cells had high levels of nucleosome components while the supernatant of necroptotic cells had high levels of lysosomal proteins [105]. How this translates into the different inflammatory potential between apoptosis and necroptosis is unknown. Furthermore, it is conceivable to think that only a limited number of factors, differentially released from necroptotic cells with respect to apoptosis, have the potential to trigger inflammation. More refined proteomic analysis to be conducted on more clinically related settings will be needed to try to determine the inflammatory potential of factors specifically secreted by dying cells. Proteomic analysis could be coupled to transcriptomic-based and mass-cytometry-based approaches with the purpose to examine cell death-specific signatures. Reliable animal models that recapitulate human mutation-driven, cell death-dependent diseases would be again the starting point for this combinatorial approach.

8. Conclusions

The last decade saw an impressive body of work that enabled us to understand how cell death modulates inflammatory diseases and how this was associated with human genetics. It is not always expected that mouse models completely recapitulate the human settings. However, model organisms spanning from 2D cells/organoids to invertebrates and mice bring us closer to the identification of aetiological factors of chronic inflammation and autoimmune disorders in humans. Excitingly, the gap between mice and men is becoming smaller with the advancement in technologies and preclinical animal models. Currently, mouse work has become the springboard for human studies with the ultimate purpose to design novel therapies that improve the care of patients affected by inflammatory/autoimmune diseases.

Author Contributions: N.P. and A.A. wrote the text and drew the figures. Both authors approved the submitted version. All authors have read and agreed to the published version of the manuscript.

Funding: This work was funded by the Deutsche Forschungsgemeinschaft (DFG, German Research Foundation, project no. AN 1717/1-1), the Jürgen Manchot Foundation, and the CMMC Junior Research Group Program for A.A.; N.P. is supported by Deutsche Forschungsgemeinschaft (DFG, German Research Foundation; project SFB1399, ID 413326622) and the Jürgen Manchot Foundation. Both N.P. and A.A. are supported by the collaborative research centres SFB1430, ID. 414786233 (A10 for N.P. and associated project for A.A.) and SFB1503, ID 455784452 (B02 for N.P. and A05 for A.A.).

Institutional Review Board Statement: Not applicable.

Informed Consent Statement: Not applicable.

Acknowledgments: We would like to thank members of the Annibaldi, Peltzer, and Liccardi laboratories for helpful discussions. We would like to apologise to the many authors whose work we could not cite due to space restrictions.

Conflicts of Interest: The authors declare no conflict of interest.

References

1. Strasser, A.; Vaux, D.L. Cell Death in the Origin and Treatment of Cancer. *Mol. Cell* **2020**, *78*, 1045–1054. [CrossRef] [PubMed]
2. Delbridge, A.R.; Grabow, S.; Strasser, A.; Vaux, D.L. Thirty years of BCL-2: Translating cell death discoveries into novel cancer therapies. *Nat. Rev. Cancer* **2016**, *16*, 99–109. [CrossRef] [PubMed]
3. Legrand, A.J.; Konstantinou, M.; Goode, E.F.; Meier, P. The Diversification of Cell Death and Immunity: Memento Mori. *Mol. Cell* **2019**, *76*, 232–242. [CrossRef] [PubMed]
4. Kayagaki, N.; Dixit, V.M. Rescue from a fiery death: A therapeutic endeavor. *Science* **2019**, *366*, 688–689. [CrossRef]
5. Najjar, M.; Saleh, D.; Zelic, M.; Nogusa, S.; Shah, S.; Tai, A.; Finger, J.N.; Polykratis, A.; Gough, P.J.; Bertin, J.; et al. RIPK1 and RIPK3 Kinases Promote Cell-Death-Independent Inflammation by Toll-like Receptor 4. *Immunity* **2016**, *45*, 46–59. [CrossRef]
6. Gitlin, A.D.; Heger, K.; Schubert, A.F.; Reja, R.; Yan, D.; Pham, V.C.; Suto, E.; Zhang, J.; Kwon, Y.C.; Freund, E.C.; et al. Integration of innate immune signaling by caspase-8 cleavage of N4BP1. *Nature* **2020**, *587*, 275–280. [CrossRef]
7. Orozco, S.L.; Daniels, B.P.; Yatim, N.; Messmer, M.N.; Quarato, G.; Chen-Harris, H.; Cullen, S.P.; Snyder, A.G.; Ralli-Jain, P.; Frase, S.; et al. RIPK3 Activation Leads to Cytokine Synthesis that Continues after Loss of Cell Membrane Integrity. *Cell Rep.* **2019**, *28*, 2275.e5–2287.e5. [CrossRef]
8. Meier, P.; Banreti, A. Tissue Repair: How to Inflame Your Neighbours. *Curr. Biol.* **2016**, *26*, R192–R194. [CrossRef]
9. Peltzer, N.; Walczak, H. Cell Death and Inflammation—A Vital but Dangerous Liaison. *Trends Immunol.* **2019**, *40*, 387–402. [CrossRef]
10. Annibaldi, A.; Meier, P. Checkpoints in TNF-Induced Cell Death: Implications in Inflammation and Cancer. *Trends Mol. Med.* **2017**, *24*, 49–65. [CrossRef]
11. Walczak, H. Death receptor-ligand systems in cancer, cell death, and inflammation. *Cold Spring Harb. Perspect. Biol.* **2013**, *5*, a008698. [CrossRef] [PubMed]
12. Walczak, H.; Krammer, P.H. The CD95 (APO-1/Fas) and the TRAIL (APO-2L) apoptosis systems. *Exp.Cell Res.* **2000**, *256*, 58–66. [CrossRef] [PubMed]
13. Micheau, O.; Tschopp, J. Induction of TNF receptor I-mediated apoptosis via two sequential signaling complexes. *Cell* **2003**, *114*, 181–190. [CrossRef]
14. Kelliher, M.A.; Grimm, S.; Ishida, Y.; Kuo, F.; Stanger, B.Z.; Leder, P. The death domain kinase RIP mediates the TNF-induced NF-kappaB signal. *Immunity* **1998**, *8*, 297–303. [CrossRef]
15. Rothe, M.; Pan, M.G.; Henzel, W.J.; Ayres, T.M.; Goeddel, D.V. The TNFR2-TRAF signaling complex contains two novel proteins related to baculoviral inhibitor of apoptosis proteins. *Cell* **1995**, *83*, 1243–1252. [CrossRef]
16. Annibaldi, A.; John, S.W.; Berghe, T.V.; Swatek, K.N.; Ruan, J.; Liccardi, G.; Bianchi, K.; Elliott, P.R.; Choi, S.M.; Coillie, S.V.; et al. Ubiquitin-Mediated Regulation of RIPK1 Kinase Activity Independent of IKK and MK2. *Mol. Cell* **2018**, *69*, 566.e5–580.e5. [CrossRef]

17. Dynek, J.N.; Goncharov, T.; Dueber, E.C.; Fedorova, A.V.; Izrael-Tomasevic, A.; Phu, L.; Helgason, E.; Fairbrother, W.J.; Deshayes, K.; Kirkpatrick, D.S.; et al. c-IAP1 and UbcH5 promote K11-linked polyubiquitination of RIP1 in TNF signalling. *EMBO J.* **2010**, *29*, 4198–4209. [CrossRef]
18. Haas, T.L.; Emmerich, C.H.; Gerlach, B.; Schmukle, A.C.; Cordier, S.M.; Rieser, E.; Feltham, R.; Vince, J.; Warnken, U.; Wenger, T.; et al. Recruitment of the linear ubiquitin chain assembly complex stabilizes the TNF-R1 signaling complex and is required for TNF-mediated gene induction. *Mol.Cell* **2009**, *36*, 831–844. [CrossRef]
19. Ikeda, F.; Deribe, Y.L.; Skanland, S.S.; Stieglitz, B.; Grabbe, C.; Franz-Wachtel, M.; van Wijk, S.J.L.; Goswami, P.; Nagy, V.; Terzic, J.; et al. SHARPIN forms a linear ubiquitin ligase complex regulating NF-kappaB activity and apoptosis. *Nature* **2011**, *471*, 637–641. [CrossRef]
20. Gerlach, B.; Cordier, S.M.; Schmukle, A.C.; Emmerich, C.H.; Rieser, E.; Haas, T.L.; Webb, A.I.; Rikard, J.A.; Anderton, H.; Wong, W.W.-L.; et al. Linear ubiquitination prevents inflammation and regulates immune signalling. *Nature* **2011**, *471*, 591–596. [CrossRef]
21. Draber, P.; Kupka, S.; Reichert, M.; Draberova, H.; Lafont, E.; de Miguel, D.; Spilgies, L.; Surinova, S.; Taraborrelli, L.; Hartwig, T.; et al. LUBAC-Recruited CYLD and A20 Regulate Gene Activation and Cell Death by Exerting Opposing Effects on Linear Ubiquitin in Signaling Complexes. *Cell Rep.* **2015**, *13*, 2258–2272. [CrossRef] [PubMed]
22. Ori, D.; Kato, H.; Sanjo, H.; Tartey, S.; Mino, T.; Akira, S.; Takeuchi, O. Essential roles of K63-linked polyubiquitin-binding proteins TAB2 and TAB3 in B cell activation via MAPKs. *J. Immunol.* **2013**, *190*, 4037–4045. [CrossRef] [PubMed]
23. Rahighi, S.; Ikeda, F.; Kawasaki, M.; Akutsu, M.; Suzuki, N.; Kato, R.; Kensche, T.; Uejima, T.; Bloor, S.; Komander, D.; et al. Specific recognition of linear ubiquitin chains by NEMO is important for NF-kappaB activation. *Cell* **2009**, *136*, 1098–1109. [CrossRef] [PubMed]
24. Jaco, I.; Annibaldi, A.; Lalaoui, N.; Wilson, R.; Tenev, T.; Laurien, L.; Kim, C.; Jamal, K.; Wicky Hohn, S.; Liccardi, G.; et al. MK2 Phosphorylates RIPK1 to Prevent TNF-Induced Cell Death. *Mol. Cell* **2017**, *66*, 698–710. [CrossRef]
25. Lafont, E.; Draber, P.; Rieser, E.; Reichert, M.; Kupka, S.; de Miguel, D.; Draberova, H.; von Massenhausen, A.; Bhamra, A.; Henderson, S.; et al. TBK1 and IKKepsilon prevent TNF-induced cell death by RIPK1 phosphorylation. *Nat. Cell Biol.* **2018**, *12*, 1389–1399. [CrossRef] [PubMed]
26. Hayden, M.S.; Ghosh, S. Regulation of NF-κB by TNF family cytokines. *Semin. Immunol.* **2014**, *26*, 253–266. [CrossRef]
27. Kupka, S.; de Miguel, D.; Draber, P.; Martino, L.; Surinova, S.; Rittinger, K.; Walczak, H. SPATA2-Mediated Binding of CYLD to HOIP Enables CYLD Recruitment to Signaling Complexes. *Cell Rep.* **2016**, *16*, 2271–2280. [CrossRef]
28. Elliott, P.R.; Leske, D.; Wagstaff, J.; Schlicher, L.; Berridge, G.; Maslen, S.; Timmermann, F.; Ma, B.; Fischer, R.; Freund, S.M.V.; et al. Regulation of CYLD activity and specificity by phosphorylation and ubiquitin-binding CAP-Gly domains. *Cell Rep.* **2021**, *37*, 109777. [CrossRef]
29. Ting, A.T.; Bertrand, M.J. More to Life than NF-kappaB in TNFR1 Signaling. *Trends Immunol.* **2016**, *37*, 535–545. [CrossRef]
30. Wang, L.; Du, F.; Wang, X. TNF-alpha induces two distinct caspase-8 activation pathways. *Cell* **2008**, *133*, 693–703. [CrossRef]
31. Dondelinger, Y.; Declercq, W.; Montessuit, S.; Roelandt, R.; Goncalves, A.; Bruggeman, I.; Hulpiau, P.; Weber, K.; Sehon, C.A.; Marquis, R.W.; et al. MLKL Compromises Plasma Membrane Integrity by Binding to Phosphatidylinositol Phosphates. *Cell Rep.* **2014**, *7*, 971–981. [CrossRef] [PubMed]
32. Degterev, A.; Hitomi, J.; Germscheid, M.; Ch'en, I.L.; Korkina, O.; Teng, X.; Abbott, D.; Cuny, G.D.; Yuan, C.; Wagner, G.; et al. Identification of RIP1 kinase as a specific cellular target of necrostatins. *Nat. Chem. Biol.* **2008**, *4*, 313–321. [CrossRef] [PubMed]
33. Sun, L.; Wang, H.; Wang, Z.; He, S.; Chen, S.; Liao, D.; Wang, L.; Yan, J.; Liu, W.; Lei, X.; et al. Mixed lineage kinase domain-like protein mediates necrosis signaling downstream of RIP3 kinase. *Cell* **2012**, *148*, 213–227. [CrossRef] [PubMed]
34. Sarhan, J.; Liu, B.C.; Muendlein, H.I.; Li, P.; Nilson, R.; Tang, A.Y.; Rongvaux, A.; Bunnell, S.C.; Shao, F.; Green, D.R.; et al. Caspase-8 induces cleavage of gasdermin D to elicit pyroptosis during Yersinia infection. *Proc. Natl. Acad. Sci. USA* **2018**, *115*, e10888–e10897. [CrossRef]
35. Chen, K.W.; Demarco, B.; Heilig, R.; Shkarina, K.; Boettcher, A.; Farady, C.J.; Pelczar, P.; Broz, P. Extrinsic and intrinsic apoptosis activate pannexin-1 to drive NLRP3 inflammasome assembly. *Embo. J.* **2019**, *38*, e101638. [CrossRef] [PubMed]
36. Peltzer, N.; Darding, M.; Walczak, H. Holding RIPK1 on the Ubiquitin Leash in TNFR1 Signaling. *Trends Cell Biol.* **2016**, *11*, 445–461. [CrossRef] [PubMed]
37. Zinngrebe, J.; Montinaro, A.; Peltzer, N.; Walczak, H. Ubiquitin in the immune system. *EMBO Rep.* **2014**, *15*, 28–45. [CrossRef]
38. Yabal, M.; Muller, N.; Adler, H.; Knies, N.; Gross, C.J.; Damgaard, R.B.; Kanegane, H.; Ringelhan, M.; Kaufmann, T.; Heikenwalder, M.; et al. XIAP Restricts TNF- and RIP3-Dependent Cell Death and Inflammasome Activation. *Cell Rep.* **2014**, *28*, 1796–1808. [CrossRef]
39. Peltzer, N.; Darding, M.; Montinaro, A.; Draber, P.; Draberova, H.; Kupka, S.; Rieser, E.; Fisher, A.; Hutchinson, C.; Taraborrelli, L.; et al. LUBAC is essential for embryogenesis by preventing cell death and enabling haematopoiesis. *Nature* **2018**, *557*, 112–117. [CrossRef]
40. Seymour, R.E.; Hasham, M.G.; Cox, G.A.; Shultz, L.D.; Hogenesch, H.; Roopenian, D.C.; Sundberg, J.P. Spontaneous mutations in the mouse Sharpin gene result in multiorgan inflammation, immune system dysregulation and dermatitis. *Genes Immun.* **2007**, *8*, 416–421. [CrossRef]
41. Rickard, J.A.; Anderton, H.; Etemadi, N.; Nachbur, U.; Darding, M.; Peltzer, N.; Lalaoui, N.; Lawlor, K.E.; Vanyai, H. TNFR1-dependent cell death drives inflammation in Sharpin-deficient mice. *Elife* **2014**, *3*, e03464. [CrossRef] [PubMed]

42. Kumari, S.; Redouane, Y.; Lopez-Mosqueda, J.; Shiraishi, R.; Romanowska, M.; Lutzmayer, S.; Kuiper, J.; Martinez, C.; Dikic, I.; Pasparakis, M.; et al. Sharpin prevents skin inflammation by inhibiting TNFR1-induced keratinocyte apoptosis. *Elife* **2014**, *3*, 03422. [CrossRef] [PubMed]
43. Peltzer, N.; Rieser, E.; Taraborrelli, L.; Draber, P.; Darding, M.; Pernaute, B.; Shimizu, Y.; Sarr, A.; Draberova, H.; Montinaro, A.; et al. HOIP Deficiency Causes Embryonic Lethality by Aberrant TNFR1-Mediated Endothelial Cell Death. *Cell Rep.* **2014**, *9*, 153–165. [CrossRef]
44. Taraborrelli, L.; Peltzer, N.; Montinaro, A.; Kupka, S.; Rieser, E.; Hartwig, T.; Sarr, A.; Darding, M.; Draber, P.; Haas, T.L.; et al. LUBAC prevents lethal dermatitis by inhibiting cell death induced by TNF, TRAIL and CD95L. *Nat. Commun.* **2018**, *9*, 3910. [CrossRef] [PubMed]
45. Tang, Y.; Tu, H.; Liu, G.; Zheng, G.; Wang, M.; Li, L.; Zhao, X.; Lin, X. RNF31 Regulates Skin Homeostasis by Protecting Epidermal Keratinocytes from Cell Death. *J. Immunol.* **2018**, *200*, 4117–4124. [CrossRef] [PubMed]
46. Shimizu, Y.; Peltzer, N.; Sevko, A.; Lafont, E.; Sarr, A.; Draberova, H.; Walczak, H. The linear ubiquitin chain assembly complex acts as a liver tumor suppressor and inhibits hepatocyte apoptosis and hepatitis. *Hepatology* **2017**, *24*, 1963–1978. [CrossRef]
47. Teh, C.E.; Lalaoui, N.; Jain, R.; Policheni, A.N.; Heinlein, M.; Alvarez-Diaz, S.; Sheridan, J.; Rieser, E.; Deuser, S.; Darding, M.; et al. Linear ubiquitin chain assembly complex coordinates late thymic T-cell differentiation and regulatory T-cell homeostasis. *Nat. Commun.* **2016**, *7*, 13353. [CrossRef]
48. Sasaki, Y.; Sano, S.; Nakahara, M.; Murata, S.; Kometani, K.; Aiba, Y.; Sakamoto, S.; Watanabe, Y.; Tanaka, K.; Kurosaki, T.; et al. Defective immune responses in mice lacking LUBAC-mediated linear ubiquitination in B cells. *Embo J.* **2013**, *32*, 2463–2476. [CrossRef]
49. Annibaldi, A.; Meier, P. Ripk1 and haematopoiesis: A case for LUBAC and Ripk3. *Cell Death Differ.* **2018**, *25*, 1361. [CrossRef]
50. Stieglitz, B.; Morris-Davies, A.C.; Koliopoulos, M.G.; Christodoulou, E.; Rittinger, K. LUBAC synthesizes linear ubiquitin chains via a thioester intermediate. *EMBO Rep.* **2012**, *13*, 840–846. [CrossRef]
51. Fuseya, Y.; Fujita, H.; Kim, M.; Ohtake, F.; Nishide, A.; Sasaki, K.; Saeki, Y.; Tanaka, K.; Takahashi, R.; Iwai, K. The HOIL-1L ligase modulates immune signalling and cell death via monoubiquitination of LUBAC. *Nat. Cell Biol.* **2020**, *22*, 663–673. [CrossRef] [PubMed]
52. Kelsall, I.R.; McCrory, E.H.; Xu, Y.; Scudamore, C.L.; Nanda, S.K.; Mancebo-Gamella, P.; Wood, N.T.; Knebel, A.; Matthews, S.J.; Cohen, P. HOIL-1-catalysed ubiquitylation of unbranched glucosaccharides and its activation by ubiquitin oligomers. *bioRxiv* **2021**. [CrossRef]
53. Boisson, B.; Laplantine, E.; Prando, C.; Giliani, S.; Israelsson, E.; Xu, Z.; Abhyankar, A.; Israël, L.; Trevejo-Nunez, G.; Bogunovic, D.; et al. Immunodeficiency, autoinflammation and amylopectinosis in humans with inherited HOIL-1 and LUBAC deficiency. *Nat. Immunol.* **2012**, *13*, 1178–1186. [CrossRef] [PubMed]
54. Wang, K.; Kim, C.; Bradfield, J.; Guo, Y.; Toskala, E.; Otieno, F.G.; Hou, C.; Thomas, K.; Cardinale, C.; Lyon, G.J.; et al. Whole-genome DNA/RNA sequencing identifies truncating mutations in RBCK1 in a novel Mendelian disease with neuromuscular and cardiac involvement. *Genome. Med.* **2013**, *5*, 67. [CrossRef]
55. Krenn, M.; Salzer, E.; Simonitsch-Klupp, I.; Rath, J.; Wagner, M.; Haack, T.B.; Strom, T.M.; Schänzer, A.; Kilimann, M.W.; Schmidt, R.L.J.; et al. Mutations outside the N-terminal part of RBCK1 may cause polyglucosan body myopathy with immunological dysfunction: Expanding the genotype-phenotype spectrum. *J. Neurol.* **2018**, *265*, 394–401. [CrossRef]
56. Boisson, B.; Laplantine, E.; Dobbs, K.; Cobat, A.; Tarantino, N.; Hazen, M.; Lidov, H.G.W.; Hopkins, G.; Du, L.; Belkadi, A.; et al. Human HOIP and LUBAC deficiency underlies autoinflammation, immunodeficiency, amylopectinosis, and lymphangiectasia. *J. Exp. Med.* **2015**, *212*, 939–951. [CrossRef]
57. Oda, H.; Beck, D.B.; Kuehn, H.S.; Moura, N.S.; Hoffmann, P.; Ibarra, M.; Stoddard, J.; Tsai, W.L.; Gutierrez-Cruz, G.; Gadina, M.; et al. Second Case of HOIP Deficiency Expands Clinical Features and Defines Inflammatory Transcriptome Regulated by LUBAC. *Front. Immunol.* **2019**, *10*, 479. [CrossRef]
58. Asanomi, Y.; Shigemizu, D.; Miyashita, A.; Mitsumori, R.; Mori, T.; Hara, N.; Ito, K.; Niida, S.; Ikeuchi, T.; Ozaki, K. A rare functional variant of SHARPIN attenuates the inflammatory response and associates with increased risk of late-onset Alzheimer's disease. *Mol. Med.* **2019**, *25*, 20. [CrossRef]
59. Asanomi, Y.; Shigemizu, D.; Akiyama, S.; Miyashita, A.; Mitsumori, R.; Hara, N.; Ikeuchi, T.; Niida, S.; Ozaki, K. A functional variant of SHARPIN confers increased risk of late-onset Alzheimer's disease. *J. Hum. Genet.* **2022**, *67*, 203–208. [CrossRef]
60. Rivkin, E.; Almeida, S.M.; Ceccarelli, D.F.; Juang, Y.-C.; MacLean, T.A.; Srikumar, T.; Huang, H.; Dunham, W.H.; Fukumura, R.; Xie, G.; et al. The linear ubiquitin-specific deubiquitinase gumby regulates angiogenesis. *Nature* **2013**, *498*, 318–324. [CrossRef]
61. Damgaard, R.B.; Walker, J.A.; Marco-Casanova, P.; Morgan, N.V.; Titheradge, H.L.; Elliott, P.R.; McHale, D.; Maher, E.; McKenzie, A.; Komander, D. The Deubiquitinase OTULIN Is an Essential Negative Regulator of Inflammation and Autoimmunity. *Cell* **2016**, *166*, 1215–1230. [CrossRef] [PubMed]
62. Zinngrebe, J.; Moepps, B.; Monecke, T.; Gierschik, P.; Schlichtig, F.; Barth, T.F.E.; Strauß, G.; Boldrin, E.; Posovszky, C.; Schulz, A.; et al. Compound heterozygous variants in OTULIN are associated with fulminant atypical late-onset ORAS. *EMBO Mol. Med.* **2022**, *14*, e14901. [CrossRef] [PubMed]
63. Zhou, Q.; Yu, X.; Demirkaya, E.; Deuitch, N.; Stone, D.; Tsai, W.L.; Kuehn, H.S.; Wang, H.; Yang, D.; Park, Y.H.; et al. Biallelic hypomorphic mutations in a linear deubiquitinase define otulipenia, an early-onset autoinflammatory disease. *Proc. Natl. Acad. Sci. USA* **2016**, *113*, 10127–10132. [CrossRef] [PubMed]

64. Heger, K.; Wickliffe, K.E.; Ndoja, A.; Zhang, J.; Murthy, A.; Dugger, D.L.; Maltzman, A.; de Sousa e Melo, F.; Hung, J.; Zeng, Y.; et al. OTULIN limits cell death and inflammation by deubiquitinating LUBAC. *Nature* **2018**, *559*, 120–124. [CrossRef]
65. Hoste, E.; Lecomte, K.; Annusver, K.; Vandamme, N.; Roels, J.; Maschalidi, S.; Verboom, L.; Vikkula, H.-K.; Sze, M.; Van Hove, L.; et al. OTULIN maintains skin homeostasis by controlling keratinocyte death and stem cell identity. *Nat. Commun.* **2021**, *12*, 5913. [CrossRef]
66. Schünke, H.; Göbel, U.; Dikic, I.; Pasparakis, M. OTULIN inhibits RIPK1-mediated keratinocyte necroptosis to prevent skin inflammation in mice. *Nat. Commun.* **2021**, *12*, 5912. [CrossRef]
67. Damgaard, R.B.; Jolin, H.E.; Allison, M.E.D.; Davies, S.E.; Titheradge, H.L.; McKenzie, A.N.J.; Komander, D. OTULIN protects the liver against cell death, inflammation, fibrosis, and cancer. *Cell Death Differ.* **2020**, *27*, 1457–1474. [CrossRef]
68. Wertz, I.E.; O'Rourke, K.M.; Zhou, H.; Eby, M.; Aravind, L.; Seshagiri, S.; Wu, P.; Wiesmann, C.; Baker, R.; Boone, D.L.; et al. De-ubiquitination and ubiquitin ligase domains of A20 downregulate NF-kappaB signalling. *Nature* **2004**, *430*, 694–699. [CrossRef]
69. Zhou, Q.; Wang, H.; Schwartz, D.M.; Stoffels, M.; Park, Y.H.; Zhang, Y.; Yang, D.; Demirkaya, E.; Takeuchi, M.; Tsai, W.L.; et al. Loss-of-function mutations in TNFAIP3 leading to A20 haploinsufficiency cause an early-onset autoinflammatory disease. *Nat. Genet.* **2016**, *48*, 67–73. [CrossRef]
70. Graham, R.R.; Cotsapas, C.; Davies, L.; Hackett, R.; Lessard, C.J.; Leon, J.M.; Burtt, N.P.; Guiducci, C.; Parkin, M.; Gates, C.; et al. Genetic variants near TNFAIP3 on 6q23 are associated with systemic lupus erythematosus. *Nat. Genet.* **2008**, *40*, 1059–1061. [CrossRef]
71. Plenge, R.M.; Cotsapas, C.; Davies, L.; Price, A.L.; Bakker, P.I.W.D.; Maller, J.; Pe'Er, I.; Burtt, N.P.; Blumenstiel, B.; DeFelice, M.; et al. Two independent alleles at 6q23 associated with risk of rheumatoid arthritis. *Nat. Genet.* **2007**, *39*, 1477–1482. [CrossRef] [PubMed]
72. Thomson, W.; Wellcome Trust Case Control Consortium; Barton, A.; Walker, N.; Eyre, S.; Hinks, A.; Bowes, J.; Donn, R.; Symmons, D.; Hider, S.; et al. Rheumatoid arthritis association at 6q23. *Nat. Genet.* **2007**, *39*, 1431–1433. [CrossRef] [PubMed]
73. Musone, S.L.; Taylor, K.E.; Lu, T.T.; Nititham, J.; Ferreira, R.C.; Ortmann, W.; Shifrin, N.; Petri, M.A.; Kamboh, M.I.; Manzi, S.; et al. Multiple polymorphisms in the TNFAIP3 region are independently associated with systemic lupus erythematosus. *Nat. Genet.* **2008**, *40*, 1062–1064. [CrossRef] [PubMed]
74. Nair, R.P.; Duffin, K.C.; Helms, C.; Ding, J.; Stuart, P.E.; Goldgar, D.; Gudjonsson, J.E.; Li, Y.; Tejasvi, T.; Feng, B.; et al. Genome-wide scan reveals association of psoriasis with IL-23 and NF-kappaB pathways. *Nat. Genet.* **2009**, *41*, 199–204. [CrossRef]
75. Fung, E.Y.; Smyth, D.J.; Howson, J.M.; Cooper, J.D.; Walker, N.M.; Stevens, H.; Wicker, L.S.; Todd, J.A. Analysis of 17 autoimmune disease-associated variants in type 1 diabetes identifies 6q23/TNFAIP3 as a susceptibility locus. *Genes Immun.* **2009**, *10*, 188–191. [CrossRef]
76. Polykratis, A.; Martens, A.; Eren, R.O.; Shirasaki, Y.; Yamagishi, M.; Yamaguchi, Y.; Uemura, S.; Miura, M.; Holzmann, B.; Kollias, G.; et al. A20 prevents inflammasome-dependent arthritis by inhibiting macrophage necroptosis through its ZnF7 ubiquitin-binding domain. *Nat. Cell Biol.* **2019**, *21*, 731–742. [CrossRef]
77. Wahida, A.; Müller, M.; Hiergeist, A.; Popper, B.; Steiger, K.; Branca, C.; Tschurtschenthaler, M.; Engleitner, T.; Donakonda, S.; De Coninck, J.; et al. XIAP restrains TNF-driven intestinal inflammation and dysbiosis by promoting innate immune responses of Paneth and dendritic cells. *Sci. Immunol.* **2021**, *6*, eabf7235. [CrossRef]
78. Strigli, A.; Gopalakrishnan, S.; Zeissig, Y.; Basic, M.; Wang, J.; Schwerd, T.; Doms, S.; Peuker, K.; Hartwig, J.; Harder, J.; et al. Deficiency in X-linked inhibitor of apoptosis protein promotes susceptibility to microbial triggers of intestinal inflammation. *Sci. Immunol.* **2021**, *6*, eabf7473. [CrossRef]
79. Rigaud, S.; Fondanèche, M.-C.; Lambert, N.C.; Pasquier, B.; Mateo, V.; Soulas-Sprauel, P.; Galicier, L.; Le Deist, F.; Rieux-Laucat, F.; Revy, P.; et al. XIAP deficiency in humans causes an X-linked lymphoproliferative syndrome. *Nature* **2006**, *444*, 110–114. [CrossRef]
80. Crowley, E.; Warner, N.; Pan, J.; Khalouei, S.; Elkadri, A.; Fiedler, K.; Foong, J.; Turinsky, A.L.; Bronte-Tinkew, D.; Zhang, S.; et al. Prevalence and Clinical Features of Inflammatory Bowel Diseases Associated With Monogenic Variants, Identified by Whole-Exome Sequencing in 1000 Children at a Single Center. *Gastroenterology* **2020**, *158*, 2208–2220. [CrossRef]
81. Hrdinka, M.; Gyrd-Hansen, M. The Met1-Linked Ubiquitin Machinery: Emerging Themes of (De)regulation. *Mol. Cell* **2017**, *68*, 265–280. [CrossRef] [PubMed]
82. Dixit, V.M.; Green, S.; Sarma, V.; Holzman, L.B.; Wolf, F.W.; O'Rourke, K.; Ward, P.A.; Prochownik, E.V.; Marks, R.M. Tumor necrosis factor-alpha induction of novel gene products in human endothelial cells including a macrophage-specific chemotaxin. *J. Biol. Chem.* **1990**, *265*, 2973–2978. [CrossRef]
83. Opipari, A.W., Jr.; Hu, H.M.; Yabkowitz, R.; Dixit, V.M. The A20 zinc finger protein protects cells from tumor necrosis factor cytotoxicity. *J. Biol. Chem.* **1992**, *267*, 12424–12427. [CrossRef]
84. Lee, E.G.; Boone, D.L.; Chai, S.; Libby, S.L.; Chien, M.; Lodolce, J.P.; Ma, A. Failure to regulate TNF-induced NF-kappaB and cell death responses in A20-deficient mice. *Science* **2000**, *289*, 2350–2354. [CrossRef] [PubMed]
85. De Dainichi, A.T.; Rathinam, C.V.; Ghosh, S. The deubiquitinase activity of A20 is dispensable for NF-kappaB signaling. *EMBO Rep.* **2014**, *15*, 775–783.
86. Lu, T.T.; Onizawa, M.; Hammer, G.; Turer, E.E.; Yin, Q.; Damko, E.; Agelidis, A.; Shifrin, N.; Advincula, R.; Barrera, J.; et al. Dimerization and ubiquitin mediated recruitment of A20, a complex deubiquitinating enzyme. *Immunity* **2013**, *38*, 896–905. [CrossRef]
87. Wertz, I.E.; Newton, K.; Seshasayee, D.; Kusam, S.; Lam, C.; Zhang, J.; Popovych, N.; Helgason, E.; Schoeffler, A.; Jeet, S.; et al. Phosphorylation and linear ubiquitin direct A20 inhibition of inflammation. *Nature* **2015**, *528*, 370–375. [CrossRef]

88. Skaug, B.; Chen, J.; Du, F.; He, J.; Ma, A.; Chen, Z.J. Direct, noncatalytic mechanism of IKK inhibition by A20. *Mol. Cell* **2011**, *44*, 559–571. [CrossRef]
89. Tokunaga, F.; Nishimasu, H.; Ishitani, R.; Goto, E.; Noguchi, T.; Mio, K.; Kamei, K.; Ma, A.; Iwai, K.; Nureki, O. Specific recognition of linear polyubiquitin by A20 zinc finger 7 is involved in NF-κB regulation. *Embo J.* **2012**, *31*, 3856–3870. [CrossRef]
90. Martens, A.; van Loo, G. A20 at the Crossroads of Cell Death, Inflammation, and Autoimmunity. *Cold Spring Harb. Perspect. Biol.* **2020**, *12*, a036418. [CrossRef]
91. Walle, L.V.; Van Opdenbosch, N.; Jacques, P.; Fossoul, A.; Verheugen, E.; Vogel, P.; Beyaert, R.; Elewaut, D.; Kanneganti, T.-D.; Van Loo, G.; et al. Negative regulation of the NLRP3 inflammasome by A20 protects against arthritis. *Nature* **2014**, *512*, 69–73. [CrossRef] [PubMed]
92. Ma, A.; Malynn, B.A. A20: Linking a complex regulator of ubiquitylation to immunity and human disease. *Nat. Rev. Immunol.* **2012**, *12*, 774–785. [CrossRef] [PubMed]
93. Gyrd-Hansen, M.; Meier, P. IAPs: From caspase inhibitors to modulators of NF-kappaB, inflammation and cancer. *Nat. Rev. Cancer* **2010**, *10*, 561–574. [CrossRef] [PubMed]
94. Chai, J.; Shiozaki, E.; Srinivasula, S.M.; Wu, Q.; Datta, P.; Alnemri, E.S.; Shi, Y. Structural basis of caspase-7 inhibition by XIAP. *Cell* **2001**, *104*, 769–780. [CrossRef]
95. Shiozaki, E.N.; Chai, J.; Rigotti, D.J.; Riedl, S.J.; Li, P.; Srinivasula, S.M.; Alnemri, E.S.; Fairman, R.; Shi, Y. Mechanism of XIAP-mediated inhibition of caspase-9. *Mol Cell.* **2003**, *11*, 519–527. [CrossRef]
96. Fesik, S.W.; Shi, Y. Structural biology. Controlling the caspases. *Science* **2001**, *294*, 1477–1478. [CrossRef]
97. Scott, F.L.; Denault, J.B.; Riedl, S.J.; Shin, H.; Renatus, M.; Salvesen, G.S. XIAP inhibits caspase-3 and -7 using two binding sites: Evolutionarily conserved mechanism of IAPs. *Embo J.* **2005**, *24*, 645–655. [CrossRef]
98. Damgaard, R.B.; Nachbur, U.; Yabal, M.; Wong, W.W.-L.; Fiil, B.K.; Kastirr, M.; Rieser, E.; Rickard, J.A.; Bankovacki, A.; Peschel, C.; et al. The ubiquitin ligase XIAP recruits LUBAC for NOD2 signaling in inflammation and innate immunity. *Mol Cell.* **2012**, *46*, 746–758. [CrossRef]
99. Lawlor, K.E.; Feltham, R.; Yabal, M.; Conos, S.A.; Chen, K.W.; Ziehe, S.; Graß, C.; Zhan, Y.; Nguyen, T.A.; Hall, C.; et al. XIAP Loss Triggers RIPK3- and Caspase-8-Driven IL-1beta Activation and Cell Death as a Consequence of TLR-MyD88-Induced cIAP1-TRAF2 Degradation. *Cell Rep.* **2017**, *20*, 668–682. [CrossRef]
100. Mudde, A.C.A.; Booth, C.; Marsh, R.A. Evolution of Our Understanding of XIAP Deficiency. *Front. Pediatr.* **2021**, *9*, 660520. [CrossRef]
101. Filipovich, A.H.; Zhang, K.; Snow, A.L.; Marsh, R.A. X-linked lymphoproliferative syndromes: Brothers or distant cousins? *Blood* **2010**, *116*, 3398–3408. [CrossRef] [PubMed]
102. Marsh, R.A.; Madden, L.; Kitchen, B.J.; Mody, R.; McClimon, B.; Jordan, M.B.; Bleesing, J.J.; Zhang, K.; Filipovich, A.H. XIAP deficiency: A unique primary immunodeficiency best classified as X-linked familial hemophagocytic lymphohistiocytosis and not as X-linked lymphoproliferative disease. *Blood* **2010**, *116*, 1079–1082. [CrossRef] [PubMed]
103. Zeissig, Y.; Petersen, B.-S.; Milutinovic, S.; Bosse, E.; Mayr, G.; Peuker, K.; Hartwig, J.; Keller, A.; Kohl, M.; Laass, M.W.; et al. XIAP variants in male Crohn's disease. *Gut* **2015**, *64*, 66–76. [CrossRef] [PubMed]
104. Clemens, M.J.; Bushell, M.; Jeffrey, I.W.; Pain, V.M.; Morley, S.J. Translation initiation factor modifications and the regulation of protein synthesis in apoptotic cells. *Cell Death Differ.* **2000**, *7*, 603–615. [CrossRef] [PubMed]
105. Tanzer, M.C.; Frauenstein, A.; Stafford, C.A.; Phulphagar, K.; Mann, M.; Meissner, F. Quantitative and Dynamic Catalogs of Proteins Released during Apoptotic and Necroptotic Cell Death. *Cell Rep.* **2020**, *30*, 1260.e5–1270.e5. [CrossRef]

biomedicines

MDPI

Article

Comparison of Transcriptional Signatures of Three Staphylococcal Superantigenic Toxins in Human Melanocytes

Nabarun Chakraborty [1,*,†], Seshamalini Srinivasan [1,2,†], Ruoting Yang [1], Stacy-Ann Miller [1], Aarti Gautam [1], Leanne J. Detwiler [1,2], Bonnie C. Carney [3,4,5], Abdulnaser Alkhalil [3], Lauren T. Moffatt [3,4,5], Marti Jett [1], Jeffrey W. Shupp [3,4,5,6] and Rasha Hammamieh [1]

1 Medical Readiness Systems Biology, Walter Reed Army Institute of Research, Silver Spring, MD 20910, USA; saissesh@gmail.com (S.S.); ruoting.yang.civ@mail.mil (R.Y.); stacyann.m.miller.civ@mail.mil (S.-A.M.); aarti.gautam.civ@mail.mil (A.G.); leanne_detwiler@yahoo.com (L.J.D.); marti.jett-tilton.civ@mail.mil (M.J.); rasha.hammamieh1.civ@mail.mil (R.H.)
2 The Geneva Foundation, Walter Reed Army Institute of Research, Silver Spring, MD 20910, USA
3 Firefighters' Burn and Surgical Research Laboratory, MedStar Health Research Institute, Washington, DC 20010, USA; bonnie.c.carney@medstar.net (B.C.C.); abdulnaser.alkhalil@medstar.net (A.A.); lauren.t.moffatt@medstar.net (L.T.M.); jeffrey.w.shupp@medstar.net (J.W.S.)
4 Department of Surgery, Georgetown University School of Medicine, Washington, DC 20057, USA
5 Department of Biochemistry and Molecular & Cellular Biology, Georgetown University School of Medicine, Washington, DC 20057, USA
6 The Burn Center, MedStar Washington Hospital Center, Washington, DC 20010, USA
* Correspondence: nabarun.chakraborty2.civ@mail.mil; Tel.: +1-301-452-8940 or +1-301-319-7363
† These authors contributed equally to this work.

Citation: Chakraborty, N.; Srinivasan, S.; Yang, R.; Miller, S.-A.; Gautam, A.; Detwiler, L.J.; Carney, B.C.; Alkhalil, A.; Moffatt, L.T.; Jett, M.; et al. Comparison of Transcriptional Signatures of Three Staphylococcal Superantigenic Toxins in Human Melanocytes. *Biomedicines* **2022**, *10*, 1402. https://doi.org/10.3390/biomedicines10061402

Academic Editor: Marianna Christodoulou

Received: 20 April 2022
Accepted: 6 May 2022
Published: 14 June 2022

Abstract: *Staphylococcus aureus*, a gram-positive bacterium, causes toxic shock through the production of superantigenic toxins (sAgs) known as Staphylococcal enterotoxins (SE), serotypes A-J (SEA, SEB, etc.), and toxic shock syndrome toxin-1 (TSST-1). The chronology of host transcriptomic events that characterizes the response to the pathogenesis of superantigenic toxicity remains uncertain. The focus of this study was to elucidate time-resolved host responses to three toxins of the superantigenic family, namely SEA, SEB, and TSST-1. Due to the evolving critical role of melanocytes in the host's immune response against environmental harmful elements, we investigated herein the transcriptomic responses of melanocytes after treatment with 200 ng/mL of SEA, SEB, or TSST-1 for 0.5, 2, 6, 12, 24, or 48 h. Functional analysis indicated that each of these three toxins induced a specific transcriptional pattern. In particular, the time-resolved transcriptional modulations due to SEB exposure were very distinct from those induced by SEA and TSST-1. The three superantigens share some similarities in the mechanisms underlying apoptosis, innate immunity, and other biological processes. Superantigen-specific signatures were determined for the functional dynamics related to necrosis, cytokine production, and acute-phase response. These differentially regulated networks can be targeted for therapeutic intervention and marked as the distinguishing factors for the three sAgs.

Keywords: superantigens; gene expression; transcriptional dynamics; staphylococcal enterotoxins; SEB; SEA; TSST-1; toxins; biological networks; clustering; functional pathways; time–course analysis; cDNA microarray; human melanocytes

1. Introduction

Staphylococcus aureus (*S. aureus*) is widely circulated in nature and carried by 25–33% of normal individuals in the anterior nares and skin [1,2]. The extreme penetrance of this bacteria and its ability to colonize skin, open wounds, and other surfaces makes it a serious threat in facilities that provide health care [3,4]. The myriad of exotoxins synthesized and secreted by *S. aureus* include the Streptococcal enterotoxins (SEs), such as SEA-SEE, SEG-SEI, SEK-SET, and SEY, and the toxic shock syndrome toxin (TSST-1). As SEA is the most common toxin in food poisoning, SEB is recognized for its potent toxicity as a biological weapon, and TSST-1 is known for being the causative agent of lethal toxic shock [5–7], they remain the primary focus of *S. aureus* toxins research [8].

Staphylococcal enterotoxins and TSST-1 are superantigens (sAgs) that bind as an intact molecule to the major histocompatibility complex II (MHC) and interact directly with the variable region of the beta chain of T-cell receptors (TCRs) without the need for processing or presentation by the antigen presenting cells (APC). These interactions activate T-cells, resulting in massive production of cytokines and chemokines, activation-induced apoptosis, and T-cell anergy [9].

The interaction of sAgs with immune cells and the ensuing pathogenesis have been well documented [10,11]. Previous work from our lab identified a set of genes in human peripheral blood mononuclear cells (PBMCs) that were expressed as early as 2 h post-SEB treatment [11] and played important roles in tissue repair, inflammation, and increased vascular permeability. Supporting studies reported SEB-induced proinflammatory mediators contribute to vasodilation, vascular leak, and edema [12–14].

The immunologic barrier raised by the skin is a concerted effort from different cell types. Keratinocytes, melanocytes, and Langerhans cells actively contributed to the innate immune response to sAgs [15]. We presently focused on the melanocytes, which are dendritic cells of neuroectodermal origin and an integral part of the epidermis [16–18]. The dendritic nature and strategic location of melanocytes in the epidermal layer of the skin allow for an ideal milieu to interact with the extra-skin environment and build response coordination among neighboring shallow skin cells.

The immunological responses of melanocytes have been attributed to their ability to express MHC molecules and other various adhesion molecules, including intercellular adhesion molecule (ICAM)-1 and vascular cell adhesion molecule (VCAM)-1 [19–21]. In addition, melanocytes can produce several cytokines, tumor necrosis factor alpha (TNF-α), and transforming growth factor beta (TGF-β1) with potential functions in phagocytosis and antigen processing and presentation [20,22,23]. The immunomodulatory cytotoxic properties of melanocytes were highlighted in a recent in vitro study, where melanocytes were exposed to *C. albicans* infection [24]. Despite the wide coverage of melanocyte research and the increasing knowledge of their role in immune response, minimal information is available about the role of melanocytes in response to sAgs. The selection of melanocytes for the present study was further justified by the differential host-sAgs responses that were essentially determined by the significant structural differences of the three sAgs, thus affecting their interactions with the host cells [8,25]. Major findings include the toxin-specific melanocyte response dynamics enabling the distinction of toxin pathogenesis; in particular, we elucidated later-stage molecular events that could have the potential for common or customized therapeutic targets for the three toxins of choice.

2. Materials and Methods

2.1. Cell Culture and Toxin Treatment

Normal human epidermal melanocytes (NHEM) and the reagents required for culturing the cells were purchased from Clonetics® (Lonza, Walkersville, MD, USA). Cells were maintained in Melanocyte Growth Medium (MGM) BulletKit® according to the supplier's instructions (Lonza, Walkersville, MD, USA). Cell cultures were established at the recommended starting cell density of 10,000 cells per cm^2 and maintained in 150 cm^2 flasks at 37 °C and 5% CO_2 in a humidified incubator.

SEA, SEB, and TSST-1 were purchased from Toxin Technology (Toxin Technology, Sarasota, FL, USA). The toxins were diluted from the stock solution to 25 µg/mL in the MGM growth media (Lonza, Walkersville, MD, USA). On the day of the assay, the cells were treated with the appropriate amount of toxins to reach a final concentration of 200 ng/mL. The toxins were inactivated by adding TRIzol (Invitrogen, Carlsbad, CA, USA) at 0.5, 2, 6, 12, 24, and 48 h post-exposure (p.e.). As controls, untreated melanocytes were grown in parallel and harvested at the same time points. Each time point for each toxin was represented by a single culture.

2.2. RNA Isolation

Total RNA was isolated with TRIzol reagent (Invitrogen, Carlsbad, CA, USA) using the manufacturer's procedure, followed by a cleanup procedure using the RNeasy MinElute Cleanup kit (QIAGEN, Germantown, MD, USA). The integrity of the extracted RNA was assessed using the 2100 Bioanalyzer instrument (Agilent Technologies, Santa Clara, CA, USA), and RNA integrity number (RIN) values were recorded.

2.3. Transcriptomic Assay and Analysis

The dual dye microarray hybridization was carried out using the SurePrint 4 × 44 K v2 Microarray Kit (Agilent Technologies, Inc., Santa Clara, CA, USA) following the vendor's protocol. Cy-5-labelled 200 ng of purified RNA was co-hybridized with Cy-3-labelled reference RNA (Agilent Technologies, Inc., Santa Clara, CA, USA) and bound to Agilent 4 × 44 k slides (Design ID: 026652). These arrays contain 41,000 unique probes targeting 27,958 Entrez gene RNAs. Following standard protocol, overnight hybridization at 55 °C was followed by a series of washes. The slides were scanned with an Agilent DNA microarray scanner and the features were extracted using the default setting of the Feature Extraction software (Feature Extraction software v.10.7, Agilent, CA, USA). The genes that displayed transcriptomic expressions at a fold change higher than 2 (fold change $\geq$ 2) were selected for further analysis.

Gene expression analysis used functions available in the Bioconductor Project [26] and functional heatmap tool (https://bioinfo-abcc.ncifcrf.gov/Heatmap/ (accessed on 26 August 2021). GeneSpring v.10.1 (Agilent Technologies, Inc., Santa Clara, CA, USA) was used for data visualization. Enrichment analysis was performed using Ingenuity Pathways Analysis (IPA, QIAGEN, Inc., Germantown, MD, USA). The data from this study was submitted to GEO under accession number GSE124756.

2.4. Gene Expression Validation by Nanostring Assays

A custom NanoString panel (NanoString Technologies, Seattle, WA, USA) was designed for genes deemed functionally important for the current study. The results and discussion section justify our choice of genes listed in Table S1. Six genes—*GIGYF2*, *INO80*, *USF2*, *WDR89*, *PPIA*, and *EIF2B1*—were selected as housekeeping genes based on their stable expression levels in melanocytes [27]. We followed the standard nCounter instructions [28], a master-mix containing hybridization buffer, Reporter ProbeSet, and Capture ProbeSet (volume:volume ratio of 1:1:0.5) was prepared, of which 25 μL was added to 5 μL target RNA. The GEN2 Prep Station incubation time was set at the higher sensitivity setting (3 h) and 280 fields of view (FOV) were routinely captured. Analysis and normalization of the raw NanoString data was conducted using nSolver Analysis Software v3.0 (NanoString Technologies, Seattle, WA, USA).

3. Results

3.1. Genomic Responses to the Three Toxins Are Characterized by Unique Host Expression Patterns

Principal component analysis (PCA) of transcriptomic expression data showed time-resolved clustering patterns of melanocytes exposed to three toxins for six treatment sequels (Figure 1). PC1 and PC2 represented 21.7% and 16.1% of the total variance; thus, together, PC1 and PC2 represent nearly 38% of the total variance. Within the transcriptomic variance defined by PC1 and PC2, we found three distinct clusters for each of the toxin types. These time points emerged clustered following longitudinal trends. For example, 30 min and 2 h SEA p.e. time points clustered together, and this combination was labelled as the early treatment phase. The early treatment phase was distantly located in the PCA plot from the middle treatment phase and was defined by 6 h and 12 h SEA p.e. time points. Finally, the late treatment phase was defined by 24 h and 48 h SEA p.e. time points, which were juxtaposed in the PCA landscape and distally located from the middle phase. A hypothetical line connecting these three treatment phases showed a potential temporal trend. A very similar picture emerged from TSST-1. The genes responding to SEB treatment, however, showed a different clustering pattern, which was more apparent in the late treatment phase of SEB. A considerable Euclidian distance was observed between 24 h and 48 h SEB p.e. Therefore, unlike SEA and TSST-1, we included 24 h SEB p.e. in the middle treatment phase along with its original two members, namely 6 h and 12 h SEB p.e. This arrangement automatically labelled 48 h SEB p.e. as the sole candidate of the SEB late-treatment phase. Interestingly, the middle-to-late treatment phases (12 h, 24 h, and 48 h) of SEA p.e. clustered closely to the middle treatment phases (6 h, 12 h, 24 h) of SEB p.e.

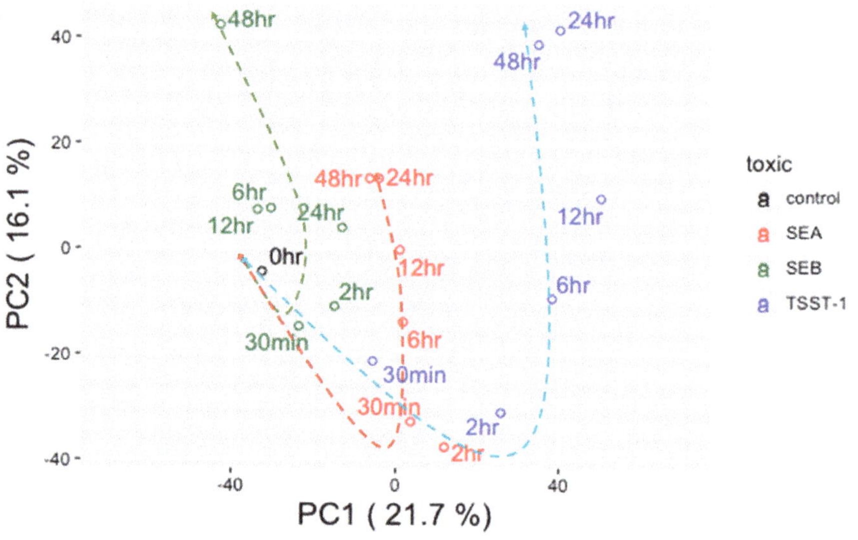

Figure 1. Principal components analysis (PCA) of time-resolved gene expression values. Black-, red-, green-, and blue-colored open circles represent control, SEA, SEB, and TSST-1, respectively. Dotted lines trace the temporal shifts caused by different toxins; here red, green, and blue dotted lines represent control, SEA, SEB, and TSST-1, respectively.

Since neither of the time point experiments have technical or biological replicates, the present strategy of grouping time sequela into the early, middle, and late treatment phases essentially enhanced the statistical confidence of the overall results. Using the longitudinal patterns of transcriptomic expressions, we sub-grouped the genes in three sets: (i) the 'Early' gene group, in which the transcriptomic fold changes were greater than |2| for at least one of the two time points (30 m and 2 h p.e.) of the early treatment phase; (ii) the 'Consistent' gene group, in which the transcriptomic fold changes were greater than |2| in all time points, and (iii) the 'Late' gene group, in which the transcriptomic fold changes were greater than |2| for at least one of the two time points (24 h and 48 h p.e.) of the late treatment phase. The exception was the SEB treatment, for which the late treatment phase included only 48 h p.e. Next, we combined (i) and (ii) to form 'Early–Consistent' gene groups; similarly, (ii) and (iii) were combined to form 'Late–Consistent' gene groups. These gene groups were used for functional analysis.

Figures S1A, S2A, and S3A depict Early–Consistent gene profiles of SEA, SEB, and TSST-1, respectively. Likewise, Figures S1B, S2B, and S3A depict Late–Consistent gene profiles of SEA, SEB, and TSST-1, respectively. A total of 445, 123, and 376 transcripts emerged, and these time clusters were called 'SEA—Early–Consistent' (SEA-E), 'SEB—Early–Consistent' (SEB–E), and 'TSST—Early–Consistent' (TSST-1-E), respectively. As explained above, the clustering for late-phase SEB exposure was performed differently than late-phase SEA and TSST-1 exposures. Hence, genes responding exclusively at 48 h p.e. (for SEB, Figure S2C) or in one of the two late p.e. phases (24 h or 48 h p.e. for SEA, Figure S1C, and TSST-1, Figure S3C) were combined with their respective consistently expressed genes (i.e., Figure S2A for SEB, Figure S1A for SEA, and Figure S3A for TSST-1). A total of 555, 1071, and 661 genes emerged, and they were called 'SEA—Late–Consistent' (SEA-L), 'SEB—Late–Consistent' (SEB-L) and 'TSST—Late–Consistent' (TSST-1-L), respectively.

3.2. Differences in Transcriptional Regulation in Response to the Three Toxins

In agreement with the PCA trend, the number of genes showing altered transcription varied greatly in response to the three toxins (Figure S4). Comparisons of early and late genomic responses to each of the toxins showed differences that were at their maximum after SEB treatment in both up- and downregulated genes. The largest number of genes responding with fold change (FC > |2|, nearly 1100 genes) were observed in SEB-L, whereas nearly 100 genes showed FC > |2| in SEB-E. In contrast, SEA-E and SEA-L comprised the least number of genes with FC > |2|. Nearly 450 genes showed transcriptomic modulations at early time points and nearly 600 genes were modulated during the late time points. Treatment with TSST-1 toxin elicited a response somewhat like SEA p.e. Interestingly, there was a common trend among all three toxins: the number of perturbed genes increased with the progression of treatment time, indicating the transcriptomic storm typically augmented by this family of sAgs [29–32] (Figure S4).

3.3. Biological Networks and Functions That Were Differentially Regulated by the Three Toxins

Functional analysis was performed using the genes listed under SEA-E, SEA-L, SEB-E, SEB-L, TSST-1-E, and TSST-1-L, respectively, to elucidate the time-dependent, toxin-specific enrichment profiles of biological and canonical functions. Table S2 lists the top biological functions ($p < 0.001$) and canonical networks ($p < 0.01$) associated with the three early treatment categories, SEA-E, SEB-E, and TSST-1-E. The list was filtered to include only those biological processes which were significantly enriched and functionally relevant to cell survival and the defense and maintenance of skin cells. In a similar fashion, genes belonging to the late treatment phase were probed to generate a list of significant biological and canonical processes that were enriched due to the prolonged toxin exposure (Table S3).

Table 1 lists the top biological pathways ($p < 0.001$) and canonical functions ($p < 0.01$) that represent melanocytes' dendritic cell-like (DC like) or macrophage-like property. 'Antigen presentation pathways', 'dendritic cell maturation', 'IL17 signalling', and 'chemokine signalling', among others, emerged as the top functions that are related to melanocytes' immunogenicity.

ILK signalling emerged as a significant network that was conserved between the early and late treatment phases in response to all three sAg. Functional annotation of the 36 genes (Table S4) associated with the ILK signalling pathway demonstrated association with two cellular processes, namely the cell death and tight junction signalling. Other networks that responded in common to at least two toxins and were conserved throughout the time–course of the study include acute phase response signaling, the antigen presentation pathway, the complement system, and agranulocyte adhesion and diapedesis.

The Venn diagram in Figure S5A elucidated those biological networks that were common among as well as exclusive to SEA-E, SEB-E, and TSST-1-E. Nine networks related to cell survival and maintenance were affected by all three toxins. SEA-E and TSST-1-E shared the largest number (28) of networks, including those, which were associated with endometriosis, proliferation of connective tissue cells, and angiogenesis. SEA-E and SEB-E shared the smallest group of networks (2), which were related to skin disorders such as chronic skin disorder and chronic psoriasis.

A Venn diagram of the functional annotation enriched by the three late treatment phases, namely SEA-L, SEB-L, and TSST-1-L, (Figure S5B), demonstrated a cohesive picture of the early treatment phase (Figure S5A). The number of overall annotated networks was greater for the late phase (87 as compared to 66 networks for the early treatment phase), as described in Tables S2 and S3. The largest number of networks was shared between SEA-L and TSST-1-L as in the early treatment phase, with similar enriched networks, namely endometriosis and proliferation of connective tissue cells. A total of 19 networks were commonly enriched for SEA-L and SEB-L; hence, the late treatment phase was associated with a higher number of significantly enriched gene networks than those associated with the early treatment phase.

Table 1. Biological pathways ($p < 0.001$) and canonical functions ($p < 0.01$) that represent melanocytes' dendritic cell-like (DC-like) or macrophage-like property. Networks which are perturbed by the toxins are double tick (√√) marked. In addition, the association of the networks with DC-like and/or macrophage-like properties are noted by single tick (√) mark.

Biological or Canonical Functions	Toxin			Biofunction Relevant to Which Melanocyte Character?	
	SEA	TSST	SEB	DC-Like	Macrophage-Like
Early					
Adhesion of blood cells	√√	√√		√	√
Antigen Presentation Pathway	√√	√√		√	√
Cdc42 Signalling	√√	√√		√	
cell movement of leukocytes	√√	√√		√	√
cell movement of phagocytes	√√	√√		√	√
Chemokine Signaling		√√	√√		√
chemotaxis of phagocytes	√√			√	√
Complement System	√√			√	√
Crosstalk between Dendritic Cells and Natural Killer Cells	√√			√	√
Dendritic Cell Maturation	√√			√	√
Differential Regulation of Cytokine Production in Macrophages and T Helper Cells by IL-17A and IL-17F	√√			√	√
ERK5 Signalling	√√				√
HMGB1 Signalling	√√			√	√
IL-17 Signalling	√√			√	√
IL-17A Signalling in Fibroblasts	√√			√	
IL-8 Signalling	√√				√
Immune response of cells	√√			√	√
Immune response of leukocytes	√√			√	√
Immune response of phagocytes			√√	√	√
Inflammatory response			√√	√	√
MAPKKK cascade			√√	√	√
Migration of phagocytes	√√			√	√
Oxidative Phosphorylation	√√			√	√
PDGF Signalling	√√				√
Proliferation of immune cells	√√			√	√
synthesis of prostaglandin	√√			√	
synthesis of prostaglandin E2	√√			√	

Table 1. *Cont.*

Biological or Canonical Functions	Toxin			Biofunction Relevant to Which Melanocyte Character?	
	SEA	TSST	SEB	DC-Like	Macrophage-Like
T-cell lymphoproliferative disorder	√√			√	√
Late					
Activation of blood cells	√√	√√		√	√
Adhesion of blood cells	√√	√√		√	√
Aggregation of blood cells	√√			√	√
Antigen Presentation Pathway	√√			√	√
Autophagy of cells	√√			√	√
Cell movement of connective tissue cells	√√				√
Cell movement of leukocytes	√√			√	√
Chemokine Signalling	√√			√	√
Chemotaxis of neutrophils	√√			√	√
Chemotaxis of phagocytes	√√			√	√
Complement System	√√			√	√
Crosstalk between Dendritic Cells and Natural Killer Cells	√√			√	√
Dendritic Cell Maturation	√√			√	√
Differentiation of hematopoietic progenitor cells	√√			√	√
eNOS Signalling	√√			√	√
IL-17 Signalling	√√			√	√
Immune response of cells	√√			√	√
Immune response of leukocytes	√√			√	√
Metabolism of eicosanoid	√√			√	√
Metabolism of prostaglandin			√√	√	
Migration of antigen presenting cells			√√	√	√
Migration of phagocytes			√√	√	√
Phagosome Maturation			√√	√	√
PI3K/AKT Signalling		√√		√	
Signalling by Rho Family GTPases		√√		√	√
Superoxide Radicals Degradation		√√		√	√
Synthesis of prostaglandin		√√		√	
Transmigration of phagocytes		√√		√	√

All three sAgs induced responses highly enriched for three biological processes: necrosis, skin diseases, and inflammation. Separate hierarchical clustering was performed using three gene sets, namely 217 genes from the necrosis network (Figure 2), 53 genes from the inflammation network (Figure S6), and 167 genes from the skin diseases network (Figure S7). The clustering analysis in Figure 2 identified four distinct groups of genes (indicated within yellow borders and labeled as groups A–D in Figure 2), which could be exclusive necrosis markers for TSST-1-L, TSST-1-E, SEA-L, and SEA-E, respectively. This hierarchical analysis failed to mine any exclusive signatures for SEB-E and SEB-L, respectively. Both the inflammation (Figure S6) and the skin diseases (Figure S7) clusters were mined as a single set, each under SEB-L (labelled group A in Figures S6 and S7, respectively). The complete list of all six gene sets is compiled in Table S5.

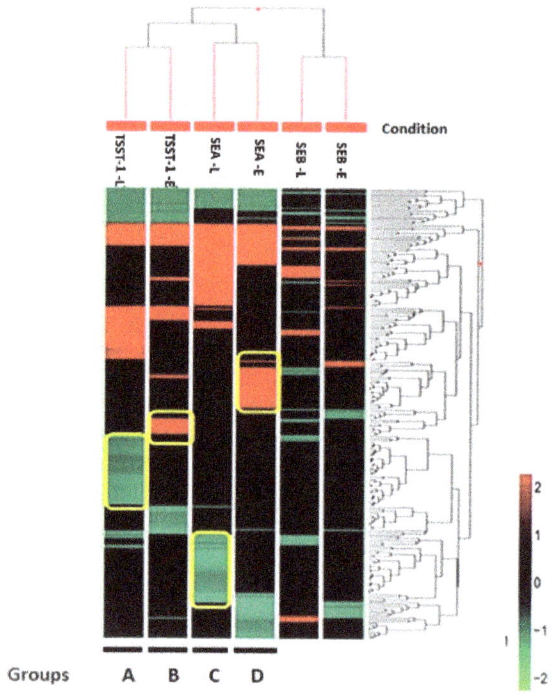

Figure 2. Hierarchical clustering analysis using of 217 genes with a $\log_2$ fold change > |2| enriching the necrosis pathway. The Euclidian algorithm was used to sort both conditions and genes. Each block represents one gene, and its color code is at the bottom right. Clusters bordered by yellow lines represent those genes which were potentially unique signatures of the particular condition. The conditions from left to right are named as TSST-1-L, TSST-1-E, SEA-L, SEA-E, SEB-L, and SEB-E, which represent TSST-1 at the late time point, TSST-1 at the early time point, SEA at the late time point, SEA at the early time point, SEB at the late time point, and SEB at the early time point, respectively.

3.4. *Confirmation of Expression Pattern for Select Genes from the Necrosis Clusters*

We performed validation of gene expression levels by NanoString nCounter® technology. Table S1 lists the top thirteen highly perturbed genes (up- and downregulated) grouped under necrosis. This list is limited to genes responding only to SEA and TSST-1 for two reasons: first, none of the genes responding in the SEB-E phase were grouped in the three clusters discussed above (Table S5), and second, a lack of sufficient RNA samples for the SEB 48 h treatment point forced us to exclude genes that belong to the SEB-L treatment phase.

Overall, a positive correlation was observed between the NanoString and microarrays results. Of the 13 genes tested, 12 genes followed the same directionality of fold changes for the NanoString and the microarray results (Figure S8) with the exception of one gene (PLCB1).

4. Discussion

The present study investigated in vitro host gene expression patterns induced by SEA, SEB, and TSST-1 during six time points ranging from 0.5 h to 48 h post-toxin exposure. A less frequently tested human skin cell type, but a major component of skin cell-mediated immunology, namely melanocytes, were selected as the target cells. The hybrid character of melanocytes was highlighted as we mined those biofunctions which were linked to the dendritic cell activities and/or the macrophage-based immune responses. This study could have benefited from incorporating additional time points to enhance the resolution of sequential biological events. For instance, our data suggested that the dosages of SEA and TSST-1 used for melanocyte treatments were potentially exhaustive within 24 h p.e.; in this context, extended time points could be highly informative. Furthermore, additional replicates in this study would result in better statistically significant gene identification. To mitigate this drawback to some extent, we mined the networks that met the cut off $p < 0.05$ using hypergeometric tests.

4.1. Distinct Temporal Trend of Pathogenesis Initiated by sAgs

The three toxins SEA, SEB, and TSST-1 of the sAg family are distinct in their structural, functional, and mechanistic properties [7,8,33]. Present literature not only lacks an understanding of molecular pathogenesis underlying the sAgs' toxicity, but also fails to fully comprehend the role of melanocytes in response to sAgs. The melanocytes' dendritic-like nature and their strategic location in the superficial layers of skin qualify them to be excellent mediators of initial immune defense against the sAgs [16–22]. We presented a whole genome-level investigation to compare the melanocytes temporal responses to SEA, SEB, and TSST-1.

A striking observation when comparing SEA and TSST-1 was the similarity in their gene expression patterns across the p.e. time course. Although SEA and TSST-1 share weak overall structural homology, TSST-1 can be displaced by SEA due to shared MHC class II binding sites [33]. This sheds light on the similarities in their mode of action as evidenced by the maximum number of shared networks for both early and late treatment phases.

Compared to SEA and TSST-1, the magnitude of transcriptional response perturbed by SEB was relatively smaller during the early treatment phase. However, the number of genes perturbed by SEB sequentially ramped up. This sort of delayed response is typical for any tissue that is not enriched with lymphocytes, as they are not the direct cellular targets of SEB [14]. Subsequently, SEB caused considerable genomic perturbations between 24 h and 48 h p.e. This trend is to be expected, as SEB typically causes a rapid neutrophil cell death accompanied by vascular congestion and leakage 24 h p.e., causing a shift to a predominantly adaptive immune response [30]. A perturbation in eNOS signalling pathways, potent vasodilators, was reported in the current study.

Another important observation was the temporal differences between SEA- and SEB-induced pathogenesis, particularly during their middle-to-late treatment phases (Figure 1). Nevertheless, a certain cohesiveness emerged between these two sAgs at the functional level. There were 11 and 19 networks that were synchronously enriched by both SEA and SEB at the early and late treatment phases (Figure S3A,B). This fact may demonstrate an underlying similarity in their mode of pathogenesis. Early pathogenesis caused by SEA- and SEB-perturbed genes manifested in skin disorders. In concurrence, SEB exclusively targeted genes linked to T-lymphocytes and their related functions, whereas SEA targeted glucose and protein metabolism networks. The consequences may include dysregulation of immune functions, apoptosis and cell death.

All three toxins enriched several networks related to cell death at early exposure phase and this response continues throughout the time course of the study. This response could be attributed to the moderately high doses of toxins used in the present study. Even though the three toxins perturbed the similar networks during the early exposure phase, as time progressed, each toxin had its unique mode of action in achieving the outcome manifested by cell death and apoptosis. One of the networks that was consistently perturbed by all three toxins across the p.e. time-course was ILK signalling. ILK functions as a kinase and signal transmitter or as a scaffold protein to facilitate cell–matrix interactions, cell signalling, and cytoskeletal organization [34]. These signals control processes related to survival, proliferation, differentiation, adhesion, migration, contractility, and neovascularization. Inhibition of ILK arrests the cell cycle and promotes apoptosis [35]. This is a key observation to support the following argument.

Early perturbation of genes associated with superoxide radical degradation in SEA indicates an oxidative stress-driven early onset of cell death [36]. TSST also perturbed this mechanism at later time points. Treatment with SEA and TSST down regulated the transcriptional levels of SOD1 and TYRP1, which potentially diminished the synthesis of different isoforms of superoxide dismutase (SODs). The potential loss of SODs highlighted the onset of oxidative stress initiated by the toxins [37], ultimately leading to onset of apoptosis during the late phase p.e.

Additional aspects of the apoptotic network, such as ERK/MAPK, were enriched by SEA and SEB at early p.e. phases, which appears to show a SE-induced apoptotic pathway distinct from that induced by TSST-1 [38,39]. We observed increased expression levels of FOS and NFAT genes during early p.e. SEA and SEB treatments. The FOS gene encodes the proto-oncogene c-FOS protein and NFATs, which are known widely for their cytokine gene expression properties and have been increasingly shown to regulate other genes related to cell cycle progression, cell differentiation, and apoptosis [39,40]. Late phase, SEB p.e. up-regulated genes that encode oncoproteins, such as Rho GTPase, which is also linked with ERK/MAPK [41]. Consequently, the G2/M DNA Damage Checkpoint Regulation, a critical biofunction closely linked with apoptosis, was highly perturbed. At late p.e. phase, SEA cross activated PI3K/AKT signaling, a critical pathway which affects many intracellular processes, including cell survival, growth, and migration.

4.2. Late Phase SEB Is Associated with Certain Dermatological Disorders

sAgs have long been implicated in the development of various inflammatory skin diseases such as psoriasis, atopic dermatitis, Kawasaki Syndrome, etc. [42,43] We observed that all three toxins modulated genes associated with the pathogenesis of psoriasis and chronic psoriasis starting from the early treatment phase. Psoriasis is often associated with functions like cell death, inflammation, autoimmune syndrome, and the production of ROS and nitric oxide [44]. From early to late treatment phases, SEA and TSST-1 shifted the expression of the gene enriching networks that are linked to lichen planus and endometriosis. During the late treatment phase, SEB regulated two unique set of genes that are closely linked to psoriasis and dermatomyositis, respectively. These genes are listed under their respective disease names in Table S3. Concurrent enrichment of oxidative stress networks could be related to the NRF2-mediated oxidative stress response and eNOS signaling pathways. Together these networks typically compromise the host's antioxidant defense mechanisms, a hallmark indicator of psoriasis [45].

4.3. Several Genes of Immunological Networks Are Differentially Modulated by Toxins

The skin exhibits a highly specialized innate immune response to invading pathogens and external stimuli. The major immune players—keratinocytes, Langerhans cells, dendritic cells, resident T-cells, and innate lymphoid cells—act in a coordinated fashion, from sensing the external stimuli to communicating through inflammatory signalling cascades, to ultimately regulating immune homeostasis [46,47]. Accumulating evidence uncovered a hybrid role of melanocytes in regulating innate and adaptive immunity [16–22,24,48–54]. Similar to keratinocytes, melanocytes express several types of toll-like receptors (TLRs) and have the ability to produce several pro-inflammatory cytokines and chemokines [48,52,54]. Melanocytes also regulate the adaptive immunity through their functional similarities to lysosomes, such as capability to phagocytose and their antigen presentation and processing aptitudes [20,48,55]. In this context, we listed those networks (Table 1) which are associated with melanocytes' hybrid role in responding to sAgs.

All of the toxin-induced adaptive immune responses could be attributed to the networks associated with leukocyte (granulocyte/agranulocyte) adhesion, a marker for second tier responses to inflammation induced by infection. Although all toxins contributed to adaptive immunity simulation, the patterns of cytokine production and acute-phase responses differed among the three toxins. For instance, during the early treatment phases of both SEA and SEB, the cytokine and chemokine signalling networks were comprised of CXCL1, CXCL12, and PLCB1, which control leukocyte trafficking; CCL2 and CCL7, which are involved in monocyte migration and macrophage recruitment; and CFL1, which regulates cell morphology and cytoskeletal organization. Early host responses to SEB and TSST-1 included an acute phase response signal that typically triggers non-specific inflammation, leukocytosis, complement activation, protease inhibition, clotting, etc. These responses persisted until 48 h p.e.

All three toxins perturbed IL-17 signalling, a pro-inflammatory signal that bridges innate and adaptive immune responses by playing critical roles in T-cell activation and in promoting the expansion and recruitment of innate immune cells, such as neutrophils [56]. The IL-17 signalling pathway was implicated in response to toxins via alterations of the transcription of several genes in this network, including CXCL1, CXCL5, CXCL8, CCL2, CCL20, and MAP2K6.

5. Conclusions

To our knowledge, this is the first mRNA-level study describing the temporal response of human melanocytes to three staphylococcal superantigenic toxins, namely SEA, SEB, and TSST-1. We observed distinct temporal patterns of transcriptomic regulation for the three individual toxins. The majority of the identified networks were related to necrosis and inflammation, in agreement with previous publications [38–40], although most of the past studies targeted different cells than melanocytes. Pathways related to innate immunity, such as the patterns of cytokine production and acute-phase response, showed toxin-specific regulation. The time-resolved response to SEB assault took a more differential pattern than SEA and TSST-1. In conclusion, these three toxins followed distinguishable pathways to achieve a common endpoint manifested by the cell death coordinated with apoptosis and necrosis. Hence, the temporal knowledge of their pathogenesis could be the key to customized intervention.

Supplementary Materials: The following supporting information can be downloaded at: https://www.mdpi.com/article/10.3390/biomedicines10061402/s1, Figure S1: Temporal profile of overall gene expression patterns in response to SEA: (A) Genes that follow the same expression pattern independent of the treatment time. (B) Genes with altered expression patterns only in one of the two early time points, i.e., 30 min or 2 h post-exposure. (C) Genes with altered expression patterns only toward the end of the exposure period, i.e., either 24 h or 48 h post-exposure; Figure S2: Temporal profile of the overall gene expression patterns in response to SEB: (A) genes that follow the same expression pattern independent of the treatment time. (B) Genes with altered expression patterns only in one of the two early time points, i.e., 30 min or 2 h post-exposure. (C) Genes with altered expression patterns only toward the end of the exposure period, i.e., 48 h post-exposure; Figure S3: Temporal profile of the overall gene expression patterns in response to TSST-1. (A) Genes that follow the same expression pattern independent of the treatment time. (B) Genes with altered expression patterns only in one of the two early time points, i.e., 30 min or 2 h post-exposure. (C) Genes with altered expression patterns only toward the end of the exposure period, i.e., either 24 h or 48 h post-exposure; Figure S4: Number of genes altered in the different experimental conditions. The stacked bar chart shows the number of over-expressed (Fold change > 2) and under-expressed (Fold change < -2) that are marked by black and white color; Figure S5: Venn diagram showing the temporal profiles of non-canonical pathways. (A) Common and unique non-canonical pathways enriched by the three sAgs at Early phases of pathogenesis. For instance, there are 17, 4, and 12 networks which are uniquely perturbed by SEA, SEB, and TSST-1 at early time points. There are 2 networks commonly perturbed by SEA and SEB. Likewise, 19 and 3 networks are commonly perturbed by SEA and TSST-1, and SEB and TSST-1, respectively. There are 9 non-canonical networks that were perturbed by all three sAgs. All of these networks are listed in the diagram. (B) Common and unique non-canonical pathways enriched by the three sAgs at late phases of pathogenesis. For instance, there are 24, 30 and 3 networks are uniquely perturbed by SEA, SEB and TSST-1 at early time points. There are 10 networks commonly perturbed by SEA and SEB. Likewise, 11 and 0 networks are commonly perturbed by SEA and TSST-1, and SEB and TSST-1, respectively. There are 9 non-canonical networks that were perturbed by all three sAgs. All of these networks are listed in the diagram; Figure S6: Hierarchical clustering analysis using of 53 genes with log2 fold change > |2| enriching the inflammation pathway. Euclidian algorithm is used to sort both conditions and genes. Each block represents one gene, and its color code is at the bottom right. The conditions from left to right are named as TSST-1-L, TSST-1-E, SEA-L, SEA-E, SEB-L and SEB-E, which represent TSST-1-L at late time point, TSST-1-E at early time point, SEA-L at late time point, SEA-E at early time point, SEB-L at late time point and SEB-E at early time point, respectively; Figure S7: Hierarchical clustering analysis using of 167 genes with log2 fold change > |2| enriching the skin disease pathway. The Euclidian algorithm is used to sort both conditions and genes. Each block represents one gene, and its color code is at the bottom right. The conditions from left to right are named as TSST-1-L, TSST-1-E, SEA-L, SEA-E, SEB-L and SEB-E; Figure S8: Targeted gene expression analysis using the NanoString platform to validate the microarray data. Validation study includes a set of 13 genes from three different conditions, namely SEA at early time point, SEA at late time point and TSST-1 at late time point. The color code profile is at the bottom left; Table S1: Top 13 highly perturbed genes grouped under the necrosis cluster; Table S2: Top biological functions and diseases ($p < 0.001$) and canonical functions ($p < 0.01$) identified through IPA for early post-exposure SEA, SEB, and TSST-1 treatments; Table S3: Top biological functions and diseases ($p < 0.001$) and canonical functions ($p < 0.01$) identified through IPA for late post-exposure SEA, SEB, and TSST-1 treatments ($p < 0.001$); Table S4: A list of 36 genes highly perturbed by one of the three sAgs during the early and late post-exposure phases; Table S5: List of significantly different genes that enrich necrosis, inflammation, and the skin diseases pathways, respectively. Genes are sorted by their fold changes, that is, log_2 transformed. The necrosis network is perturbed by TSST-1 at the early (TSST-1-E) and late (TSST-1-L) time points, and SEA at the early (SEA-E) and late (SEA-L) time points. Likewise, networks linked to skin disease and inflammation, respectively, are perturbed by SEB at the late time point (SEB-L).

Author Contributions: Conceptualization, R.H., J.W.S. and M.J.; methodology, R.Y. and N.C.; software, R.Y.; validation, L.J.D.; formal analysis, N.C., R.Y., and S.S.; investigation, S.S., A.G., S.-A.M., L.T.M. and B.C.C.; resources, R.H., J.W.S., R.Y. and M.J.; data curation, R.Y., N.C., S.S. and A.A.; writing—original draft preparation, N.C. and S.S.; writing—review and editing, N.C., A.G., R.Y., A.A. and J.W.S.; visualization, N.C. and S.S.; supervision, R.H., J.W.S., and M.J.; project administration, S.S., N.C. and A.G.; funding acquisition, R.H., J.W.S. and M.J. All authors have read and agreed to the published version of the manuscript.

Funding: This research was funded by the Defense Threat Reduction Agency, Project Number: G0020_04_WR_B.

Institutional Review Board Statement: Not Applicable.

Informed Consent Statement: Not Applicable.

Data Availability Statement: Data is contained within the article and supplementary files. Array data is available in GEO under accession number GSE124756.

Acknowledgments: All of the individual support from the Integrative Systems Biology Program, the US Army Center for Environmental Health Research, and the Geneva Foundation, and the editing assistance of Julia Scheerer, Derese Getnet, and Linda Brennan are deeply appreciated. We also thank Joshua Williams for his assistance with the Functional Heatmap tool.

Conflicts of Interest: The authors declare no conflict of interest.

Disclaimer: Material has been reviewed by the Walter Reed Army Institute of Research. There is no objection to its presentation and/or publication. The opinions or assertions contained herein are the private views of the author, and are not to be construed as official, or as reflecting the true views of the Department of the Army or the Department of Defense.

References

1. Kluytmans, J.; van Belkum, A.; Verbrugh, H. Nasal carriage of Staphylococcus aureus: Epidemiology, underlying mechanisms, and associated risks. *Clin. Microbiol. Rev.* **1997**, *10*, 505–520. [CrossRef] [PubMed]
2. Brown, A.F.; Leech, J.M.; Rogers, T.R.; McLoughlin, R.M. Staphylococcus aureus Colonization: Modulation of Host Immune Response and Impact on Human Vaccine Design. *Front. Immunol.* **2014**, *4*, 507. [CrossRef] [PubMed]
3. Ramarathnam, V.; De Marco, B.; Ortegon, A.; Kemp, D.; Luby, J.; Sreeramoju, P. Risk factors for development of methicillin-resistant Staphylococcus aureus infection among colonized patients. *Am. J. Infect. Control* **2013**, *41*, 625–628. [CrossRef]
4. McKinnell, J.A.; Huang, S.S.; Eells, S.J.; Cui, E.; Miller, L.G. Quantifying the impact of extranasal testing of body sites for methicillin-resistant Staphylococcus aureus colonization at the time of hospital or intensive care unit admission. *Infect. Control Hosp. Epidemiol.* **2013**, *34*, 161–170. [CrossRef] [PubMed]
5. Regenthal, P.; Hansen, J.S.; Andre, I.; Lindkvist-Petersson, K. Thermal stability and structural changes in bacterial toxins responsible for food poisoning. *PLoS ONE* **2017**, *12*, e0172445. [CrossRef]
6. Xu, S.X.; McCormick, J.K. Staphylococcal superantigens in colonization and disease. *Front. Cell. Infect. Microbiol.* **2012**, *2*, 52. [CrossRef]
7. Dinges, M.M.; Orwin, P.M.; Schlievert, P.M. Exotoxins of Staphylococcus aureus. *Clin. Microbiol. Rev.* **2000**, *13*, 16–34. [CrossRef]
8. Pinchuk, I.V.; Beswick, E.J.; Reyes, V.E. Staphylococcal Enterotoxins. *Toxins* **2010**, *2*, 2177–2197. [CrossRef]
9. Fraser, J.; Arcus, V.; Kong, P.; Baker, E.; Proft, T. Superantigens—Powerful modifiers of the immune system. *Mol. Med. Today* **2000**, *6*, 125–132. [CrossRef]
10. Krakauer, T. Immune response to staphylococcal superantigens. *Immunol. Res.* **1999**, *20*, 163–173. [CrossRef]
11. Mendis, C.; Das, R.; Hammamieh, R.; Royaee, A.; Yang, D.; Peel, S.; Jett, M. Transcriptional response signature of human lymphoid cells to staphylococcal enterotoxin B. *Genes Immun.* **2005**, *6*, 84–94. [CrossRef]
12. Vial, T.; Descotes, J. Immune-mediated side-effects of cytokines in humans. *Toxicology* **1995**, *105*, 31–57. [CrossRef]
13. Mattsson, E.; Herwald, H.; Egesten, A. Superantigens from Staphylococcus aureus induce procoagulant activity and monocyte tissue factor expression in whole blood and mononuclear cells via IL-1beta. *J. Thromb. Haemost.* **2003**, *1*, 2569–2576. [CrossRef] [PubMed]
14. Ferreyra, G.A.; Elinoff, J.M.; Demirkale, C.Y.; Starost, M.F.; Buckley, M.; Munson, P.J.; Krakauer, T.; Danner, R.L. Late Multiple Organ Surge in Interferon-Regulated Target Genes Characterizes Staphylococcal Enterotoxin B Lethality. *PLoS ONE* **2014**, *9*, e88756. [CrossRef] [PubMed]
15. Frieri, M.; Capetandes, A. Vascular Endothelial Cell Growth Factor (VEGF) Production by Staphylococcal Enterotoxin B (SEB) and IL-1β-Stimulated Human Keratinocytes is Inhibited by Pimecrolimus. *J. Allergy Clin. Immunol.* **2009**, *123*, S225. [CrossRef]
16. Sulaimon, S.S.; Kitchell, B.E. The biology of melanocytes. *Vet. Dermatol.* **2003**, *14*, 57–65. [CrossRef]
17. Plonka, P.M.; Passeron, T.; Brenner, M.; Tobin, D.J.; Shibahara, S.; Thomas, A.; Slominski, A.L.; Kadekaro, D.; Hershkovitz, E.; Peters, J.J.; et al. What are melanocytes really doing all day long . . . ? *Exp. Dermatol.* **2009**, *18*, 799–819. [CrossRef]
18. Mackintosh, J.A. The antimicrobial properties of melanocytes, melanosomes and melanin and the evolution of black skin. *J. Theor. Biol.* **2001**, *212*, 101–113. [CrossRef]
19. Al Badri, A.M.T.; Todd, P.M.; Garioch, J.J.; Gudgeon, J.E.; Stewart, D.G.; Goudie, R.B. An immunohistological study of cutaneous lymphocytes in vitiligo. *J. Pathol.* **1993**, *170*, 149–155. [CrossRef]
20. Le Poole, I.C.; Mutis, T.; van den Wijngaard, R.M.; Westerhof, W.; Ottenhoff, T.; De Vries, R.R.; Das, P.K. A novel, antigen-presenting function of melanocytes and its possible relationship to hypopigmentary disorders. *J. Immunol.* **1993**, *15*, 7284–7292.
21. Smit, N.; Le Poole, I.; van den Wijngaard, R.; Tigges, A.; Westerhof, W.; Das, P. Expression of different immunological markers by cultured human melanocytes. *Arch. Dermatol. Res.* **1993**, *285*, 356–365. [CrossRef] [PubMed]
22. Lu, Y.; Zhu, W.-Y.; Tan, C.; Yu, G.-H.; Gu, J.-X. Melanocytes are Potential Immunocompetent Cells: Evidence from Recognition of Immunological Characteristics of Cultured Human Melanocytes. *Pigment Cell Res.* **2002**, *15*, 454–460. [CrossRef] [PubMed]
23. Gasque, P.; Jaffar-Bandjee, M.C. The immunology and inflammatory responses of human melanocytes in infectious diseases. *J. Infect.* **2015**, *71*, 413–421. [CrossRef] [PubMed]
24. Tapia, C.V.; Falconer, M.; Tempio, F.; Falcón, F.; López, M.; Fuentes, M.; Alburquenque, C.; Amaro, J.; Bucarey, S.A.; Di Nardo, A. Melanocytes and melanin represent a first line of innate immunity against Candida albicans. *Med. Mycol.* **2014**, *52*, 445–454. [CrossRef] [PubMed]
25. Spaulding, A.R.; Salgado-Pabón, W.; Kohler, P.L.; Horswill, A.R.; Leung, D.Y.M.; Schlievert, P.M. Staphylococcal and Streptococcal Superantigen Exotoxins. *Clin. Microbiol. Rev.* **2013**, *26*, 422–447. [CrossRef] [PubMed]
26. Huber, W.; Carey, V.J.; Gentleman, R.; Anders, S.; Carlson, M.; Carvalho, B.S.; Bravo, H.C.; Davis, S.; Gatto, L.; Girke, T.; et al. Orchestrating high-throughput genomic analysis with Bioconductor. *Nat. Methods* **2015**, *12*, 115–121. [CrossRef] [PubMed]
27. Beard, R.E.; Abate-Daga, D.; Rosati, S.F.; Zheng, Z.; Wunderlich, J.R.; Rosenberg, S.A.; Morgan, R.A. Gene Expression Profiling using Nanostring Digital RNA Counting to Identify Potential Target Antigens for Melanoma Immunotherapy. *Clin. Cancer Res.* **2013**, *19*, 4941–4950. [CrossRef] [PubMed]
28. Veldman-Jones, M.H.; Brant, R.; Rooney, C.; Geh, C.; Emery, H.; Harbron, C.G.; Wappett, M.; Sharpe, A.; Dymond, M.; Barrett, J.C.; et al. Evaluating Robustness and Sensitivity of the NanoString Technologies nCounter Platform to Enable Multiplexed Gene Expression Analysis of Clinical Samples. *Cancer Res.* **2015**, *75*, 2587–2593. [CrossRef]

29. Dauwalder, O.; Thomas, D.; Ferry, T.; Debard, A.-L.; Badiou, C.; Vandenesch, F.; Etienne, J.; Lina, G.; Monneret, G. Comparative inflammatory properties of staphylococcal superantigenic enterotoxins SEA and SEG: Implications for septic shock. *J. Leukoc. Biol.* **2006**, *80*, 753–758. [CrossRef]
30. Bi, S.; Das, R.; Zelazowska, E.; Mani, S.; Neill, R.; Coleman, G.D.; Yang, D.C.; Hammamieh, R.; Shupp, J.W.; Jett, M. The Cellular and Molecular Immune Response of the Weanling Piglet to Staphylococcal Enterotoxin B. *Exp. Biol. Med.* **2009**, *234*, 1305–1315. [CrossRef]
31. Dauwalder, O.; Pachot, A.; Cazalis, M.A.; Paye, M.; Faudot, C.; Badiou, C.; Mougin, B.; Vandenesch, F.; Etienne, J.; Lina, G.; et al. Early kinetics of the transcriptional response of human leukocytes to staphylococcal superantigenic enterotoxins A and G. *Microb. Pathog.* **2009**, *47*, 171–176. [CrossRef]
32. Peterson, M.L.; Ault, K.; Kremer, M.J.; Klingelhutz, A.J.; Davis, C.C.; Squier, C.A.; Schlievert, P.M. The Innate Immune System Is Activated by Stimulation of Vaginal Epithelial Cells with *Staphylococcus aureus* and Toxic Shock Syndrome Toxin 1. *Infect. Immun.* **2005**, *73*, 2164–2174. [CrossRef] [PubMed]
33. Bavari, S.; Ulrich, R.G. Staphylococcal enterotoxin A and toxic shock syndrome toxin compete with CD4 for human major histocompatibility complex class II binding. *Infect. Immun.* **1995**, *63*, 423–429. [CrossRef] [PubMed]
34. Parsons, J.T.; Horwitz, A.R.; Schwartz, M.A. Cell adhesion: Integrating cytoskeletal dynamics and cellular tension. *Nat. Rev. Mol. Cell Biol.* **2010**, *11*, 633–643. [CrossRef] [PubMed]
35. Persad, S.; Attwell, S.; Gray, V.; Delcommenne, M.; Troussard, A.; Sanghera, J.; Dedhar, S. Inhibition of integrin-linked kinase (ILK) suppresses activation of protein kinase B/Akt and induces cell cycle arrest and apoptosis of PTEN-mutant prostate cancer cells. *Proc. Natl. Acad. Sci. USA* **2000**, *97*, 3207–3212. [CrossRef] [PubMed]
36. Nucci, L.A.; Santos, S.S.; Brunialti, M.; Sharma, N.; Machado, F.R.; Assunção, M.; De Azevedo, L.C.P.; Salomao, R. Expression of genes belonging to the interacting TLR cascades, NADPH-oxidase and mitochondrial oxidative phosphorylation in septic patients. *PLoS ONE* **2017**, *12*, e0172024. [CrossRef]
37. Miao, L.; St Clair, D.K. Regulation of superoxide dismutase genes: Implications in disease. *Free Radic. Biol. Med.* **2009**, *47*, 344–356. [CrossRef]
38. Below, S.; Konkel, A.; Zeeck, C.; Müller, C.; Kohler, C.; Engelmann, S.; Hildebrandt, J.-P. Virulence factors of *Staphylococcus aureus* induce Erk-MAP kinase activation and c-Fos expression in S9 and 16HBE14o- human airway epithelial cells. *Am. J. Physiol. Cell. Mol. Physiol.* **2009**, *296*, L470–L479. [CrossRef]
39. Viola, J.P.; Carvalho, L.D.; Fonseca, B.P.; Teixeira, L.K. NFAT transcription factors: From cell cycle to tumor development. *Braz. J. Med. Biol. Res.* **2005**, *38*, 335–344. [CrossRef]
40. Prager, G.; Hadamitzky, M.; Engler, A.; Doenlen, R.; Wirth, T.; Pacheco-Lopez, G.; Krügel, U.; Schedlowski, M.; Engler, H. Amygdaloid Signature of Peripheral Immune Activation by Bacterial Lipopolysaccharide or Staphylococcal Enterotoxin B. *J. Neuroimmune Pharmacol.* **2013**, *8*, 42–50. [CrossRef]
41. Zhang, S.; Han, J.; Sells, M.A.; Chernoff, J.; Knaus, U.G.; Ulevitch, R.J.; Bokoch, G.M. Rho Family GTPases Regulate p38 Mitogen-activated Protein Kinase through the Downstream Mediator Pak1. *J. Biol. Chem.* **1995**, *270*, 23934–23936. [CrossRef]
42. Yarwood, J.M.; Leung, D.Y.; Schlievert, P.M. Evidence for the involvement of bacterial superantigens in psoriasis, atopic dermatitis, and Kawasaki syndrome. *FEMS Microbiol. Lett.* **2000**, *192*, 1–7. [CrossRef]
43. Skov, L.; Baadsgaard, O. Bacterial superantigens and inflammatory skin diseases. *Clin. Exp. Dermatol.* **2000**, *25*, 57–61. [CrossRef] [PubMed]
44. Baz, K.; Cimen, M.B.; Kokturk, A.; Yazici, A.C.; Eskandari, H.G.; Ikizoglu, G.; Api, H.; Atik, U. Oxidant/Antioxidant Status in Patients with Psoriasis. *Yonsei Med. J.* **2003**, *44*, 987–990. [CrossRef]
45. Kadam, D.P.; Suryakar, A.N.; Ankush, R.D.; Kadam, C.Y.; Deshpande, K.H. Role of Oxidative Stress in Various Stages of Psoriasis. *Indian J. Clin. Biochem.* **2010**, *25*, 388–392. [CrossRef]
46. Pasparakis, M.; Haase, I.; Nestle, F.O. Mechanisms regulating skin immunity and inflammation. *Nat. Rev. Immunol.* **2014**, *14*, 289–301. [CrossRef] [PubMed]
47. Elias, P.M. The skin barrier as an innate immune element. *Semin. Immunopathol.* **2007**, *29*, 3–14. [CrossRef]
48. Ahn, J.H.; Park, T.J.; Jin, S.H.; Kang, H.Y. Human melanocytes express functional Toll-like receptor 4. *Exp. Dermatol.* **2008**, *17*, 412–417. [CrossRef] [PubMed]
49. Dessinioti, C.; Stratigos, A.J.; Rigopoulos, D.; Katsambas, A.D. A review of genetic disorders of hypopigmentation: Lessons learned from the biology of melanocytes. *Exp. Dermatol.* **2009**, *18*, 741–749. [CrossRef]
50. Hirobe, T. Structure and function of melanocytes: Microscopic morphology and cell biology of mouse melanocytes in the epidermis and hair follicle. *Histol. Histopathol.* **1995**, *10*, 223–237.
51. Hong, Y.; Song, B.; Chen, H.-D.; Gao, X.-H. Melanocytes and Skin Immunity. *J. Investig. Dermatol. Symp. Proc.* **2015**, *17*, 37–39. [CrossRef] [PubMed]
52. Jin, S.H.; Kang, H.Y. Activation of Toll-like Receptors 1, 2, 4, 5, and 7 on Human Melanocytes Modulate Pigmentation. *Ann. Dermatol.* **2010**, *22*, 486–489. [CrossRef] [PubMed]
53. Kruger-Krasagakes, S.; Krasagakis, K.; Garbe, C.; Diamantstein, T. Production of cytokines by human melanoma cells and melanocytes. *Recent Results Cancer Res.* **1995**, *139*, 155–168.
54. Tam, I.; Stępień, K. Secretion of proinflammatory cytokines by normal human melanocytes in response to lipopolysaccharide. *Acta Biochim. Pol.* **2011**, *58*, 507–511. [CrossRef] [PubMed]

55. Le Poole, I.C.; van den Wijngaard, R.P.; Westerhof, W.; Verkruisen, R.P.; Dutrieux, R.; Dingemans, K.P.; Das, P.K. Phagocytosis by Normal Human Melanocytes In Vitro. *Exp. Cell Res.* **1993**, *205*, 388–395. [CrossRef] [PubMed]
56. Yu, J.J.; Gaffen, S.L. Interleukin-17: A novel inflammatory cytokine that bridges innate and adaptive immunity. *Front. Biosci.* **2008**, *13*, 170–177. [CrossRef]

biomedicines

MDPI

Review

Novel Biomarkers, Diagnostic and Therapeutic Approach in Rheumatoid Arthritis Interstitial Lung Disease—A Narrative Review

Alesandra Florescu [1,†], Florin Liviu Gherghina [2,†], Anca Emanuela Mușetescu [1,*], Vlad Pădureanu [3,*], Anca Roșu [1], Mirela Marinela Florescu [4], Cristina Criveanu [1], Lucian-Mihai Florescu [5] and Anca Bobircă [6]

1 Department of Rheumatology, University of Medicine and Pharmacy of Craiova, 200349 Craiova, Romania; alesandra.florescu@umfcv.ro (A.F.); anca.rosu@umfcv.ro (A.R.); cristina.criveanu@umfcv.ro (C.C.)
2 Department of Physical Medicine and Rehabilitation, University of Medicine and Pharmacy of Craiova, 200349 Craiova, Romania; florin.gherghina@umfcv.ro
3 Department of Internal Medicine, University of Medicine and Pharmacy of Craiova, 200349 Craiova, Romania
4 Department of Pathology, University of Medicine and Pharmacy of Craiova, 200349 Craiova, Romania; mirela.florescu@umfcv.ro
5 Department of Radiology and Medical Imaging, University of Medicine and Pharmacy of Craiova, 200349 Craiova, Romania; lucian.florescu@umfcv.ro
6 Department of Rheumatology and Internal Medicine, Carol Davila University of Medicine and Pharmacy, 050474 Bucharest, Romania; anca.bobirca@umfcd.ro
* Correspondence: anca.musetescu@umfcv.ro (A.E.M.); vlad.padureanu@umfcv.ro (V.P.)
† All these authors contributed equally to this work.

Citation: Florescu, A.; Gherghina, F.L.; Mușetescu, A.E.; Pădureanu, V.; Roșu, A.; Florescu, M.M.; Criveanu, C.; Florescu, L.-M.; Bobircă, A. Novel Biomarkers, Diagnostic and Therapeutic Approach in Rheumatoid Arthritis Interstitial Lung Disease—A Narrative Review. *Biomedicines* **2022**, *10*, 1367. https://doi.org/10.3390/biomedicines10061367

Academic Editor: Marianna Christodoulou

Received: 25 April 2022
Accepted: 7 June 2022
Published: 9 June 2022

Publisher's Note: MDPI stays neutral with regard to jurisdictional claims in published maps and institutional affiliations.

Abstract: Rheumatoid arthritis (RA) is considered a systemic inflammatory disease marked by polyarthritis which affects the joints symmetrically, leading to progressive damage of the bone structure and eventually joint deformity. Lung involvement is the most prevalent extra-articular feature of RA, affecting 10–60% of patients with this disease. In this review, we aim to discuss the patterns of RA interstitial lung disease (ILD), the molecular mechanisms involved in the pathogenesis of ILD in RA, and also the therapeutic challenges in this particular extra-articular manifestation. The pathophysiology of RA-ILD has been linked to biomarkers such as anti-citrullinated protein antibodies (ACPAs), MUC5B mutation, Krebs von den Lungen 6 (KL-6), and other environmental factors such as smoking. Patients at the highest risk for RA-ILD and those most likely to advance will be identified using biomarkers. The hope is that finding biomarkers with good performance characteristics would help researchers better understand the pathophysiology of RA-ILD and, in turn, lead to the development of tailored therapeutics for this severe RA manifestation.

Keywords: rheumatoid arthritis; interstitial lung disease; biomarkers; treatment

1. Introduction

Rheumatoid arthritis (RA) is considered a systemic inflammatory disease marked by polyarthritis, which affects the joints symmetrically, leading to progressive damage of the bone structure and eventually joint deformity. This pathology affects around 1% of the population in the United States and northern Europe [1,2]. Even though arthritis is the most prevalent clinical manifestation of RA, extra-articular manifestations are often evidenced in people with the disease. Extra-articular manifestations include cardiac, ocular, lung, cutaneous, gastrointestinal, neurological, and renal involvement, but also rheumatoid vasculitis and rheumatoid nodules [3,4].

Lung involvement is the most prevalent extra-articular feature of RA, affecting 10–60% of patients with this disease. Any segment of the respiratory tract can be affected in RA patients. The involved segments include the parenchyma, which can cause ILD or rheumatoid nodules, the pleura, causing pleural effusions or inflammation, the small and large airways (bronchiolitis, bronchiectasis, and cricoarytenoid inflammation), but also the pulmonary vessels, resulting in vasculitis and pulmonary hypertension. ILD is considered to have a prevalence ranging from 5 to 58%, clinically overt RA-ILD being encountered in less than 50% of patients [2,5,6].

Pleural effusion was thought to be the most frequent feature of RA-ILD before the development of computed tomography (CT), which aids in assessing the correct diagnosis. High-resolution computed tomography (HRCT) can identify more subtle changes in the parenchyma, leading to earlier discovery of the ILD, especially in subclinical phases when the patients have not developed symptoms such as dyspnea [7–9].

The aim of this review is to present the patterns involved in RA-ILD and the molecular mechanisms described in the pathogenesis of this extra-articular manifestation. We also aim to present the diagnostic and therapeutic approach in patients with RA-ILD.

2. Pathogenesis

Rheumatoid factors (RF) and anti-citrullinated protein antibodies (ACPAs) are frequently found in the serum of RA patients. These autoantibodies are discovered in 50–80% of RA patients. They were discovered in the serum of patients with subclinical disease several years prior to clinical manifestations, thus testifying to the affirmation that genetic and environmental predispositions play an important part in the development of antibodies [10]. The production of antibodies leads to inflammation, followed by the development of clinical manifestations of the disease. Citrullination, the process through which arginine is converted to citrulline, leads to an immune response which implies the formation of ACPAs. ACPAs are significantly linked to the development of RA in those who are genetically susceptible [11,12].

Several immunopathogenic routes for RA-ILD have been proposed, although the precise location of the trigger event in the RA pathogenic cascade remains unknown. It is thought that the citrullinated proteins cross-react with the antigens in the lungs, albeit the immune response might be initiated in the synovium. This finding is reinforced by the fact that articular involvement precedes the pulmonary involvement in patients with RA. Recent literature data have shown that the microbiome plays a lead part in the development of RA due to its role in modulating the immune response. The "mucosal origins" theory posits that the development of RA begins in the mucosa of either the mouth, airway, or gastrointestinal tract. The bacterial, viral or mycobacterial antigens cross-react with antibodies, leading to the development of RA. Germs such as Proteus spp. and Porphyromonas gingivalis are thought to be involved in the pathogenesis of RA-ILD [13,14].

The genetic background of a patient might have either a predisposing (HLADRB1*15, HLADRB1*16, DQB1*06, and HLA-A*31:01 alleles) or protecting (HLA-DRB1 SE) role in the establishment of RA-ILD. Environmental conditions have a critical impact in genetically predisposed individuals. Tobacco usage has been identified as a probable cause of RA-ILD development. Smoking can harm pulmonary epithelial and vascular endothelial cells directly and increase citrullination of proteins in the lungs by activating PAD enzymes locally. Citrullinated proteins act as antigen targets, even in the preclinical stage, leading to a local immune response. This process leads to the formation of ACPAs, followed by the generation of RA and ILD. This stage is characterized by increased citrullination [15–17].

These formed antibodies lead to the development of an inflammatory response such as the production of pro-inflammatory cytokines, tumor necrosis factor (TNF)-α being one of the most important. B-cells are activated, and their differentiation is promoted by T-lymphocytes after antigen exposure. $CD4^+$ T cells infiltrates are more prominent in RA-IL than $CD3^+$ T lymphocytes, in contrast with idiopathic pulmonary fibrosis (IPF) infiltrates. Other researchers have speculated that $CD8^+$ T cells might have a role in the progression of pulmonary fibrosis in RA. Certain data attest to the fact that $CD8^+$ lymphocytes also have an important role in the development of ILD associated with RA, although this affirmation is portrayed in a study which suggests that smoking leads to an increase in $CD8^+$ T lymphocytes in the lungs. [15,18–20].

The lung cellular infiltrates in RA-ILD have proven to be complex in SKG mouse models, consisting of $CD4^+$ T lymphocytes, B lymphocytes, neutrophils, and macrophages. Cytokines and chemokines are of great importance in interstitial lung involvement in RA patients. TNF-α is a pro-inflammatory cytokine generated primarily by activated lymphocytes, macrophages, endothelial, and epithelial cells involved in the pathophysiology of ILD. TNF-α is important in the early stages and preservation of the cytokine and chemokine generation cascade and the induction of cell–cell adhesion and trans-endothelial migration [6,21].

The proliferation of fibroblasts is stimulated by TNF-α. Additionally, TNF-α promotes their capacity to degrade the extracellular matrix and to trigger the appearance of growth factors (GFs). GFs implied in the pathogenesis of ILD are platelet-derived growth factor (PDGF-β), transforming growth factor (TGF-β), but also vascular endothelial cell growth factor (VEGF). Nevertheless, the expression of cytokines such as IL-4 and IL-13, and chemokines (CXCL5, 8, 12, and 13) is also important. The GFs, cytokines and chemokines stimulate the fibroblasts to differentiate and proliferate, thus connecting the inflammatory and fibrotic stages. Macrophages, fibroblasts, epithelial, and endothelial cells all generate PDGF-β. PDGF-β is one of the pro-fibrotic, and pro-inflammatory molecules recognized to be important in the pathophysiology of ILD, such as TGF-β and TNF-α [5,22].

TGF-β's profibrotic effect is mediated via monocyte and fibroblast recruitment and activation and the stimulation of extracellular matrix deposition. TGF-β also causes fibroblasts to differentiate into myofibroblasts, which are the primary source of the extracellular matrix in the process of fibrosis of the lungs. Chemokines do not have a well-defined role in the formation of the inflammatory infiltrates in the lungs of patients with RA-ILD. These chemokines are produced by macrophages, fibroblasts, and epithelial cells, and they function by recruiting and activating fibroblasts.

The pro-fibrotic and/or pro-inflammatory cytokines and GFs are known to activate the Janus kinase (JAK)/ signal transducer and activator of transcription (STAT) pathway. JAK/STAT activation leads to the polarization of macrophages into pro-inflammatory M1 type macrophages, with increases the secretion of cytokines such as IL-6, CXCL10 and TNF-α. These pro-inflammatory cytokines promote inflammation and/or fibrotic changes.

Other mediators included in the pathogenesis are matrix metalloproteinases (MMPs) produced by damaged epithelia. MMPs maintain the crosstalk between inflammation and fibrosis by increasing the recruitment of cells such as B and T lymphocytes, macrophages, and neutrophils and producing additional pro-fibrotic mediators. The inflammatory process promotes the generation of the VEGF which aids the angiogenetic process. The exact mechanism of the generation of VEGF is still not well determined (Figure 1).

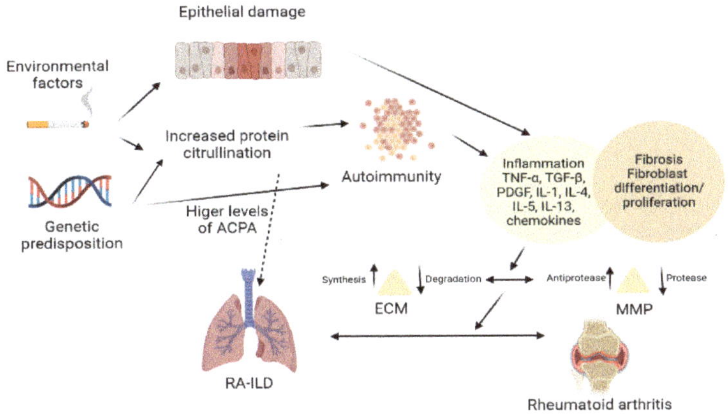

Figure 1. Pathogenesis of RA-ILD.

3. Biomarkers

3.1. Antibody Biomarkers

Patients with RA are known to have a preclinical stage in which autoantibodies such as RF and/or ACPA are detected in the serum, before the appearance of clinical synovial inflammation. However, the presence of the serological markers in the serum in situations when ILD develops before the articular manifestations might somehow be confusing [23].

RF are autoantibodies oriented against modified Fc segments of immunoglobulin (Ig) G. RFs are found in the bloodstream of up to 80% of patients with RA. The majority of RF consists of IgM antibodies, which are linked to the development of interstitial lung involvement in RA. IgA RFs have also been linked to ILD [24].

ACPAs have specificity for proteins in which peptidyl arginine deaminases (PAD) have transformed the arginine residues and are seen in the sera of 70–80% of RA patients. Anti-PAD3 antibodies have been associated with interstitial lung disease. ACPAs are reported to have greater specificity for RA than RFs. ACPA levels over a certain threshold are linked to ILD in RA. It has been claimed that IPF is related to the generation of IgA type ACPAs, albeit this has not been linked to ILD as a RA consequence. Circulating secretory IgA-ACPAs have also been found in the serum of RA patients with ILD [25,26].

Other RA-ILD ACPAs have been discovered in RA-ILD patients. Antibodies against the citrullinated alpha-enolase peptide 1 (anti-CEP1) were linked to RA-ILD in an Italian investigation. In a Chinese study, it was also shown that increased levels of anti-CEP antibodies contributed to the development of ILD in RA patients. Antibodies against anticitrullinated heat shock protein 90 (cit-Hsp90) α or β have also been linked to RA-ILD, with low sensitivity but high specificity. Patients with RA-ILD produced more interferon g (IFN-γ) than those without ILD when their peripheral blood mononuclear cells (PBMC) were grown in contact with cit-Hsp90 beta. IFN- γ was not discovered in the PBMC of patients with other connective tissue diseases (CTD)-ILD. This is due to the fact that IFN- γ production is increased by cit-Hsp90 T lymphocytes specific for RA-ILD [19,27,28].

Antibodies targeting additional post-translationally modified proteins have been reported in addition to citrullinated proteins. Antibodies against anti-carbamylated proteins (anti-CarP) have recently been linked to the development of RA-ILD. Four anti-CarP antibodies were often discovered with high serum titers: IgG anti-fetal calf serum (FCS), antichimeric fibrin/filaggrin homocitrullinated peptide, anti-fibrinogen, and IgA anti-FCS. Finally, antibodies against malondialdehyde-acetaldehyde (anti-MAA) have been linked to lung involvement in RA. Anti-MAA antibodies have been linked with increased disease activity and response to ACPA [29–31].

Patients with RA-ILD had greater plasma levels of IgA and IgM anti-MAA antibodies than those with RA without ILD. Levels of IgM anti-MAA antibodies were also higher in patients with RA-ILD than in those with lung disease not related to RA, such as chronic obstructive pulmonary disease (COPD) [32,33].

To summarize, there is no evidence that ACPAs have a role in RA-ILD risk. In clinical practice, ACPA positivity should not be used as a predictor for the risk of development of RA-ILD. However, high anti-CCP antibody titers and rheumatoid factor titers may assist in identifying individuals with RA who are at high risk of ILD [34].

3.2. Genetic Biomarkers

Not many papers have reported the genetic connections involving interstitial lung involvement in RA, even though genetic risk factors for RA or IPF have been thoroughly researched. A single nucleotide variation (SNV) in the promoter region of the MUC5B gene, rs35705950, has been linked to familial and sporadic IPF. Additionally, a link has been established between RA-ILD and the mutation of the MUC5B gene [35,36].

The MUC5B gene is overexpressed when this risk allele is present. MUC5B overproduction may impede alveolar repair. On the other hand, this risk allele has been linked to a better prognosis in IPF patients, depicting its relevance in moderate IPF. In order to attest the influent aspect of common variations in disease predisposition, genome-wide association studies (GWASs) have been developed. In a Japanese GWAS, the SNV rs12702634 in the RPA3-UMAD1 gene proved to have a major association with the development of RA-ILD. This genetic polymorphism was mostly associated with the UIP pattern [37–39].

A study conducted by Jönsson et al. on 1466 RA patients from northern Sweden analyzed 571151 SNVs, finding that 4 of the tested SNPs were associated with interstitial lung involvement in RA, as follows: rs35705950 (MUC5B gene), rs2609255 (FAM13A), rs111521887 (TOLLIP gene), and rs2736100 (TERT gene). However, more extensive studies on a larger number of patients are yet to be conducted [40].

The antigens are provided to the T-cell receptors by HLA molecules; thus, HLA alleles are connected to a wide range of diseases. IPF is linked to HLA-B*15, HLA-B*40, HLA-DR2 (DRB1*15 and DRB1*16), and MICA*001. RA is linked to HLA-DRB1*04:01, *04:04, *04:05, *01:01, and *10:01. These RA risk alleles are known as "shared epitope" (SE) alleles because they share amino acid sequences at positions 70–74 of the HLA-DR protein (QKRAA, RRRAA, or QRRAA). In RA, DR2 alleles have been found to predispose to ILD, whilst SE alleles have been found to protect against ILD. Even though SE alleles are closely linked to ACPA-positive RA, the frequency of these alleles is lower in RA patients with interstitial lung involvement [41].

Micro RNAs (miRNAs) control the expression of genes that code proteins and are non-coding RNAs formed from about 22 nucleotides. Circulating miRNAs are rapidly emerging as disease biomarkers in several illnesses. Plasma levels of hsa-miR-214-5p and hsa-miR-7-5p are elevated in RA and IPF. Additionally, the potential of long non-coding RNAs has been tested. They are transcripts that are longer than 200 nucleotides, but do not have the capacity to be translated into proteins. The levels of several of these long non-coding RNAs were likewise shown to be higher in RA-ILD patients' PBMCs [42,43].

3.3. Other Biomarkers

Krebs von den Lungen 6/MUC1 (KL-6) is a mucin-like glycoprotein which stimulates fibrosis and inhibits apoptosis of pulmonary fibroblasts. Serum KL-6 levels were shown to be higher in those with RA lung involvement, suggesting that it might help detect ILD development early on. In a study of 47 RA patients, the findings on lung computed tomography proved to be related to higher levels of serum KL-6 levels and increased disease severity. Severity was defined as extensive lung fibrosis on HRCT (>30%) or forced vital capacity (FVC) on PFT less than 50% and also the need of oxygen supplementation. Increased levels of KL-6 were also found in a study by Lee and colleagues in the serum of patients with CTD-ILD [44]. Type II pneumocytes and bronchiolar epithelial cells both express KL-6. KL-6 is expected to leak into the vascular system after epithelium breakdown caused by lung damage, indicating that it might be employed as a marker of epithelial injury. KL-6 might be used as a diagnostic marker in CTD-ILD. According to Oguz et al. in a study conducted on 113 CTD patients and 45 healthy controls, median KL-6 readings were significantly higher in the CTD-ILD group [44–48].

The pathophysiology of IPF is influenced by matrix metalloproteinases (MMPs) and tissue inhibitors of metalloproteinases (TIMPs), but also by cytokines and chemokines. MMP-7 levels were regularly observed to be higher in IPF patients. Several studies have looked at the involvement of these proteins in interstitial lung involvement in RA patients. High levels of MMP-7, soluble programmed death-ligand 1, C-X-C motif chemokine ligand 10 (CXCL10), interleukin (IL)-13, and IL-18 were discovered in the serum of patients with lung involvement in RA [49,50].

Chen et al. proved that MMP-7 and CXCL10 serum levels were more elevated in patients than those with RA without ILD [51]. Doyle et al. conducted a study which might help diagnose RA-ILD in the subclinical phase by discovering that a biomarker profile consisting of MMP-7, activation-regulation chemokines, and surfactant protein D (SP-D) is consistent with the development of ILD in RA patients [49,52,53].

Fu et al. discovered that lysyl oxidase-like 2 (LOXL2) levels in RA patients with or without ILD were higher in comparison with healthy controls. LOXL2 levels were substantially increased in subjects with RA-ILD who had ILD for ≤3 months than those who had ILD for >3 months [52]. The main candidates for biomarkers in RA-ILD are presented in Table 1.

Table 1. Value of biomarkers in RA-ILD.

Category	Biomarkers	Value	Evidence
Autoantibodies	ACPA	Diagnosis	High specificity—high titers associated with higher incidence of RA-ILD [54]
	Anticitrullinated HSP90	Diagnosis	Relatively high specificty and sensitivity [55]
	PAD	Severity	No utility for diagnosis—possible link to disease severity [56]
	Rheumatoid factor	Diagnosis	Low specificity—high titers associated with higher incidence of RA-ILD [57]
	Anti-CEP1	Diagnosis	More specific for synovial disease [58]
	Anti-CarP	Diagnosis	Relatively high specificity [59]
	Anti-MAA	Diagnosis	Not specific for RA-ILD—also found in RA without ILD [60]
Genetic biomarkers	MUC5B gene	Diagnosis	Highly suggestive, specifically in those with UIP pattern [61,62]
	microRNAs (has-miR-214-5p, has-miR-7-5p)	Diagnosis	High specificity, lower sensitivity—higher levels in patients with RA-ILD, not RA without ILD [63]
	HLA haplotypes (HLA-DR2, HLA-DQB1*04, *06, HLA-DR4, HLA-DRB1*14:06, HLA-DRB1*16:02-DQB1*05:02)	Diagnosis	Relatively low frequencies in RA-ILD [64]
Other biomarkers	KL-6	Severity	Highly suggestive for severity on HRCT [65,66]
	MMP-7	Diagnosis	Suggestive for fibrotic ILD—elevated in patients with ILD, not RA without ILD [41]
	CXCL10	Diagnosis	Elevated in patients with ILD, not RA without ILD [49]
	sPD-L1	Predictive	Relatively high specificity, lower sensitivity [67]
	IL-18	Diagnosis	Relatively high sensitivity and specificity [68]
	IL-13	Severity	Higher levels in RA-ILD, not RA without ILD [69]
	SP-D	Diagnosis	High specificity, lower sensitivity—influenced by bacterial lung infections [70]
	LOLX2	Diagnosis	High specificity for diagnosis [71]

ACPA—anticitrullinated protein antibodies; Anticitrullinated HSP90—heat shock protein 90; PAD—peptidyl arginine deaminases; anti-CEP1—citrullinated alpha-enolase peptide 1; anti-CarP—anti-carbamylated proteins; anti-MAA—anti- malondialdehyde-acetaldehyde; KL-6—Krebs von den Lungen 6/MUC1; MMP-7—matrix metalloproteinases 7; CXCL10—C-X-C motif chemokine ligand 10; sPD-1—soluble programmed death ligand 1; IL—interleukin; SP-D—surfactant protein D; LOLX2—lysyl oxidase-like 2.

In conclusion, each of these potential compounds, such as RF and ACPA, have some evidence of a link to RA-ILD. If any of these relationships are to be regarded as clinically effective biomarkers for RA-ILD, more research is needed to explain them and establish their validity. There are multiple ongoing clinical studies which aim to investigate biomarkers in RA-ILD, as presented in Table 2.

Table 2. Clinical study in RA-ILD biomarkers.

Clinical Study	Type	Primary Outcome	Secondary Outcome	Biomarkers	Status/Results
Soluble Programmed Death 1 (sPD1) is a Diagnostic Biomarker of ILD in Patients With Rheumatoid Arthritis Disease NCT05105230 Year: 2021	Observational, Retrospective, Case-control, NP—66	Evaluation of the levels of serum (sPD1) in RA patients and its association with ILD	Detection of subclinical RA-ILD for early diagnosis and management of this extra-articular manifestation	sPD1	Not yet recruiting
Deciphering Rheumatoid Arthritis -associated Interstitial Lung Disease Pathogenesis 2 (TRANSLATE2) Year: 2020	Observational, Prospective, Case-control, NP—500	Identification of genetic factors involved in RA-ILD using whole exome sequencing (WES).	Identification of genetic factors implicated in RA-ILD using genome wide association study. Description of RA-ILD natural history by a 5 years annual follow up of RA-ILD patients. Description of the effect of disease modifying Anti Rheumatic Drugs (DMARDs) on ILD course and mortality.	Genetic	Recruiting
Rheumatoid Arthritis Patients at Risk for Interstitial Lung Disease (RAPID) NCT03297775 Year: 2017	Observational, Prospective, Cohort, NP—750	Presence or absence of ILD on HRCT imaging. The study is designed to evaluate individuals affected by RA and explore associated lung disease so that the investigators can better understand the clinical phenotype and genetic and molecular endotypes of this disease.		Multiple biomarkers (genetic, serum, sputum)	Recruiting
Lysophosphatidic Acid (LPA)/Autotaxin (ATX) Axis in Rheumatoid Lung Disease (LYSLUNG) NCT04284735 Year:2020	Interventional, Non-randomized NP—40	ATX and LPA levels in sputum from RA patients with ILD in comparison with sputum from RA patients without ILD	Correlation between ATX and LPA levels and severity of RA-ILD estimated by tomodensitometry	LPA/ATX	Recruiting
Rheumatoid Arthritis-Associated Interstitial Lung Disease: Characterization of Lung Disease Progression (BERTHA) NCT04136223 Year: 2019	Observational, Prospective, Cohort NP—100	Interstitial Lung Disease progression—FVC	Interstitial Lung Disease progression—imaging, death, FVC dichotomous variable	Genetic	Recruiting

Table 2. *Cont.*

Clinical Study	Type	Primary Outcome	Secondary Outcome	Biomarkers	Status/Results
Factors of ILD in Newly Diagnosed Rheumatoid Arthritis (FIND-RA) NCT04002765 Year: 2019	Interventional, Single group assignment NP—300	Presence of an interstitial lung disease	Proportion of different ILD subtypes Proportion of patients with a non-ILD lung involvement of RA rs35705950 variant of the MUC5B promoter Anti-CCP antibodies Relevant exposure	rs35705950 variant of the MUC5B promoter anti-CCP antibodies	Recruiting
Effects of Tofacitinib vs. Methotrexate on Clinical and Molecular Disease Activity Markers in Joints and Lungs in Early Rheumatoid Arthritis (PULMORA)—A Randomized, Controlled, Open-label, Assessor-blinded, Phase IV Trial NCT04311567 Year: 2020	Interventional, Randomized, Parallel assignment NP—48	Change in total interstitial disease score of pulmonary abnormalities by HRCT	Change in extent of parenchymal lung disease by HRCT pattern Change in Forced Vital Capacity (FVC) Change in Diffusion Capacity of Carbon Monoxide (DLCO) Change in walking distance (meters) Change in blood oxygen saturation (SpO2) after 6 min walking Patient reported outcome of breathing and airway symptoms	Exploratory sub-study: Gene expression of bulk tissue and sorted cells of synovial biopsies and broncho-alveolar lavage samples by RNA sequencing. Levels of cytokines (IL-1β, IFN-α2, IFN-γ, TNF-α, IL-6, IL-8 (CXCL8), IL-10, IL-12p70, IL-17A, IL-18, IL-23, and IL-33), chemokines (CCL2, CCL3, CCL4, CCL5, CCL11, CCL17, CCL20, CXCL1, CXCL5, CXCL8, CXCL9, CXCL10, CXCL11) and growth factors (GM-CSF, PDGF and TGFbeta1) of synovial fluid, blood and broncho-alveolar lavage will be determined by bead-based immunoassay.	Recruiting
Change in Serum and Sputum Biomarkers Over Time in the Development of Rheumatoid Arthritis-associated Lung Disease NCT03616158 Year: 2019	Observational, NP—340	Define which patients with RA (or pre-RA) are at greatest risk for developing RA-related lung disease (LD)		Multiple biomarkers (genetic, serum, sputum)	Recruiting

4. Similarities between RA-ILD and IPF

RA-ILD has certain phenotypic similarities with IPF, unlike other CTD-associated ILD. First, some risk variables are shared by RA-ILD and IPF, the most important being smoking, followed by age and male sex. On the second hand, they have a similar imaging and pathology phenotype, with an apparent prevalence of the usual interstitial pneumonia (UIP) pattern, which is the most prevalent pattern of interstitial lung involvement in RA [72].

ACPAs have recently been discovered in patients with IPF. ACPA positivity was shown to be more common in two different IPF cohorts. In these two IPF cohorts, IgA-ACPA positivity was higher than in the general population control group. The concept of a common genetic foundation in RA interstitial lung involvement and IPF is supported by phenotypic resemblance and shared environmental risk factors [73]. An increase in rare variations in genes associated with familial pulmonary fibrosis has been identified in RA-ILD. The functional MUC5B rs35705950 promoter mutation has recently been described as a risk factor for RA-ILD, in addition to being a significant risk factor for IPF. Strong MUC5B staining was evidenced in lung samples from individuals with RA-ILD, located in the areas with alveolar epithelium hyperplasia in the fibrotic regions, comparable to that seen in IPF. According to immunohistochemistry, IgA-ACPA positivity was higher than IgG-ACPA positivity in patients with IPF, whereas IgG-ACPA positivity was higher than IgA-ACPA positivity in patients with RA [74,75].

5. Diagnosis of RA-ILD

The diagnosis of ILD in patients with diagnosed or suspected RA demands a coordinated multidisciplinary approach involving radiology, pathology, rheumatology, and pulmonology expertise, as well as consideration of other possible causes of ILD. Each specialist has a well determined role in the diagnosis and treatment of ILD. After a diagnosis of RA is established by the rheumatologist, the patients have to be thoroughly evaluated. A HRCT has to be performed and interpreted by a specialized radiologist and if alterations in the lung parenchyma are detected, a complete evaluation with pulmonary function tests (PFTs) has to be conducted by a pulmonologist. Regarding the treatment, collaboration between the rheumatologist and pulmonologist is of great importance, since the therapeutic arsenal is different in each specialty. Thus, frequent meetings and conferences, or even the formation of multidisciplinary teams, are of great importance in the diagnosis and treatment of RA-ILD [76].

5.1. Clinical Presentation

Exertional dyspnea, cough, chest discomfort, and exhaustion are symptoms of ILD that are similar to those of a variety of more frequent lung disorders.

In individuals with fibrotic ILD, a clinical evaluation might reveal digital clubbing and/or Velcro-crackles on lung auscultation. Up to 15% of patients with RA-ILD have been reported to present clubbing [77,78].

Patients with RA-ILD have been found to exhibit bilateral basal crackles in almost 90% of cases. Crackles were found in individuals with RA who did not have ILD, albeit to a lesser level. The complexity of the illness and the diversity in HRCT patterns are most likely responsible for the clinical variability [34].

5.2. Imaging

The use of a chest X-ray to detect ILD in RA patients is ineffective. On a thoracic radiograph, up to 64% of individuals with ILD on HRCT will have no visible interstitial abnormalities. As a result, if ILD is suspected, HRCT must be performed as part of the diagnostic process.

The UIP pattern is the most frequently encountered in RA-ILD, although all types of interstitial pneumonia have been described. UIP, obliterative bronchiolitis, nonspecific interstitial pneumonia (NSIP), and organizing pneumonia (OP) were identified as the four primary HRCT patterns in individuals with RA-ILD [79,80].

5.3. Phenotypes of RA-ILD

The most prevalent type of ILD is usual interstitial pneumonia, evidenced in up to 70% of cases. It is associated with worse outcomes in comparison with other RA-ILD patterns. UIP typical HRCT features include a subpleural distribution with a basal predominance, honeycombing, which is highly specific, reflecting the stage and the severity of the disease, reticular opacities associated with honeycombing and traction bronchiectasis, ground-glass opacities, which are usually less extensive than the reticular pattern, architectural distortion, and lobar volume loss [81].

Non-specific interstitial pneumonia is less prevalent than UIP. NSIP has two main subtypes: fibrotic and cellular, with lung involvement being mostly subpleural with an apicobasal gradient. NSIP typical HRCT features include ground-glass opacities with immediate subpleural sparing, mostly bilateral and symmetric, reticular opacities and irregular linear opacities, thickening of bronchovascular bundles, traction bronchiectasis, and lung volume loss, particularly in the lower lobes. It is associated with a lower risk of disease progression and a better response to treatment in comparison with UIP (Figure 2) [82].

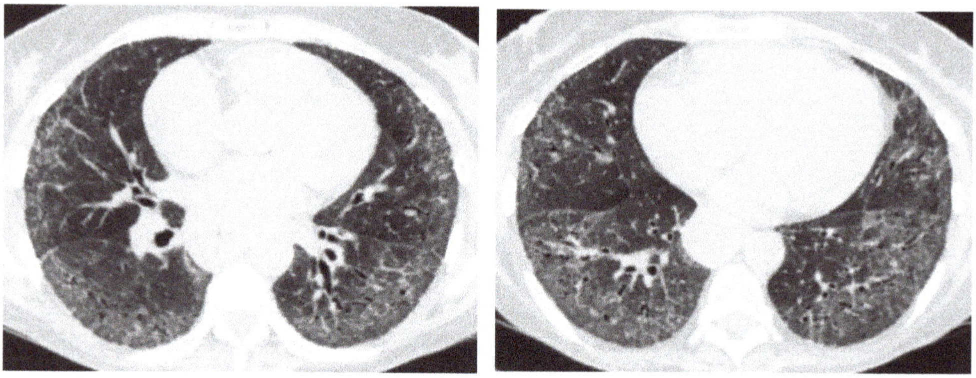

Figure 2. CT of the thorax—lung window—showing bilateral fine interstitial thickening and ground glass opacities with a basal predominance, minimal traction bronchiectasis and relative subpleural sparing (NSIP pattern).

Organizing pneumonia is a less frequent pattern encountered in RA-ILD. HRCT typical features include focal ground-glass opacities, consolidation and reversed halo sign.

Other less common patterns are lymphocytic interstitial pneumonia (LIP) and desquamative interstitial pneumonia (DIP) [83].

LIP may present HRCT features such as diffuse with mid to lower lobe predominance, interstitial thickening along lymph channels, thickening of the bronchovascular bundles pulmonary nodules, either centrilobular or subpleural, ground-glass opacities, and thin wall cysts.

DIP is characterized on HRCT by ground-glass opacities, irregular linear opacities, and small cystic spaces [84,85].

5.4. Pulmonary Function Tests

PFTs, especially the lung's carbon monoxide diffusing capacity (DLCO), are able to detect subclinical pulmonary disease. The presence of concomitant emphysema and the variability of the PFTs within the normal values, may restrict the use of this diagnostic method. PFT outcomes in individuals with RA-ILD vary depending on the research groups and severity of the illness. PFT abnormalities are present in 45–65% of individuals with RA, whether or not they have respiratory symptoms.

Restrictive patterns, but also airway obstruction, and decreased DLCO are among the patterns. The incidence of a restricted pattern ranges from 5 to 25%. Approximately 20–45% of people with RA have a DLCO that is impaired. Although many people have abnormal PFTs, most of these abnormalities are clinically inconsequential and silent [86,87].

5.5. Bronchoalveolar Lavage

In individuals with RA-ILD, the cellular characteristics of bronchoalveolar lavage (BAL) fluid are frequently aberrant but nonspecific. Lymphocytosis tends to be more common in the non-UIP pattern, while increased neutrophil levels are characteristic of the UIP pattern. BAL is not always required. Usually, it is conducted to rule out other causes of lung disease. The nonspecific results prevent this method from being a useful diagnostic tool [88].

5.6. Histopathology

Insights into the histopathological structure of interstitial pneumonia obtained through surgical lung biopsy (SLB) may help to clarify the diagnosis and might also have prognostic significance. However, the risks outweigh the benefits in some cases, and the decision to perform a SLB needs to be carefully considered [89].

In RA-ILD, the histological patterns are varied, and any kind of interstitial pneumonia can occur and even overlap. Idiopathic interstitial pneumonias are classified according to a variety of distinct histological characteristics that are also observed in RA-ILD. Patches of fibrosis with honeycombing and fibroblast foci alternate with patches of normal lung tissue in a UIP pattern marked by heterogeneity. The appearance of NSIP is uniform, with thickening of the alveolar septa and various degrees of inflammatory and fibrotic changes. DIP and follicular bronchiolitis include peribronchiolar inflammation and fibrosis, while intra-alveolar connective tissue plugs characterize OP [90].

6. Treatment

It is critical to carry out a baseline evaluation of disease severity in patients diagnosed with RA-ILD and closely monitor patients to identify those who develop disease progression. When selecting whether to start or continue therapy in individuals with RA-ILD, the severity and progression of the illness are two significant variables to consider.

The best therapeutic plan for RA-ILD patients has yet to be determined. There have been no randomized controlled trials (RCTs) comparing drugs for the therapeutic options of RA-ILD to date [91].

6.1. Corticosteroids, Synthetic, Biological, and Targeted Therapy

In patients with refractory disease the most frequently utilized therapeutic strategies consist of corticosteroids, azathioprine, and mycophenolate, with rituximab or TNF-α inhibitors. In RA-ILD with an inflammatory pattern, treatment response is frequently better. Fibrotic lung disease, for example RA-UIP, is usually less responsive to treatment and disease progression is similar to IPF [92].

Current therapy is primarily centered on immunosuppression and is based on empirical information. Corticosteroids are usually administered either in a daily oral dose or as pulse therapy. The dose is tapered over several months according to tolerance and clinical response. In inflammatory types of RA-ILD, such as NSIP and OP, corticosteroids have proved to have a limited effect on disease progression. TNF-alpha inhibitors, methotrexate (MTX), azathioprine (AZA), mycophenolate mofetil (MMF), and cyclophosphamide (CYC) are among the immunosuppressive medications used as maintenance therapy or in corticosteroid-resistant cases [93].

Therapy with corticosteroids alone or in combination with DMARDs alleviated or stabilized the disease in almost half of the 84 patients with RA-UIP, according to a retrospective search by Song et al., but there was no substantial difference in lifespan compared to the untreated group [94].

In rapidly advancing, severe ILD and RA-ILD with substantial UIP, cyclophosphamide in conjunction with methylprednisolone have shown potential efficacy; however, the data is based on a limited retrospective case series [22].

In RA patients, methotrexate is recommended as the first-line therapy, since it successfully reduces disease progression, disability, and mortality. MTX, on the other hand, has been linked to the development or worsening ILD in RA patients [95]. Kiely et al. intended to see whether treatment with MTX is linked to RA-ILD diagnosis and delays RA-ILD development. They found that MTX exposure was linked with a substantially lower incidence of RA-ILD in a multicenter prospective early RA cohort analysis involving 2701 participants. Furthermore, they discovered that therapy may help RA patients postpone the onset of ILD. This research offers us reason to believe that MTX may be helpful in the prevention and treatment of RA-ILD [96].

In a study conducted by Yusof et al., rituximab (RTX) was administered in 700 individuals with RA, 56 of whom already had RA-ILD. After receiving rituximab, 68% of these patients had improved or maintained pulmonary function. Rituximab was shown to have a good safety profile, only three individuals (0.4%) having developed following therapy [96,97].

Interstitial lung involvement induced by medication has been cited for most TNF agents, including infliximab, adalimumab, etanercept, certolizumab pegol, golimumab, and IL-6 receptor antagonist tocilizumab. The majority of evidence for TNF inhibitors-related ILD comes from case reports. A thorough literature quest revealed that establishing a causal link between RA treatment and the beginning or progression of ILD is extremely challenging [98].

However, due to the lack of a dedicated RCT, the effect of bDMARDs on RA-ILD is uncertain. Rituximab, tocilizumab, and abatacept have all been shown to have favorable results in recent studies, with the disease in treated individuals maintaining constant or improving as measured by PFTs. However, most of these investigations are small, uncontrolled retrospective studies, and their findings must be confirmed in RCTs [99–101].

The JAK/STAT pathway is incriminated in the development of ILD. The beneficial effect of JAK inhibitors on CTD-ILD has been reported in a number of case reports presented in recent literature, in mouse models and in a few clinical studies. An open-label trial conducted by Chen Z. et al. evaluated the efficiency of tofacitinib in amyopathic dermatomyositis associated with ILD in patients with anti-melanoma differentiation-associated gene 5 (MDA5). The study involved 18 patients treated with GC and tofacitinib in doses of 10mg/day, while 32 patients treated with GC alone were included as historical controls. The 6 month survival rate was significantly higher in the group treated with tofacitinib than in the control group. Favorable outcomes were also noted in the case of FVC, DLCO and findings on the HRCT in the study group [102].

D'Alessandro et al. conducted a study on 15 patients (out of which 4 were diagnosed with RA-ILD) with RA in order to evaluate the adipokine levels in RA patients after 6 months of baricitinib treatment. The study showed a significant decrease in KL-6 levels in the patients with ILD, also showing an improvement in DLCO. Although the RA-ILD group was too small to have statistical significance, the results of this study may be a cornerstone for the development of other trials [103]. Other case reports on ruxolitinib have shown improvement in PFTs and HRCT in patients with ILD [104–106]. However, more expensive RCTs have to be conducted in order to establish the beneficial effect of JAK inhibitors in RA-ILD and the potential adverse events.

6.2. Antifibrotic Therapy

Due to the mechanistic similarities between RA-related UIP and IPF, antifibrotic medication may have a beneficial effect on progressive fibrotic RA-ILD, particularly with UIP patterns. Antifibrotic drugs are not known to be beneficial for articular symptoms of the condition, thus immunomodulating therapy may be needed in addition. When a varied group of patients with progressive fibrotic ILD (PF-ILD) (other than IPF) were placed together as a single entity, the results of the INBUILD study suggested a therapeutic advantage with nintedanib in individuals displaying pulmonary disease progression. A post hoc assessment of all diagnostic categories (including some autoimmune-ILDs) revealed a treatment advantage (particularly, the rate of FVC decline) [107–110].

Another treatment option is represented by pirfenidone which lowers serum concentrations of IL-6 and TNF-alpha, two important cytokines in RA pathogenesis. A recent discovery suggested that pirfenidone prevents the transition from fibroblast to myofibroblast in the lung tissues of patients with ILD. Due to this fact, treatment with pirfenidone may be considered in the case of UIP patterns. According to recent studies, pirfenidone has a beneficial effect on disease progression by slowing it in patients with unclassifiable PF-ILD [111,112].

The main limitation of our review is the fact that it is a narrative review, therefore eligibility criteria for studies, search strategy, selection process, study risk of bias assessment, and data collection are not explained.

7. Conclusions

For RA patients, ILD is a frequent and sometimes fatal consequence. Unfortunately, the precise etiology of RA-ILD is not fully understood yet. The pathophysiology of RA-ILD has been linked to biomarkers such as ACPA, MUC5B mutation, KL-6, and other environmental factors such as smoking. Patients at the highest risk for RA-ILD and those most likely to advance will be identified using biomarkers. The hope is that finding biomarkers with good performance characteristics would help researchers better understand the pathophysiology of RA-ILD and, in turn, lead to the development of tailored therapeutics for this severe RA manifestation. Although multiple biomarkers have been studied, none have proven performance characteristics in order to reliably identify interstitial lung disease in RA patients. More studies have to be performed in order to establish and validate the clinical implications, sensitivity, specificity, utility in diagnosis, prognosis and disease severity.

Author Contributions: Conceptualization, A.F. and F.L.G.; methodology, A.E.M., V.P. and A.B.; validation, C.C. and A.R.; investigation, L.-M.F. and M.M.F.; data curation, A.F. and A.E.M.; writing—original draft preparation, L.-M.F., M.M.F. and C.C.; writing—review and editing, A.R., F.L.G. and A.F.; supervision, A.E.M., V.P. and A.B. All authors have read and agreed to the published version of the manuscript.

Funding: The Article Processing Charges were funded by the University of Medicine and Pharmacy of Craiova, Romania.

Institutional Review Board Statement: Not applicable.

Informed Consent Statement: Not applicable.

Data Availability Statement: Not applicable.

Conflicts of Interest: The authors declare no conflict of interest.

References

1. Huang, S.; Kronzer, V.L.; Dellaripa, P.F.; Deane, K.D.; Bolster, M.B.; Nagaraja, V.; Khanna, D.; Doyle, T.J.; Sparks, J.A. Rheumatoid Arthritis–Associated Interstitial Lung Disease: Current Update on Prevalence, Risk Factors, and Pharmacologic Treatment. *Curr. Treat. Options Rheumatol.* **2020**, *6*, 337–353. [CrossRef] [PubMed]
2. Marcucci, E.; Bartoloni, E.; Alunno, A.; Leone, M.C.; Cafaro, G.; Luccioli, F.; Valentini, V.; Valentini, E.; la Paglia, G.M.C.; Bonifacio, A.F.; et al. Extra-Articular Rheumatoid Arthritis. *Reumatismo* **2018**, *70*, 212–224. [CrossRef] [PubMed]
3. Marigliano, B.; Soriano, A.; Margiotta, D.; Vadacca, M.; Afeltra, A. Lung Involvement in Connective Tissue Diseases: A Comprehensive Review and a Focus on Rheumatoid Arthritis. *Autoimmun. Rev.* **2013**, *12*, 1076–1084. [CrossRef] [PubMed]
4. Cojocaru, M.; Cojocaru, I.M.; Silosi, I.; Vrabie, C.D.; Tanasescu, R. Extra-Articular Manifestations in Rheumatoid Arthritis. *Maedica* **2010**, *5*, 286. [CrossRef]
5. Ha, Y.J.; Lee, Y.J.; Kang, E.H. Lung Involvements in Rheumatic Diseases: Update on the Epidemiology, Pathogenesis, Clinical Features, and Treatment. *BioMed. Res. Int.* **2018**, *2018*, 6930297. [CrossRef]

6. Dai, Y.; Wang, W.; Yu, Y.; Hu, S. Rheumatoid Arthritis–Associated Interstitial Lung Disease: An Overview of Epidemiology, Pathogenesis and Management. *Clin. Rheumatol.* **2021**, *40*, 1211–1220. [CrossRef]
7. Spagnolo, P.; Lee, J.S.; Sverzellati, N.; Rossi, G.; Cottin, V. The Lung in Rheumatoid Arthritis: Focus on Interstitial Lung Disease. *Arthritis Rheumatol.* **2018**, *70*, 1544–1554. [CrossRef]
8. Azhar, M.; Muhammad, Q.; Halilu, F.; Woodford, M. Interstitial lung disease as the only manifestation of rheumatoid arthritis: Can lungs be the site of origin for rheumatoid arthritis-related inflammation? *Chest* **2021**, *160*, A2090. [CrossRef]
9. Esposito, A.J.; Chu, S.G.; Madan, R.; Doyle, T.J.; Dellaripa, P.F. Thoracic Manifestations of Rheumatoid Arthritis. *Clin. Chest Med.* **2019**, *40*, 545–560. [CrossRef]
10. Petrovská, N.; Prajzlerová, K.; Vencovský, J.; Šenolt, L.; Filková, M. The Pre-Clinical Phase of Rheumatoid Arthritis: From Risk Factors to Prevention of Arthritis. *Autoimmun. Rev.* **2021**, *20*, 102797. [CrossRef]
11. Bongartz, T.; Cantaert, T.; Atkins, S.R.; Harle, P.; Myers, J.L.; Turesson, C.; Ryu, J.H.; Baeten, D.; Matteson, E.L. Citrullination in Extra-Articular Manifestations of Rheumatoid Arthritis. *Rheumatology* **2007**, *46*, 70–75. [CrossRef] [PubMed]
12. Darrah, E.; Andrade, F. Rheumatoid Arthritis and Citrullination. *Curr. Opin. Rheumatol.* **2018**, *30*, 72. [CrossRef] [PubMed]
13. Manfredi, A.; Luppi, F.; Cassone, G.; Vacchi, C.; Salvarani, C.; Sebastiani, M. Pathogenesis and Treatment of Idiopathic and Rheumatoid Arthritis-Related Interstitial Pneumonia. The Possible Lesson from COVID-19 Pneumonia. *Expert Rev. Clin. Immunol.* **2020**, *16*, 751–770. [CrossRef] [PubMed]
14. Holers, V.M.; Demoruelle, M.K.; Kuhn, K.A.; Buckner, J.H.; Robinson, W.H.; Okamoto, Y.; Norris, J.M.; Deane, K.D. Rheumatoid Arthritis and the Mucosal Origins Hypothesis: Protection Turns to Destruction. *Nat. Rev. Rheumatol.* **2018**, *14*, 542–557. [CrossRef]
15. Johnson, C. Recent Advances in the Pathogenesis, Prediction, and Management of Rheumatoid Arthritis-Associated Interstitial Lung Disease. *Curr. Opin. Rheumatol.* **2017**, *29*, 254–259. [CrossRef]
16. Bendstrup, E.; Møller, J.; Kronborg-White, S.; Prior, T.S.; Hyldgaard, C. Interstitial Lung Disease in Rheumatoid Arthritis Remains a Challenge for Clinicians. *J. Clin. Med.* **2019**, *8*, 2038. [CrossRef]
17. Khan, T.; Jose, R.J.; Renzoni, E.A.; Mouyis, M. A Closer Look at the Role of Anti-CCP Antibodies in the Pathogenesis of Rheumatoid Arthritis-Associated Interstitial Lung Disease and Bronchiectasis. *Rheumatol. Ther.* **2021**, *8*, 21.
18. McDermott, G.C.; Doyle, T.J.; Sparks, J.A. Interstitial Lung Disease throughout the Rheumatoid Arthritis Disease Course. *Curr. Opin. Rheumatol.* **2021**, *33*, 284–291. [CrossRef]
19. Pruijn, G.J.M. Citrullination and Carbamylation in the Pathophysiology of Rheumatoid Arthritis. *Front. Immunol.* **2015**, *6*, 192. [CrossRef]
20. Cavagna, L.; Monti, S.; Grosso, V.; Boffini, N.; Scorletti, E.; Crepaldi, G.; Caporali, R. The Multifaceted Aspects of Interstitial Lung Disease in Rheumatoid Arthritis. *BioMed Res. Int.* **2013**, *2013*, 13. [CrossRef]
21. Mizutani, S.; Nishio, J.; Kondo, K.; Motomura, K.; Yamada, Z.; Masuoka, S.; Yamada, S.; Muraoka, S.; Ishii, N.; Kuboi, Y.; et al. Treatment with an Anti-Cx3cl1 Antibody Suppresses M1 Macrophage Infiltration in Interstitial Lung Disease in Skg Mice. *Pharmaceuticals* **2021**, *14*, 474. [CrossRef] [PubMed]
22. Kadura, S.; Raghu, G. Rheumatoid Arthritis-Interstitial Lung Disease: Manifestations and Current Concepts in Pathogenesis and Management. *Eur. Respir. Rev.* **2021**, *30*, 210011. [CrossRef] [PubMed]
23. Kuwana, M.; Gil-Vila, A.; Selva-O'Callaghan, A. Role of Autoantibodies in the Diagnosis and Prognosis of Interstitial Lung Disease in Autoimmune Rheumatic Disorders. *Ther. Adv. Musculoskelet. Dis.* **2021**, *13*, 1759720X211032457. [CrossRef]
24. Furukawa, H.; Oka, E.; Toma, S. Biomarkers for Rheumatoid Arthritis Associated Lung Disease. *Clin. Rheumatol. Relat. Res.* **2020**, *32*, 318–322. [CrossRef]
25. Lucchino, B.; di Paolo, M.; Gioia, C.; Vomero, M.; Diacinti, D.; Mollica, C.; Alessandri, C.; Diacinti, D.; Palange, P.; di Franco, M. Identification of Subclinical Lung Involvement in ACPA-Positive Subjects through Functional Assessment and Serum Biomarkers. *Int. J. Mol. Sci.* **2020**, *21*, 5162. [CrossRef] [PubMed]
26. Solomon, J.J.; Matson, S.; Kelmenson, L.B.; Chung, J.H.; Hobbs, S.B.; Rosas, I.O.; Dellaripa, P.F.; Doyle, T.J.; Poli, S.; Esposito, A.J.; et al. IgA Antibodies Directed Against Citrullinated Protein Antigens Are Elevated in Patients with Idiopathic Pulmonary Fibrosis. *Chest* **2020**, *157*, 1513–1521. [CrossRef] [PubMed]
27. Tanguy, J.; Pommerolle, L.; Garrido, C.; Kolb, M.; Bonniaud, P.; Goirand, F.; Bellaye, P.S. Extracellular Heat Shock Proteins as Therapeutic Targets and Biomarkers in Fibrosing Interstitial Lung Diseases. *Int. J. Mol. Sci.* **2021**, *22*, 9316. [CrossRef] [PubMed]
28. Samara, K.D.; Trachalaki, A.; Tsitoura, E.; Koutsopoulos, A.V.; Lagoudaki, E.D.; Lasithiotaki, I.; Margaritopoulos, G.; Pantelidis, P.; Bibaki, E.; Siafakas, N.M.; et al. Upregulation of Citrullination Pathway: From Autoimmune to Idiopathic Lung Fibrosis. *Respir. Res.* **2017**, *18*, 218. [CrossRef]
29. Mohamed, A.; Enein, A.; Abdelsalam, N.; Balata, M.; Abdellatif, S.; Rizk, E.; Fathy, A. Utility of Anti-Carbamylated Protein Antibodies in the Diagnosis of Early Rheumatoid Arthritis. *Indian J. Rheumatol.* **2019**, *14*, 37. [CrossRef]
30. Nakabo, S. Clinical and Etiological Meaning of Anti-Carbamylated Protein Antibodies in Rheumatoid Arthritis. *Immunol. Med.* **2018**, *41*, 147–153. [CrossRef]
31. Wu, C.Y.; Yang, H.Y.; Luo, S.F.; Lai, J.H. From Rheumatoid Factor to Anti-Citrullinated Protein Antibodies and Anti-Carbamylated Protein Antibodies for Diagnosis and Prognosis Prediction in Patients with Rheumatoid Arthritis. *Int. J. Mol. Sci.* **2021**, *22*, 686. [CrossRef] [PubMed]

32. Mikuls, T.R.; Edison, J.; Meeshaw, E.; Sayles, H.; England, B.R.; Duryee, M.J.; Hunter, C.D.; Kelmenson, L.B.; Kay Moss, L.; Feser, M.L.; et al. Autoantibodies to Malondialdehyde–Acetaldehyde Are Detected Prior to Rheumatoid Arthritis Diagnosis and After Other Disease Specific Autoantibodies. *Arthritis Rheumatol.* **2020**, *72*, 2025–2029. [CrossRef] [PubMed]
33. Thiele, G.M.; Duryee, M.J.; Anderson, D.R.; Klassen, L.W.; Mohring, S.M.; Young, K.A.; Benissan-Messan, D.; Sayles, H.; Dusad, A.; Hunter, C.D.; et al. Malondialdehyde-Acetaldehyde Adducts and Anti-Malondialdehyde-Acetaldehyde Antibodies in Rheumatoid Arthritis. *Arthritis Rheumatol.* **2015**, *67*, 645–655. [CrossRef] [PubMed]
34. Brito, Y.; Glassberg, M.K.; Ascherman, D.P. Rheumatoid Arthritis-Associated Interstitial Lung Disease: Current Concepts. *Curr. Rheumatol. Rep.* **2017**, *19*, 79. [CrossRef]
35. Furukawa, H.; Oka, S.; Higuchi, T.; Shimada, K.; Hashimoto, A.; Matsui, T.; Tohma, S. Biomarkers for Interstitial Lung Disease and Acute-Onset Diffuse Interstitial Lung Disease in Rheumatoid Arthritis. *Ther. Adv. Musculoskelet. Dis.* **2021**, *13*, 1–12. [CrossRef]
36. Oliveira, R.P.; Ribeiro, R.; Melo, L.; Grima, B.; Oliveira, S.; Alves, J.D. Connective Tissue Disease-Associated Interstitial Lung Disease. *Pulmonology* **2020**, *28*, 113–118. [CrossRef]
37. Saad, M.N.; Mabrouk, M.S.; Eldeib, A.M.; Shaker, O.G. Identification of Rheumatoid Arthritis Biomarkers Based on Single Nucleotide Polymorphisms and Haplotype Blocks: A Systematic Review and Meta-Analysis. *J. Adv. Res.* **2016**, *7*, 1–16. [CrossRef]
38. Deane, K.D.; Demoruelle, M.K.; Kelmenson, L.B.; Kuhn, K.A.; Norris, J.M.; Holers, V.M. Genetic and Environmental Risk Factors for Rheumatoid Arthritis. *Best Pract. Res. Clin. Rheumatol.* **2017**, *31*, 3–18. [CrossRef]
39. Shirai, Y.; Honda, S.; Ikari, K.; Kanai, M.; Takeda, Y.; Kamatani, Y.; Morisaki, T.; Tanaka, E.; Kumanogoh, A.; Harigai, M.; et al. Association of the RPA3-UMAD1 Locus with Interstitial Lung Diseases Complicated with Rheumatoid Arthritis in Japanese. *Ann. Rheum. Dis.* **2020**, *79*, 1305–1309. [CrossRef]
40. Jönsson, E.; Ljung, L.; Norrman, E.; Freyhult, E.; Lisbeth, L.; Dahlqvist, J.; Rantapää-Dahlqvist, S. Pulmonary Fibrosis in Relation to Genetic Loci in an Inception Cohort of Patients with Early Rheumatoid Arthritis from Northern Sweden. *Rheumatology* **2022**, *61*, 943–952. [CrossRef]
41. Chen, J.; Doyle, T.J.; Liu, Y.; Aggarwal, R.; Wang, X.; Shi, Y.; Ge, S.X.; Huang, H.; Lin, Q.; Liu, W.; et al. Biomarkers of Rheumatoid Arthritis-Associated Interstitial Lung Disease. *Arthritis Rheumatol.* **2015**, *67*, 28–38. [CrossRef] [PubMed]
42. Moran-Moguel, M.C.; del Rio, S.P.; Mayorquin-Galvan, E.E.; Zavala-Cerna, M.G. Rheumatoid Arthritis and MiRNAs: A Critical Review through a Functional View. *J. Immunol. Res.* **2018**, *2018*, 2474529. [CrossRef] [PubMed]
43. Konigsberg, I.R.; Borie, R.; Walts, A.D.; Cardwell, J.; Rojas, M.; Metzger, F.; Hauck, S.M.; Fingerlin, T.E.; Yang, I.V.; Schwartz, D.A. Molecular Signatures of Idiopathic Pulmonary Fibrosis. *Am. J. Respir. Cell Mol. Biol.* **2021**, *65*, 430–441. [CrossRef] [PubMed]
44. Lee, J.S.; Lee, E.Y.; Ha, Y.J.; Kang, E.H.; Lee, Y.J.; Song, Y.W. Serum Kl-6 Levels Reflect the Severity of Interstitial Lung Disease Associated with Connective Tissue Disease. *Arthritis Res. Ther.* **2019**, *21*, 58. [CrossRef]
45. Amigues, I.; Ramadurai, D.; Swigris, J.J. Current Perspectives on Emerging Biomarkers for Rheumatoid Arthritis-Associated Interstitial Lung Disease. *Open Access Rheumatol. Res. Rev.* **2019**, *11*, 229. [CrossRef]
46. Wang, Y.; Chen, S.; Zheng, S.; Lin, J.; Hu, S.; Zhuang, J.; Lin, Q.; Xie, X.; Zheng, K.; Zhang, W.; et al. The Role of Lung Ultrasound B-Lines and Serum KL-6 in the Screening and Follow-up of Rheumatoid Arthritis Patients for an Identification of Interstitial Lung Disease: Review of the Literature, Proposal for a Preliminary Algorithm, and Clinical Application to Cases. *Arthritis Res. Ther.* **2021**, *23*, 212. [CrossRef]
47. Ma, H.; Lu, J.; Song, Y.; Wang, H.; Yin, S. The Value of Serum Krebs von Den Lungen-6 as a Diagnostic Marker in Connective Tissue Disease Associated with Interstitial Lung Disease. *BMC Pulm. Med.* **2020**, *20*, 6. [CrossRef]
48. Oguz, E.O.; Kucuksahin, O.; Turgay, M.; Yildizgoren, M.T.; Ates, A.; Demir, N.; Kumbasar, O.O.; Kinikli, G.; Duzgun, N. Association of Serum KL-6 Levels with Interstitial Lung Disease in Patients with Connective Tissue Disease: A Cross-Sectional Study. *Clin. Rheumatol.* **2016**, *35*, 663–666. [CrossRef]
49. Doyle, T.J.; Patel, A.S.; Hatabu, H.; Nishino, M.; Wu, G.; Osorio, J.C.; Golzarri, M.F.; Traslosheros, A.; Chu, S.G.; Frits, M.L.; et al. Detection of Rheumatoid Arthritis-Interstitial Lung Disease Is Enhanced by Serum Biomarkers. *Am. J. Respir. Crit. Care Med.* **2015**, *191*, 1403–1412. [CrossRef]
50. Mohan, C.; Assassi, S. Biomarkers in Rheumatic Diseases: How Can They Facilitate Diagnosis and Assessment of Disease Activity? *BMJ* **2015**, *351*, h5079. [CrossRef]
51. Chen, Z.; Bozec, A.; Ramming, A.; Schett, G. Anti-Inflammatory and Immune-Regulatory Cytokines in Rheumatoid Arthritis. *Nat. Rev. Rheumatol.* **2019**, *15*, 9–17. [CrossRef] [PubMed]
52. Kass, D.J.; Nouraie, M.; Glassberg, M.K.; Ramreddy, N.; Fernandez, K.; Harlow, L.; Zhang, Y.; Chen, J.; Kerr, G.S.; Reimold, A.M.; et al. Comparative Profiling of Serum Protein Biomarkers in Rheumatoid Arthritis–Associated Interstitial Lung Disease and Idiopathic Pulmonary Fibrosis. *Arthritis Rheumatol.* **2020**, *72*, 409–419. [CrossRef] [PubMed]
53. Mahalanobish, S.; Saha, S.; Dutta, S.; Sil, P.C. Matrix Metalloproteinase: An Upcoming Therapeutic Approach for Idiopathic Pulmonary Fibrosis. *Pharmacol. Res.* **2020**, *152*, 104591. [CrossRef] [PubMed]
54. Zhu, J.; Zhou, Y.; Chen, X.; Li, J. A Metaanalysis of the Increased Risk of Rheumatoid Arthritis-Related Pulmonary Disease as a Result of Serum Anticitrullinated Protein Antibody Positivity. *J. Rheumatol.* **2014**, *41*, 1282–1289. [CrossRef]
55. Harlow, L.; Rosas, I.O.; Gochuico, B.R.; Mikuls, T.R.; Dellaripa, P.F.; Oddis, C.V.; Ascherman, D.P. Identification of Citrullinated Hsp90 Isoforms as Novel Autoantigens in Rheumatoid Arthritis-Associated Interstitial Lung Disease. *Arthritis Rheum.* **2013**, *65*, 869–879. [CrossRef]

56. Giles, J.T.; Darrah, E.; Danoff, S.; Johnson, C.; Andrade, F.; Rosen, A.; Bathon, J.M. Association of Cross-Reactive Antibodies Targeting Peptidyl-Arginine Deiminase 3 and 4 with Rheumatoid Arthritis-Associated Interstitial Lung Disease. *PLoS ONE* **2014**, *9*, e98794. [CrossRef]
57. Mori, S.; Koga, Y.; Sugimoto, M. Different Risk Factors between Interstitial Lung Disease and Airway Disease in Rheumatoid Arthritis. *Respir. Med.* **2012**, *106*, 1591–1599. [CrossRef]
58. Li, H.; Li, L.; Liu, C.; Cheng, L.; Yan, S.; Chen, H.; Li, Y. Diagnostic Value of Anti-citrullinated A-enolase Peptide 1 Antibody in Patients with Rheumatoid Arthritis: A Systematic Review and Meta-analysis. *Int. J. Rheum. Dis.* **2021**, *24*, 633. [CrossRef]
59. Castellanos-Moreira, R.; Rodríguez-García, S.C.; Gomara, M.J.; Ruiz-Esquide, V.; Cuervo, A.; Casafont-Solé, I.; Ramírez, J.; Holgado, S.; Gómez-Puerta, J.A.; Cañete, J.D.; et al. Anti-Carbamylated Proteins Antibody Repertoire in Rheumatoid Arthritis: Evidence of a New Autoantibody Linked to Interstitial Lung Disease. *Ann. Rheum. Dis.* **2020**, *79*, 587–594. [CrossRef]
60. England, B.R.; Duryee, M.J.; Roul, P.; Mahajan, T.D.; Singh, N.; Poole, J.A.; Ascherman, D.P.; Caplan, L.; Demoruelle, M.K.; Deane, K.D.; et al. Malondialdehyde-Acetaldehyde Adducts and Antibody Responses in Rheumatoid Arthritis-Associated Interstitial Lung Disease. *Arthritis Rheumatol.* **2019**, *71*, 1483–1493. [CrossRef]
61. Peljto, A.L.; Zhang, Y.; Fingerlin, T.E.; Shwu-Fan, M.; Garcia, J.G.N.; Richards, T.J.; Silveira, L.J.; Lindell, K.O.; Steele, M.P.; Loyd, J.E.; et al. Association between the MUC5B Promoter Polymorphism and Survival in Patients with Idiopathic Pulmonary Fibrosis. *JAMA* **2013**, *309*, 2232–2239. [CrossRef] [PubMed]
62. Juge, P.A.; Borie, R.; Kannengiesser, C.; Gazal, S.; Revy, P.; Wemeau-Stervinou, L.; Debray, M.P.; Ottaviani, S.; Marchand-Adam, S.; Nathan, N.; et al. Shared Genetic Predisposition in Rheumatoid Arthritis-Interstitial Lung Disease and Familial Pulmonary Fibrosis. *Eur. Respir. J.* **2017**, *49*, 1602314. [CrossRef] [PubMed]
63. Oka, S.; Furukawa, H.; Shimada, K.; Hashimoto, A.; Komiya, A.; Fukui, N.; Tsuchiya, N.; Tohma, S. Plasma MiRNA Expression Profiles in Rheumatoid Arthritis Associated Interstitial Lung Disease. *BMC Musculoskelet. Disord.* **2017**, *18*, 1–7. [CrossRef] [PubMed]
64. Migita, K.; Nakamura, T.; Koga, T.; Eguchi, K. HLA-DRB1 Alleles and Rheumatoid Arthritis-Related Pulmonary Fibrosis. *J. Rheumatol.* **2010**, *37*, 205–207. [CrossRef] [PubMed]
65. Deng, K.; Fan, Q.; Yang, Y.; Deng, X.; He, R.; Tan, Y.; Lan, Y.; Deng, X.; Pan, Y.; Wang, Y.; et al. Prognostic Roles of KL-6 in Disease Severity and Lung Injury in COVID-19 Patients: A Longitudinal Retrospective Analysis. *J. Med. Virol.* **2021**, *93*, 2505–2512. [CrossRef] [PubMed]
66. Ko, U.W.; Cho, E.J.; Oh, H.B.; Koo, H.J.; Do, K.H.; Song, J.W. Serum Krebs von Den Lungen-6 Level Predicts Disease Progression in Interstitial Lung Disease. *PLoS ONE* **2020**, *15*, e0244114. [CrossRef] [PubMed]
67. Wu, X.; Xu, L.; Cheng, Q.; Nie, L.; Zhang, S.; Du, Y.; Xue, J. Increased Serum Soluble Programmed Death Ligand 1(SPD-L1) Is Associated with the Presence of Interstitial Lung Disease in Rheumatoid Arthritis: A Monocentric Cross-Sectional Study. *Respir. Med.* **2020**, *166*, 105948. [CrossRef]
68. Matsuo, T.; Hashimoto, M.; Ito, I.; Kubo, T.; Uozumi, R.; Furu, M.; Ito, H.; Fujii, T.; Tanaka, M.; Terao, C.; et al. Interleukin-18 Is Associated with the Presence of Interstitial Lung Disease in Rheumatoid Arthritis: A Cross-Sectional Study. *Scand. J. Rheumatol.* **2019**, *48*, 87–94. [CrossRef]
69. Hussein, M.S.; El-Barbary, A.M.; Nada, D.W.; Gaber, R.A.; Elkolaly, R.M.; Aboelhawa, M.A. Identification of Serum Interleukin-13 and Interleukin-13 Receptor Subunit Expressions: Rheumatoid Arthritis–Associated Interstitial Lung Disease. *Int. J. Rheum. Dis.* **2021**, *24*, 591–598. [CrossRef]
70. Ohnishi, H.; Yokoyama, A.; Kondo, K.; Hamada, H.; Abe, M.; Nishimura, K.; Hiwada, K.; Kohno, N. Comparative Study of KL-6, Surfactant Protein-A, Surfactant Protein-D, and Monocyte Chemoattractant Protein-1 as Serum Markers for Interstitial Lung Diseases. *Am. J. Respir. Crit. Care Med.* **2002**, *165*, 378–381. [CrossRef]
71. Fu, Q.; Bai, Y.; Liu, Y.; Zhou, J.; Zheng, Y. The Serum Level and Significance of Lysyl Oxidase-like 2 in Patients with Rheumatoid Arthritis-Associated Interstitial Lung Disease. *Clin. Rheumatol.* **2018**, *37*, 193–198. [CrossRef] [PubMed]
72. De Lauretis, A.; Veeraraghavan, S.; Renzoni, E. Connective Tissue Disease-Associated Interstitial Lung Disease: How Does It Differ from IPF? How Should the Clinical Approach Differ? *Chronic Respir. Dis.* **2011**, *8*, 53–82. [CrossRef] [PubMed]
73. Paulin, F.; Doyle, T.J.; Fletcher, E.A.; Ascherman, D.P.; Rosas, I.O. Rheumatoid Arthritis-Associated Interstitial Lung Disease and Idiopathic Pulmonary Fibrosis: Shared Mechanistic and Phenotypic Traits Suggest Overlapping Disease Mechanisms. *Rev. Investig. Clin.* **2015**, *67*, 280–286.
74. Matson, S.; Lee, J.; Eickelberg, O. Two Sides of the Same Coin? A Review of the Similarities and Differences between Idiopathic Pulmonary Fibrosis and Rheumatoid Arthritis-Associated Interstitial Lung Disease. *Eur. Respir. J.* **2021**, *57*, 2002533. [CrossRef] [PubMed]
75. Perry, E.; Kelly, C.; Eggleton, P.; de Soyza, A.; Hutchinson, D. The Lung in ACPA-Positive Rheumatoid Arthritis: An Initiating Site of Injury? *Rheumatology* **2014**, *53*, 1940–1950. [CrossRef] [PubMed]
76. Tirelli, C.; Morandi, V.; Valentini, A.; La Carrubba, C.; Dore, R.; Zanframundo, G.; Morbini, P.; Grignaschi, S.; Franconeri, A.; Oggionni, T.; et al. Multidisciplinary Approach in the Early Detection of Undiagnosed Connective Tissue Diseases in Patients with Interstitial Lung Disease: A Retrospective Cohort Study. *Front. Med.* **2020**, *7*, 11. [CrossRef]
77. Fischer, A.; du Bois, R. Interstitial Lung Disease in Connective Tissue Disorders. *Lancet* **2012**, *380*, 689–698. [CrossRef]
78. Oldham, J.M.; Adegunsoye, A.; Valenzi, E.; Lee, C.; Witt, L.; Chen, L.; Husain, A.N.; Montner, S.; Chung, J.H.; Cottin, V.; et al. Characterisation of Patients with Interstitial Pneumonia with Autoimmune Features. *Eur. Respir. J.* **2016**, *47*, 1767. [CrossRef]

79. Zamora-Legoff, J.A.; Krause, M.L.; Crowson, C.S.; Ryu, J.H.; Matteson, E.L. Patterns of Interstitial Lung Disease and Mortality in Rheumatoid Arthritis. *Rheumatology* **2017**, *56*, 344–350. [CrossRef]
80. Yunt, Z.X.; Chung, J.H.; Hobbs, S.; Fernandez-Perez, E.R.; Olson, A.L.; Huie, T.J.; Keith, R.C.; Janssen, W.J.; Goldstein, B.L.; Lynch, D.A.; et al. High Resolution Computed Tomography Pattern of Usual Interstitial Pneumonia in Rheumatoid Arthritis-Associated Interstitial Lung Disease: Relationship to Survival. *Respir. Med.* **2017**, *126*, 100–104. [CrossRef]
81. Salaffi, F.; Carotti, M.; di Carlo, M.; Tardella, M.; Giovagnoni, A.; Adamek, M. High-Resolution Computed Tomography of the Lung in Patients with Rheumatoid Arthritis: Prevalence of Interstitial Lung Disease Involvement and Determinants of Abnormalities. *Medicine* **2019**, *98*, e17088. [CrossRef] [PubMed]
82. Lan, S.; Huayang, D.; Ruxuan, C.; Xin, S.; Wei, S.; Zhengyu, J. Chest High-Resolution Computed Tomography Imaging Features of Lung Involvement in Rheumatoid Arthritis. *Acta Acad. Med. Sin.* **2020**, *42*, 202–208. [CrossRef]
83. Huang, S.; Doyle, T.J.; Hammer, M.M.; Byrne, S.C.; Huang, W.; Marshall, A.A.; Iannaccone, C.K.; Huang, J.; Feathers, V.; Weinblatt, M.E.; et al. Rheumatoid Arthritis-Related Lung Disease Detected on Clinical Chest Computed Tomography Imaging: Prevalence, Risk Factors, and Impact on Mortality. *Semin. Arthritis Rheum.* **2020**, *50*, 1216–1225. [CrossRef] [PubMed]
84. Balbir-Gurman, A.; Guralnik, L.; Yigla, M.; Braun-Moscovici, Y.; Hardak, E. Imaging Aspects of Interstitial Lung Disease in Patients with Rheumatoid Arthritis: Literature Review. *Autoimmun. Rev.* **2018**, *17*, 87–93. [CrossRef]
85. Gautam, M.; Masood, M.J.; Arooj, S.; Mahmud, M.-H.; Mukhtar, M.U. Rheumatoid Arthritis Related Interstitial Lung Disease: Patterns of High-Resolution Computed Tomography. *Cureus* **2020**, *12*, e6875. [CrossRef]
86. Zamora-Legoff, J.A.; Krause, M.L.; Crowson, C.S.; Ryu, J.H.; Matteson, E.L. Progressive Decline of Lung Function in Rheumatoid Arthritis–Associated Interstitial Lung Disease. *Arthritis Rheumatol.* **2017**, *69*, 542–549. [CrossRef]
87. Robles-Pérez, A.; Luburich, P.; Bolivar, S.; Dorca, J.; Nolla, J.M.; Molina-Molina, M.; Narváez, J. A Prospective Study of Lung Disease in a Cohort of Early Rheumatoid Arthritis Patients. *Sci. Rep.* **2020**, *10*, 15640. [CrossRef]
88. Hyldgaard, C.; Hilberg, O.; Pedersen, A.B.; Ulrichsen, S.P.; Løkke, A.; Bendstrup, E.; Ellingsen, T. A Population-Based Cohort Study of Rheumatoid Arthritis-Associated Interstitial Lung Disease: Comorbidity and Mortality. *Ann. Rheum. Dis.* **2017**, *76*, 1700–1706. [CrossRef]
89. Kim, E.J.; Collard, H.R.; King, T.E. Rheumatoid Arthritis-Associated Interstitial Lung Disease: The Relevance of Histopathologic and Radiographic Pattern. *Chest* **2009**, *136*, 1397–1405. [CrossRef]
90. Manfredi, A.; Cassone, G.; Luppi, F.; Atienza-Mateo, B.; Cavazza, A.; Sverzellati, N.; González-Gay, M.A.; Salvarani, C.; Sebastiani, M. Rheumatoid Arthritis Related Interstitial Lung Disease. *Expert. Rev. Clin. Immunol.* **2021**, *17*, 485–497. [CrossRef]
91. Kelly, C.; Saravanan, V. Treatment Strategies for a Rheumatoid Arthritis Patient with Interstitial Lung Disease. *Expert Opin. Pharmacother.* **2008**, *9*, 3221–3230. [CrossRef] [PubMed]
92. Vij, R.; Strek, M.E. Diagnosis and Treatment of Connective Tissue Disease-Associated Interstitial Lung Disease. *Chest* **2013**, *143*, 814–824. [CrossRef] [PubMed]
93. Iqbal, K.; Kelly, C. Treatment of Rheumatoid Arthritis-Associated Interstitial Lung Disease: A Perspective Review. *Ther. Adv. Musculoskelet. Dis.* **2015**, *7*, 247–267. [CrossRef] [PubMed]
94. Song, L.N.; Kong, X.D.; Wang, H.J.; bin Zhan, L. Establishment of a Rat Adjuvant Arthritis-Interstitial Lung Disease Model. *Biomed. Res. Int.* **2016**, *2016*, 2970783. [CrossRef]
95. Juge, P.A.; Lee, J.S.; Lau, J.; Kawano-Dourado, L.; Serrano, J.R.; Sebastiani, M.; Koduri, G.; Matteson, E.; Bonfiglioli, K.; Sawamura, M.; et al. Methotrexate and Rheumatoid Arthritis Associated Interstitial Lung Disease. *Eur. Respir. J.* **2021**, *57*, 2000337. [CrossRef]
96. Kiely, P.; Busby, A.D.; Nikiphorou, E.; Sullivan, K.; Walsh, D.A.; Creamer, P.; Dixey, J.; Young, A. Is Incident Rheumatoid Arthritis Interstitial Lung Disease Associated with Methotrexate Treatment? Results from a Multivariate Analysis in the ERAS and ERAN Inception Cohorts. *BMJ Open* **2019**, *9*, e028466. [CrossRef]
97. Md Yusof, M.Y.; Kabia, A.; Darby, M.; Lettieri, G.; Beirne, P.; Vital, E.M.; Dass, S.; Emery, P. Effect of Rituximab on the Progression of Rheumatoid Arthritis–Related Interstitial Lung Disease: 10 Years' Experience at a Single Centre. *Rheumatology* **2017**, *56*, 1348. [CrossRef]
98. Bonniaud, P.; Georges, M.; Favrolt, N.; Camus, P. Drug-Induced Interstitial Lung Disease. *Eur. Respir. J.* **2022**, *64*, 951–956. [CrossRef]
99. Fernández-Díaz, C.; Castañeda, S.; Melero-González, R.B.; Ortiz-Sanjuán, F.; Juan-Mas, A.; Carrasco-Cubero, C.; Casafont-Solé, I.; Olivé, A.; Rodríguez-Muguruza, S.; Almodóvar-González, R.; et al. Abatacept in Interstitial Lung Disease Associated with Rheumatoid Arthritis: National Multicenter Study of 263 Patients. *Rheumatology* **2020**, *59*, 3906–3916. [CrossRef]
100. Mena-Vázquez, N.; Redondo-Rodríguez, R.; Rojas-Gimenez, M.; Romero-Barco, C.M.; Manrique-Arija, S.; Ortega-Castro, R.; Hidalgo Conde, A.; Arnedo Díez de Los Ríos, R.; Cabrera César, E.; Espildora, F.; et al. Efficacy and Safety of Rituximab in Autoimmune Disease-Associated Interstitial Lung Disease: A Prospective Cohort Study. *J. Clin. Med.* **2022**, *11*, 927. [CrossRef]
101. Zhang, J.; Wang, D.; Wang, L.; Wang, S.; Roden, A.C.; Zhao, H.; Li, X.; Prakash, X.Y.S.; Matteson, E.L.; Tschumperlin, D.J.; et al. Profibrotic Effect of IL-17A and Elevated IL-17RA in Idiopathic Pulmonary Fibrosis and Rheumatoid Arthritis-Associated Lung Disease Support a Direct Role for IL-17A/IL-17RA in Human Fibrotic Interstitial Lung Disease. *Am. J. Physiol. Lung Cell. Mol. Physiol.* **2019**, *316*, L487–L497. [CrossRef] [PubMed]
102. Chen, Z.; Wang, X.; Ye, S. Tofacitinib in Amyopathic Dermatomyositis–Associated Interstitial Lung Disease. *N. Engl. J. Med.* **2019**, *381*, 291–293. [CrossRef] [PubMed]

103. D'Alessandro, M.; Perillo, F.; Metella Refini, R.; Bergantini, L.; Bellisai, F.; Selvi, E.; Cameli, P.; Manganelli, S.; Conticini, E.; Cantarini, L.; et al. Efficacy of Baricitinib in Treating Rheumatoid Arthritis: Modulatory Effects on Fibrotic and Inflammatory Biomarkers in a Real-Life Setting. *Int. Immunopharmacol.* **2020**, *86*, 106748. [CrossRef] [PubMed]
104. Silva-Carmona, M.; Vogel, T.P.; Marchal, S.; Guesmi, M.; Jean-Christophe, D.; Leroy, S.; Fabre, A.; Barlogis, V.; Forbes, L.R.; Giovannini-Chami, L. Successful Treatment of Interstitial Lung Disease in STAT3 Gain-of-Function Using JAK Inhibitors. *Am. J. Respir. Crit. Care Med.* **2020**, *202*, 893–897. [CrossRef]
105. Balci, S.; Ekinci, R.M.K.; de Jesus, A.A.; Goldbach-Mansky, R.; Yilmaz, M. Baricitinib Experience on STING-Associated Vasculopathy with Onset in Infancy: A Representative Case from Turkey. *Clin. Immunol.* **2020**, *212*, 108273. [CrossRef]
106. Bader-Meunier, B.; Hadchouel, A.; Berteloot, L.; Polivka, L.; Béziat, V.; Casanova, J.L.; Lévy, R. Effectiveness and Safety of Ruxolitinib for the Treatment of Refractory Systemic Idiopathic Juvenile Arthritis like Associated with Interstitial Lung Disease: A Case Report. *Ann. Rheum. Dis.* **2022**, *81*, E20. [CrossRef]
107. Makino, S. Progressive Fibrosing Interstitial Lung Diseases: A New Concept and Indication of Nintedanib. *Mod. Rheumatol.* **2021**, *31*, 13–19. [CrossRef]
108. Miura, Y.; Ohkubo, H.; Niimi, A.; Kanazawa, S. Suppression of Epithelial Abnormalities by Nintedanib in Induced-Rheumatoid Arthritis-Associated Interstitial Lung Disease Mouse Model. *ERJ Open Res.* **2021**, *7*, 00345. [CrossRef]
109. Wollin, L.; Distler, J.H.W.; Redente, E.F.; Riches, D.W.H.; Stowasser, S.; Schlenker-Herceg, R.; Maher, T.M.; Kolb, M. Potential of Nintedanib in Treatment of Progressive Fibrosing Interstitial Lung Diseases. *Eur. Respir. J.* **2019**, *54*, 1900161. [CrossRef]
110. Wells, A.U.; Flaherty, K.R.; Brown, K.K.; Inoue, Y.; Devaraj, A.; Richeldi, L.; Moua, T.; Crestani, B.; Wuyts, W.A.; Stowasser, S.; et al. Nintedanib in Patients with Progressive Fibrosing Interstitial Lung Diseases-Subgroup Analyses by Interstitial Lung Disease Diagnosis in the INBUILD Trial: A Randomised, Double-Blind, Placebo-Controlled, Parallel-Group Trial. *Lancet Respir. Med.* **2020**, *8*, 453–460. [CrossRef]
111. Behr, J.; Prasse, A.; Kreuter, M.; Johow, J.; Rabe, K.F.; Bonella, F.; Bonnet, R.; Grohe, C.; Held, M.; Wilkens, H.; et al. Pirfenidone in Patients with Progressive Fibrotic Interstitial Lung Diseases Other than Idiopathic Pulmonary Fibrosis (RELIEF): A Double-Blind, Randomised, Placebo-Controlled, Phase 2b Trial. *Lancet Respir. Med.* **2021**, *9*, 476–486. [CrossRef]
112. Behr, J.; Neuser, P.; Prasse, A.; Kreuter, M.; Rabe, K.; Schade-Brittinger, C.; Wagner, J.; Günther, A. Exploring Efficacy and Safety of Oral Pirfenidone for Progressive, Non-IPF Lung Fibrosis (RELIEF)—A Randomized, Double-Blind, Placebo-Controlled, Parallel Group, Multi-Center, Phase II Trial. *BMC Pulm. Med.* **2017**, *17*, 122. [CrossRef] [PubMed]

biomedicines

Article

Gelsolin as a Potential Biomarker for Endoscopic Activity and Mucosal Healing in Ulcerative Colitis

Keiko Maeda [1,*], Masanao Nakamura [2], Takeshi Yamamura [2], Tsunaki Sawada [1], Eri Ishikawa [2], Akina Oishi [2], Shuji Ikegami [2], Naomi Kakushima [2], Kazuhiro Furukawa [2], Tadashi Iida [2], Yasuyuki Mizutani [2], Takuya Ishikawa [2], Eizaburo Ohno [2], Takashi Honda [2], Masatoshi Ishigami [2] and Hiroki Kawashima [2]

1 Department of Endoscopy, Graduate School of Medicine, Nagoya University, 65 Tsurumai-cho, Showa-ku, Nagoya 466-8550, Japan; t.sawada@med.nagoya-u.ac.jp
2 Department of Gastroenterology and Hepatology, Graduate School of Medicine, Nagoya University, 65 Tsurumai-cho, Showa-ku, Nagoya 466-8550, Japan; mnakamura@med.nagoya-u.ac.jp (M.N.); tyamamu@med.nagoya-u.ac.jp (T.Y.); erieri-i@med.nagoya-u.ac.jp (E.I.); baron@med.nagoya-u.ac.jp (A.O.); s.ikegami@med.nagoya-u.ac.jp (S.I.); nk202004@med.nagoya-u.ac.jp (N.K.); kazufuru@med.nagoya-u.ac.jp (K.F.); iidatyuw@med.nagoya-u.ac.jp (T.I.); y-mizu@med.nagoya-u.ac.jp (Y.M.); ishitaku@med.nagoya-u.ac.jp (T.I.); eono@med.nagoya-u.ac.jp (E.O.); honda@med.nagoya-u.ac.jp (T.H.); masaishi@med.nagoya-u.ac.jp (M.I.); h-kawa@med.nagoya-u.ac.jp (H.K.)
* Correspondence: kmaeda@med.nagoya-u.ac.jp; Tel.: +81-52-744-2172

Abstract: The therapeutic goal in ulcerative colitis is mucosal healing, which requires improved non-invasive biomarkers to evaluate disease activity. Gelsolin is associated with several autoimmune diseases, and here, we aimed to analyze its usefulness as a serological biomarker for clinical and endoscopic activities in ulcerative colitis. Patients with ulcerative colitis (n = 138) who had undergone blood tests and colonoscopy were included. Serum gelsolin was measured using enzyme-linked immunosorbent assay, and correlation between the gelsolin level and clinical and endoscopic activities was examined. The serum gelsolin level in patients with ulcerative colitis was significantly lower than that in healthy subjects, and it decreased in proportion to increasing Mayo score and Mayo endoscopic subscore. The area under the curve for correlation between clinical and endoscopic remission and serum gelsolin level was higher than that for C-reactive protein. Furthermore, in C-reactive protein-negative patients, the serum gelsolin level was lower in the active phase than in remission. Our findings indicate that the serum gelsolin level correlates with clinical and endoscopic activities in ulcerative colitis, has a higher sensitivity and specificity than C-reactive protein, and can detect mucosal healing, suggesting that gelsolin can be used as a biomarker for ulcerative colitis.

Keywords: biomarker; ulcerative colitis; gelsolin; mucosal healing

Citation: Maeda, K.; Nakamura, M.; Yamamura, T.; Sawada, T.; Ishikawa, E.; Oishi, A.; Ikegami, S.; Kakushima, N.; Furukawa, K.; Iida, T.; et al. Gelsolin as a Potential Biomarker for Endoscopic Activity and Mucosal Healing in Ulcerative Colitis. *Biomedicines* **2022**, *10*, 872. https://doi.org/10.3390/biomedicines10040872

Academic Editor: Marianna Christodoulou

Received: 17 March 2022
Accepted: 6 April 2022
Published: 9 April 2022

Publisher's Note: MDPI stays neutral with regard to jurisdictional claims in published maps and institutional affiliations.

1. Introduction

Inflammatory bowel disease (IBD), represented by ulcerative colitis (UC) and Crohn's disease, is a chronic inflammatory ailment of the gastrointestinal tract with an increasing incidence worldwide [1,2]. With recent advances in medical therapies, such as the development of immunomodulators and biologics, the goal of IBD treatment has shifted from alleviating clinical symptoms to achieving endoscopic mucosal healing. Mucosal healing reduces subsequent recurrence, surgery, and carcinogenesis rates, [3–5] and the concept of treat-to-target, aimed at achieving endoscopic mucosal healing, is being widely accepted [6,7]. The gold standard for determining disease activity and mucosal healing in IBD is endoscopy; however, this method is associated with physical, time, and economic burdens. Therefore, in clinical practice, non-invasive, and repeatable blood and stool-based biomarkers that reflect disease activity and mucosal healing are necessary. In addition to fecal markers such as fecal calprotectin and fecal immunochemical test, serum C-reactive protein (CRP) [8,9] and serum leucine rich glycoprotein (LRG) [10–12] have been reported as useful.

CRP is produced by hepatocytes in the acute phase upon IL-6 stimulation and is used as a biomarker for various inflammatory diseases. In terms of correlation with disease activity, CRP is associated with endoscopic activity in CD, but only with histologically severe inflammation in UC. Therefore, in UC, a low CRP does not necessarily mean the absence of endoscopic activity, which is problematic owing to its low sensitivity. Leucine-rich glycoprotein is also expressed on neutrophils, macrophages, and intestinal epithelial cells and is induced by interleukin (IL)-22, tumor necrosis factor-α, and IL-1 independent of IL-6. In CD, the LRG level is strongly correlated with disease activity, and in UC, it correlates with endoscopic activity, but more evidence is needed in this regard. Therefore, biomarkers that more accurately reflect clinical and endoscopic activities and predict mucosal healing than conventional markers are needed. The noninvasive assessment of accurate disease activity would enable optimal therapeutic choices and improve patient prognosis.

In this study, we performed proteasome analysis of colon mucus samples from patients with UC and focused on GSN, which is significantly downregulated in active UC compared with that in remission. Gelsolin (GSN) is an 82–84 kDa protein consisting of 730 amino acids organized into six homologous domains and expressed in both extracellular fluids and cytoplasm of most human cells [13,14]. GSN is a multifunctional protein; because of its strong effect on the cytoskeleton and inflammation-related biological processes, it shows potential as a biomarker for inflammation-associated medical conditions, such as for predicting illness severity, treatment efficacy, and clinical outcomes [15,16]. Reduced GSN level has been observed in patients with chronic autoimmune diseases such as rheumatoid arthritis and psoriasis [17,18]. GSN localization is also altered in patients with Crohn's disease. Moreover, GSN is an actin-depolymerizing protein that regulates actin dynamics and is involved in cytoskeletal remodeling [16,17]. Its extracellular isoform, plasma GSN, is expressed in the blood, urine, and other extracellular fluids, such as lymph, burn wound fluid, cerebrospinal fluid, and airway surface fluids [19,20]. The secreted GSN also functions in the extracellular actin scavenger system, where it is responsible for the severance and removal of actin filaments from dead cells into the bloodstream [13]. In addition, the secreted GSN binds to lipopolysaccharides, which are compounds derived from the cell wall of Gram-negative bacilli, and inhibits the activation of Toll-like receptors, thereby regulating immune responses [21–24]. The secreted GSN has anti-inflammatory properties, and decreased GSN levels in the blood have been reported in chronic inflammatory diseases. Although the mechanism by which the GSN levels in the blood are reduced remains unclear, the re-distribution of GSN to inflammatory sites, binding to some plasma factors secreted in association with inflammation, and decreased GSN production have previously been reported [23,24]. Therefore, in this study, we examined whether GSN can be used as a biomarker for the clinical and endoscopic activities of UC.

2. Materials and Methods

2.1. Study Subjects and Sample Collection

In total, 138 patients with UC and 16 healthy controls were enrolled in this study at the Department of Gastroenterology and Hepatology, Nagoya University Hospital, between April 2016 and April 2021. The healthy controls comprised 10 women, and their median age was 45 (range 36–66) years. Patients were diagnosed with UC based on clinical, endoscopic, and histological criteria and received medical therapy. Clinical and endoscopic activity scores were reviewed from their medical records. Blood sampling and endoscopy were performed within a maximum interval of 1 month. Serum was obtained from the blood samples and stored at −80 °C until GSN analysis. Patients with UC comprised 84 women and 54 men, and their median age was 47 (range 20–82) years. The median duration of disease was 143 (range 7–372) months. The median C-reactive protein level was 0.08 (range 0–8.4) mg/dL. The median albumin level was 4.1 (range 1.8-4.9) g/dL. The median Mayo score was 3 (range 0–12). Here, 74.6% (103/138) patients were administered 5-aminosalicylic acids, 13% (18/138) patients were administered corticosteroids, and 32.6% (45/138) patients were administered biologic agents. Patients with UC were classified according to the extent of disease involvement as those with proctitis, left-sided colitis, or pancolitis, as described in the Montreal classification.

The proportion of patients with proctitis, left-sided colitis, and extensive colitis was 5.7% (8/138), 26% (36/138), and 68.3% (94/138), respectively. Clinical activity was determined using the Mayo score, and remission was defined by a score of $\leq$2. The endoscopic Mayo score was used to determine endoscopic activity, and endoscopic remission was defined by a score of 0. The proportion of clinically and endoscopically active patients was 56.5% (70/138) and 63.6% (88/138), respectively. The patient characteristics are presented in Table 1.

Table 1. Characteristics of patients with ulcerative colitis.

Patient Characteristic	*N* = 138
Sex, female/male, *N*	84/54
Age, years, median (range)	47 (20–82)
Duration of disease, months (range)	143 (7–372)
Disease location, *N* Extensive/left-sided/proctitis	94/36/8
Treatment, *N*	
Oral 5-aminosalicylic acid, *N* (%)	103 (74.6)
Corticosteroids, *N* (%)	18 (13.0)
Biologic agents, *N* (%)	45 (32.6)
Immunomodulators, *N* (%)	27 (19.6)
Calcineurin inhibitors, *N* (%)	3 (2.2)
Topical agents, *N* (%)	45 (32.6)
C-reactive protein, mg/dL, median (range)	0.08 (0–8.4)
Albumin, g/dL, median (range)	4.1 (1.8–4.9)
Mayo score median (range)	3 (0–12)

2.2. Measurement of Serum GSN Level

Serum GSN level was measured using an enzyme-linked immunosorbent assay (ELISA) kit (Abcam, Cambridge, UK), according to the manufacturer's instructions. The absorbance of each sample was measured at 450 and 570 nm using a PowerScan4 microplate reader (DS Pharma Medical Co., Osaka, Japan). The level of GSN was calculated using a standard curve.

2.3. Mass Spectrometry

Lower rectum intestinal mucus samples of 3 patients with active UC and 3 patients with UC in remission were collected through colonoscopy. Colon mucus from the anterior and right rectal walls was collected using brush catheters (Colonoscope Cytology Brush®; Cook Medical, Winston-Salem, NC, USA).

Patients were diagnosed with UC based on clinical, endoscopic, and histological criteria. These samples were lysed using a Minute Total Protein Extraction Kit for mass spectrometry (Funakoshi, Tokyo, Japan), and the specimens were adjusted to the same protein level before mass spectrometry (MS).

The proteins were digested using trypsin for 16 h at 37 °C after reduction and alkylation, and the peptides were analyzed using LC−MS on an Orbitrap Fusion mass spectrometer (Thermo Fisher Scientific Inc., Waltham, MA, USA) coupled to an UltiMate3000 RSLC nano LC system (Dionex Co., Amsterdam, The Netherlands), using a nano HPLC capillary column (Nikkyo Technos Co., Tokyo, Japan) with a nano electrospray ion source. Reverse-phase chromatography was performed with a linear gradient (0 min, 5% B, 100 min, 40% B) of solvent A (2% acetonitrile with 0.1% formic acid) and solvent B (95% acetonitrile with 0.1% formic acid). A precursor ion scan was carried out using a 400–1600 mass to charge ratio (m/z) before MS/MS analysis.

2.4. Data Analysis

The raw data were processed using Proteome Discoverer 1.4 (Thermo Fisher Scientific) in conjunction with the MASCOT search engine, version 2.6.0 (Matrix Science Inc., Boston, MA, USA) for protein identification. The peptides and proteins were identified using the human protein database in UniProt (release 2020_03) with a precursor mass tolerance of 10 ppm and a fragment ion mass tolerance of 0.8 Da. Fixed modification was set to carbamidomethylation of cysteine, and variable modification was set to oxidation of methionine.

2.5. Statistical Analysis

All analyses were performed using Prism software (GraphPad prism version 8 Software, GraphPad Software, San Diego, CA, USA). Differences between groups were compared using Mann–Whitney U-test and Kruskal–Wallis test. The area under the receiver operating characteristic (ROC) curve (AUC) was calculated by plotting sensitivity on the y axis against 1—specificity on the x axis for each value. The correlation analysis was performed using Pearson coefficients. Statistical significance was defined as $p < 0.05$ (*, $p < 0.05$; **, $p < 0.01$; and ***, $p < 0.001$).

2.6. Ethical Considerations

This study was approved by the ethics committee of the Nagoya University Hospital, Japan (Protocol number 2015-0420, August 2016). Written informed consent was obtained from all patients before their enrollment in accordance with tenets of the Declaration of Helsinki.

3. Results

3.1. Downregulation of Serum GSN in Patients with Clinically Active UC

We first conducted proteomic analysis of the specimens from patients with UC in the active phase and in remission. The specimens were collected from the anterior and right walls of the rectal mucosa using brush samples. We identified 460 proteins with a score of ≥30 from the brush samples in patients with active UC (Table S1). Inflammatory protein markers (protein S100-A9) and a neurotrophic protein (myeloperoxidase) presented high scores. We compared protein expression in patients with UC in remission and in the active phase. Consistent with previous study results, Mucin-5B and Mucin-13 were downregulated in active UC compared with those in remission UC [25,26]. In addition, we found that GSN was downregulated in patients with active UC compared with that in patients with remission UC (Table 2).

Table 2. List of genes downregulated in active UC compared with those in remission UC.

Accession	Description	p
Q9HC84	Mucin-5B	0.03349
Q9H3R2	Mucin-13	0.004309
P06396	Gelsolin	0.03639

We then compared the serum GSN level between patients with UC and healthy controls. We analyzed samples from 138 patients (54 males and 84 females) whose median age was 47 (20–82) years. Of all patients, 68.1% had extensive colitis, 26% had left-sided colitis, and 5.8% had proctitis. The proportion of clinically and endoscopically active patients was 56.5% (70/138) and 63.6% (88/138), respectively. The serum GSN level was lower in patients with UC than in the healthy controls (138 patients with UC and 16 healthy controls, $p < 0.001$, Figure 1a). In addition, the serum GSN level was significantly lower in clinically active patients with UC than in those in remission (138 patients with UC, $p < 0.001$, Figure 1b). The correlation between the GSN levels and Mayo scores was determined using

Pearson coefficients, and a significant correlation was found (r = −0.70229, $p < 0.001$) (Figure S1).

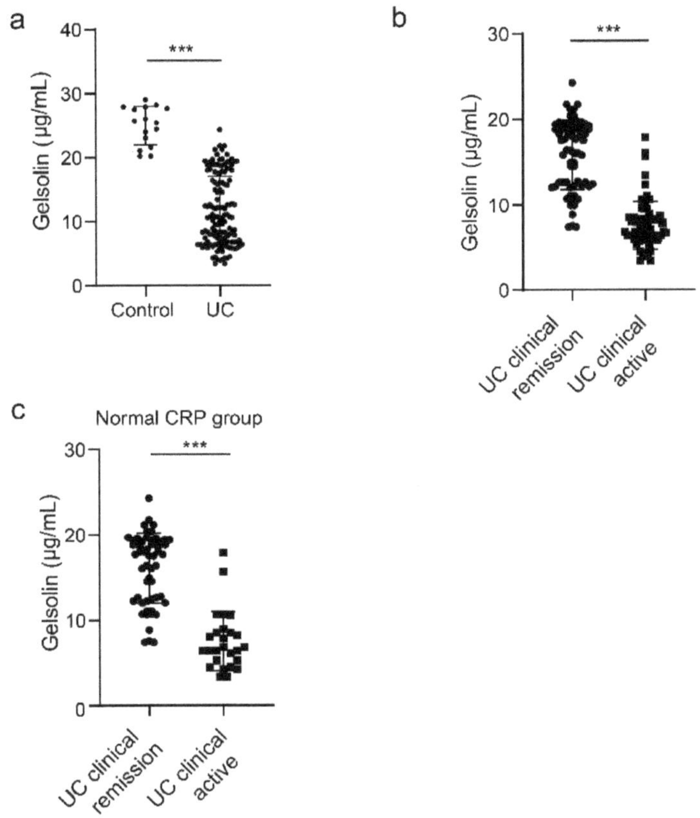

Figure 1. Serum gelsolin (GSN) level decreased in clinically active patients with ulcerative colitis (UC). Serum GSN level was measured in (**a**) 138 patients with UC and 16 healthy subjects (control); (**b**) 68 patients with UC in clinical remission and 70 patients with clinically active UC; and (**c**) 56 patients with UC in clinical remission and 26 patients with clinically active UC and normal C-reactive protein (CRP) level (<0.14 mg/dL). Statistical significance was defined as $p < 0.05$ (*** $p < 0.001$) using Mann–Whitney U-test.

The expression of CRP, which is used as a serum biomarker for UC, is induced by IL-6, but the expression of GSN is downregulated by a mechanism different from that of CRP. The correlation between the sensitivity of CRP and disease activity is low; therefore, we tested whether GSN is effective for patients in whom activity was difficult to assess with CRP. Among the 82 patients with UC whose CRP level was normal (<0.14 mg/dL), the GSN level was significantly lower in patients with clinically active disease than in those in the remission phase (82 patients with UC, $p < 0.001$, Figure 1c). These findings indicate that the GSN level correlates with clinical activity, even in cases with a normal CRP level.

3.2. Inverse Relationship between the Serum GSN Level and Endoscopic Activity in Patients with UC

We analyzed whether the serum GSN level is associated with endoscopic activity in patients with UC. A decreased GSN level correlated with an increased endoscopic activity score (Mayo 0 vs. 1, $p = 0.999$; Mayo 0 vs. 2, $p = 0.0113$: Mayo 0 vs. 3, $p < 0.01$; Mayo 1 vs. 2, $p = 0.0549$; Mayo 1 vs. 3, $p < 0.001$; and Mayo 2 vs. 3, $p < 0.001$). Patients with MES 2 had a lower GSN level than those with MES 0 (54 patients (Mayo = 2), 50 patients (Mayo = 0), $p = 0.0113$), possibly reflecting minor mucosal changes (Figure 2a).

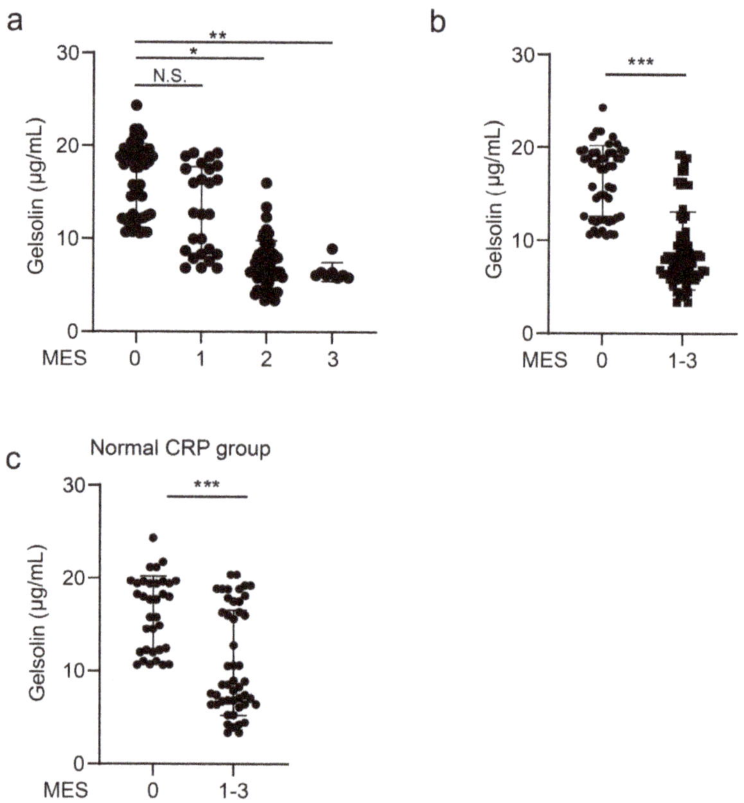

Figure 2. GSN level correlates with the endoscopic activity score in patients with UC. (**a**) Serum GSN level in patients with UC categorized according to disease activity (MES 0 ($n = 50$), 1 ($n = 26$), 2 ($n = 54$), and 3 ($n = 8$)). (**b**) Serum GSN level in 50 patients with UC in endoscopic remission (Mayo endoscopic score (MES) = 0) and 88 patients with endoscopically active UC (MES > 0). (**c**) Serum GSN level was measured in 34 patients with UC in endoscopic remission and 48 patients with endoscopically active UC and normal CRP level (CRP < 0.14 mg/dL). Statistical significance was defined as $p < 0.05$ (* $p < 0.05$; ** $p < 0.01$; *** $p < 0.001$; and N.S., not significant) using Mann–Whitney U-test and Kruskal–Wallis test.

Recently, mucosal healing has been reported to reduce operative and relapse rates, and therapeutic goals have shifted from symptom relief to mucosal healing [3–5]. The detection of mucosal healing is important when using a serum biomarker that correlates with UC activity. We defined mucosal healing using the Mayo endoscopic score of 0 and tested whether mucosal healing could be detected by the GSN level.

The GSN level was lower in patients with endoscopically active UC (Mayo endoscopic score (MES) > 0) than in those with endoscopic remission (MES = 0) ($p < 0.001$, Figure 2a). The correlation between the GSN levels and Mayo endoscopic scores was measured using Pearson coefficients, and a significant correlation was found ($r = -0.7585$, $p < 0.01$) (Figure S2). Pearson coefficient was also used to determine the correlation between the GSN and CRP levels, and it was found that they had a low correlation ($r = -0.287$, $p = 0.006$) (Figure S3A).

The correlation between the GSN levels and Mayo endoscopic scores was measured using Pearson coefficients, and it was found that the GSN level and albumin had a low correlation ($r = 0.44755$, $p < 0.001$) (Figure S3B).

In addition, we tested whether the GSN level could detect mucosal healing in cases with a normal CRP level. Among the 82 patients with UC whose CRP level was within the normal level (<0.14 mg/dL), the GSN level was significantly lower in patients in the endoscopically active phase than in those in the remission phase ($p < 0.001$, Figure 2c). These findings indicate that the GSN level could detect clinical and endoscopic activities in UC patients with high sensitivity. Furthermore, even in patients with a normal CRP level, it correlated with clinical and endoscopic activities, making CRP useful for patients whose activity is difficult to assess with conventional blood tests.

3.3. GSN as a Serological Biomarker of Clinically and Endoscopically Active UC

Given that the GSN level correlated with clinical and endoscopic activities in patients with UC, we next investigated its diagnostic potential to detect clinical remission and mucosal healing in order to use it as a serum biomarker.

We compared the sensitivity and specificity of GSN with those of CRP using ROC curve and AUC analyses. The sensitivity and specificity of GSN were 91.43% and 89.71%, respectively, for the detection of clinical remission at a cut-off of 10.67 μg/mL (Figure 3a). The AUC of GSN was 0.874 and that of CRP was 0.78. For the detection of endoscopic remission, the sensitivity and specificity of GSN were 78.41% and 86.54%, respectively (Figure 3c), whereas those of CRP were 56.82% and 82.00%, respectively (Figure 3d). The AUC of GSN was 0.835, and that of CRP was 0.692. The AUC of GSN was higher than that of CRP for identifying both clinical and endoscopic remission (Figure 3a–d). These data suggest that GSN is a biomarker that reflects clinical and endoscopic activities and that it can detect mucosal healing.

To determine whether GSN can be used as a biomarker for assessing the clinical and endoscopic activities of UC, we analyzed its sensitivity and specificity using the ROC curve and AUC analyses and compared the results of GSN and CRP, an existing UC marker. The AUC of GSN was higher than that of CRP for identifying both clinical and endoscopic remission (Figure 3a–d).

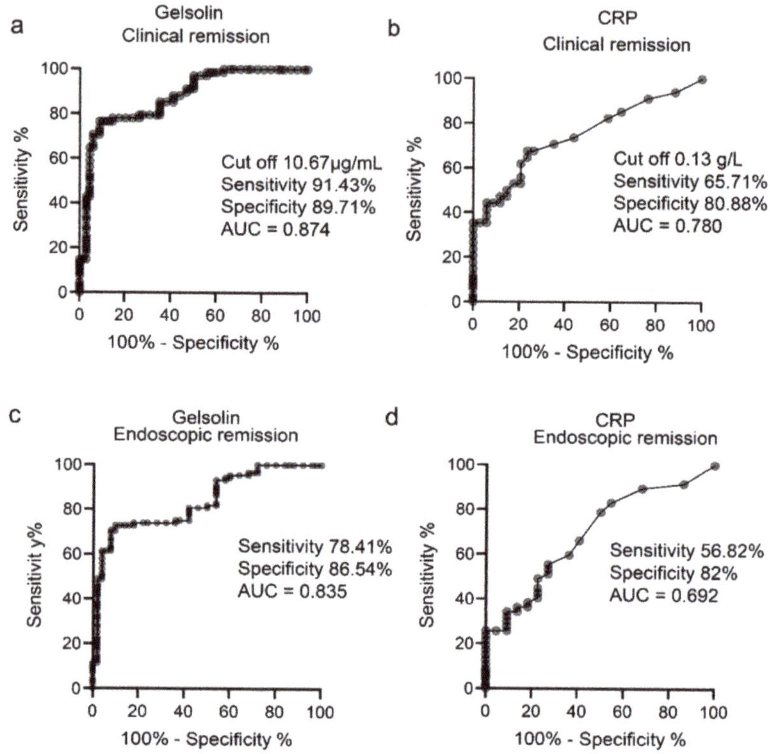

Figure 3. GSN level reflects clinical and endoscopic activities in patients with UC. Receiver operating characteristic (ROC) curves for GSN and CRP indicating their sensitivity and specificity in discriminating (**a**,**b**) clinical remission and (**c**,**d**) endoscopic remission.

4. Discussion

In this study, we showed that the GSN level correlates with the clinical and endoscopic activities of UC. GSN also showed high sensitivity and specificity in predicting the achievement of mucosal healing in patients with UC.

Currently, CRP and LRG are used as blood-based biomarkers to evaluate the activity of IBD. CRP expression is induced by IL-6 and is used to evaluate various inflammatory diseases [27]. CRP is a useful marker for the diagnosis of IBD, evaluation of disease activity, and prediction of therapeutic efficacy [8,9,28]. However, CRP correlates mostly with severe inflammation and does not reflect mild inflammation [29].

We performed proteomic analysis of the specimens from patients with UC in the active phase and remission phase. We identified 460 proteins in patients with active UC, as in previous studies, the inflammatory markers such as S100-A9 and myeloperoxitase were detected. In addition, IgG-Fc, which is required for the stabilization of Mucin-2 was also detected. Among them, we focused on GSN, the expression of which was downregulated in patients with active UC compared with that in patents with remission UC, and its expression in patents with remission UC was higher than that in patients with active UC.

The GSN level decreases with endoscopic activity, and patients with MES 2 have significantly lower GSN level than patients with MES 0, suggesting that GSN may also reflect mild intestinal inflammation. In addition, the fact that the GSN level reflects clinical and endoscopic activities even in a group of patients with normal CRP levels suggests that the GSN level may be useful for patients whose activity has been difficult to assess with conventional biomarkers. Furthermore, in this study, we demonstrated that the GSN level reflects clinical and endoscopic remission with higher sensitivity and specificity than CRP. We believe that GSN can help detect IL-6-independent inflammation and mucosal healing because it reflects even mild inflammation. Moreover, it can be used to assess endoscopic activity even in CRP-negative cases and could be a new biomarker with an underlying mechanism of action that is different from that of CRP. LRG, a newly identified serum marker for UC, has also been reported to correlate with clinical and endoscopic activities of the disease [10,12]. In the future, it will be necessary to compare the sensitivity and specificity of LRG and GSN, and utilize them according to disease activity and stage. Furthermore, it has been suggested that the measurement of both LRG and GSN may allow more accurate assessment of disease activity and predict mucosal healing with higher sensitivity and specificity.

Mucosal healing was previously defined as MES 0 and MES 1; however, as patients with MES 1 have a higher relapse rate than those with MES 0 [4,30], several studies have considered only MES 0 to reflect mucosal healing. Therefore, mucosal healing was defined as MES 0 in this study. As the operation and relapse rates are low in patients who have achieved mucosal healing, mucosal healing has become the therapeutic goal in UC. GSN presented higher sensitivity and specificity than CRP in detecting mucosal healing with a cut off of 10.67 g/mL. UC is a chronic inflammatory disease with recurrent remissions and relapses, and optimization of treatment based on more accurate assessment of disease activity is needed. Optimal therapeutic options may improve the prognosis of patients by enabling long-term maintenance of mucosal healing. Therefore, using GSN as a biomarker will enable accurate assessment of mucosal healing and treatment optimization.

GSN is a multifunctional protein with altered blood levels in chronic inflammatory and autoimmune disorders such as rheumatoid arthritis [17,31], ankylosing spondylitis [32], systemic lupus erythematosus [33], and Henoch–Schoenlein purpura [34]. However, there have been no reports on the association between GSN level and disease activity in patients with IBD. In the gastrointestinal tract, GSN, along with the GSN superfamily protein villin-1, regulates actin dynamics, intestinal epithelial cell death, and intestinal inflammation [35], but its function in IBD is unknown, and the mechanism of its decreased expression in the intestinal tissues and blood requires further analysis. GSN binds to lipopolysaccharides (LPS), a bacterial cell wall component, and inhibits the activation of Toll-like receptors on the surface of innate immune system cells, such as macrophages and dendritic cells. In IBD, the intestinal epithelial barrier, including mucus production and tight junction formation, is disrupted, and this disruption induces bacterial translocation of LPS from the intestinal tract into the bloodstream. The progressive disruption of the intestinal epithelial barrier mechanism associated with inflammation in IBD may induce an increase in the blood levels of LPS and decrease the level of GSN. In addition, GSN has been reported to be associated with multiple immune cell functions, such as neutrophil migration, suggesting that the abnormal activation of immune cells associated with chronic inflammation may be related to the mechanism of decreased GSN level. The proteins that we identified using proteomic analysis included calprotectin, which is currently used as a stool marker, and could have comprised proteins that can be used as blood- or stool-based markers of disease activity.

Our study had some limitations. For instance, it was a retrospective, single-center study with a small number of patients with heterogeneous backgrounds and treatments. Future studies should involve the recruitment of a prospective cohort to ascertain whether GSN reflects endoscopic activity and mucosal healing in patients with UC. As mentioned earlier, GSN is affected by other inflammatory and autoimmune disorders, and therefore, it may not be useful when other inflammations are involved. Prospective correlations between the GSN level and clinical and endoscopic activities should be carefully examined for the presence of intestinal and other infections or other autoimmune complications.

Nevertheless, we believe that GSN has the potential to be developed into a biomarker to assess UC disease activity and mucosal healing, and can contribute to the realization of treatment targets aimed at achieving mucosal healing. Our findings may lead to a reduction in the number of endoscopic procedures that are needed to assess UC disease activity, reducing patient stress and medical costs. Furthermore, non-invasive markers for disease activity will enable us to accurately assess UC disease activity and adjust treatments appropriately, as well as enable the use of treat-to-target approaches to achieve mucosal healing.

5. Patents

This work has been submitted for a patent application.

Supplementary Materials: The following supporting information can be downloaded at: https://www.mdpi.com/article/10.3390/biomedicines10040872/s1, Table S1: Proteins detected in colonic samples from active UC patients; Figure S1: Correlation between the GSN level and Mayo score determined using Pearson coefficients; Figure S2: Correlation between the GSN level and Mayo endoscopic score determined using Pearson coefficients; Figure S3: GSN level and C-reactive protein or albumin had a low correlation.

Author Contributions: Conceptualization, K.M. and M.N.; methodology, K.M.; formal analysis, K.M. and A.O.; investigation, T.Y., T.S. and S.I.; resources, K.M.; data curation, K.M. and E.I.; writing—original draft preparation, K.M.; writing—review and editing, N.K., K.F., T.I. (Takuya Ishikawa), Y.M., T.I. (Tadashi Iida) and E.O.; supervision, M.I., T.H. and H.K.; project administration, K.M.; funding acquisition, K.M. All authors have read and agreed to the published version of the manuscript.

Funding: This research was funded by a Grant-in-Aid for Young Scientists (to K.M), grant number 19K20667 and 21K15920).

Institutional Review Board Statement: The study was conducted in accordance with the tenets of the Declaration of Helsinki and approved by the Ethics Committee of the Nagoya University Hospital, Japan (protocol code 04667146 and 1 April in 2016 of approval).

Informed Consent Statement: Written informed consent was obtained from all participants.

Data Availability Statement: The original data and the materials generated in this study are available from the corresponding author on request.

Acknowledgments: We thank members of the Department of Gastroenterology at Nagoya University Graduate School of Medicine for careful reading the manuscript and for helpful discussions.

Conflicts of Interest: The authors declare no conflict of interest. The funders had no role in the design of the study; in the collection, analyses, or interpretation of data; in the writing of the manuscript; or in the decision to publish the results.

References

1. Ng, S.C.; Shi, H.Y.; Hamidi, N.; Underwood, F.E.; Tang, W.; Benchimol, E.I.; Panaccione, R.; Ghosh, S.; Wu, J.C.Y.; Chan, F.K.L.; et al. Worldwide incidence and prevalence of inflammatory bowel disease in the 21st century: A systematic review of population-based studies. *Lancet* **2017**, *390*, 2769–2778. [CrossRef]
2. Kobayashi, T.; Siegmund, B.; Le Berre, C.; Wei, S.C.; Ferrante, M.; Shen, B.; Bernstein, C.N.; Danese, S.; Peyrin-Biroulet, L.; Hibi, T. Ulcerative colitis. *Nat. Rev. Dis. Prim.* **2020**, *6*, 1–20. [CrossRef] [PubMed]
3. Rutgeerts, P.; Vermeire, S.; Van Assche, G. Mucosal healing in inflammatory bowel disease: Impossible ideal or therapeutic target? *Gut* **2007**, *56*, 453–455. [CrossRef] [PubMed]
4. Yokoyama, K.; Kobayashi, K.; Mukae, M.; Sada, M.; Koizumi, W. Clinical Study of the Relation between Mucosal Healing and Long-Term Outcomes in Ulcerative Colitis. *Gastroenterol. Res. Pract.* **2013**, *2013*, 192794. [CrossRef]
5. Carvalho, P.B.; de Castro, F.D.; Rosa, B.; Moreira, M.J.; Cotter, J. Mucosal Healing in Ulcerative Colitis—When Zero is Better. *J. Crohn's Colitis* **2015**, *10*, 20–25. [CrossRef]
6. Colombel, J.-F.; D'Haens, G.; Lee, W.-J.; Petersson, J.; Panaccione, R. Outcomes and Strategies to Support a Treat-to-target Approach in Inflammatory Bowel Disease: A Systematic Review. *J. Crohn's Colitis* **2019**, *14*, 254–266. [CrossRef]
7. Turner, D.; Ricciuto, A.; Lewis, A.; D'Amico, F.; Dhaliwal, J.; Griffiths, A.M.; Bettenworth, D.; Sandborn, W.J.; Sands, B.E.; Reinisch, W.; et al. STRIDE-II: An Update on the Selecting Therapeutic Targets in Inflammatory Bowel Disease (STRIDE) Initiative of the International Organization for the Study of IBD (IOIBD): Determining Therapeutic Goals for Treat-to-Target strategies in IBD. *Gastroenterology* **2021**, *160*, 1570–1583. [CrossRef]
8. Fagan, E.A.; Dyck, R.F.; Maton, P.N.; Hodgson, H.J.F.; Chadwick, V.S.; Petrie, A.; Pepys, M.B. Serum levels of C-reactive protein in Crohn's disease and ulcerative colitis. *Eur. J. Clin. Investig.* **1982**, *12*, 351–359. [CrossRef]

9. Menees, S.B.; Powell, C.; Kurlander, J.; Goel, A.; Chey, W.D. A Meta-Analysis of the Utility of C-Reactive Protein, Erythrocyte Sedimentation Rate, Fecal Calprotectin, and Fecal Lactoferrin to Exclude Inflammatory Bowel Disease in Adults With IBS. *Am. J. Gastroenterol.* **2015**, *110*, 444–454. [CrossRef]
10. Shinzaki, S.; Matsuoka, K.; Iijima, H.; Mizuno, S.; Serada, S.; Fujimoto, M.; Arai, N.; Koyama, N.; Morii, E.; Watanabe, M.; et al. Leucine-rich Alpha-2 Glycoprotein is a Serum Biomarker of Mucosal Healing in Ulcerative Colitis. *J. Crohn's Colitis* **2016**, *11*, 84–91. [CrossRef]
11. Shinzaki, S.; Matsuoka, K.; Tanaka, H.; Takeshima, F.; Kato, S.; Torisu, T.; Ohta, Y.; Watanabe, K.; Nakamura, S.; Yoshimura, N.; et al. Leucine-rich alpha-2 glycoprotein is a potential biomarker to monitor disease activity in inflammatory bowel disease receiving adalimumab: PLANET study. *J. Gastroenterol.* **2021**, *56*, 560–569. [CrossRef] [PubMed]
12. Yasutomi, E.; Inokuchi, T.; Hiraoka, S.; Takei, K.; Igawa, S.; Yamamoto, S.; Ohmori, M.; Oka, S.; Yamasaki, Y.; Kinugasa, H.; et al. Leucine-rich alpha-2 glycoprotein as a marker of mucosal healing in inflammatory bowel disease. *Sci. Rep.* **2021**, *11*, 1–11. [CrossRef]
13. Epstein, F.H.; Lee, W.M.; Galbraith, R.M. The Extracellular Actin-Scavenger System and Actin Toxicity. *N. Engl. J. Med.* **1992**, *326*, 1335–1341. [CrossRef] [PubMed]
14. Sun, H.Q.; Yamamoto, M.; Mejillano, M.; Yin, H.L. Gelsolin, a Multifunctional Actin Regulatory Protein. *J. Biol. Chem.* **1999**, *274*, 33179–33182. [CrossRef] [PubMed]
15. Feldt, J.; Schicht, M.; Garreis, F.; Welss, J.; Schneider, U.W.; Paulsen, F. Structure, regulation and related diseases of the actin-binding protein gelsolin. *Expert Rev. Mol. Med.* **2018**, *20*, e7. [CrossRef] [PubMed]
16. Piktel, E.; Levental, I.; Durnaś, B.; Janmey, P.A.; Bucki, R. Plasma Gelsolin: Indicator of Inflammation and Its Potential as a Diagnostic Tool and Therapeutic Target. *Int. J. Mol. Sci.* **2018**, *19*, 2516. [CrossRef] [PubMed]
17. Osborn, T.M.; Verdrengh, M.; Stossel, T.P.; Tarkowski, A.; Bokarewa, M. Decreased levels of the gelsolin plasma isoform in patients with rheumatoid arthritis. *Arthritis Res. Ther.* **2008**, *10*, R117. [CrossRef]
18. Esawy, M.; Makram, W.K.; Albalat, W.; Shabana, M.A. Plasma gelsolin levels in patients with psoriatic arthritis: A possible novel marker. *Clin. Rheumatol.* **2020**, *39*, 1881–1888. [CrossRef]
19. Silacci, P.; Mazzolai, L.; Gauci, C.; Stergiopulos, N.; Yin, H.L.; Hayoz, D. Gelsolin Superfamily Proteins: Key Regulators of Cellular Functions. *Cell Mol. Life Sci.* **2004**, *61*, 2614–2623. [CrossRef]
20. Kwiatkowski, D.J.; Stossel, T.P.; Orkin, S.H.; Mole, J.E.; Coltens, H.R.; Yin, H.L. Plasma and cytoplasmic gelsolins are encoded by a single gene and contain a duplicated actin-binding domain. *Nature* **1986**, *323*, 455–458. [CrossRef]
21. Lind, S.E.; Janmey, P.A. Human plasma gelsolin binds to fibronectin. *J. Biol. Chem.* **1984**, *259*, 13262–13266. [CrossRef]
22. Osborn, T.M.; Dahlgren, C.; Hartwig, J.H.; Stossel, T.P. Modifications of cellular responses to lysophosphatidic acid and platelet-activating factor by plasma gelsolin. *Am. J. Physiol. Physiol.* **2007**, *292*, C1323–C1330. [CrossRef] [PubMed]
23. Bucki, R.; Byfield, F.J.; Kulakowska, A.; McCormick, M.E.; Drozdowski, W.; Namiot, Z.; Hartung, T.; Janmey, P.A. Extracellular Gelsolin Binds Lipoteichoic Acid and Modulates Cellular Response to Proinflammatory Bacterial Wall Components. *J. Immunol.* **2008**, *181*, 4936–4944. [CrossRef] [PubMed]
24. Bucki, R.; Kułakowska, A.; Byfield, F.J.; Żendzian-Piotrowska, M.; Baranowski, M.; Marzec, M.; Winer, J.P.; Ciccarelli, N.J.; Górski, J.; Drozdowski, W.; et al. Plasma gelsolin modulates cellular response to sphingosine 1-phosphate. *Am. J. Physiol. Physiol.* **2010**, *299*, C1516–C1523. [CrossRef]
25. Lavelle, A.; Lennon, G.; O'Sullivan, O.; Docherty, N.; Balfe, A.; Maguire, A.; Mulcahy, H.E.; Doherty, G.; O'Donoghue, D.; Hyland, J.; et al. Spatial variation of the colonic microbiota in patients with ulcerative colitis and control volunteers. *Gut* **2015**, *64*, 1553–1561. [CrossRef] [PubMed]
26. van der Post, S.; Jabbar, K.S.; Birchenough, G.; Arike, L.; Akhtar, N.; Sjovall, H.; Johansson, M.E.V.; Hansson, G.C. Structural weakening of the colonic mucus barrier is an early event in ulcerative colitis pathogenesis. *Gut* **2019**, *68*, 2142–2151. [CrossRef]
27. Pepys, M.B.; Hirschfield, G. C-reactive protein: A critical update. *J. Clin. Investig.* **2003**, *111*, 1805–1812. [CrossRef]
28. Shine, B.; Berghouse, L.; Jones, J.; Landon, J. C-Reactive protein as an aid in the differentiation of functional and inflammatory bowel disorders. *Clin. Chim. Acta* **1985**, *148*, 105–109. [CrossRef]
29. A Solem, C.; Loftus, E.V.; Tremaine, W.J.; Harmsen, W.S.; Zinsmeister, A.R.; Sandborn, W.J. Correlation of C-Reactive Protein with Clinical, Endoscopic, Histologic, and Radiographic Activity in Inflammatory Bowel Disease. *Inflamm. Bowel Dis.* **2005**, *11*, 707–712. [CrossRef]
30. Acosta, M.B.-D.; Vallejo, N.; De La Iglesia, D.; Uribarri, L.; Bastón, I.; Ferreiro-Iglesias, R.; Lorenzo, A.; Dominguez-Munoz, J.E. Evaluation of the Risk of Relapse in Ulcerative Colitis According to the Degree of Mucosal Healing (Mayo 0 vs 1): A Longitudinal Cohort Study. *J. Crohn's Colitis* **2015**, *10*, 13–19. [CrossRef]
31. Liao, H.; Wu, J.; Kuhn, E.; Chin, W.; Chang, B.; Jones, M.D.; O'Neil, S.; Clauser, K.R.; Karl, J.; Hasler, F.; et al. Use of mass spectrometry to identify protein biomarkers of disease severity in the synovial fluid and serum of patients with rheumatoid arthritis. *Arthritis Care Res.* **2004**, *50*, 3792–3803. [CrossRef] [PubMed]
32. Genre, F.; López-Mejías, R.; Miranda-Filloy, J.A.; Ubilla, B.; Carnero-López, B.; Gómez-Acebo, I.; Blanco, R.; Ochoa, R.; Rueda-Gotor, J.; González-Juanatey, C.; et al. Gelsolin levels are decreased in ankylosing spondylitis patients undergoing an-ti-TNF-alpha therapy. *Clin. Exp. Rheumatol.* **2014**, *32*, 218–224. [PubMed]

33. Hu, Y.; Li, H.; Li, W.H.; Meng, H.X.; Fan, Y.Z.; Li, W.J.; Ji, Y.T.; Zhao, H.; Zhang, L.; Jin, X.M.; et al. The value of decreased plasma gelsolin levels in patients with systemic lupus erythematosus and rheumatoid arthritis in diagnosis and disease activity evaluation. *Lupus* **2013**, *22*, 1455–1461. [CrossRef] [PubMed]
34. Kandur, Y.; Celik, A.; Gözübenli, F.; Çetinkaya, A.; Olgar, Ş. Plasma gelsolin as a potential biomarker for Henoch–Schoenlein purpura. *Scand. J. Rheumatol.* **2016**, *46*, 166–168. [CrossRef]
35. Roy, S.; Esmaeilniakooshkghazi, A.; Patnaik, S.; Wang, Y.; George, S.P.; Ahrorov, A.; Hou, J.K.; Herron, A.J.; Sesaki, H.; Khurana, S. Villin-1 and Gelsolin Regulate Changes in Actin Dynamics That Affect Cell Survival Signaling Pathways and Intestinal Inflammation. *Gastroenterology* **2018**, *154*, 1405–1420.e2. [CrossRef]

biomedicines

Article

Periodontal Disease Augments Cardiovascular Disease Risk Biomarkers in Rheumatoid Arthritis

Jeneen Panezai [1,2,3,*], Ambereen Ghaffar [4], Mohammad Altamash [5], Mikael Åberg [6], Thomas E. Van Dyke [3,7], Anders Larsson [6] and Per-Erik Engström [1]

1 Section of Oral Health and Periodontology, Division of Oral Diseases, Department of Dental Medicine, Karolinska Institutet, 14104 Huddinge, Sweden; per-erik.engstrom@ki.se
2 Department of Microbiology, Faculty of Life Sciences and Informatics, Balochistan University of Information Technology, Engineering and Management Sciences (BUITEMS), Quetta 87300, Pakistan
3 Department of Applied Oral Sciences, The Forsyth Institute, Cambridge, MA 02142, USA; tvandyke@forsyth.org
4 Rheumatology Clinic, Habib Medical Centre, Karachi 75950, Pakistan; drambereen@gmail.com
5 Department of Periodontology, Altamash Institute of Dental Medicine, Karachi 75500, Pakistan; drmaltamash@gmail.com
6 Clinical Chemistry, Department of Medical Sciences, Uppsala University, 75236 Uppsala, Sweden; mikael.aberg@medsci.uu.se (M.Å.); anders.larsson@akademiska.se (A.L.)
7 Department of Oral Medicine, Infection and Immunity, Harvard Faculty of Medicine, Boston, MA 02115, USA
* Correspondence: jpanezai@forsyth.org or Jeneen.panezai@buitms.edu.pk

Abstract: Objectives: Periodontal disease (PD) and rheumatoid arthritis (RA) are known chronic conditions with sustained inflammation leading to osteolysis. Cardiovascular diseases (CVD) are frequent comorbidities that may arise from sustained inflammation associated with both PD and RA. In order to determine CVD risk, alterations at the molecular level need to be identified. The objective of this study, therefore, was to assess the relationship of CVD associated biomarkers in RA patients and how it is influenced by PD. Methods: The study consisted of patient (26 RA with PD, 21 RA without PD, 51 patients with PD only) and systemically and periodontally healthy control (n = 20) groups. Periodontal parameters bleeding on probing, probing pocket depth, and marginal bone loss were determined to characterize the patient groups. Proteomic analysis of 92 CVD-related protein biomarkers was performed using a multiplex proximity extension assay. Biomarkers were clustered using the search tool for retrieval of interacting genes (STRING) to determine protein–protein interaction (PPI) networks. Results: RA patients with PD had higher detection levels for 47% of the measured markers (ANGPT1, BOC, CCL17, CCL3, CD4, CD84, CTRC, FGF-21, FGF-23, GLO1, HAOX1, HB-EGF, hOSCAR, HSP 27, IL16, IL-17D, IL18, IL-27, IL6, LEP, LPL, MERTK, MMP12, MMP7, NEMO, PAPPA, PAR-1, PARP-1, PD-L2, PGF, PIgR, PRELP, RAGE, SCF, SLAMF7, SRC, THBS2, THPO, TNFRSF13B, TRAIL-R2, VEGFD, VSIG2, and XCL1) as compared to RA without PD. Furthermore, a strong biological network was identified amongst these proteins (clustering coefficient = 0.52, PPI enrichment p-value < 0.0001). Coefficients for protein clusters involved in CVD (0.59), metabolic (0.53), and skeletal (0.51) diseases were strongest in the PD group. Conclusion: Periodontal disease augments CVD-related biomarkers in RA through shared pathological clusters, concurrently enhancing metabolic and skeletal disease protein interactions, independent of autoimmune status.

Keywords: inflammation; proteins; proteomics; rheumatoid arthritis; periodontal disease; cardiovascular disease

Citation: Panezai, J.; Ghaffar, A.; Altamash, M.; Åberg, M.; Van Dyke, T.E.; Larsson, A.; Engström, P.-E. Periodontal Disease Augments Cardiovascular Disease Risk Biomarkers in Rheumatoid Arthritis. *Biomedicines* **2022**, *10*, 714. https://doi.org/10.3390/biomedicines10030714

Academic Editor: Marianna Christodoulou

Received: 15 February 2022
Accepted: 16 March 2022
Published: 19 March 2022

Publisher's Note: MDPI stays neutral with regard to jurisdictional claims in published maps and institutional affiliations.

1. Introduction

Chronic inflammation stems from persistent acute inflammation due to the failure to resolve the acute phase, often associated with the inability to remove the inducing agent or stimulus. Several diseases that acquire such chronicity due to a dysregulated immune response include atherosclerosis, type 2 diabetes, rheumatoid arthritis (RA), and periodontal disease (PD) [1].

Cardiovascular diseases (CVD) are the leading cause of global mortality with over 75% of cases in low- and middle-income countries [2]. CVD comprises a group of disorders that involve the heart muscle and blood vessels. The most common pathogenic pathway that leads to CVD is atherosclerosis [3]. Risk factors such as smoking, diabetes, hypertension, and obesity are transduced into atherosclerotic events via complex interactions between endothelial adhesion molecules and inflammatory cells including macrophages and T lymphocytes. The inflammatory response also has an autoimmune component as low-density lipoprotein (LDL) cholesterol, one of the retained lipids in atherosclerotic plaques, is antigenic resulting in production of high affinity antibodies [4].

PD is an independent risk factor for the development of CVD [5]. Systematic reviews have shown a consistent association between CVD and PD which may be partially attributed to shared risk factors and the dissemination of periodontal pathogens into the bloodstream or an increase in systemic inflammation [6].

RA is an autoimmune disease characterized by synovial inflammation and destruction due to immune mediated inflammation. This sustained inflammation in RA promotes cardiovascular pathology to such an extent that it remains the leading cause of mortality in RA patients [7]. The overall increased CVD risk in RA has been attributed less to traditional CVD risk factors and more to underlying autoimmunity and inflammatory burden [8].

The influence of PD and RA combined may increase the menacing effects of inflammation and raise an individual's risk of developing CVD even further. This can be evaluated through exploration of emergent biomarkers involved in CVD initiation and pathology and, therefore, the aim of this study is to assess CVD related biomarkers in RA patients and how it is influenced by PD.

2. Materials and Methods

The study was performed at the Department of Periodontology, the Altamash Institute of Dental Medicine, between October 2012 and August 2017 in Karachi, Pakistan. Upon obtaining informed consent, a detailed questionnaire was used to acquire information pertaining to medical and dental history. The minimum sample size was calculated using the Epitools Epidemiological Calculators [9] with the assumptions of a power of 80% and a confidence interval (CI) of 95%. The parameters reported in the literature pertaining to South Asian population were used [10]: (1) frequency of 60% for PD among RA patients and (2) frequency of 28% for PD without RA.

2.1. Rheumatoid Arthritis Group

A total of 47 RA patients were recruited via consecutive sampling from the Department of Rheumatology at the Habib Medical Centre in Karachi, Pakistan. These patients were established RA cases diagnosed by a rheumatologist (AG) using current ACR/EULAR classification criteria [11]. All patients were receiving disease-modifying anti-rheumatic drugs (DMARDs), corticosteroids, non-steroidal anti-inflammatory drugs (NSAIDs), or a biologic DMARD (Rituximab) at the time of the examination. Based on their periodontal status, the patients were divided into two groups: RA with PD ($n = 26$) and RA without PD ($n = 21$).

2.2. Periodontal Disease Group

A group of 51 participants diagnosed with PD, but exhibiting no signs of RA, gout, or osteoarthritis were also included. Individuals with a history of treatment for PD during the last six months and/or treatment with antibiotics in the last three months were excluded.

We used twenty controls for comparison and better characterization of the groups. All controls had clinically healthy periodontium and no systemic disease. Blood samples were drawn from all participants and prepared sera were stored at −80 °C until the time of analyses.

2.3. Periodontal Examination

Periodontal examination was carried out for all teeth except for the third molars by a single examiner (JP). The defining criteria for PD was probing pocket depth (PPD) of $\geq$5 mm (to the nearest millimeter) in at least three different sites using a periodontal probe (Hu-Friedy manufacturing, Chicago, IL, USA). Pockets measuring $\geq$5 mm were added to calculate the sum of deep pockets representing PD affected sites. Our group has designed continuous periodontal indices to gauge the severity of PD as a continuous variable rather than dichotomous. These parameters were used to assess the severity of PD:

- Bleeding on probing (BOP)
- ΣPPD Total
- ΣPPD Disease
- Adjusted PPD Total
- Adjusted PPD Diseased sites
- Σ Marginal bone loss (MBL)
- Adjusted ΣMBL

Details of the parameters and their measurements are described in our previous publications [12,13].

2.4. Anthropometric Measures

Body weight was measured to the nearest kg. Using non-stretchable measuring tape, height and waist were measured to the nearest cm. Waist was measured in the horizontal plane at the midpoint between the lowest rib and the iliac crest. Body mass index (BMI) was calculated from weight and height measurements (kg/m^2). Anthropometric measurements were recorded for all four groups.

2.5. Glycated Hemoglobin (HbA1c)

Glycated hemoglobin levels were determined for all four groups after collecting four milliliters of whole blood into spray-coated EDTA tubes (lavender top, Becton, Dickinson, Franklin Lakes, NJ, USA). The samples were analyzed on the same day using the ion-exchange high-performance liquid chromatography system Bio-Rad D-10 Hemoglobin Testing System (Bio-Rad Laboratories, Hercules, CA, USA). The HbA1c values are standardized according to the National Glycohemoglobin Standardization Program (NGSP) system [14].

2.6. Proteomic Profiling

Serum samples were analyzed at the SciLifeLab Affinity Proteomics Uppsala, Uppsala University, (Uppsala, Sweden) using proximity extension assay (PEA) technology (Olink Proteomics, Uppsala, Sweden). Levels of 92 proteins from the Olink Target96 CVD II panel were measured (Supplementary File Table S1). The PEA technology utilize pairs of antibodies equipped with DNA reporter molecules [15]. When binding in close proximity to their correct targets, the antibody pairs give rise to new DNA amplicons each ID-barcoding their respective antigens. The amplicons are subsequently quantified using the Fluidigm BioMark™ HD real-time PCR platform (South San Francisco, CA, USA). For data analysis and quality control Olink NPX Manager Software was used and the inter-plate variability was adjusted by intensity normalization. The final protein values are expressed as Normalized Protein eXpression (NPX) values which are on a log2 scale and one unit higher NPX represents a doubling of the measured protein concentration. Data quality was controlled and normalized using an internal and an interpolate control. Assay validation data for all proteins from the panel are available (www.olink.com).

2.7. Protein–Protein Interaction (PPI) Network Analysis

The Protein–Protein Interactions (PPI) Network analysis was performed using the search tool for retrieval of interacting genes (STRING) (https://string-db.org, accessed on 21 February 2021). The STRING database interaction evidence is thematically grouped into 'channels' (such as text mining, co-expression, and lab experiments) and limited to *Homo sapiens*. An interaction score > 0.4 was applied to construct the PPI networks. STRING performed identifier mapping (test the proteins of each known pathway for any nonrandom skew within the user-provided input values, and report statistically significant pathways) and displayed a network with all of the mapped proteins and their interconnections [16]. In the networks, the nodes correspond to the proteins and the edges represent the interactions. STRING was employed to seek potential interactions among proteins. The clustering coefficient, where 0 represents the absence of connections and 1 a fully connected network, was calculated quantifying the abundance of connected nodes in a PPI network. PPI enrichment p-value is used to indicate that the nodes are not random and that the observed number of edges is significant.

2.8. Statistical Analyses

All analyses were performed test using GraphPad Prism version 9.0. for Windows (GraphPad Software, San Diego, CA, USA). Patient characteristics were analyzed using one-way ANOVA and Kruskal–Wallis tests depending on the normality of the data to identify group wise differences. Inter-group differences in biomarker distributions were analyzed using the Mann–Whitney U test. The relationship between each of the 92 protein biomarkers and periodontal parameters was assessed using Spearman correlation analyses. To control the false discovery rate (FDR), the Benjamini–Hochberg procedure was applied to adjust p-values from multiple testing [17]. The significance level was defined at $p \leq 0.05$.

For exploration of biomarker patterns within disease groups, principal component analysis (PCA) was performed. PCA is a powerful exploratory model statistically used for data exploration and simplification. The technique is based on generating principal components (latent variable) from the original dataset. The relationship of the principal components to the samples is referred to as 'scores', and that to the variables is called 'loadings'. A threshold of 0.5 was deemed significant for variable loadings.

3. Results

3.1. Characteristics of Study Groups

The characteristics for the disease groups (RA with PD, RA without PD, and PD only) and controls are shown in Table 1. There were no differences in age amongst the four groups. The number of females were higher in disease groups as compared to controls. The clinical status comprised self-reported hypertension and diabetes confirmed by medication and prescription. The frequency of both conditions were similar amongst the disease groups. Periodontal parameters, waist circumference and HbA1c differed significantly amongst the groups with the highest medians in PD patients. The median for HbA1c value in the PD group classifies them as pre-diabetes overall (Supplementary File Table S3).

Table 1. Characteristics for RA and PD (disease) groups and controls.

Characteristics	Disease Groups			Control	*p*-Value
	RA with PD (*n* = 26)	**RA without PD** (*n* = 21)	**PD** (*n* = 51)	(*n* = 20)	
Age (years) [a]	48.5 (8.8)	43.1 (13.3)	47 (9.5)	43 (6.3)	0.11
Female sex, *n* (%) [b]	21 (81)	20 (95)	33 (65)	8 (40)	<0.001
Clinical Status, *n* [b]					
- Hypertension	9	7	8	-	0.14
- Diabetes	2	6	10	-	
BOP [c]	23 (57)	43 (60)	77 (56)	15 (32)	<0.0001
PPD Total [c]	301(81)	276 (86)	384 (113)	191 (24)	<0.0001
PPD Disease [c]	107.5 (104)	0 (2.5)	229 (136)	0 (5)	<0.0001
Adjusted PPD Total [c]	11.6 (2.9)	10.8 (2.8)	15.5 (4.2)	6.8 (1)	<0.0001
Adjusted PPD Disease [c]	8 (4.3)	0 (0)	10.4 (4)	0 (0)	<0.0001
∑MBL [c]	27.4 (10.8) [d]	13.5 (12.9) [e]	34.2 (15.4)	8.8 (17.5) [f]	<0.0001
Adjusted ∑MBL [c]	4.57 (1)	3.02 (0.9)	5.24 (2)	2.88 (0.8)	<0.0001
Body mass index (kg/m^2) [c]	24.2 (5)	24.1 (6.2)	25.2 (4)	23.9 (4.6)	0.35
Waist circumference (cm) [c]	102 (30)	97 (23)	109 (19)	86 (17)	<0.0001
HbA1c % [c]	5.0 (1)	5.0 (2)	5.7 (1.2)	4.5 (0.8)	<0.0001

BOP = bleeding on probing, PPD = probing pocket depth, MBL = marginal bone loss, HbA1c = glycated hemoglobin. [a] Differences in means were tested using one-way ANOVA test (testing overall difference among the three groups). [b] Differences in frequency were tested using χ^2 (chi-squared) test (testing overall difference among the three groups). [c] Differences in medians were tested using Kruskal–Wallis test (testing overall difference among the three groups). [d] Missing data (*n* = 5) was excluded in the analyses. [e] Missing data (*n* = 1) was excluded in the analyses. [f] Missing data (*n* = 1) was excluded in the analyses.

3.2. Group-Wise Biomarker Distribution

The distribution of 92 CVD biomarkers was assessed among the four groups. Two samples from the PD group were excluded due to unacceptable technical variations. Biomarkers with significantly increased levels in RA with PD groups as compared to RA without PD are shown in Figure 1. Higher NPX values were noted for 47% (43/92) of the markers which were: ANGPT1, BOC, CCL17, CCL3, CD4, CD84, CTRC, FGF-21, FGF-23, GLO1, HAOX1, HB-EGF, hOSCAR, HSP 27, IL16, IL-17D, IL18, IL-27, IL6, LEP, LPL, MERTK, MMP12, MMP7, NEMO, PAPPA, PAR-1, PARP-1, PD-L2, PGF, PIgR, PRELP, RAGE, SCF, SLAMF7, SRC, THBS2, THPO, TNFRSF13B, TRAIL-R2, VEGFD, VSIG2 and XCL1. For 32 of these biomarkers (BOC, CCL17, CCL3, CD84, CTRC, FGF-21, GLO1, HAOX1, HB-EGF, hOSCAR, HSP 27, IL-16, IL-17D, IL-18, IL-27, IL-6, LEP, LPL, MERTK, MMP12, NEMO, PAR-1, PARP-1, PD-L2, PRELP, RAGE, SCF, SLAMF7, THBS2, TNFRSF13B, TRAIL-R2, and XCL1) PD and RA with PD groups exhibited no differences (Supplementary File Table S2).

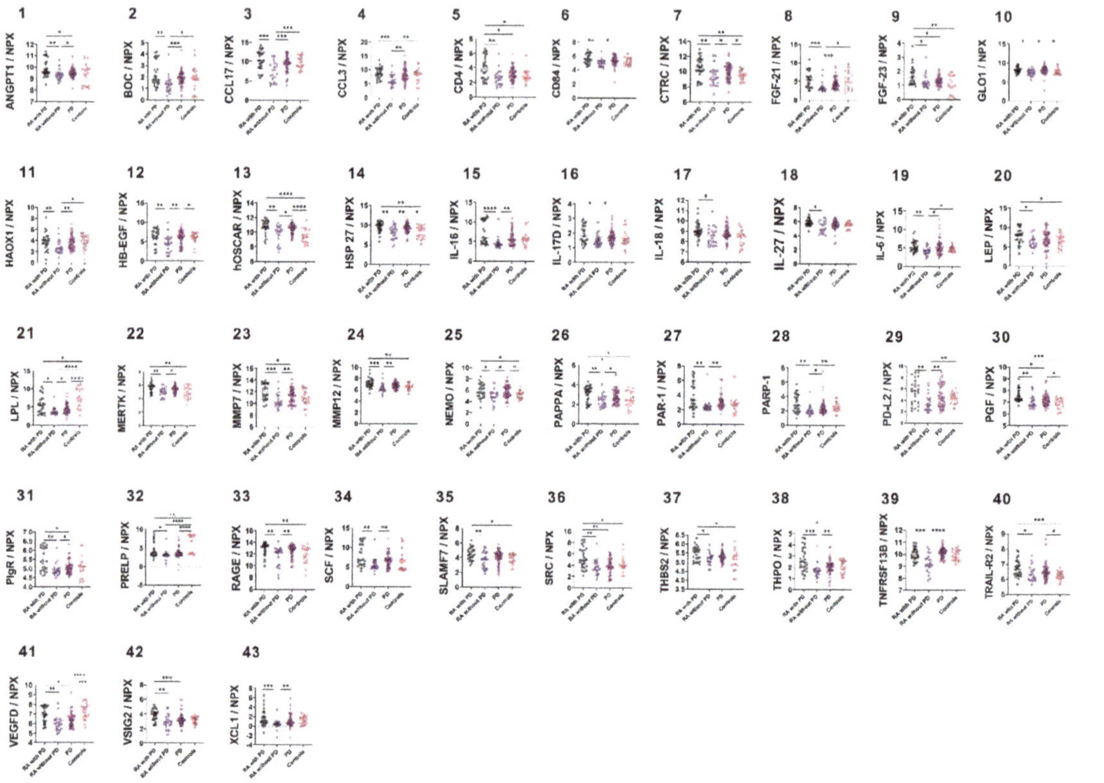

Figure 1. Group wise analyses for CVD-related biomarkers. Graphs 1–43 showing higher detection levels in RA with PD as compared to RA without PD. (1) ANGPT1, (2) BOC, (3) CCL17, (4) CCL3, (5) CD4, (6) CD84, (7) CTRC, (8) FGF-21, (9) FGF-23, (10) GLO1, (11) HAOX1, (12) HB-EGF, (13) hOSCAR (14) HSP 27, (15) IL-16, (16) IL-17D, (17) IL-18 (18) IL-27, (19) IL-6, (20) LEP, (21) LPL, (22) MERTK, (23) MMP7, (24), MMP12, (25) NEMO, (26) PAPPA, (27) PAR-1, (28) PARP-1, (29) PD-L2, (30) PGF, (31) PIgR, (32) PRELP, (33) RAGE, (34) SCF, (35) SLAMF7, (36) SRC, (37) THBS2, (38) THPO, (39) TNFRSF13B, (40) TRAIL-R2 (41) VEGFD, (42) VSIG2, and (43) XCL1. Data are presented as median with interquartile range. Group differences were calculated using Mann–Whitney U test. * p value $\leq$ 0.05, ** p value < 0.01, *** p value < 0.001, **** p value < 0.0001.

3.3. Correlation of CVD Biomarkers with Periodontal Parameters

The correlation between periodontal parameters and CVD-related biomarkers are shown in Table 2. The highest frequency of significant correlations was seen in the PD group for all parameters except for adjusted MBL. Anti-inflammatory marker IL-4RA was inversely related with three out of five indices for inflammation and pocketing. The Proto-oncogene tyrosine-protein kinase Src (SRC) was inversely correlated with four out of five indices for inflammation and pocketing. All correlations were direct amongst the RA with PD group, except for ACE-2. The most frequent and moderately strong correlations were noted with adjusted MBL. Least frequent correlations were noted in the RA without PD group. Dickkopf-related protein 1 (Dkk-1) and thrombospondin 2 (THBS2) were directly associated with Adjusted PPD Total. There was no overlap between the associated biomarkers amongst the three groups.

Table 2. Correlations of CVD risk biomarkers with periodontal pocketing and marginal bone loss.

	Periodontal Pocketing and Inflammation										Marginal Bone Loss			
	BOP		PPD Total		PPD Disease		Adj. PPD Total		Adj. PPD Disease		∑MBL		Adj. MBL	
	Analyte	r	Analyte	r	Analyte	r	Analyte	r	Analyte	r	Analyte	r	Analyte	r
RA with PD			ACE-2	−0.42	LOX-1	0.41			PTX3	0.44	ANGPT1 PGF	0.47 0.45	LEP TNFRSF13B IL-27	0.48 0.48 0.46
PD	CXCL1 SRC XCL1	−0.31 −0.32 −0.36	IL-4RA	−0.29	IL-4RA MERTK SRC	−0.37 −0.28 −0.29	IL-4RA MMP-12 SRC	−0.33 −0.31 −0.38	MMP-12 SRC	−0.34 −0.42	ADAM-TS13	−0.29		
RA without PD	FGF-23	−0.52					Dkk-1 THBS2	0.47 0.49			CD40-L TGM2	−0.47 −0.45		

Spearman rank correlation was used to identify correlations. All coefficients show biomarkers with adjusted p-values ≤ 0.05 after using the Benjamini–Hochberg procedure for multiple testing.

3.4. PCA

PCA was performed using standardized data and PC selection via parallel analysis. In the initial PCA output, selected component PC1 with loading structure >0.5 are shown for all disease groups (Table 3). The individual values show the correlation between the specific biomarker and the PC 1 for which the loading is calculated for. For RA with PD group, 64 biomarkers exceeded the 0.5 threshold of loading significance. Similarly, RA without PD had 65 biomarkers exceeding the threshold whereas PD group showed 55 biomarkers exceeding the threshold.

Table 3. PC loadings for disease groups.

RA with PD			PD			RA without PD		
Variable	**PC1**	**PC2**	**Variable**	**PC1**	**PC2**	**Variable**	**PC1**	**PC2**
C22D4	**0.98**	−0.08	PDGF subunit B	**0.87**	−0.12	PDGF subunit B	**0.92**	0.02
SCF	**0.96**	0.05	SOD2	**0.87**	−0.05	CD84	**0.90**	−0.14
IL-17D	**0.95**	0.00	MMP7	**0.86**	0.12	SCF	**0.90**	−0.31
PAR-1	**0.93**	−0.16	CD4	**0.85**	−0.28	BOC	**0.90**	−0.32
BOC	**0.93**	−0.16	hOSCAR	**0.84**	0.08	CXCL1	**0.90**	0.11
PIgR	**0.93**	−0.02	CCL17	**0.84**	−0.08	PD-L2	**0.89**	0.21
VEGFD	**0.93**	0.20	IL16	**0.83**	−0.24	MERTK	**0.87**	0.22
IL16	**0.93**	0.07	HB-EGF	**0.83**	0.05	VEGFD	**0.87**	0.06
MMP7	**0.92**	0.15	CCL3	**0.83**	0.11	VSIG2	**0.87**	0.20
SPON2	**0.91**	−0.14	PIgR	**0.80**	−0.30	MMP7	**0.86**	0.09
PDGF subunit B	**0.91**	0.19	VEGFD	**0.80**	0.01	THBS2	**0.86**	−0.11
THPO	**0.91**	−0.18	RAGE	**0.79**	0.07	BNP	**0.85**	−0.06
CD84	**0.90**	0.21	SCF	**0.78**	−0.41	PIgR	**0.84**	−0.33
hOSCAR	**0.90**	0.25	HSP 27	**0.78**	−0.21	PARP-1	**0.82**	−0.36
LPL	**0.90**	−0.18	IL-17D	**0.77**	0.04	HO-1	**0.82**	0.16
FABP2	**0.89**	−0.23	THBS2	**0.77**	−0.02	hOSCAR	**0.82**	0.27
FGF-21	**0.89**	−0.30	CD84	**0.76**	−0.14	CD4	**0.82**	−0.26
THBS2	**0.88**	−0.06	ADAM-TS13	**0.75**	0.17	CCL17	**0.82**	0.28
CCL17	**0.88**	0.28	PD-L2	**0.75**	0.05	IL-17D	**0.82**	−0.20
CXCL1	**0.88**	0.30	FGF-21	**0.74**	−0.06	DECR1	**0.81**	−0.17
PARP-1	**0.88**	−0.10	HO-1	**0.73**	−0.22	FABP2	**0.81**	−0.41
CTRC	**0.87**	0.26	Dkk-1	**0.72**	−0.22	GDF-2	**0.81**	0.15
ANGPT1	**0.86**	−0.04	BOC	**0.72**	−0.36	RAGE	**0.80**	0.17
PRELP	**0.85**	−0.44	PAPPA	**0.71**	0.03	SORT1	**0.80**	0.20
MERTK	**0.85**	−0.06	VSIG2	**0.70**	0.44	PGF	**0.80**	0.28
TM	**0.85**	−0.28	SORT1	**0.70**	0.32	ADAM-TS13	**0.80**	0.23
CCL3	**0.84**	0.18	MERTK	**0.69**	0.07	CCL3	**0.80**	−0.23
BMP-6	**0.84**	−0.36	GDF-2	**0.68**	0.33	FGF-21	**0.80**	−0.31
SOD2	**0.83**	0.02	TNFRSF13B	**0.67**	0.11	HB-EGF	**0.78**	0.48

Table 3. *Cont.*

RA with PD			PD			RA without PD		
Variable	**PC1**	**PC2**	**Variable**	**PC1**	**PC2**	**Variable**	**PC1**	**PC2**
SORT1	**0.83**	0.36	CXCL1	**0.66**	0.07	THPO	**0.78**	−0.39
STK4	**0.81**	−0.07	GLO1	**0.66**	0.03	SLAMF7	**0.77**	0.39
VSIG2	**0.81**	−0.04	NEMO	**0.66**	0.02	LEP	**0.77**	0.24
SRC	**0.81**	0.10	CTRC	**0.66**	−0.02	Dkk-1	**0.77**	−0.22
RAGE	**0.81**	0.24	SPON2	**0.65**	−0.11	TF	**0.76**	−0.01
PGF	**0.81**	−0.31	CTSL1	**0.65**	0.32	AGRP	**0.76**	−0.14
HO-1	**0.81**	−0.06	IL-1ra	**0.63**	0.16	IL-1ra	**0.76**	0.18
FGF-23	**0.80**	−0.28	AGRP	**0.61**	−0.19	NEMO	**0.75**	−0.06
IL18	**0.79**	−0.18	FGF-23	**0.60**	−0.29	TNFRSF13B	**0.74**	0.19
Dkk-1	**0.79**	0.01	IL1RL2	**0.60**	0.08	PAR-1	**0.74**	−0.57
AGRP	**0.78**	−0.27	THPO	**0.60**	−0.37	PSGL-1	**0.73**	−0.04
XCL1	**0.76**	−0.50	DCN	**0.59**	−0.05	IL16	**0.73**	−0.39
HB-EGF	**0.76**	**0.53**	IL-27	**0.57**	0.18	DCN	**0.73**	0.25
HSP 27	**0.76**	0.20	TNFRSF11A	**0.56**	**0.61**	HSP 27	**0.72**	−0.01
TNFRSF13B	**0.75**	−0.15	PTX3	**0.56**	0.08	PAPPA	**0.72**	−0.22
PD-L2	**0.74**	0.50	BNP	**0.56**	−0.05	SOD2	**0.71**	−0.38
NEMO	**0.74**	0.39	IL6	**0.55**	0.37	TNFRSF11A	**0.71**	**0.56**
BNP	**0.74**	−0.08	TGM2	**0.54**	0.13	TM	**0.70**	0.32
LEP	**0.73**	0.08	PGF	**0.54**	**0.63**	LPL	**0.70**	−0.62
PAPPA	**0.73**	0.42	LPL	**0.54**	−0.67	FGF-23	**0.69**	−0.40
IL-1ra	**0.72**	−0.17	PRSS27	**0.54**	0.08	GLO1	**0.66**	0.16
TNFRSF11A	**0.70**	−0.24	FABP2	**0.53**	−0.53	SPON2	**0.65**	0.05
DCN	**0.70**	**0.50**	PAR-1	**0.52**	−0.66	SRC	**0.64**	−0.33
IL1RL2	**0.69**	0.04	TIE2	**0.52**	0.36	CTRC	**0.64**	0.10
PSGL-1	**0.68**	−0.05	ANGPT1	**0.51**	−0.44	IL1RL2	**0.64**	−0.06
TF	**0.67**	−0.05	IDUA	**0.50**	0.04	IL-27	**0.63**	0.47
ADM	**0.66**	−0.30	IL18	**0.50**	0.33	PRSS27	**0.62**	0.38
ADAM-TS13	**0.66**	**0.55**	SRC	0.49	−0.39	BMP-6	**0.62**	−0.49
TNFRSF10A	**0.59**	−0.53	LEP	0.48	0.07	IL18	**0.61**	0.30
MARCO	**0.57**	−0.24	TM	0.47	**0.64**	TIE2	**0.60**	**0.52**
CTSL1	**0.57**	0.04	CEACAM8	0.47	0.09	SERPINA12	**0.57**	0.25
GDF-2	**0.57**	0.29	STK4	0.47	−0.49	ANGPT1	**0.55**	−0.48
PRSS27	**0.56**	0.35	PARP-1	0.47	−0.40	STK4	**0.54**	−0.53
TRAIL-R2	**0.51**	−0.61	HAOX1	0.47	−0.04	ITGB1BP2	**0.54**	−0.20
IL-27	**0.50**	−0.30	DECR1	0.43	0.33	REN	**0.53**	0.31
GLO1	**0.50**	0.28	TRAIL-R2	0.42	**0.72**	PRELP	**0.51**	−0.77
IgG Fc receptor II-b	0.44	−0.31	TF	0.42	**0.56**	IL6	0.49	0.01
SERPINA12	0.39	0.17	TNFRSF10A	0.40	**0.63**	GH	0.47	0.15
HAOX1	0.36	−0.32	ITGB1BP2	0.38	−0.10	GIF	0.46	−0.03
IL6	0.33	−0.18	PRELP	0.37	−0.63	CTSL1	0.46	0.25
DECR1	0.31	0.26	LOX-1	0.36	0.08	TGM2	0.44	0.39
GH	0.30	0.20	BMP-6	0.36	−0.58	TNFRSF10A	0.42	**0.54**
FS	0.29	−0.21	XCL1	0.34	−0.56	XCL1	0.38	−0.79
SLAMF7	0.29	−0.35	GH	0.34	−0.28	TRAIL-R2	0.38	**0.56**
KIM1	0.25	−0.46	Gal-9	0.33	0.24	PTX3	0.36	−0.33
Gal-9	0.22	−0.33	MARCO	0.30	−0.14	LOX-1	0.36	−0.18
ITGB1BP2	0.20	0.36	ADM	0.29	−0.38	KIM1	0.34	**0.68**
PTX3	0.18	0.12	AMBP	0.26	**0.64**	CD40-L	0.32	0.07
TGM2	0.15	0.27	CA5A	0.25	0.34	IgG Fc receptor II-b	0.32	−0.02
GIF	0.12	−0.18	KIM1	0.24	0.74	GT	0.31	0.42
MMP12	0.11	−0.54	PSGL-1	0.19	0.23	CA5A	0.29	0.26
CEACAM8	0.09	0.25	SERPINA12	0.18	0.15	MARCO	0.29	−0.01
ACE2	0.08	−0.49	SLAMF7	0.17	0.10	HAOX1	0.27	−0.29
AMBP	0.06	0.26	GIF	0.17	0.36	IDUA	0.25	0.36
TIE2	−0.03	**0.58**	REN	0.15	0.35	AMBP	0.23	**0.63**

Table 3. *Cont.*

	RA with PD			PD			RA without PD		
	Variable	PC1	PC2	Variable	PC1	PC2	Variable	PC1	PC2
	CA5A	−0.04	−0.44	ACE2	0.13	**0.54**	PRSS8	0.14	**0.81**
	LOX-1	−0.04	0.37	IL-4RA	0.12	0.31	MMP12	0.11	0.34
	IDUA	−0.10	−0.03	CD40-L	0.08	0.02	Gal-9	0.10	0.35
	REN	−0.10	−0.21	IgG Fc receptor II-b	0.03	0.24	CEACAM8	0.06	0.09
	GT	−0.11	−0.66	PRSS8	0.02	**0.81**	ADM	0.03	−0.25
	CD40-L	−0.20	−0.04	MMP12	−0.04	0.32	FS	0.01	0.23
	PRSS8	−0.44	−0.17	FS	−0.06	0.34	ACE2	0.00	0.16
	IL-4RA	−0.46	−0.76	GT	−0.13	0.16	IL-4RA	−0.01	0.26
Proportion of variance		**47.0%**	**8.9%**		**32.9%**	**11.6%**		**43.2%**	**11.3%**
Cumulative proportion of variance		**47.0%**	**55.9%**		**32.9%**	**44.5%**		**43.2%**	**54.5%**

PC = principal component. All loadings > 0.5 are in bold. The variance represented by two principal components in proportion and cumulatively are shown as percentages.

Visual representation of PC loadings plot (Figure 2) shows how the biomarkers are clustered closely together. In the disease groups, the majority of the biomarkers not only correlated strongly with each other but also with PC1 as most of the values were close to 1. The clustering pattern was more similar between the PD and RA without PD groups as some biomarkers showed stronger correlation with PC2. Using controls for reference, the loading plot showed weaker correlations between the biomarkers themselves and PC 1 and 2. The PC score plots reveal the variation in the dimensionality of the four groups.

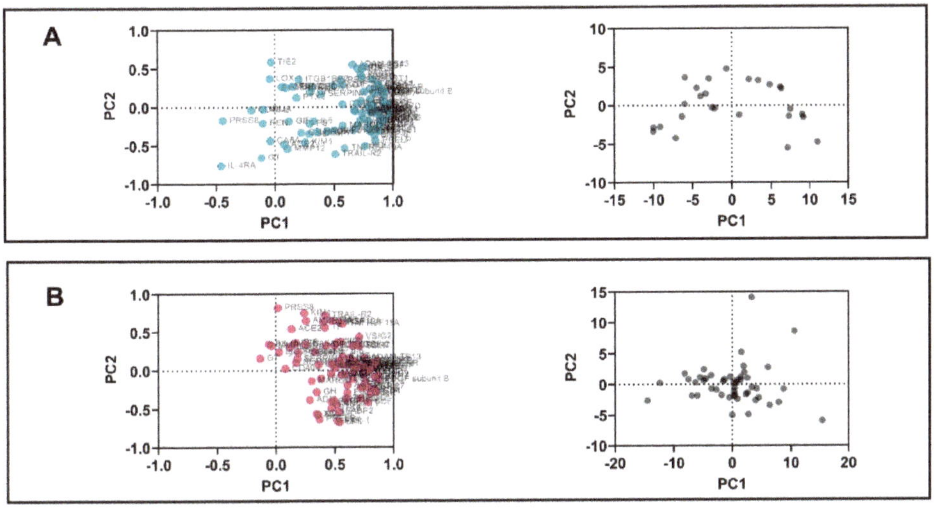

Figure 2. *Cont.*

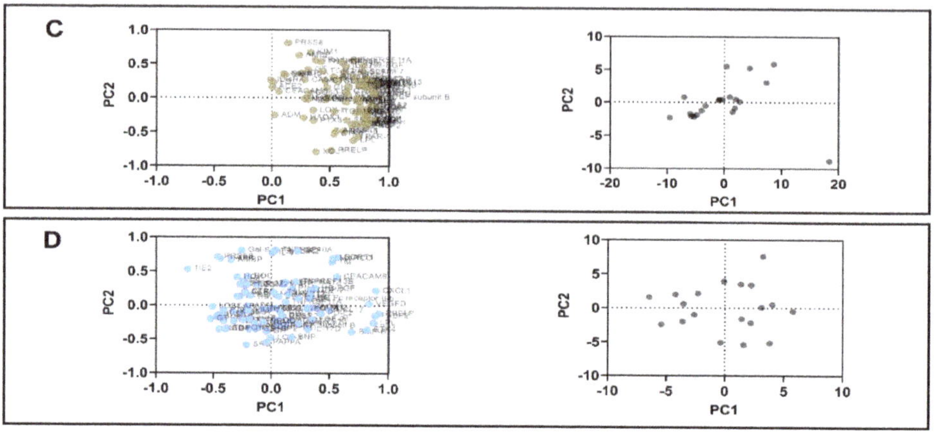

Figure 2. PCA analysis. The principal component analysis showing loadings (left side) and scores (right side) RA with PD (panel **A**), PD (panel **B**), RA without PD (panel **C**), and controls (panel **D**).

3.5. Protein–Protein Interaction Network

The Protein–Protein interaction (PPI) network analysis of 43 proteins discriminating RA with PD from RA without PD is shown in Figure 3. The potential interactions between ANGPT1, BOC, CCL17, CCL3, CD4, CD84, FGF-21, FGF-23, HB-EGF, hOSCAR, HSP 27, IL16, IL-17D, IL18, IL-27, IL6, LEP, LPL, MMP12, MMP7, NEMO, PAPPA, PAR-1, PARP-1, PD-L2, PGF, RAGE, SCF, SRC, THBS2, THPO, TNFRSF13B, TRAIL-R2, VEGFD, and XCL1 yielded a clustering coefficient of 0.52, with a PPI enrichment p value < 0.0001. Markers from the CVD panel that also play a significant role in metabolic and skeletal disease areas were identified from the PC1 results for each disease group based on bioinformatic databases, including UniProt, Human Protein Atlas, Gene Ontology (GO), and DisGeNET. PPI network analysis was performed for three disease areas per group. These results are shown in Figure 4. The clustering coefficient was strongest for the PD group in all three disease areas when compared to the RA groups. The metabolic disease proteins were identical in clustering strength in both RA groups, uninfluenced by PD status.

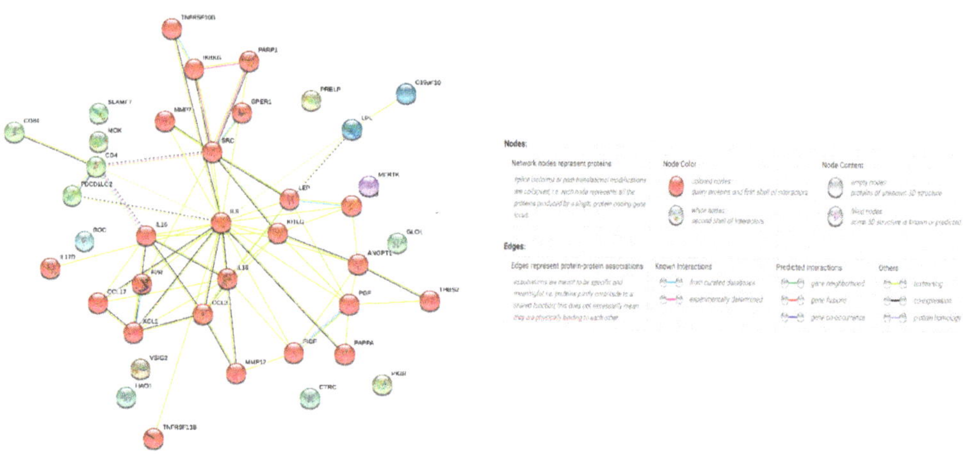

Figure 3. Protein–Protein interactions (PPI) showing networking of 43 CVD related biomarkers identified to be increased in RA with PD patients. The cluster shows frequent and strong interactions (represented by the same color of the nodes).

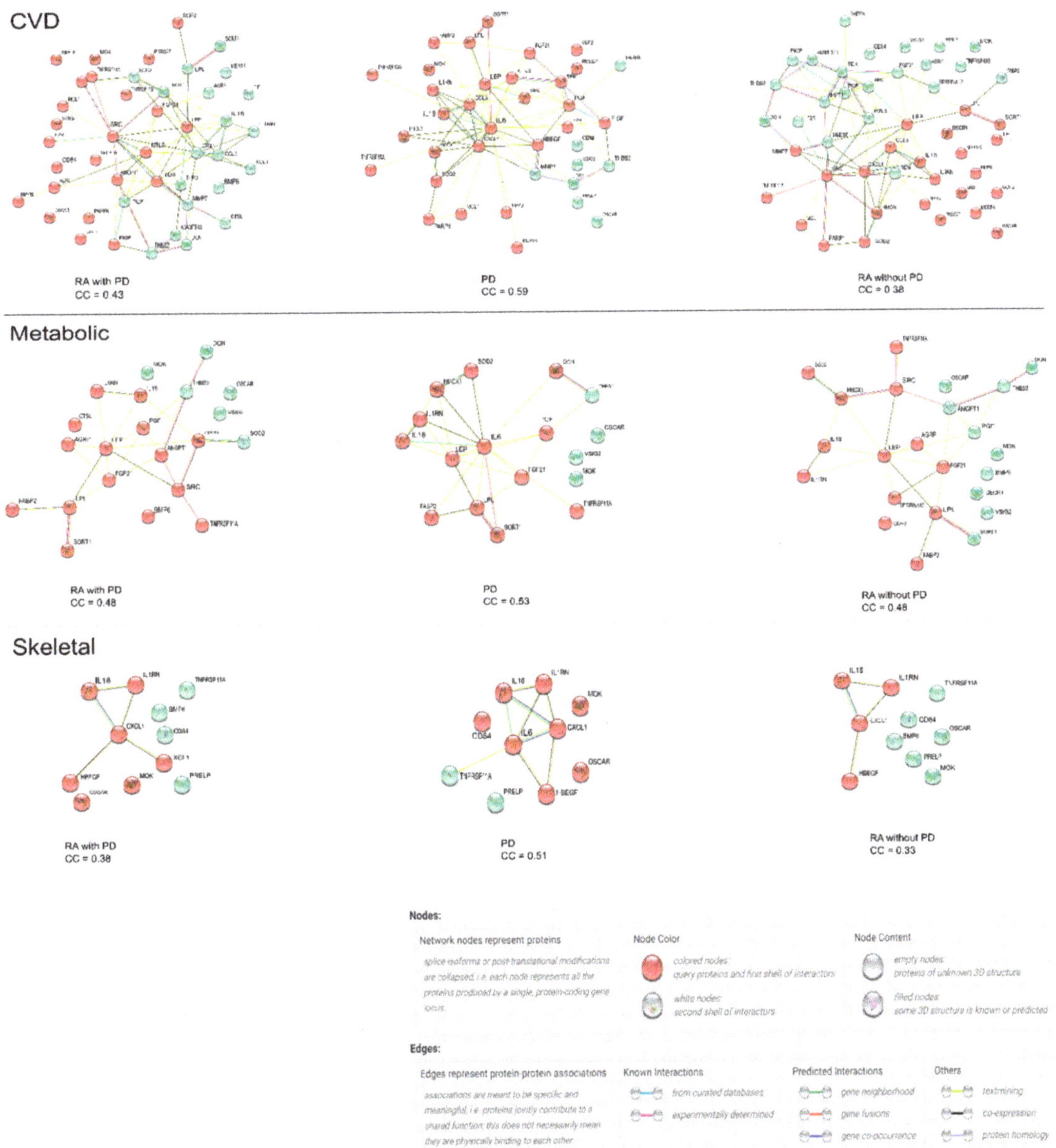

Figure 4. PC1 biomarkers and their protein network analysis according to disease area in RA with PD, PD, and RA without PD groups. The network nodes represent proteins with red colored nodes denoting first shell interactors and green color showing second shell of interactors. All cluster coefficients (CC) have a PPI enrichment value of <0.0001.

4. Discussion

In this report, we identified 43 markers with a strong interactive network in patients suffering from PD, with and without RA. The risk of CVD exists in both PD and RA through shared pathological clusters. Several markers also increase associated metabolic and skeletal disease risk, independent of autoimmune status. In order to prevent CVD related morbidity and mortality in chronic inflammatory conditions, it is crucial to identify and study CVD risk biomarkers in the early stages of inflammatory disease. Studying a vast array of biomarkers that are significant in CVD development is an advantage offered by protein profiling using proteomic techniques. The biological mechanisms can be better understood with identification of early stage biomarkers which predispose RA and PD independently or combined to risk of CVD.

In our study, we examined an array of 92 biomarkers related to cardiovascular dysfunction and inflammation in RA patients with or without PD and PD patients alone. The disease groups showed a higher number of women of a relatively young age (<50 years). The gender dominance of women was expected since they are affected more by RA and seek dental care more frequently as compared to men [18]. In young women, being affected with RA is a risk for CVD [19]. RA female patients are 2.6-times more likely to develop CVD as compared to the general population. Our findings in relation to the age of the present cohort are, therefore, relevant.

An overall dysregulated level of HbA1c and increased waist circumference, a measure of central obesity, in PD patients has been confirmed previously as well [20]. Periodontal parameters were less severe in RA patients with PD and attributable to the use of disease modifying anti-rheumatic drugs (DMARDs) by the former group.

For direct comparisons, CVD biomarker distribution was assessed in all groups. Based on biological processes, the frequency of PD relevant biomarkers represented immune response (47%), cell adhesion (40%), intracellular mitogen-activated protein kinase (MAPK) signaling cascade (35%), inflammation (30%), catabolic process (23%), and proteolysis (19%). MAPKs are implicated as key regulators of inflammatory cytokines like IL-6 and TNF, thus transducing inflammation [21]. One of the contributing factors to CVD is endothelial dysfunction which is brought about by over expression of adhesion molecules due to inflammatory mediators [22].

The association of periodontal parameters with CVD biomarkers was also examined per disease group. In RA with PD, the associated biomarkers for periodontal pocketing spanned from enzymes (ACE-2) and membrane proteins (LOX-1) to plasma proteins (PTX3). The inverse relationship between ACE-2 levels and PPD Total scores reflect a pro-atherogenic state as ACE-2 levels have been detected in RA patients with a negative correlation with intima media thickness of carotid arteries [23]. Diseased probing sites correlated moderately with PTX3, also a pro-atherogenic inflammatory marker expressed by vascular endothelium known to modify angiogenesis and atherosclerotic lesion development [24]. Oxidized low density lipoproteins (ox-LDL) have been directly implicated in the pathogenesis of RA through signaling via the lectin-like ox-LDL receptor 1 (LOX-1) in the joint synovium [25]. LOX-1 activates downstream pathways that enhance atherosclerosis via endothelial dysfunction. LOX-1 is also expressed in platelets, where it enhances platelet activation and adhesion to endothelial cells [26]. Both LOX-1 and PTX3 associations with PPD Disease were moderately strong suggesting that PD contributes to a pro-atherosclerotic milieu in RA.

In PD patients only, three biomarkers (IL-4RA, SRC, MMP-12) conveyed a consistent pattern associated with deep pocketing. They reflect an unbalanced state with low anti-inflammatory IL-4RA levels confirming previous findings [27]. These findings corroborate a defect in the regulatory involvement of SRC and MMP-12 with phagocytosis and host defense mechanisms in PD patients. Low-MMP12 levels in periodontal tissues may be a risk factor underlying excessive pro-inflammatory IFN-γ macrophage activation in disease [28].

FGF-23, a bone-derived hormone, can also drive an increased production of pro-inflammatory cytokines [29]. Dkk-1 is known to play a pathophysiological role in bone erosion and joint remodeling in RA patients [30]. It negatively regulates the function of the Wnt pathway which is involved in the differentiation of osteoblasts. Thrombospondin 2 (THBS2) a matricellular protein, has been demonstrated as an endogenous regulator of angiogenesis and inflammation in the RA synovium [31].

High levels of LEP (leptin) associated with increased MBL in RA with PD patients further enforce previous findings of increased LEP levels found in dysfunctional immune phenotype including insulin resistance, inflammation, and disturbances in hemostatic factors [32]. TNSFR13B and its association with MBL reflects an increased B-cell proliferative and surviving capacity via its receptor BAFF (B-cell-activating factor). BAFF are up-regulated in RA synovial joints as well as early stages of PD [33,34]. IL-27′s enhancement of TNF-α mediated upregulation of adhesion molecules and pro-inflammatory IL-6 in blood monocytes of patients with acute myocardial infarction (MI) makes it high CVD risk associated [35]. The TGM2 levels in RA without PD correlate inversely with the total sum of MBL which aligns with previous findings that TGM2 correlates with RANKL production in human periodontal ligament cells as part of the inflammatory response in PD [36].

Protein biomarkers with high loadings (>0.5) on PC1 exceeded 50% of the total biomarkers analyzed in all disease groups. The biomarkers contributing to the greatest variance were similar in all three groups. Based on their disease–gene associations, these biomarkers are involved in vascular inflammation (HO-1, LPL, PAPPA, ADAMTS13, ADM, PGF, and GDF-2), hypertension, and arterial disease (PAPPA, ADAMTS13, ADM, and PGF). The underlying gene ontology represents upregulation of chemotaxis (XCL1 and CCL3), T helper 1 cytokines (SLAMF1, IL-18, XCL1, and IL-1ra), T helper 2 cytokines (XCL1), negative regulation of vasoconstriction (ADM and LEP), hematopoietic stem cell proliferation (THPO and ATXN) and increased bone loss (TNFSF11A and TF) [37].

The additional analyses of PPI networking for PC1 markers in CVD, metabolic and skeletal disease areas was performed as osseous and metabolic disturbances, especially insulin resistance, are highly frequent co-morbidities in both PD and RA [38–41]. The clustering coefficients displayed by PD group PC 1 biomarkers reflect a greater involvement of disease related proteins that make them a group with the highest risk for developing CVD, insulin resistance and skeletal diseases. The dampening of inflammatory circuits due to the use of NSAIDs and DMARDs in RA groups are to have some impact on the level of engagement amongst these proteins. Future studies are required to identify and validate markers of diagnostic and therapeutic relevance that may enhance the 'treat-to-target' strategy for RA and, hopefully, PD.

The limitations to our study pertain to the limited size of samples and the exploration of proteins which have been associated with cardiovascular diseases. Due to the exploratory nature of our study and the low prevalence of RA (~1%), we used a non-probability sampling method in which groups were not sex-matched. Despite these limitations, our findings have identified a direction for the exploration of other pathways in order to understand molecular alterations responsible for increased risk of CVD development in RA and PD.

5. Conclusions

We identified 43 markers with a strong interactive network in patients suffering from PD, with and without RA. In addition, several of these markers also increase associated metabolic and skeletal disease risk, independent of autoimmune status.

Supplementary Materials: The following supporting information can be downloaded at: https://www.mdpi.com/article/10.3390/biomedicines10030714/s1, Table S1, Olink Cardiovascular II panel; Table S2; Table S3, Clinical Parameters.

Author Contributions: Conceptualization, J.P., M.Å., T.E.V.D., A.L. and P.-E.E.; methodology, J.P., A.G., M.A., M.Å., A.L. and P.-E.E.; software, J.P., M.Å. and A.L.; validation, J.P., A.G., M.A., M.Å., T.E.V.D., A.L. and P.-E.E.; formal analysis, J.P., M.Å. and A.L.; investigation, J.P., A.G., M.A., M.Å. and A.L.; resources, A.G., M.A. and P.-E.E.; data curation, J.P.; writing—original draft preparation, J.P.; writing—review and editing, J.P., A.G., M.A., M.Å., T.E.V.D., A.L. and P.-E.E.; visualization, J.P., T.E.V.D., A.L. and P.-E.E.; supervision, J.P., M.A., T.E.V.D., A.L. and P.-E.E.; project administration, J.P., A.G., M.A., T.E.V.D., A.L. and P.-E.E. All authors have read and agreed to the published version of the manuscript.

Funding: This research received no external funding.

Institutional Review Board Statement: The study protocol was approved by the regional research ethical committees in Karachi, Pakistan (26 September 2012, 30 September 2016) and Stockholm, Sweden (2016/296-31/1).

Informed Consent Statement: Informed consent was obtained from all subjects involved in the study.

Data Availability Statement: All relevant data are provided as Supporting Information in Tables S1–S3.

Acknowledgments: The authors would like to acknowledge support of SciLifeLab Affinity Proteomics Uppsala, (Uppsala University, Sweden) for aiding in protein analyses. All authors gave their final approval and agree to be accountable for all aspects of the work.

Conflicts of Interest: The authors declare no conflict of interest. Van Dyke is supported by USPHS, NIH, NIDCR DE025020.

References

1. Hajishengallis, G.; Moutsopoulos, N.M.; Hajishengallis, E.; Chavakis, T. Immune and regulatory functions of neutrophils in inflammatory bone loss. *Semin. Immunol.* **2016**, *28*, 146–158. [CrossRef]
2. Li, Z.; Lin, L.; Wu, H.; Yan, L.; Wang, H.; Yang, H.; Li, H. Global, Regional, and National Death, and Disability-Adjusted Life-Years (DALYs) for Cardiovascular Disease in 2017 and Trends and Risk Analysis from 1990 to 2017 Using the Global Burden of Disease Study and Implications for Prevention. *Public Health Front.* **2021**, *9*, 559751. [CrossRef]
3. Gallino, A.; Aboyans, V.; Diehm, C.; Cosentino, F.; Stricker, H.; Falk, E.; Schouten, O.; Lekakis, J.; Amann-Vesti, B.; Siclari, F.; et al. Non-coronary atherosclerosis. *Eur. Heart J.* **2014**, *35*, 1112–1119. [CrossRef]
4. Kobiyama, K.; Ley, K. Atherosclerosis. *Circ. Res.* **2018**, *123*, 1118–1120. [CrossRef]
5. Beukers, N.G.; van der Heijden, G.J.; van Wijk, A.J.; Loos, B.G. Periodontitis is an independent risk indicator for atherosclerotic cardiovascular diseases among 60174 participants in a large dental school in the Netherlands. *J. Epidemiol. Community Health* **2017**, *71*, 37–42. [CrossRef]
6. Makkar, H.; Reynolds, M.A.; Wadhawan, A.; Dagdag, A.; Merchant, A.T.; Postolache, T.T. Periodontal, metabolic, and cardiovascular disease: Exploring the role of inflammation and mental health. *Pteridines* **2018**, *29*, 124–163. [CrossRef]
7. Aviña-Zubieta, J.A.; Choi, H.K.; Sadatsafavi, M.; Etminan, M.; Esdaile, J.M.; Lacaille, D. Risk of cardiovascular mortality in patients with rheumatoid arthritis: A meta-analysis of observational studies. *Arthritis Care Res.* **2008**, *59*, 1690–1697. [CrossRef]
8. Santos, M.J.; Cordeiro, A.C.; Gil, V.M. *Biomarkers of Cardiometabolic Risk, Inflammation and Disease*; Springer: Cham, Switzerland, 2015.
9. Sergeant, ESG, 2018. Epitools Epidemiological Calculators. Ausvet. Available online: http://epitools.ausvet.com (accessed on 20 May 2021).
10. Potikuri, D.; Dannana, K.C.; Kanchinadam, S.; Agrawal, S.; Kancharla, A.; Rajasekhar, L.; Pothuraju, S.; Gumdal, N. Periodontal disease is significantly higher in non-smoking treatment-naive rheumatoid arthritis patients: Results from a case-control study. *Ann. Rheum. Dis.* **2012**, *71*, 1541–1544. [CrossRef] [PubMed]
11. Aletaha, D.; Neogi, T.; Silman, A.J.; Funovits, J.; Felson, D.T.; Bingham, C.O., III; Birnbaum, N.S.; Burmester, G.R.; Bykerk, V.P.; Cohen, M.D.; et al. 2010 rheumatoid arthritis classification criteria: An American College of Rheumatology/European League Against Rheumatism collaborative initiative. *Arthritis Rheum* **2010**, *62*, 2569–2581. [CrossRef] [PubMed]
12. Panezai, J.; Ali, A.; Ghaffar, A.; Benchimol, D.; Altamash, M.; Klinge, B.; Engström, P.E.; Larsson, A. Upregulation of circulating inflammatory biomarkers under the influence of periodontal disease in rheumatoid arthritis patients. *Cytokine* **2020**, *131*, 155117. [CrossRef] [PubMed]
13. Panezai, J.; Ghaffar, A.; Altamash, M.; Sundqvist, K.G.; Engström, P.E.; Larsson, A. Correlation of serum cytokines, chemokines, growth factors and enzymes with periodontal disease parameters. *PLoS ONE* **2017**, *12*, e0188945. [CrossRef] [PubMed]
14. Little, R.R.; Rohlfing, C.L.; Sacks, D.B. National Glycohemoglobin Standardization Program (NGSP) Steering Committee Status of hemoglobin A1c measurement and goals for improvement: From chaos to order for improving diabetes care. *Clin. Chem.* **2011**, *57*, 205–214. [CrossRef] [PubMed]
15. Assarsson, E.; Lundberg, M.; Holmquist, G.; Bjorkesten, J.; Thorsen, S.B.; Ekman, D.; Eriksson, A.; Rennel Dickens, E.; Ohlsson, S.; Edfeldt, G.; et al. Homogenous 96-plex PEA immunoassay exhibiting high sensitivity, specificity, and excellent scalability. *PLoS ONE* **2014**, *9*, e95192. [CrossRef] [PubMed]
16. Szklarczyk, D.; Gable, A.L.; Nastou, K.C.; Lyon, D.; Kirsch, R.; Pyysalo, S.; Doncheva, N.T.; Legeay, M.; Fang, T.; Bork, P.; et al. The STRING database in 2021, customizable protein–protein networks, and functional characterization of user-uploaded gene/measurement sets. *Nucleic Acids Res.* **2021**, *49*, D605–D612. [CrossRef]
17. Benjamini, Y.; Hochberg, Y. Controlling the false discovery rate: A practical and powerful approach to multiple testing. *J. R Stat. Soc. Ser. B Stat. Methodol.* **1995**, *1*, 289–300. [CrossRef]

18. Lipsky, M.S.; Su, S.; Crespo, C.J.; Hung, M. Men and Oral Health: A Review of Sex and Gender Differences. *Am. J. Men's Health* **2021**, *15*, 15579883211016361. [CrossRef]
19. Fransen, J.; Kazemi-Bajestani, S.M.; Bredie, S.J.; Popa, C.D. Rheumatoid Arthritis Disadvantages Younger Patients for Cardiovascular Diseases: A Meta-Analysis. *PLoS ONE* **2016**, *11*, e0157360. [CrossRef]
20. Lindström, J.; Tuomilehto, J. The diabetes risk score: A practical tool to predict type 2 diabetes risk. *Diabetes Care* **2003**, *26*, 725–731. [CrossRef]
21. Thalhamer, T.; McGrath, M.A.; Harnett, M.M. MAPKs and their relevance to arthritis and inflammation. *Rheumatology* **2008**, *47*, 409–414. [CrossRef]
22. Theofilis, P.; Sagris, M.; Oikonomou, E.; Antonopoulos, A.S.; Siasos, G.; Tsioufis, C.; Tousoulis, D. Inflammatory Mechanisms Contributing to Endothelial Dysfunction. *Biomedicines* **2021**, *9*, 781. [CrossRef]
23. Braz, N.F.; Pinto, M.R.; Vieira, É.L.; Souza, A.J.; Teixeira, A.L.; Simoes-e-Silva, A.C.; Kakehasi, A.M. Renin–angiotensin system molecules are associated with subclinical atherosclerosis and disease activity in rheumatoid arthritis. *Mod. Rheumatol.* **2021**, *31*, 119–126. [CrossRef] [PubMed]
24. CieŚlik, P.; Hrycek, A. Long pentraxin 3 (PTX3) in the light of its structure, mechanism of action and clinical implications. *Autoimmunity* **2012**, *45*, 119–128. [CrossRef] [PubMed]
25. Ishikawa, M.; Ito, H.; Akiyoshi, M.; Kume, N.; Yoshitomi, H.; Mitsuoka, H.; Tanida, S.; Murata, K.; Shibuya, H.; Kasahara, T.; et al. Lectin-like oxidized low-density lipoprotein receptor 1 signal is a potent biomarker and therapeutic target for human rheumatoid arthritis. *Arthritis Rheum.* **2012**, *64*, 1024–1034. [CrossRef] [PubMed]
26. Barreto, J.; Karathanasis, S.K.; Remaley, A.; Sposito, A.C. Role of LOX-1 (lectin-like oxidized low-density lipoprotein receptor 1) as a cardiovascular risk predictor: Mechanistic insight and potential clinical use. *Arter. Thromb. Vasc. Biol.* **2021**, *41*, 153–166. [CrossRef]
27. Pradeep, A.R.; Roopa, Y.; Swati, P.P. Interleukin-4, a T-helper 2 cell cytokine, is associated with the remission of periodontal disease. *J. Periodontal Res.* **2008**, *43*, 712–716. [CrossRef]
28. Dufour, A.; Bellac, C.L.; Eckhard, U.; Solis, N.; Klein, T.; Kappelhoff, R.; Fortelny, N.; Jobin, P.; Rozmus, J.; Mark, J.; et al. C-terminal truncation of IFN-γ inhibits proinflammatory macrophage responses and is deficient in autoimmune disease. *Nature Commun.* **2018**, *9*, 2416. [CrossRef]
29. Czaya, B.; Faul, C. The role of fibroblast growth factor 23 in inflammation and anemia. *Int. J. Mol. Sci.* **2019**, *20*, 4195. [CrossRef]
30. Rossini, M.; Viapiana, O.; Adami, S.; Fracassi, E.; Idolazzi, L.; Dartizio, C.; Povino, M.R.; Orsolini, G.; Gatti, D. In patients with rheumatoid arthritis Dickkopf-1 serum levels are correlated with parathyroid hormone, bone erosions and bone mineral density. *Clin. Exp. Rheumatol.* **2015**, *33*, 77–83.
31. Park, Y.W.; Kang, Y.M.; Butterfield, J.; Detmar, M.; Goronzy, J.J.; Weyand, C.M. Thrombospondin 2 functions as an endogenous regulator of angiogenesis and inflammation in rheumatoid arthritis. *Am. J. Pathol.* **2004**, *165*, 2087–2098. [CrossRef]
32. Wannamethee, S.G.; Tchernova, J.; Whincup, P.; Lowe, G.D.; Kelly, A.; Rumley, A.; Wallace, A.M.; Sattar, N. Plasma leptin: Associations with metabolic, inflammatory and haemostatic risk factors for cardiovascular disease. *Atherosclerosis* **2007**, *191*, 418–426. [CrossRef]
33. Abe, T.; AlSarhan, M.; Benakanakere, M.R.; Maekawa, T.; Kinane, D.F.; Cancro, M.P.; Korostoff, J.M.; Hajishengallis, G. The B cell-stimulatory cytokines BLyS and APRIL are elevated in human periodontitis and are required for B cell-dependent bone loss in experimental murine periodontitis. *J. Immunol.* **2015**, *195*, 1427–1435. [CrossRef] [PubMed]
34. Wei, F.; Chang, Y.; Wei, W. The role of BAFF in the progression of rheumatoid arthritis. *Cytokine.* **2015**, *76*, 537–544. [CrossRef] [PubMed]
35. Al Shahi, H.; Shimada, K.; Miyauchi, K.; Yoshihara, T.; Sai, E.; Shiozawa, T.; Naito, R.; Aikawa, T.; Ouchi, S.; Kadoguchi, T.; et al. Elevated circulating levels of inflammatory markers in patients with acute coronary syndrome. *Int. J. Vasc. Med.* **2015**, *2015*, 805375. [CrossRef] [PubMed]
36. Matarese, G.; Currò, M.; Isola, G.; Caccamo, D.; Vecchio, M.; Giunta, M.L.; Ramaglia, L.; Cordasco, G.; Williams, R.C.; Ientile, R. Transglutaminase 2 up-regulation is associated with RANKL/OPG pathway in cultured HPDL cells and THP-1-differentiated macrophages. *Amino Acids* **2015**, *47*, 2447–2455. [CrossRef] [PubMed]
37. Carbon, S.; Ireland, A.; Mungall, C.J.; Shu, S.; Marshall, B.; Lewis, S. AmiGO Hub, Web Presence Working Group. AmiGO: Online access to ontology and annotation data. *Bioinformatics* **2009**, *25*, 288–289. [CrossRef]
38. Song, I.S.; Han, K.; Park, Y.M.; Ji, S.; Jun, S.H.; Ryu, J.J.; Park, J.B. Severe periodontitis is associated with insulin resistance in non-abdominal obese adults. *J. Clin. Endocrinol. Metab.* **2016**, *101*, 4251–4259. [CrossRef] [PubMed]
39. Llorente, I.; García-Castañeda, N.; Valero, C.; González-Álvaro, I.; Castañeda, S. Osteoporosis in rheumatoid arthritis: Dangerous liaisons. *Front. Med.* **2020**, *7*, 802. [CrossRef]
40. Avouac, J.; Elhai, M.; Forien, M.; Sellam, J.; Eymard, F.; Molto, A.; Banal, F.; Damiano, J.; Dieudé, P.; Larger, E.; et al. Influence of inflammatory and non-inflammatory rheumatic disorders on the clinical and biological profile of type-2 diabetes. *Rheumatology* **2021**, *60*, 3598–3606. [CrossRef]
41. Hong, S.J.; Yang, B.E.; Yoo, D.M.; Kim, S.J.; Choi, H.G.; Byun, S.H. Analysis of the relationship between periodontitis and osteoporosis/fractures: A cross-sectional study. *BMC Oral Health* **2021**, *21*, 125. [CrossRef]

biomedicines

MDPI

Systematic Review

Frequency and Clinical Significance of Elevated IgG4 in Rheumatoid Arthritis: A Systematic Review

Rajalingham Sakthiswary [1,*], Syahrul Sazliyana Shaharir [1] and Asrul Abdul Wahab [2]

[1] Department of Medicine, Universiti Kebangsaan Malaysia Medical Centre, Kuala Lumpur 56000, Malaysia; sazliyana@hotmail.com

[2] Department of Immunology, Universiti Kebangsaan Malaysia Medical Centre, Kuala Lumpur 56000, Malaysia; saw@ppukm.ukm.edu.my

* Correspondence: sakthis5@hotmail.com

Abstract: Immunoglobulin (Ig)G4 is a unique protein molecule and its role in autoimmune diseases remains elusive and controversial. Accumulating evidence suggests a pathogenic role of IgG4 in rheumatoid arthritis (RA). Rheumatoid factors (RF) in RA can recognize the Fc domains of IgG4 to form RF-IgG4 immune complexes that may activate the complement system leading to synovial injury. The aim of this article was to systematically review the literature from the past 2 decades to determine the frequency of elevated IgG4 and its clinical significance in RA. We comprehensively searched the Pubmed, Scopus, and Web of Science databases with the following terms: "IgG4", "rheumatoid arthritis", and "immunoglobulin G4", and scrutinized all of the relevant publications. Based on the selection criteria, 12 studies were incorporated, which involved a total of 1715 RA patients. Out of 328 subjects from three studies, the pooled frequency of elevated non-specific IgG4 was 35.98%. There was a significant positive correlation between the IgG4 levels and the RA disease activity based on DAS-28 measurements (r = 0.245–0.253) and inflammatory markers, i.e., erythrocyte sedimentation rate (ESR) and C-reactive protein (CRP) levels (r = 0.262–0.389). Longitudinal studies that measured the serial levels of IgG4 consistently showed a decline in the concentrations (up to 48% less than baseline) with disease modifying anti-rheumatic drug (DMARD) treatment. Current evidence suggests that serum IgG4 levels are significantly elevated in RA compared to the general population. This review indicates that IgG4 is a promising biomarker of disease activity and tends to decline in response to DMARD therapies. Biologic therapies have revolutionized the therapeutic armamentarium of RA in the recent decade, and IgG4 appears to be a potential treatment target.

Keywords: arthritis; rheumatoid; immunoglobulin G; immune system

Citation: Sakthiswary, R.; Shaharir, S.S.; Wahab, A.A. Frequency and Clinical Significance of Elevated IgG4 in Rheumatoid Arthritis: A Systematic Review. *Biomedicines* **2022**, *10*, 558. https://doi.org/10.3390/biomedicines10030558

Academic Editor: Marianna Christodoulou

Received: 4 January 2022
Accepted: 17 February 2022
Published: 26 February 2022

1. Introduction

Immunoglobulin (Ig)G accounts for 80% of the total immunoglobulins in human serum, and can be divided into four subclasses, i.e., IgG1 (60–70%), IgG2 (15–20%), IgG3 (5–10%), and IgG4 (4–6%). Each of these has different immunological properties and functions [1]. Immunoglobulins play a pivotal role in autoimmune diseases such as rheumatoid arthritis (RA), systemic lupus erythematosus (SLE), and myasthenia gravis. RA is a chronic inflammatory joint disease with a complex pathogenesis. The orchestrated interaction of a wide array of cytokines, autoreactive B cells, and T cells underpin the mechanisms in RA. The sera of RA patients tend to typically exhibit a wide variety of autoantibodies [2]. Rheumatoid factors (RF), which are the predominant autoantibodies in RA, have Fab segments, which react with the Fc portion of the IgG molecule to generate IgM (RF)-IgG immune complexes, which can stimulate the complement system and trigger a cascade of events in the synovial microenvironment [3]. IgG4 molecules have stirred much interest among researchers in the past decade, ever since IgG4-related disease (IgG4-RD) was endorsed in 2011 [4]. The striking difference between IgG4-RD and RA is the presence and frequency of joint involvement. IgG4-RD is a multisystemic disorder, with arthritis being reported in only 10% of patients [5]. The common forms of presentations of IgG4-RD include autoimmune pancreatitis, sclerosing cholangitis, sclerosing sialedenitis, orbital disease, and retroperitoneal fibrosis [6]. RA, on the other hand, is primarily a disease of the joints. The presence of arthritis is mandatory to diagnose RA [7].

IgG4 is a poorly understood molecule with controversial roles in the immune system. Traditionally, IgG4 has been viewed as a "non-inflammatory" molecule, which dampens rather than incites immune activation. This is due to the unique molecular structure of IgG4, whereby the heavy chains in each IgG4 molecule have inefficient disulphide bridges due to a single amino acid difference in the hinge region [8]. In hemi-IgG4 molecules, one heavy chain may covalently bind with one light chain, and then dissociate from each other and re-associate randomly with other hemi-IgG4 molecules. This phenomenon is known as the "Fab arm exchange", which exclusively occurs in the IgG4 subclass [9]. This half-antibody exchange generates antibodies that are capable of binding two different antigens, but are rarely able to form large immune complexes [10]. Based on these theories, IgG4 have a limited ability to form immune responses.

Nevertheless, accumulating evidence suggests a pathogenic role of IgG4 based on its correlations with disease activity and the severity of certain disease entities [4,11]. IgG4 may theoretically bind to Fc receptors on macrophages and eosinophils, and facilitate the presentation of extracellular antigens to CD4+ T lymphocytes [12,13]. Some recent publications have implicated IgG4 autoantibodies in the pathogenesis of RA. Histopathological findings of IgG4 infiltration in the rheumatoid synovium and elevated serum levels of IgG4 in RA patients lend further credence to the above notion [3,14,15].

As far as we know, there are no published systematic reviews focusing on IgG4 in RA. Hence, the purpose of this systematic review was to gather and scrutinize all available literature in the past few decades to determine the pooled frequency of elevated IgG4 and its clinical significance.

2. Materials and Methods

2.1. Search Strategy

We comprehensively searched the Pubmed, Scopus, and Web of Science databases with the following terms: "IgG4", "immunoglobulin G4", and "rheumatoid arthritis", and tracked all of the publications. All three authors independently performed a literature search by title and abstract screening using the Endnote software. In the event of uncertainty, the full text of the article was obtained and assessed. Disagreements were resolved by a consensus-based discussion. Only articles that were approved after much scrutiny by all were finally included in the review. To minimize the selection, information, and confounding biases, the PICOT (patient/population, intervention, control, outcome, time) approach was employed to develop the inclusion and exclusion criteria [16]. The population in this review referred to patients with RA, intervention in most studies included treatment with disease modifying anti-rheumatic drugs, and the outcome was the serum IgG4 levels. A clear search protocol reduced the ambiguity in the selection process of the articles. In order to achieve extensive coverage without missing any relevant articles, the references of all retrieved articles were reviewed. This systematic review was conducted in accordance with the standards set by the Preferred Reporting Item for Systematic Review and Meta-Analysis (PRISMA) Statement [17]. Figure 1 summarizes our search strategy.

2.2. Inclusion Criteria

All adult human studies written in English that looked into IgG4 in RA were included. Conference abstracts with sufficient data were considered eligible.

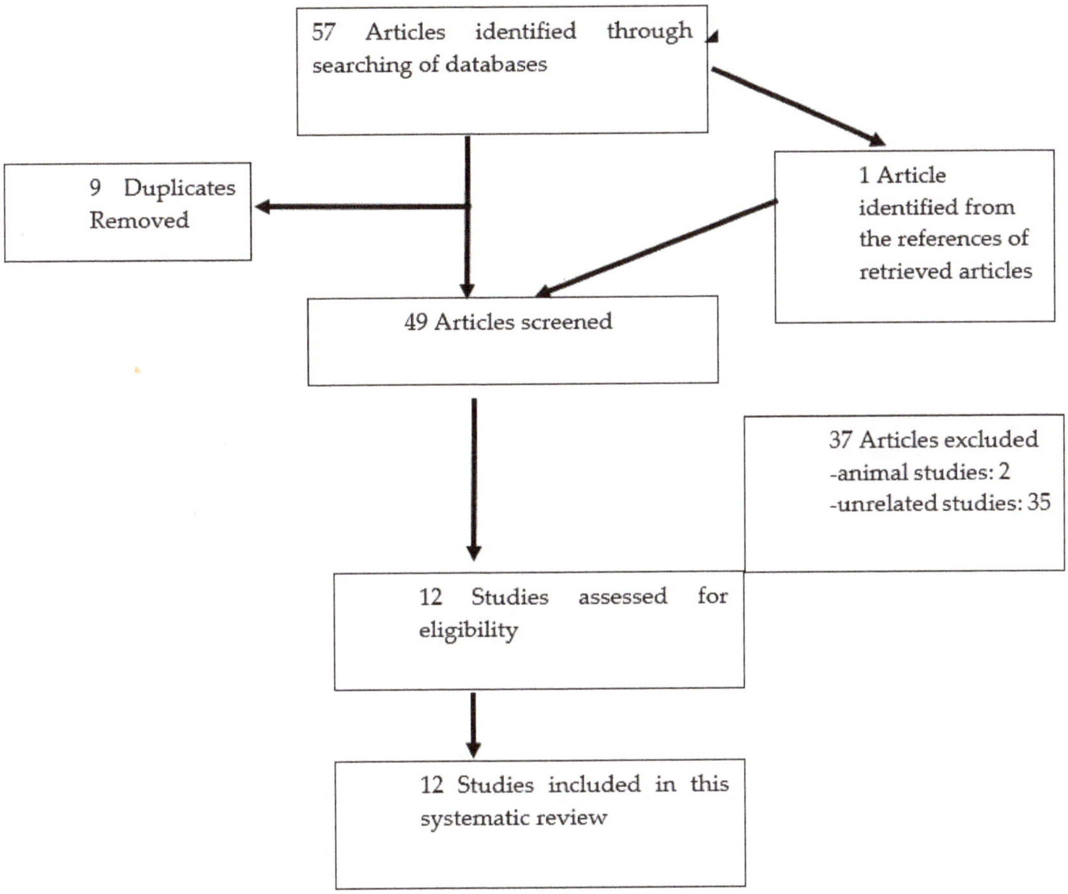

Figure 1. The algorithm for the selection of studies in this systematic review.

2.3. Exclusion Criteria

We excluded studies published before 2000. Furthermore, articles in other languages, case reports, case series, animal studies, editorials, and review articles were excluded.

2.4. Data Extraction

After compiling the relevant studies, the authors extracted the relevant data from each paper, including year of publication, country, study design, study population, frequency of subjects with elevated IgG4, mean/median IgG4 levels in RA, and the correlations with clinical and biochemical markers. The Newcastle–Ottawa Scale [18] (Table 1) was used for the quality assessment of the 11 included observational studies. The above scale is not applicable for randomized trials. Scores of $\geq$3 were considered as low risk of bias, whereas $<$3 were judged as high risk. Disagreements among the authors were solved through discussions and a consensus was reached.

Table 1. The Newcastle–Ottawa Scale for quality assessment of observational studies.

Studies	Selection				Comparability	Outcome			
	Exposed Truly Representative of Average	Selection of Non Exposed from the Same Community	Exposure Ascertained by Secure Record or Interview	Demonstration of Outcome of Interest Not Present at the Start of the Study	Study Controls for Other Variables	Follow Up Long Enough for Outcome to Occur	Complete Follow Up for All Subjects Accounted for	Subjects Lost to Follow Up Unlikely to Introduce Bias	Total Score
Chen et al. (2014)	✓	✓	✓	x	x	x	x	x	3
Lin et al. (2010)	✓	x	✓	x	x	x	x	x	2
Kim et al. (2020)	✓	x	✓	x	x	x	x	x	2
Chen et al. (2021)	✓	✓	✓	x	x	x	x	x	3
K. Martinsson et al. (2016)	✓	x	x	x	x	x	x	x	1
Carbone et al. (2013)	✓	x	✓	x	x	✓	✓	x	4
Engelmann et al. (2008)	✓	x	x	x	x	x	x	x	1
Bos et al. (2009)	✓	x	✓	x	x	✓	✓	x	4
Engelmann et al. (2018)	✓	x	✓	x	x	x	✓	x	3
Chapuy-Regaud et al. (2005)	✓	x	x	x	x	x	x	x	1
van Schouwenburg et al. (2012)	✓	x	✓	x	x	✓	x	x	3

3. Results

3.1. Study Characteristics Proteomic Analysis of IgG4

Based on the selection criteria, 12 studies were incorporated, which involved a total of 1715 RA patients. Among the included twelve research works, four were from Asia [14,15,19,20], seven from Europe [21–27], and one from North America [28]. All of the studies in this series were observational, except for one randomized trial [23]. There were seven cross-sectional studies [14,15,19,20,22,24,27] and five longitudinal studies [21,23,25,26,28] included in this review. The quality assessment of the observational studies based on the Newcastle–Ottawa scale revealed that six articles were of low-risk bias ($\geq$3 points) and the remaining five were of high-risk bias (<3 points).

IgG4 levels were detected using three methods, i.e., ELISA in seven studies [21–24,26–28], immunonephelometry in four studies [14,15,19,20], and radioimmunoassay in a single study [25]. Of note, all Asian studies used the immunonephelometry method of testing IgG4. The studies that performed the immunonephelometric quantification of IgG4 stored the samples between -80 to -70 degrees Celsius after the samples were processed in a centrifugal separator. The total levels of IgG and IgG4 were determined with liquid reagent kits [15,19]. The levels of IgG4 specific-anti-citrulinated cyclic peptide (CCP) antibodies were determined using the ELISA kit containing a CCP-coated plate with horseperoxidase-conjugated anti-human IgG4 antibodies [28].

3.2. Frequency of Elevated IgG4 in Rheumatoid Arthritis

There were five studies that analyzed the non-specific IgG4 levels in RA [14,15,19,20,23] (Table 2). There were two studies that did not provide data on the frequency of subjects with raised levels of IgG4 [20,23]. Out of 328 subjects from three studies, the pooled frequency of elevated IgG4 was 35.98%. The studies used different kits with variable units of measurements, i.e., g/L, mg/L and mg/dL. Calculation of effect size was not performed as there were only two studies [15,19] that provided the mean values of IgG4.

There were four studies that investigated the levels of IgG4 specific to citrullinated cyclic peptide (CCP) [21,22,27,28] and two studies on citrullinated cyclic fibrinogen (CCF) [24,26] (Table 3). The pooled frequency of elevated IgG4 anti-CCP was 330 out of 581 subjects (56.79%).

3.3. Clinical Significance of IgG4 in Rheumatoid Arthritis

3.3.1. IgG4 and Disease Activity

There were four studies that investigated the association of serum IgG4 levels with the RA disease activity [14,15,19,20]. Of note, two of the studies were from the same group of researchers [14,15]. Kim et al. [19] found significant correlations between serum IgG4 levels and DAS28-ESR ($r = 0.245$; $p = 0.016$), and with ESR ($r = 0.262$; $p = 0.010$). In keeping with these findings, Chen et al. [15] revealed that IgG4 levels correlated positively with CRP ($r = 0.373$), ESR ($r = 0.389$), and DAS28 ($r = 0.253$; all $p < 0.05$) [4]. The Pearson correlation coefficient r value from these studies for the correlation between IgG4 levels and the RA disease activity based on DAS-28 measurements ranged from 0.245–0.253, whereas for inflammatory markers, i.e., ESR and CRP levels, it was 0.262–0.389. The r values that fell between 0.2–0.4, in general, reflected a weak to moderate strength in the relationships of the aforementioned variables [29]. There was a trend towards higher IgG4 levels in the high disease activity group compared to the moderate, low, and remission groups, although statistical significance was not achieved.

In one of the studies, the synovial samples of RA patients had a median IgG4 positive(+) plasma cells count of 83 (10–192)/mm^2 and a median ratio of IgG4+/IgG+ plasma cells of 19.1 (8.4–31.5). Both of them were positively correlated with ESR, CRP, and serum IgG4 ($r = 0.216$–0.394, all $p < 0.05$) [14].

Table 2. Summary of rheumatoid arthritis studies on non-specific IgG4.

Year	Authors	Country	Study Design & Study Population	Frequency of RA Patients with Raised IgG4	Mean/Median IgG4 Levels in RA	IgG4 Detection Method	Key Findings
2014	Chen et al. [15]	China	Cross sectional 136 RA patients	46%	1.52 ± 1.27 g/L	Immunonephelometry	51% patients had elevated IgG4/IgG ratio. The mean IgG4 of the untreated patients was 1.82 ± 1.39 g/L, which was significantly higher than that of the treated patients (1.39 ± 1.20 g/L; p = 0.044) IgG4 of high disease activity group was significantly higher than that of the remission group (p = 0.003). CRP and ESR, of the elevated IgG4 group were significantly higher than those of the normal IgG4 group (CRP: 38 ± 42 mg/L vs. 25 ± 33 mg/L; ESR: 70 ± 42 mm/h vs. 48 ± 32 mm/h; all $p < 0.05$). IgG4 level correlated positively with CRP (r = 0.426) and ESR (r = 0.315; both $p < 0.05$). RF and anti-CCP Ab levels of the elevated IgG4 group were significantly higher than those of the normal IgG4 group (RF: 513 ± 636 IU/mL vs. 245 ± 392 IU/mL; anti-CCP Ab: 256 ± 243 U/mL vs. 162 ± 199 U/mL; both $p < 0.05$), IgG4 correlated positively and significantly with synovial IgG4-positive plasma cells but no significant correlation of IgG4 with total synovitis score or subscores.
2021	Chen et al. [14]	China	Cross sectional 96 active RA patients	49 (51.0%)	1.38 (0.86–2.42) g/L (median)	Immunonephelometry	RA patients with elevated IgG4 had significantly higher levels of ESR [90 (64–116) mm/h vs. 61 (38–75) mm/h], CRP [46.20 (17.20–74.20) mg/L vs. 18.90 (9.46–49.20) mg/L], DAS28-ESR [6.3 (5.6–7.4) vs. 5.7 (4.7–6.4)], SDAI [34.2 (25.3–48.8) vs. 27.8 (18.9–35.9)] all $p < 0.05$]. They also showed significantly higher synovial counts of IgG4+ plasma cells [106 (39–249) /mm^2 vs. 68 (3–123) /mm^2], and higher ratio of IgG4+/IgG+ plasma cells [26.3 (15.5–38.0)% vs. 15.2 (0.9–24.7)%, all $p <$ 0.05]. There were 10 (10.4%) patients showing elevated serum IgG4 and IgG4-related synovitis based on synovial biopsy.
2010	Lin et al. [20]	China	Cross sectional 72 RA patients		365.5 (72.85–1377.5) mg/L (median)	Immunonephelometry	When the patients were divided according to the clinical activity, the IgG subclass concentrations were similar between the two groups ($p > 0.05$).
2020	Kim et al. [19]	Korea	Cross sectional 128 participants (RA: 96; healthy controls: 17; Osteoarthritis: 11; and IgG4-related disease: 4)	6.3%	48.0 ± 45.4 mg/dL (mean)	Immunonephelometry	The mean serum IgG4/IgG ratio in patients with RA was 3.5 ± 2.8% (range 0.2–16.9%). None of the healthy controls or patients with osteoarthritis had elevated serum IgG4 levels. A significant correlation was found between serum IgG4 levels and the Disease Activity Score-28 with erythrocyte sedimentation rate (r = 0.245; p = 0.016).
2010	Farboud et al. [23]	United Kingdom	Longitudinal (24 weeks) 55 RA patients			ELISA	For IgG1, IgG2, IgG3 and IgG4 subclasses, in the 0.5-mg of oral bovine type II collagen group in which the best clinical response was observed, there was statistically significant decreases observed in the IgG2 and IgG3 subclasses (p = 0.047, p = 0.046) but not in IgG4.

RA: rheumatoid arthritis, ESR: erythrocyte sedimentation rate, CRP: C-reactive protein, RF: rheumatoid factor, CCP: citrulinated cyclic peptide.

3.3.2. IgG4 and Treatment Response

There were four longitudinal studies [21,23,26,28] that evaluated the changes in the levels of IgG4 with therapy. The therapies used included biologic disease modifying anti-rheumatic drugs (DMARDs) such as tocilizumab [28], adalimumab [26], conventional DMARDs [21], and an experimental agent that was oral bovine type II collagen [23]. All of these studies except for one [23] consistently showed a decline in the IgG4 levels with treatment. There was a parallel decrease in the disease activity of the subjects. Bos et al. [26] disclosed that although all types of IgG (IgG1–4) decreased with treatment, the good responders based on European League Against Rheumatism (EULAR) response criteria [30] had the greatest decline in antibody levels, and this effect was most pronounced for IgG4 (48% reduction). Similarly, Carbone et al. [28] found a 2–3-fold reduction in IgG4 levels with tocilizumab therapy, but not in IgG1 levels, despite IgG1 being the most frequent IgG subtype.

Among the subjects who were treated with adalimumab, secondary failure to this biologic therapy was due to the formation of anti-drug antibody, which was IgG4 in up to 29% of the subjects [25].

Table 3. Summary of Rheumatoid Arthritis studies on specific types of IgG4.

Year	Authors	Country	Study Design & Study Population	Frequency of RA Patients with Raised IgG4	Type of IgG4	IgG4 Detection Method	Key Findings
2016	K. Martinsson et al. [27]	Sweden	Cross sectional 504 with recent onset RA (untreated)	59%	IgG4 anti-CCP	ELISA	Among those who were RF positive, 79% subjects tested positive for IgG4. IgG anti-CCP subclasses that associate with HLA-DRB1/SE are IgG1 and IgG4. Increased proportion of IgG4 anti-CCP-positive cases that were not associated with smoking. The fractions of IgG4 anti-CCP did not differ significantly between ever and never-smokers.
2013	Carbone et al. [28]	USA	Longitudinal 8 patients with active RA were treated with tocilizumab (TCZ) monotherapy or in combination with non-biologic DMARDs over 6 months; serum samples were obtained at (0 month), 1 month, 3 months, and 6 months		IgG4 anti-CCP	ELISA	Over the 6 months of treatment, there was a prominent four-fold reduction in the levels of IgG4 (p = 0.06). The levels of IgG4 were markedly decreased in all but one patient. Pronounced reduction (2–3 fold) in the serum levels of IgG4-specific anti-CCP Abs in all patients (p = 0.011), but no statistically significant reduction in the levels of IgG1-anti-CCP Abs (p = 0.185).
2014	Engelmann et al. [22]	Germany	Cross sectional 77 RA patients	33 (42.86%) patients with anti-CCP antibodies are positive for the IgG4 subclass	IgG4 anti-CCP	ELISA	Even though IgG1 is the predominant subclass among antibodies against CCP and anti-citrullinated vimentin (MCV) in RA, IgG4 was conspicuously elevated. Elevated IgG4 titers among auto-antibodies in RA are indicative of a Th2-biased environment.
2018	Engelmann et al. [21]	Germany	Longitudinal 34 ACPA-positive RA were monitored for 3 months after therapy		IgG4 anti-CCP	ELISA	3 months after therapy, the responders showed a significant decrease in IgG4 ACPA levels, and this was independent of the individual treatment regimen.
2009	Bos et al. [26]	Netherlands	Longitudinal 180 patients treated with adalimumab for 28 weeks		IgG4 ACF	ELISA	The median reduction in anti-citrullinated fibrinogen (ACF) levels was 31% for total IgG, 29% for IgG1, 40% for IgG4, and 22% for the IgG4/IgG1 ACF ratio in the infliximab cohort. In adalimumab-treated patients, ACF levels declined 14% for total IgG and IgG1, and 36% for IgG4 ACF; the IgG4: IgG1 ratio was reduced by 24%. European League Against Rheumatism good responders had the greatest decline in antibody levels, and this effect was most pronounced for IgG4 (48% reduction). The IgG4/IgG1 ACF ratio preferentially decreased in patients with adequate therapeutic adalimumab levels.
2005	Chapuy-Regaud et al. [24]	France	Cross sectional 186 RA	21.3% (30/141) had IgG1-AhFibA in combination with IgG4-AhFibA	IgG4 ACF	ELISA	IgG4-Antihuman fibrinogen (AhFibA) observed much more frequently and at higher titers than IgG3- or IgG2-AhFibA. AhFibA were mainly IgG1 and, to a lesser extent, IgG4.
2012	van Schouwenburg et al. [25]	Netherlands	Longitudinal 271 RA patients monitored for 3 years of adalimumab treatment	29% of the patients had detectable IgG4	IgG4 against adalimumab	Radio immunoassay	The proportion IgG4 of total IgG against adalimumab varied widely between patients. IgG4 was found to contribute significantly to the anti-drug antibody (ADA) response in some patients.

RA: rheumatoid arthritis; DMARD: disease modifying anti-rheumatic drug.

4. Discussion

To the best of our knowledge, this is the first systematic review in the literature on IgG4 in RA. Our pooled analyses showed that IgG4 levels were significantly elevated in RA patients (35.98%) compared to the frequencies reported in healthy individuals. Various studies have found that the frequency of elevated IgG4 in healthy subjects ranged from 0–2.5% [19,31]. In keeping with our findings, several studies have reported higher frequencies of elevated IgG4 in autoimmune diseases such as Sjogren syndrome, systemic lupus erythematosus, myasthenia gravis, and eosinophilic granulomatous polyangiitis [32,33]. A serum IgG4 concentration of above 135 mg/dL has been widely accepted as the cut-off value to define "elevated IgG4" and as a criterion for the diagnosis of IgG4-related disease [34].

There was a significant positive correlation between the IgG4 levels and RA disease activity based on the findings of all three studies that performed correlation analyses between the above-mentioned parameters. Of note, all three studies used the same composite clinical disease activity tool, i.e., DAS28-ESR, which may partially explain the similarity in the findings. Disease activity in RA reflects synovial inflammation, which is due to the effects of circulating cytokines, such as interleukin (IL)-1, IL-6, and tumor necrosis factor (TNF) α. The synthesis of IgG4 in vitro was regulated by IL-6. IL-6 may enhance IgG4 production through IL-21 expressed in CD4+ T cells [35], which in turn promotes the differentiation of B cells into antibody-secreting plasma cells [36]. The link between IL-6 and IgG4 may explain the relationship between the latter and RA disease activity. The pro-inflammatory nature of IL-6 is well established in RA and it plays important roles in the regulation of the immune response, inflammation, and bone metabolism [37]. The reported association in the studies need not necessarily imply causation of RA disease activity directly by IgG4. Nevertheless, elevated IgG4 levels may indicate a relapse of RA. The conventional biomarkers of disease activity widely used by clinicians in day-to-day clinical practice are ESR and CRP. Clinicians may consider IgG4 as an adjunct biomarker in this regard, but not for diagnostic purposes.

The independent role of IgG4 in RA remains elusive, although there is some supporting evidence based on the histopathological analysis by Chen et al. [14,15]. The studies demonstrated marked infiltration of RA synovium by IgG4-positive plasma cells, which were correlated with a total synovitis score, inflammatory infiltration subscore, CD3-positive T cells, CD20-positive B cells, and CD38-positive plasma cells. This finding suggested that IgG4 was potentially a culprit molecule in RA rather than an innocent bystander. It is tempting to speculate that the fibrosis observed in the RA synovium could be secondary to the upregulation of a fibrogenic cytokine, i.e., transforming growth factor (TGF)-β by IgG4 [38]. This postulation is based on our knowledge on IgG4-RD and its striking histological feature, which is fibrosis [39].

Rheumatoid factors (RF) in RA can recognize the Fc domains of IgG4 to form RF-IgG4 immune complexes that may activate the complement system, leading to synovial injury [40]. Although IgG1 is the most frequent isotype against citrullinated cyclic peptide, Bos et al. [26] proposed that prolonged exposure to autoantigens might lead to changes in the IgG4/IgG1 antibody ratio switching to an IgG4-dominated response. Figure 2 illustrates the theoretical role of IgG4 in the pathogenesis of RA.

The evidence from this systematic review suggests that IgG4 is a reliable biomarker of treatment response. The results from the studies in this regard were consistent. There are a few hypothetical explanations for the above. DMARD therapy tends to inhibit IgG4 production via TNFα inhibition [41]. Furthermore, IgG4 levels, unlike IgG1 levels tend to decline with therapy among responders due to disruptions in the chronic stimulation by citrullinated proteins [42]. Citrullination is inflammation-dependent and is hence suppressed by DMARD therapies, which have anti-inflammatory properties. IgG1 levels are stable and inflammation-independent as they are predominantly produced by long-lived plasma cells, whereas IgG4 levels are produced by short-lived plasma cells, which are driven by citrullinated proteins [43].

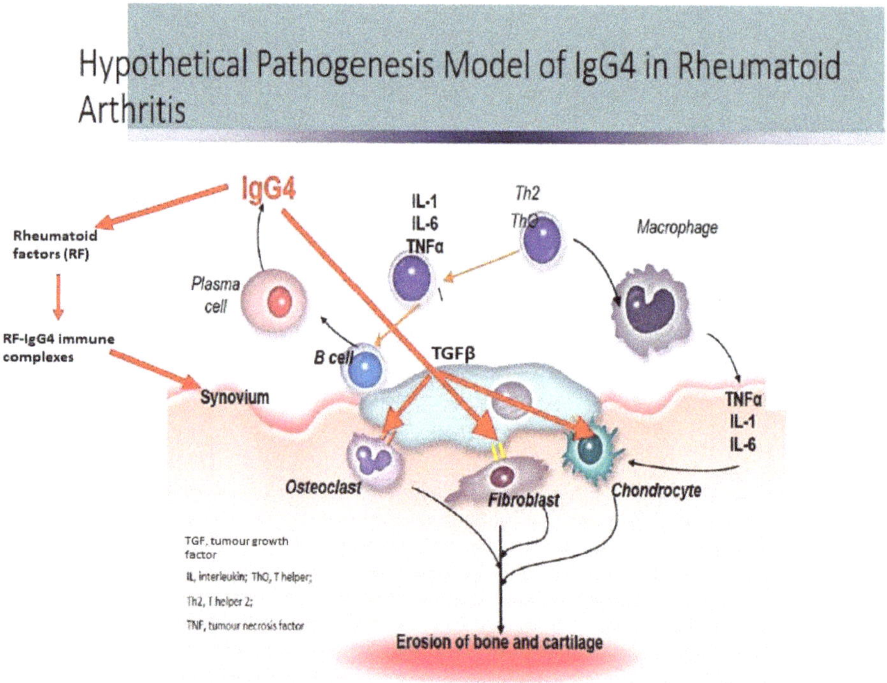

Figure 2. Hypothetical pathogenesis model on the role of IgG4 in rheumatoid arthritis.

We acknowledge the limitations of this systematic review. Most of the studies were conducted in Europe and Asia, which may limit the representativeness of the results to a certain extent. There are racial and ethnic disparities with regard to the disease characteristics and clinical outcomes in RA [44]. We were unable to calculate the effect size for the correlation between the RA disease activity and IgG4, as well as the difference in the means of IgG4 across the various categories of RA patients due to the lack of relevant numerical data in the included studies. From the limited data from a few studies, conclusions cannot be made firmly and may appear speculative. Moreover, several articles with ambiguous data description were excluded, which may affect the pooled frequency.

5. Conclusions

Current evidence suggests that the serum IgG4 levels are elevated in RA compared to the general population. This review indicates that IgG4 is a promising biomarker of disease activity, and tends to decline in response to DMARD therapies. Thus, IgG4 could serve as an alternative modality in RA to assess patients' disease severity. There are several theories with regard to the pro-inflammatory role of IgG4. Further research is necessary to substantiate these hypotheses.

Author Contributions: Conceptualization, R.S., S.S.S. and A.A.W.; methodology, R.S., S.S.S. and A.A.W.; data curation, R.S.; writing, R.S. and S.S.S.; funding acquisition, R.S. All authors have read and agreed to the published version of the manuscript.

Funding: The research was funded by the University Kebangsaan Malaysia, Kuala Lumpur, Malaysia.

Institutional Review Board Statement: Not applicable.

Informed Consent Statement: Not applicable.

Data Availability Statement: The data presented in this study are available upon request from the corresponding author.

Conflicts of Interest: The authors declare no conflict of interest.

References

1. Puissant-Lubrano, B.; Peres, M.; Apoil, P.A.; Congy-Jolivet, N.; Roubinet, F.; Blancher, A. Immunoglobulin IgA, IgD, IgG, IgM and IgG subclass reference values in adults. *Clin. Chem. Lab. Med.* **2015**, *53*, e359–e361. [CrossRef]
2. Corrigall, V.M.; Panayi, G.S. Autoantigens and immune pathways in rheumatoid arthritis. *Crit. Rev. Immunol.* **2002**, *22*, 281–293.
3. Wang, Y.; Lu, Q.; Wu, S.L.; Karger, B.L.; Hancock, W.S. Characterization and comparison of disulfide linkages and scrambling patterns in therapeutic monoclonal antibodies: Using LC-MS with electron transfer dissociation. *Anal. Chem.* **2011**, *83*, 3133–3140. [CrossRef]
4. Stone, J.H.; Khosroshahi, A.; Deshpande, V.; Chan, J.K.; Heathcote, J.G.; Aalberse, R.; Azumi, A.; Bloch, D.B.; Brugge, W.R.; Carruthers, M.N.; et al. Recommendations for the nomenclature of IgG4-related disease and its individual organ system manifestations. *Arthritis Rheum.* **2012**, *64*, 3061–3067. [CrossRef]
5. Masaki, Y.; Dong, L.; Kurose, N.; Kitagawa, K.; Morikawa, Y.; Yamamoto, M.; Takahashi, H.; Shinomura, Y.; Imai, K.; Saeki, T.; et al. Proposal for a new clinical entity, IgG4-positive multiorgan lymphoproliferative syndrome: Analysis of 64 cases of IgG4-related disorders. *Ann. Rheum. Dis.* **2009**, *68*, 1310–1315. [CrossRef]
6. Stone, J.H.; Zen, Y.; Deshpande, V. IgG4-related disease. *N. Engl. J. Med.* **2012**, *366*, 539–551. [CrossRef]
7. Cohen, S.; Emery, P. The American College of Rheumatology/European League Against Rheumatism Criteria for the classification of rheumatoid arthritis: A game changer. *Ann. Rheum. Dis.* **2010**, *69*, 1575–1576. [CrossRef]
8. Aalberse, R.C.; Stapel, S.O.; Schuurman, J.; Rispens, T. Immunoglobulin G4: An odd antibody. *Clin. Exp. Allergy* **2009**, *39*, 469–477. [CrossRef]
9. Rispens, T.; Ooijevaar-de Heer, P.; Bende, O.; Aalberse, R.C. Mechanism of immunoglobulin G4 Fab-arm exchange. *J. Am. Chem. Soc.* **2011**, *133*, 10302–10311. [CrossRef]
10. van der Neut Kolfschoten, M.; Schuurman, J.; Losen, M.; Bleeker, W.K.; Martinez-Martinez, P.; Vermeulen, E.; den Bleker, T.H.; Wiegman, L.; Vink, T.; Aarden, L.A.; et al. Anti-inflammatory activity of human IgG4 antibodies by dynamic Fab arm exchange. *Science* **2007**, *317*, 1554–1557. [CrossRef]
11. Muraki, T.; Hamano, H.; Ochi, Y.; Komatsu, K.; Komiyama, Y.; Arakura, N.; Yoshizawa, K.; Ota, M.; Kawa, S.; Kiyosawa, K. Autoimmune pancreatitis and complement activation system. *Pancreas* **2006**, *32*, 16–21. [CrossRef]
12. Padigel, U.M.; Hess, J.A.; Lee, J.J.; Lok, J.B.; Nolan, T.J.; Schad, G.A.; Abraham, D. Eosinophils act as antigen-presenting cells to induce immunity to Strongyloides stercoralis in mice. *J. Infect. Dis.* **2007**, *196*, 1844–1851. [CrossRef]
13. Rock, B.; Martins, C.R.; Theofilopoulos, A.N.; Balderas, R.S.; Anhalt, G.J.; Labib, R.S.; Futamura, S.; Rivitti, E.A.; Diaz, L.A. The pathogenic effect of IgG4 autoantibodies in endemic pemphigus foliaceus (fogo selvagem). *N. Engl. J. Med.* **1989**, *320*, 1463–1469. [CrossRef]
14. Chen, L.F.; Zhang, X.; Chen, C.; Ma, J.D.; Mo, Y.; Lin, J.; Zou, Y.Y.; Zheng, H.; Dai, L. Clinical Characteristics of Rheumatoid Arthritis Patients with Igg4-Related Synovitis. *Ann. Rheum. Dis.* **2021**, *80*, 1091–1092. [CrossRef]
15. Chen, L.F.; Mo, Y.Q.; Ma, J.D.; Luo, L.; Zheng, D.H.; Dai, L. Elevated serum IgG4 defines specific clinical phenotype of rheumatoid arthritis. *Mediat. Inflamm.* **2014**, *2014*, 635293. [CrossRef]
16. McDonagh, M.; Peterson, K.; Raina, P.; Chang, S.; Shekelle, P. Avoiding Bias in Selecting Studies. In *Methods Guide for Effectiveness and Comparative Effectiveness Reviews*; Agency for Healthcare Research and Quality: Rockville, MD, USA, 2008.
17. Moher, D.; Liberati, A.; Tetzlaff, J.; Altman, D.G.; Group, P. Preferred reporting items for systematic reviews and meta-analyses: The PRISMA statement. *J. Clin. Epidemiol.* **2009**, *62*, 1006–1012. [CrossRef]
18. Stang, A. Critical evaluation of the Newcastle-Ottawa scale for the assessment of the quality of nonrandomized studies in meta-analyses. *Eur. J. Epidemiol.* **2010**, *25*, 603–605. [CrossRef]
19. Kim, S.H.; Jeong, H.J.; Kim, J.M.; Jun, J.B.; Son, C.N. Clinical Significance of Elevated Serum Immunoglobulin G4 Levels in Patients With Rheumatoid Arthritis. *J. Rheumat. Dis.* **2020**, *27*, 96–99. [CrossRef]
20. Lin, G.G.; Li, J.M. Elevation of serum IgG subclass concentration in patients with rheumatoid arthritis. *Rheumatol. Int.* **2010**, *30*, 837–840. [CrossRef]
21. Engelmann, R.; Nekarda, S.; Kuthning, D.; Kneitz, C.; Muller-Hilke, B. Decreased IgG4 ACPA levels in responders and increased CD1c(+) classical dendritic cells in non-responders of patients with rheumatoid arthritis under therapy. *Clin. Rheumatol.* **2018**, *37*, 1783–1790. [CrossRef]
22. Engelmann, R.; Kneitz, C.; Muller-Hilke, B. The dynamics of IgG1 and IgG4 ACPA levels during therapy in rheumatoid arthritis. *Z. Rheumatol.* **2014**, *73*, 12.
23. Farboud, A.; Choy, E. Serological investigation of IgG levels and subclasses in rheumatoid arthritis patients following ingestion of bovine type II collagen: Results of a double blind, randomised controlled trial. *Clin. Rheumatol.* **2011**, *30*, 193–199. [CrossRef]
24. Chapuy-Regaud, S.; Nogueira, L.; Clavel, C.; Sebbag, M.; Vincent, C.; Serre, G. IgG subclass distribution of the rheumatoid arthritis-specific autoantibodies to citrullinated fibrin. *Clin. Exp. Immunol.* **2005**, *139*, 542–550. [CrossRef]

25. van Schouwenburg, P.A.; Krieckaert, C.L.; Nurmohamed, M.; Hart, M.; Rispens, T.; Aarden, L.; Wouters, D.; Wolbink, G.J. IgG4 production against adalimumab during long term treatment of RA patients. *J. Clin. Immunol.* **2012**, *32*, 1000–1006. [CrossRef]
26. Bos, W.H.; Bartelds, G.M.; Vis, M.; van der Horst, A.R.; Wolbink, G.J.; de Stadt, R.J.V.; van Schaardenburg, D.; Dijkmans, B.A.C.; Lems, W.F.; Nurmohamed, M.T.; et al. Preferential decrease in IgG4 anti-citrullinated protein antibodies during treatment with tumour necrosis factor blocking agents in patients with rheumatoid arthritis. *Ann. Rheum. Dis.* **2009**, *68*, 558–563. [CrossRef]
27. Martinsson, K.; Johansson, A.; Kastbom, A.; Skogh, T. Immunoglobulin (Ig)G1 and IgG4 anti-cyclic citrullinated peptide (CCP) associate with shared epitope, whereas IgG2 anti-CCP associates with smoking in patients with recent-onset rheumatoid arthritis (the Swedish TIRA project). *Clin. Exp. Immunol.* **2017**, *188*, 53–62. [CrossRef]
28. Carbone, G.; Wilson, A.; Diehl, S.A.; Bunn, J.; Cooper, S.M.; Rincon, M. Interleukin-6 Receptor Blockade Selectively Reduces IL-21 Production by CD4 T Cells and IgG4 Autoantibodies in Rheumatoid Arthritis. *Int. J. Biol. Sci.* **2013**, *9*, 279–288. [CrossRef]
29. Schober, P.; Boer, C.; Schwarte, L.A. Correlation Coefficients: Appropriate Use and Interpretation. *Anesth. Analg.* **2018**, *126*, 1763–1768. [CrossRef]
30. Fransen, J.; van Riel, P.L. The Disease Activity Score and the EULAR response criteria. *Clin. Exp. Rheumatol.* **2005**, *23*, S93–S99. [CrossRef]
31. Thewjitcharoen, Y.; Krittiyawong, S.; Porramatikul, S.; Wanothayaroj, E.; Lekpittaya, N.; Jeamjiraprasert, J.; Nakasatien, S.; Himathongkam, T. A study of serum IgG4 levels in the clinical metamorphosis of autoimmune thyroid disease. *J. Clin. Transl. Endocrinol.* **2017**, *8*, 35–40. [CrossRef]
32. Ryu, J.H.; Horie, R.; Sekiguchi, H.; Peikert, T.; Yi, E.S. Spectrum of Disorders Associated with Elevated Serum IgG4 Levels Encountered in Clinical Practice. *Int. J. Rheumatol.* **2012**, *2012*, 232960. [CrossRef]
33. Yamamoto, M.; Takahashi, H.; Suzuki, C.; Tabeya, T.; Ohara, M.; Naishiro, Y.; Yamamoto, H.; Imai, K.; Shinomura, Y. Analysis of serum IgG subclasses in Churg-Strauss syndrome–the meaning of elevated serum levels of IgG4. *Intern. Med.* **2010**, *49*, 1365–1370. [CrossRef]
34. Hao, M.; Liu, M.; Fan, G.; Yang, X.; Li, J. Diagnostic Value of Serum IgG4 for IgG4-Related Disease: A PRISMA-compliant Systematic Review and Meta-analysis. *Medicine* **2016**, *95*, e3785. [CrossRef]
35. Dienz, O.; Eaton, S.M.; Bond, J.P.; Neveu, W.; Moquin, D.; Noubade, R.; Briso, E.M.; Charland, C.; Leonard, W.J.; Ciliberto, G.; et al. The induction of antibody production by IL-6 is indirectly mediated by IL-21 produced by CD4+ T cells. *J. Exp. Med.* **2009**, *206*, 69–78. [CrossRef]
36. Ettinger, R.; Sims, G.P.; Fairhurst, A.M.; Robbins, R.; da Silva, Y.S.; Spolski, R.; Leonard, W.J.; Lipsky, P.E. IL-21 induces differentiation of human naive and memory B cells into antibody-secreting plasma cells. *J. Immunol.* **2005**, *175*, 7867–7879. [CrossRef]
37. Nishimoto, N.; Kishimoto, T. Interleukin 6: From bench to bedside. *Nat. Clin. Pract. Rheumatol.* **2006**, *2*, 619–626. [CrossRef]
38. Zen, Y.; Fujii, T.; Harada, K.; Kawano, M.; Yamada, K.; Takahira, M.; Nakanuma, Y. Th2 and regulatory immune reactions are increased in immunoglobin G4-related sclerosing pancreatitis and cholangitis. *Hepatology* **2007**, *45*, 1538–1546. [CrossRef]
39. Deshpande, V.; Zen, Y.; Chan, J.K.; Yi, E.E.; Sato, Y.; Yoshino, T.; Kloppel, G.; Heathcote, J.G.; Khosroshahi, A.; Ferry, J.A.; et al. Consensus statement on the pathology of IgG4-related disease. *Mod. Pathol.* **2012**, *25*, 1181–1192. [CrossRef]
40. Pan, Q.; Lan, Q.; Peng, Y.; Cai, J.; Zheng, J.; Dickerson, C.; Xiao, H.; Liu, H.F. Nature, functions, and clinical implications of IgG4 autoantibodies in systemic lupus erythematosus and rheumatoid arthritis. *Discov. Med.* **2017**, *23*, 169–174.
41. Aversa, G.; Punnonen, J.; de Vries, J.E. The 26-kD transmembrane form of tumor necrosis factor alpha on activated CD4+ T cell clones provides a costimulatory signal for human B cell activation. *J. Exp. Med.* **1993**, *177*, 1575–1585. [CrossRef]
42. Makrygiannakis, D.; af Klint, E.; Lundberg, I.E.; Lofberg, R.; Ulfgren, A.K.; Klareskog, L.; Catrina, A.I. Citrullination is an inflammation-dependent process. *Ann. Rheum. Dis.* **2006**, *65*, 1219–1222. [CrossRef]
43. Radbruch, A.; Muehlinghaus, G.; Luger, E.O.; Inamine, A.; Smith, K.G.; Dorner, T.; Hiepe, F. Competence and competition: The challenge of becoming a long-lived plasma cell. *Nat. Rev. Immunol.* **2006**, *6*, 741–750. [CrossRef]
44. Greenberg, J.D.; Spruill, T.M.; Shan, Y.; Reed, G.; Kremer, J.M.; Potter, J.; Yazici, Y.; Ogedegbe, G.; Harrold, L.R. Racial and ethnic disparities in disease activity in patients with rheumatoid arthritis. *Am. J. Med.* **2013**, *126*, 1089–1098. [CrossRef]

MDPI
St. Alban-Anlage 66
4052 Basel
Switzerland
www.mdpi.com

Biomedicines Editorial Office
E-mail: biomedicines@mdpi.com
www.mdpi.com/journal/biomedicines

www.ingramcontent.com/pod-product-compliance
Lightning Source LLC
Chambersburg PA
CBHW040817250926
51850CB00007BA/1188
9783036592763